Graduate Texts in Mathematics 60

Graduate Texts in Mathematics

(continued after index)

V. I. Arnold

Mathematical Methods of Classical Mechanics

Second Edition

Translated by K. Vogtmann
and A. Weinstein

With 269 Illustrations

Springer

V. I. Arnold
Department of
    Mathematics
Steklov Mathematical
    Institute
Russian Academy of
    Sciences
Moscow 117966
GSP-1
Russia

K. Vogtmann
Department of
    Mathematics
Cornell University
Ithaca, NY 14853
U.S.A.

A. Weinstein
Department of
    Mathematics
University of California
    at Berkeley
Berkeley, CA 94720
U.S.A.

Mathematics Subject Classifications (2000): 70HXX, 70D05, 58-XX

Library of Congress Cataloging-in-Publication Data
Arnold, V.I. (Vladimir Igorevich), 1937–
    [Matematicheskie metody klassicheskoĭ mekhaniki. English]
    Mathematical methods of classical mechanics / V.I. Arnold;
translated by K. Vogtmann and A. Weinstein.—2nd ed.
        p. cm.—(Graduate texts in mathematics ; 60)
    Translation of: Matematicheskie metody klassicheskoĭ mekhaniki.
    Bibliography: p.
    Includes index.
    ISBN 0-387-96890-3
    1. Mechanics, Analytic. I. Title II. Series
QA805.A6813   1989
531'.01'515—dc19   88-39823

Title of the Russian Original Edition: *Matematicheskie metody klassicheskoĭ
mekhaniki*. Nauka, Moscow, 1974.

Printed in the United States of America.   (??)

9

springeronline.com

Preface

Many different mathematical methods and concepts are used in classical mechanics: differential equations and phase flows, smooth mappings and manifolds, Lie groups and Lie algebras, symplectic geometry and ergodic theory. Many modern mathematical theories arose from problems in mechanics and only later acquired that axiomatic-abstract form which makes them so hard to study.

In this book we construct the mathematical apparatus of classical mechanics from the very beginning; thus, the reader is not assumed to have any previous knowledge beyond standard courses in analysis (differential and integral calculus, differential equations), geometry (vector spaces, vectors) and linear algebra (linear operators, quadratic forms).

With the help of this apparatus, we examine all the basic problems in dynamics, including the theory of oscillations, the theory of rigid body motion, and the hamiltonian formalism. The author has tried to show the geometric, qualitative aspect of phenomena. In this respect the book is closer to courses in theoretical mechanics for theoretical physicists than to traditional courses in theoretical mechanics as taught by mathematicians.

A considerable part of the book is devoted to variational principles and analytical dynamics. Characterizing analytical dynamics in his "Lectures on the development of mathematics in the nineteenth century," F. Klein wrote that ". . . a physicist, for his problems, can extract from these theories only very little, and an engineer nothing." The development of the sciences in the following years decisively disproved this remark. Hamiltonian formalism lay at the basis of quantum mechanics and has become one of the most often used tools in the mathematical arsenal of physics. After the significance of symplectic structures and Huygens' principle for all sorts of optimization problems was realized, Hamilton's equations began to be used constantly in engineering calculations. On the other hand, the contemporary development of celestial mechanics, connected with the requirements of space exploration, created new interest in the methods and problems of analytical dynamics.

The connections between classical mechanics and other areas of mathematics and physics are many and varied. The appendices to this book are devoted to a few of these connections. The apparatus of classical mechanics is applied to: the foundations of riemannian geometry, the dynamics of an ideal fluid, Kolmogorov's theory of perturbations of conditionally periodic motion, short-wave asymptotics for equations of mathematical physics, and the classification of caustics in geometrical optics.

These appendices are intended for the interested reader and are not part of the required general course. Some of them could constitute the basis of special courses (for example, on asymptotic methods in the theory of non-linear oscillations or on quasi-classical asymptotics). The appendices also contain some information of a reference nature (for example, a list of normal forms of quadratic hamiltonians). While in the basic chapters of the book the author has tried to develop all the proofs as explicitly as possible, avoiding · references to other sources, the appendices consist on the whole of summaries of results, the proofs of which are to be found in the cited literature.

The basis for the book was a year-and-a-half-long required course in classical mechanics, taught by the author to third- and fourth-year mathematics students at the mathematics-mechanics faculty of Moscow State University in 1966–1968.

The author is grateful to I. G. Petrovsky, who insisted that these lectures be delivered, written up, and published. In preparing these lectures for publication, the author found very helpful the lecture notes of L. A. Bunimovich, L. D. Vaingortin, V. L. Novikov, and especially, the mimeographed edition (Moscow State University, 1968) organized by N. N. Kolesnikov. The author thanks them, and also all the students and colleagues who communicated their remarks on the mimeographed text; many of these remarks were used in the preparation of the present edition. The author is grateful to M. A. Leontovich, for suggesting the treatment of connections by means of a limit process, and also to I. I. Vorovich and V. I. Yudovich for their detailed review of the manuscript.

V. ARNOLD

The translators would like to thank Dr. R. Barrar for his help in reading the proofs. We would also like to thank many readers, especially Ted Courant, for spotting errors in the first two printings.

Berkeley, 1981   K. VOGTMANN
                                                                                  A. WEINSTEIN

Preface to the second edition

The main part of this book was written twenty years ago. The ideas and methods of symplectic geometry, developed in this book, have now found many applications in mathematical physics and in other domains of applied mathematics, as well as in pure mathematics itself. Especially, the theory of short wave asymptotic expansions has reached a very sophisticated level, with many important applications to optics, wave theory, acoustics, spectroscopy, and even chemistry; this development was parallel to the development of the theories of Lagrange and Legendre singularities, that is, of singularities of caustics and of wave fronts, of their topology and their perestroikas (in Russian metamorphoses were always called "perestroikas," as in "Morse perestroika" for the English "Morse surgery"; now that the word perestroika has become international, we may preserve the Russian term in translation and are not obliged to substitute "metamorphoses" for "perestroikas" when speaking of wave fronts, caustics, and so on).

Integrable hamiltonian systems have been discovered unexpectedly in many classical problems of mathematical physics, and their study has led to new results in both physics and mathematics, for instance, in algebraic geometry.

Symplectic topology has become one of the most promising and active branches of "global analysis." An important generalization of the Poincaré "geometric theorem" (see Appendix 9) was proved by C. Conley and E. Zehnder in 1983. A sequence of works (by M. Chaperon, A. Weinstein, J.-C. Sikorav, M. Gromov, Ya. M. Eliashberg, Yu. Chekanov, A. Floer, C. Viterbo, H. Hofer, and others) marks important progress in this very lively domain. One may hope that this progress will lead to the proof of many known conjectures in symplectic and contact topology, and to the discovery of new results in this new domain of mathematics, emerging from the problems of mechanics and optics.

The present edition includes three new appendices. They represent the modern development of the theory of ray systems (the theory of singularity and of perestroikas of caustics and of wave fronts, related to the theory of Coxeter reflection groups), the theory of integrable systems (the geometric theory of elliptic coordinates, adapted to the infinite-dimensional Hilbert space generalization), and the theory of Poisson structures (which is a generalization of the theory of symplectic structures, including degenerate Poisson brackets).

A more detailed account of the present state of perturbation theory may be found in the book, *Mathematical Aspects of Classical and Celestial Mechanics* by V. I. Arnold, V. V. Kozlov, and A. I. Neistadt, Encyclopaedia of Math. Sci., Vol. 3 (Springer, 1986); Volume 4 of this series (1988) contains a survey "Symplectic geometry" by V. I. Arnold and A. B. Givental', an article by A. A. Kirillov on geometric quantization, and a survey of the modern theory of integrable systems by S. P. Novikov, I. M. Krichever, and B. A. Dubrovin.

For more details on the geometry of ray systems, see the book *Singularities of Differentiable Mappings* by V. I. Arnold, S. M. Gusein-Zade, and A. N. Varchenko (Vol. 1, Birkhäuser, 1985; Vol. 2, Birkhäuser, 1988). *Catastrophe Theory* by V. I. Arnold (Springer, 1986) (second edition) contains a long annotated bibliography.

Surveys on symplectic and contact geometry and on their applications may be found in the Bourbaki seminar (D. Bennequin, "Caustiques mystiques", February, 1986) and in a series of articles (V. I. Arnold, First steps in symplectic topology, Russian Math. Surveys, 41 (1986); Singularities of ray systems, Russian Math. Surveys, 38 (1983); Singularities in variational calculus, Modern Problems of Math., VINITI, 22 (1983) (translated in J. Soviet Math.); and O. P. Shcherbak, Wave fronts and reflection groups, Russian Math. Surveys, 43 (1988)).

Volumes 22 (1983) and 33 (1988) of the VINITI series, "Sovremennye problemy matematiki. Noveishie dostijenia," contain a dozen articles on the applications of symplectic and contact geometry and singularity theory to mathematics and physics.

Bifurcation theory (both for hamiltonian and for more general systems) is discussed in the textbook *Geometrical Methods in the Theory of Ordinary Differential Equations* (Springer, 1988) (this new edition is more complete than the preceding one). The survey "Bifurcation theory and its applications in mathematics and mechanics" (XVIIth International Congress of Theoretical and Applied Mechanics in Grenoble, August, 1988) also contains new information, as does Volume 5 of the Encyclopaedia of Math. Sci. (Springer, 1989), containing the survey "Bifurcation theory" by V. I. Arnold, V. S. Afraimovich, Yu. S. Ilyashenko, and L. P. Shilnikov. Volume 2 of this series, edited by D. V. Anosov and Ya. G. Sinai, is devoted to the ergodic theory of dynamical systems including those of mechanics.

The new discoveries in all these theories have potentially extremely wide applications, but since these results were discovered rather recently, they are discussed only in the specialized editions, and applications are impeded by the difficulty of the mathematical exposition for nonmathematicians. I hope that the present book will help to master these new theories not only to mathematicians, but also to all those readers who use the theory of dynamical systems, symplectic geometry, and the calculus of variations—in physics, mechanics, control theory, and so on. The author would like to thank Dr. T. Tokieda for his help in correcting errors in previous printings and for reading the proofs.

December 1988   V. I. Arnold

Translator's preface to the second edition

This edition contains three new appendices, originally written for inclusion in a German edition. They describe work by the author and his co-workers on Poisson structures, elliptic coordinates with applications to integrable systems, and singularities of ray systems. In addition, numerous corrections to errors found by the author, the translators, and readers have been incorporated into the text.

Contents

Contents

Contents

PART I
NEWTONIAN MECHANICS

Newtonian mechanics studies the motion of a system of point masses in three-dimensional euclidean space. The basic ideas and theorems of newtonian mechanics (even when formulated in terms of three-dimensional cartesian coordinates) are invariant with respect to the six-dimensional[1] group of euclidean motions of this space.

A newtonian potential mechanical system is specified by the masses of the points and by the potential energy. The motions of space which leave the potential energy invariant correspond to laws of conservation.

Newton's equations allow one to solve completely a series of important problems in mechanics, including the problem of motion in a central force field.

---

[1] And also with respect to the larger group of galilean transformations of space-time.

Experimental facts 1

In this chapter we write down the basic experimental facts which lie at the foundation of mechanics: Galileo's principle of relativity and Newton's differential equation. We examine constraints on the equation of motion imposed by the relativity principle, and we mention some simple examples.

1 The principles of relativity and determinacy

In this paragraph we introduce and discuss the notion of an inertial coordinate system. The mathematical statements of this paragraph are formulated exactly in the next paragraph.

A series of experimental facts is at the basis of classical mechanics.[2] We list some of them.

A *Space and time*

Our space is three-dimensional and euclidean, and time is one-dimensional.

B *Galileo's principle of relativity*

There exist coordinate systems (called inertial) possessing the following two properties:

1. All the laws of nature at all moments of time are the same in all inertial coordinate systems.
2. All coordinate systems in uniform rectilinear motion with respect to an inertial one are themselves inertial.

[2] All these "experimental facts" are only approximately true and can be refuted by more exact experiments. In order to avoid cumbersome expressions, we will not specify this from now on and we will speak of our mathematical models as if they exactly described physical phenomena.

3

In other words, if a coordinate system attached to the earth is inertial, then an experimenter on a train which is moving uniformly in a straight line with respect to the earth cannot detect the motion of the train by experiments conducted entirely inside his car.

In reality, the coordinate system associated with the earth is only approximately inertial. Coordinate systems associated with the sun, the stars, etc. are more nearly inertial.

C Newton's principle of determinacy

The initial state of a mechanical system (the totality of positions and velocities of its points at some moment of time) uniquely determines all of its motion.

It is hard to doubt this fact, since we learn it very early. One can imagine a world in which to determine the future of a system one must also know the acceleration at the initial moment, but experience shows us that our world is not like this.

2 The galilean group and Newton's equations

In this paragraph we define and investigate the galilean group of space-time transformations. Then we consider Newton's equation and the simplest constraints imposed on its right-hand side by the property of invariance with respect to galilean transformations.[3]

A Notation

We denote the set of all real numbers by $\mathbb{R}$. We denote by $\mathbb{R}^n$ an n-dimensional real vector space.

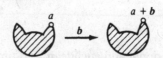

Figure 1   Parallel displacement

Affine n-dimensional space A^n is distinguished from $\mathbb{R}^n$ in that there is "no fixed origin." The group $\mathbb{R}^n$ acts on A^n as the *group of parallel displacements* (Figure 1):

$$a \to a + \mathbf{b}, \qquad a \in A^n, \mathbf{b} \in \mathbb{R}^n, a + \mathbf{b} \in A^n.$$

[Thus the sum of two points of A^n is not defined, but their difference is defined and is a vector in $\mathbb{R}^n$.]

[3] The reader who has no need for the mathematical formulation of the assertions of Section 1 can omit this section.

A *euclidean structure* on the vector space $\mathbb{R}^n$ is a positive definite symmetric bilinear form called a *scalar product*. The scalar product enables one to define the *distance*

$$\rho(x, y) = \|x - y\| = \sqrt{(x - y, x - y)}$$

between points of the corresponding *affine* space A^n. An affine space with this distance function is called a *euclidean space* and is denoted by E^n.

B *Galilean structure*

The galilean space-time structure consists of the following three elements:

1. The universe—a four-dimensional affine[4] space A^4. The points of A^4 are called *world points* or *events*. The parallel displacements of the universe A^4 constitute a vector space $\mathbb{R}^4$.
2. Time—a linear mapping $t: \mathbb{R}^4 \to \mathbb{R}$ from the vector space of parallel displacements of the universe to the real "time axis." The *time interval* from event $a \in A^4$ to event $b \in A^4$ is the number $t(b - a)$ (Figure 2). If $t(b - a) = 0$, then the events a and b are called *simultaneous*.

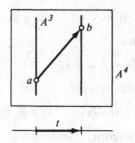

Figure 2   Interval of time t

The set of events simultaneous with a given event forms a three-dimensional affine subspace in A^4. It is called a *space of simultaneous events A^3*.

The kernel of the mapping t consists of those parallel displacements of A^4 which take some (and therefore every) event into an event simultaneous with it. This kernel is a three-dimensional linear subspace $\mathbb{R}^3$ of the vector space $\mathbb{R}^4$.

The galilean structure includes one further element.

3. The *distance between simultaneous events*

$$\rho(a, b) = \|a - b\| = \sqrt{(a - b, a - b)} \qquad a, b \in A^3$$

is given by a scalar product on the space $\mathbb{R}^3$. This distance makes every space of simultaneous events into a three-dimensional euclidean space E^3.

[4] Formerly, the universe was provided not with an affine, but with a linear structure (the geocentric system of the universe).

A space A^4, equipped with a galilean space-time structure, is called a *galilean space*.

One can speak of two events occurring simultaneously in different places, but the expression "two non-simultaneous events $a, b \in A^4$ occurring at *one and the same place in three-dimensional space*" has no meaning as long as we have not chosen a coordinate system.

The *galilean group* is the group of all transformations of a galilean space which preserve its structure. The elements of this group are called *galilean transformations*. Thus, galilean transformations are affine transformations of A^4 which preserve intervals of time and the distance between simultaneous events.

EXAMPLE. Consider the direct product[5] $\mathbb{R} \times \mathbb{R}^3$ of the t axis with a three-dimensional vector space $\mathbb{R}^3$; suppose $\mathbb{R}^3$ has a fixed euclidean structure. Such a space has a natural galilean structure. We will call this space *galilean coordinate space*.

We mention three examples of galilean transformations of this space. First, *uniform motion with velocity* $\mathbf{v}$:

$$g_1(t, \mathbf{x}) = (t, \mathbf{x} + \mathbf{v}t) \qquad \forall t \in \mathbb{R}, \mathbf{x} \in \mathbb{R}^3.$$

Next, *translation of the origin*:

$$g_2(t, \mathbf{x}) = (t + s, \mathbf{x} + \mathbf{s}) \qquad \forall t \in \mathbb{R}, \mathbf{x} \in \mathbb{R}^3.$$

Finally, *rotation of the coordinate axes*:

$$g_3(t, \mathbf{x}) = (t, G\mathbf{x}), \qquad \forall t \in \mathbb{R}, \mathbf{x} \in \mathbb{R}^3,$$

where $G: \mathbb{R}^3 \to \mathbb{R}^3$ is an orthogonal transformation.

PROBLEM. Show that every galilean transformation of the space $\mathbb{R} \times \mathbb{R}^3$ can be written in a unique way as the composition of a rotation, a translation, and a uniform motion ($g = g_1 \circ g_2 \circ g_3$) (thus the dimension of the galilean group is equal to $3 + 4 + 3 = 10$).

PROBLEM. Show that all galilean spaces are isomorphic to each other[6] and, in particular, isomorphic to the coordinate space $\mathbb{R} \times \mathbb{R}^3$.

Let M be a set. A one-to-one correspondence $\varphi_1 : M \to \mathbb{R} \times \mathbb{R}^3$ is called a *galilean coordinate system* on the set M. A coordinate system φ_2 *moves uniformly* with respect to φ_1 if $\varphi_1 \, \varphi_2^{-1} : \mathbb{R} \times \mathbb{R}^3 \to \mathbb{R} \times \mathbb{R}^3$ is a galilean transformation. The galilean coordinate systems φ_1 and φ_2 give M the same galilean structure.

---

[5] Recall that the direct product of two sets A and B is the set of ordered pairs (a, b), where $a \in A$ and $b \in B$. The direct product of two spaces (vector, affine, euclidean) has the structure of a space of the same type.

[6] That is, there is a one-to-one mapping of one to the other preserving the galilean structure.

C *Motion, velocity, acceleration*

A motion in $\mathbb{R}^N$ is a differentiable mapping $\mathbf{x}\colon I \to \mathbb{R}^N$, where I is an interval on the real axis.

The derivative

$$\dot{\mathbf{x}}(t_0) = \frac{d\mathbf{x}}{dt}\bigg|_{t=t_0} = \lim_{h \to 0} \frac{\mathbf{x}(t_0 + h) - \mathbf{x}(t_0)}{h} \in \mathbb{R}^N$$

is called the *velocity vector* at the point $t_0 \in I$.

The second derivative

$$\ddot{\mathbf{x}}(t_0) = \frac{d^2\mathbf{x}}{dt^2}\bigg|_{t=t_0}$$

is called the *acceleration vector* at the point t_0.

We will assume that the functions we encounter are continuously differentiable as many times as necessary. In the future, unless otherwise stated, mappings, functions, etc. are understood to be differentiable mappings, functions, etc. The image of a mapping $\mathbf{x}\colon I \to \mathbb{R}^N$ is called a *trajectory* or *curve* in $\mathbb{R}^N$.

PROBLEM. Is it possible for the trajectory of a differentiable motion on the plane to have the shape drawn in Figure 3? Is it possible for the acceleration vector to have the value shown?

ANSWER. Yes. No.

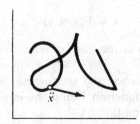

Figure 3   Trajectory of motion of a point

We now define a *mechanical system of n points moving in three-dimensional euclidean space.*

Let $\mathbf{x}\colon \mathbb{R} \to \mathbb{R}^3$ be a motion in $\mathbb{R}^3$. The graph[7] of this mapping is a curve in $\mathbb{R} \times \mathbb{R}^3$.

A curve in galilean space which appears in some (and therefore every) galilean coordinate system as the graph of a motion, is called a *world line* (Figure 4).

[7] The graph of a mapping $f\colon A \to B$ is the subset of the direct product $A \times B$ consisting of all pairs $(a, f(a))$ with $a \in A$.

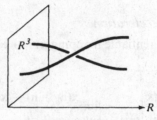

Figure 4   World lines

A motion of a system of n points gives, in galilean space, n world lines. In a galilean coordinate system they are described by n mappings $\mathbf{x}_i: \mathbb{R} \to \mathbb{R}^3$, $i = 1, \ldots, n$.

The direct product of n copies of $\mathbb{R}^3$ is called the *configuration space* of the system of n points. Our n mappings $\mathbf{x}_i: \mathbb{R} \to \mathbb{R}^3$ define one mapping

$$\mathbf{x}: \mathbb{R} \to \mathbb{R}^N \qquad N = 3n$$

of the time axis into the configuration space. Such a mapping is also called a *motion of a system of n points* in the galilean coordinate system on $\mathbb{R} \times \mathbb{R}^3$.

D *Newton's equations*

According to Newton's principle of determinacy (Section 1C) all motions of a system are uniquely determined by their initial positions ($\mathbf{x}(t_0) \in \mathbb{R}^N$) and initial velocities ($\dot{\mathbf{x}}(t_0) \in \mathbb{R}^N$).

In particular, the initial positions and velocities determine the acceleration. In other words, there is a function $\mathbf{F}: \mathbb{R}^N \times \mathbb{R}^N \times \mathbb{R} \to \mathbb{R}^N$ such that

$$(1) \qquad\qquad \ddot{\mathbf{x}} = \mathbf{F}(\mathbf{x}, \dot{\mathbf{x}}, t).$$

Newton used Equation (1) as the basis of mechanics. It is called *Newton's equation.*

By the theorem of existence and uniqueness of solutions to ordinary differential equations, the function $\mathbf{F}$ and the initial conditions $\mathbf{x}(t_0)$ and $\dot{\mathbf{x}}(t_0)$ uniquely determine a motion.[8]

For each specific mechanical system the form of the function $\mathbf{F}$ is determined experimentally. From the mathematical point of view the form of $\mathbf{F}$ for each system constitutes the *definition* of that system.

E *Constraints imposed by the principle of relativity*

Galileo's principle of relativity states that in physical space-time there is a selected galilean structure ("the class of inertial coordinate systems") having the following property.

---

[8] Under certain smoothness conditions, which we assume to be fulfilled. In general, a motion is determined by Equation (1) only on some *interval* of the time axis. For simplicity we will assume that this interval is the whole time axis, as is the case in most problems in mechanics.

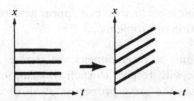

Figure 5 Galileo's principle of relativity

If we subject the world lines of all the points of any mechanical system[9] to one and the same galilean transformation, we obtain world lines of the same system (with new initial conditions) (Figure 5).

This imposes a series of conditions on the form of the right-hand side of Newton's equation written in an inertial coordinate system: Equation (1) must be invariant with respect to the group of galilean transformations.

EXAMPLE 1. Among the galilean transformations are the time translations. Invariance with respect to time translations means that "the laws of nature remain constant," i.e., if $\mathbf{x} = \varphi(t)$ is a solution to Equation (1), then for any $s \in \mathbb{R}$, $\mathbf{x} = \varphi(t + s)$ is also a solution.

From this it follows that the right-hand side of Equation (1) in an inertial coordinate system does not depend on the time:

$$\ddot{\mathbf{x}} = \Phi(\mathbf{x}, \dot{\mathbf{x}}).$$

Remark. Differential equations in which the right-hand side does depend on time arise in the following situation.

Suppose that we are studying part I of the mechanical system I + II. Then the influence of part II on part I can sometimes be replaced by a time variation of parameters in the system of equations describing the motion of part I. For example, the influence of the moon on the earth can be ignored in investigating the majority of phenomena on the earth. However, in the study of the tides this influence must be taken into account; one can achieve this by introducing, instead of the attraction of the moon, periodic changes in the strength of gravity on earth.

---

[9] In formulating the principle of relativity we must keep in mind that it is relevant only to *closed* physical (in particular, mechanical) systems, i.e., that we must include in the system all bodies whose interactions play a role in the study of the given phenomena. Strictly speaking, we should include in the system all bodies in the universe. But we know from experience that one can disregard the effect of many of them: for example, in studying the motion of planets around the sun we can disregard the attractions among the stars, etc.

On the other hand, in the study of a body in the vicinity of earth, the system is not closed if the earth is not included; in the study of the motion of an airplane the system is not closed if it does not include the air surrounding the airplane, etc. In the future, the term "mechanical system" will mean a closed system in most cases, and when there is a non-closed system in question this will be explicitly stated (cf., for example, Section 3).

Equations with variable coefficients can appear also as the result of formal operations in the solution of problems.

EXAMPLE 2. Translations in three-dimensional space are galilean transformations. Invariance with respect to such translations means that *space is homogeneous*, or "has the same properties at all of its points." That is, if $x_i = \varphi_i(t)(i = 1, \ldots, n)$ is a motion of a system of n points satisfying (1), then for any $r \in \mathbb{R}^3$ the motion $\varphi_i(t) + r\,(i = 1, \ldots, n)$ also satisfies Equation (1).

From this it follows that the right-hand side of Equation (1) in the inertial coordinate system can depend only on the "relative coordinates" $x_j - x_k$.

From invariance under passage to a uniformly moving coordinate system (which does not change $\ddot{x}_i$ or $x_j - x_k$, but adds to each $\dot{x}_i$ a fixed vector v) it follows that the right-hand side of Equation (1) in an inertial system of coordinates can depend only on the *relative* velocities

$$\ddot{x}_i = f_i(\{x_j - x_k, \dot{x}_j - \dot{x}_k\}), \qquad i, j, k = 1, \ldots, n.$$

EXAMPLE 3. Among the galilean transformations are the rotations in three-dimensional space. Invariance with respect to these rotations means that *space is isotropic*; there are no preferred directions.

Thus, if $\varphi_i \colon \mathbb{R} \to \mathbb{R}^3 (i = 1, \ldots, n)$ is a motion of a system of points satisfying (1), and $G \colon \mathbb{R}^3 \to \mathbb{R}^3$ is an orthogonal transformation, then the motion $G\varphi_i \colon \mathbb{R} \to \mathbb{R}^3 (i, \ldots, n)$ also satisfies (1). In other words.

$$F(Gx, G\dot{x}) = GF(x, \dot{x}),$$

where Gx denotes $(Gx_1, \ldots, Gx_n)$, $x_i \in \mathbb{R}^3$.

PROBLEM. Show that if a mechanical system consists of only one point, then its acceleration in an inertial coordinate system is equal to zero ("Newton's first law").

Hint. By Examples 1 and 2 the acceleration vector does not depend on x, $\dot{x}$, or t, and by Example 3 the vector F is invariant with respect to rotation.

PROBLEM. A mechanical system consists of two points. At the initial moment their velocities (in some inertial coordinate system) are equal to zero. Show that the points will stay on the line which connected them at the initial moment.

PROBLEM. A mechanical system consists of three points. At the initial moment their velocities (in some inertial coordinate system) are equal to zero. Show that the points always remain in the plane which contained them at the initial moment.

PROBLEM. A mechanical system consists of two points. Show that for any initial conditions there exists an inertial coordinate system in which the two points remain in a fixed plane.

PROBLEM. Show that mechanics "through the looking glass" is identical to ours.

Hint. In the galilean group there is a reflection transformation, changing the orientation of $\mathbb{R}^3$.

PROBLEM. Is the class of inertial systems unique?

ANSWER. No. Other classes can be obtained if one changes the units of length and time or the direction of time.

3 Examples of mechanical systems

We have already remarked that the form of the function **F** in Newton's equation (1) is determined experimentally for each mechanical system. Here are several examples.

In examining concrete systems it is reasonable not to include all the objects of the universe in a system. For example, in studying the majority of phenomena taking place on the earth we can ignore the influence of the moon. Furthermore, it is usually possible to disregard the effect of the processes we are studying on the motion of the earth itself; we may even consider a coordinate system attached to the earth as "fixed." It is clear that the principle of relativity no longer imposes the constraints found in Section 2 for equations of motion written in such a coordinate system. For example, near the earth there is a distinguished direction, the vertical.

A *Example 1: A stone falling to the earth*

Experiments show that

(2) $$\ddot{x} = -g, \quad \text{where } g \approx 9.8 \text{ m/s}^2 \text{ (Galileo)}^*$$

where x is the height of a stone above the surface of the earth.

If we introduce the "potential energy" $U = gx$, then Equation (2) can be written in the form

$$\ddot{x} = -\frac{dU}{dx}.$$

If $U: E^N \to \mathbb{R}$ is a differentiable function on euclidean space, then we will denote by $\partial U/\partial \mathbf{x}$ the gradient of the function U. If $E^N = E^{n_1} \times \cdots \times E^{n_k}$ is a direct product of euclidean spaces, then we will denote a point $\mathbf{x} \in E^N$ by $(\mathbf{x}_1, \ldots, \mathbf{x}_k)$, and the vector $\partial U/\partial \mathbf{x}$ by $(\partial U/\partial \mathbf{x}_1, \ldots, \partial U/\partial \mathbf{x}_k)$. In particular, if $x_1, \ldots, x_N$ are cartesian coordinates in E^N, then the components of the vector $\partial U/\partial \mathbf{x}$ are the partial derivatives $\partial U/\partial x_1, \ldots, \partial U/\partial x_N$.

Experiments show that the radius vector of the stone with respect to some point 0 on the earth satisfies the equation

(3) $$\ddot{\mathbf{x}} = -\frac{\partial U}{\partial \mathbf{x}}, \quad \text{where } U = -(\mathbf{g}, \mathbf{x})$$

---

* In this and other sections, the mass of a particle is taken to be 1.

11

The vector in the right-hand side is directed towards the earth. It is called the *gravitational acceleration vector* **g**. (Figure 6.)

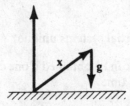

Figure 6   A stone falling to the earth

B Example 2: Falling from great height

Like all experimental facts, the law of motion (2) has a restricted domain of application. According to a more precise law of falling bodies, discovered by Newton, acceleration is inversely proportional to the square of the distance from the center of the earth:

$$\ddot{x} = -g\frac{r_0^2}{r^2},$$

where $r = r_0 + x$ (Figure 7).

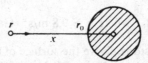

Figure 7   The earth's gravitational field

This equation can also be written in the form (3), if we introduce the potential energy

$$U = -\frac{k}{r} \qquad k = gr_0^2,$$

inversely proportional to the distance to the center of the earth.

PROBLEM. Determine with what velocity a stone must be thrown in order that it fly infinitely far from the surface of the earth.[10]

ANSWER. ≥ 11.2 km/sec.

[10] This is the so-called second cosmic velocity v_2. Our equation does not take into account the attraction of the sun. The attraction of the sun will not let the stone escape from the solar system if the velocity of the stone with respect to the earth is less than 16.6 km/sec.

C Example 3: Motion of a weight along a line
under the action of a spring

Experiments show that under small extensions of the spring the equation of motion of the weight will be (Figure 8)

$$\ddot{x} = -\alpha^2 x.$$

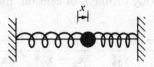

Figure 8   Weight on a spring

This equation can also be written in the form (3) if we introduce the potential energy

$$U = \frac{\alpha^2 x^2}{2}.$$

If we replace our one weight by two weights, then it turns out that, under the same extension of the spring, the acceleration is half as large.

It is experimentally established that for any two bodies the ratio of the accelerations $\ddot{x}_1/\ddot{x}_2$ under the same extension of a spring is fixed (does not depend on the extent of extension of the spring or on its characteristics, but only on the bodies themselves). The value inverse to this ratio is by definition the ratio of masses:

$$\frac{\ddot{x}_1}{\ddot{x}_2} = \frac{m_2}{m_1}.$$

For a unit of mass we take the mass of some fixed body, e.g., one liter of water. We know by experience that the masses of all bodies are positive. The product of mass times acceleration $m\ddot{x}$ does not depend on the body, and is a characteristic of the extension of the spring. This value is called the *force* of the spring acting on the body.

As a unit of force, we take the "newton." If one liter of water is suspended on a spring at the surface of the earth, the spring acts with a force of 9.8 newtons ($= 1$ kg).

D Example 4: Conservative systems

Let $E^{3n} = E^3 \times \cdots \times E^3$ be the configuration space of a system of n points in the euclidean space E^3. Let $U: E^{3n} \to \mathbb{R}$ be a differentiable function and let $m_1, \ldots, m_n$ be positive numbers.

Definition. The motion of n points, of masses $m_1, \ldots, m_n$, in the potential field with potential energy U is given by the system of differential equations

$$(4) \qquad m_i \ddot{\mathbf{x}}_i = -\frac{\partial U}{\partial \mathbf{x}_i} \qquad i = 1, \ldots, n.$$

The equations of motion in Examples 1 to 3 have this form. The equations of motion of many other mechanical systems can be written in the same form. For example, the three-body problem of celestial mechanics is problem (4) in which

$$U = -\frac{m_1 m_2}{\|\mathbf{x}_1 - \mathbf{x}_2\|} - \frac{m_2 m_3}{\|\mathbf{x}_2 - \mathbf{x}_3\|} - \frac{m_3 m_1}{\|\mathbf{x}_3 - \mathbf{x}_1\|}.$$

Many different equations of entirely different origin can be reduced to form (4), for example the equations of electrical oscillations. In the following chapter we will study mainly systems of differential equations in the form (4).

Investigation of the equations of motion

2

In most cases (for example, in the three-body problem) we can neither solve the system of differential equations nor completely describe the behavior of the solutions. In this chapter we consider a few simple but important problems for which Newton's equations can be solved.

4 Systems with one degree of freedom

In this paragraph we study the phase flow of the differential equation (1). A look at the graph of the potential energy is enough for a qualitative analysis of such an equation. In addition, Equation (1) is integrated by quadratures.

A Definitions

A *system with one degree of freedom* is a system described by one differential equation

$$(1) \qquad \ddot{x} = f(x) \qquad x \in \mathbb{R}.$$

The *kinetic energy* is the quadratic form*

$$T = \tfrac{1}{2}\dot{x}^2.$$

The *potential energy* is the function

$$U(x) = - \int_{x_0}^{x} f(\xi)d\xi.$$

The sign in this formula is taken so that the potential energy of a stone is larger if the stone is higher off the ground.

Notice that the potential energy determines f. Therefore, to specify a system of the form (1) it is enough to give the potential energy. Adding a constant to the potential energy does not change the equation of motion (1).

---

* see footnote on p. 11.

15

The *total energy* is the sum

$$E = T + U.$$

In general, the total energy is a function, $E(x, \dot{x})$, of x and $\dot{x}$.

Theorem (The law of conservation of energy). *The total energy of points moving according to the equation* (1) *is conserved*: $E(x(t), \dot{x}(t))$ *is independent of* t.

PROOF.

$$\frac{d}{dt}(T + U) = \dot{x}\ddot{x} + \frac{dU}{dx}\dot{x} = \dot{x}(\ddot{x} - f(x)) = 0. \qquad \square$$

B *Phase flow*

Equation (1) is equivalent to the system of two equations:

$$(2) \qquad \qquad \dot{x} = y, \quad \dot{y} = f(x).$$

We consider the plane with coordinates x and y, which we call the *phase plane* of Equation (1). The points of the phase plane are called *phase points*. The right-hand side of (2) determines a vector field on the phase plane, called the *phase velocity vector field*.

A solution of (2) is a motion $\boldsymbol{\varphi}: \mathbb{R} \to \mathbb{R}^2$ of a phase point in the phase plane, such that the velocity of the moving point at each moment of time is equal to the phase velocity vector at the location of the phase point at that moment.[11]

The image of $\boldsymbol{\varphi}$ is called the *phase curve*. Thus the phase curve is given by the parametric equations

$$x = \varphi(t) \qquad y = \dot{\varphi}(t).$$

PROBLEM. Show that through every phase point there is one and only one phase curve.

Hint. Refer to a textbook on ordinary differential equations.

We notice that a phase curve could consist of only one point. Such a point is called an *equilibrium position*. The vector of phase velocity at an equilibrium position is zero.

The law of conservation of energy allows one to find the phase curves easily. On each phase curve the value of the total energy is constant. Therefore, each phase curve lies entirely in one energy level set $E(x, y) = h$.

C *Examples*

EXAMPLE 1. The basic equation of the theory of oscillations is

$$\ddot{x} = -x.$$

---

[11] Here we assume for simplicity that the solution φ is defined on the whole time axis $\mathbb{R}$.

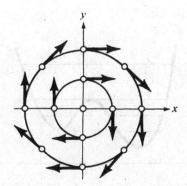

Figure 9   Phase plane of the equation $\ddot{x} = -x$

In this case (Figure 9) we have:

$$T = \frac{\dot{x}^2}{2} \qquad U = \frac{x^2}{2} \qquad E = \frac{\dot{x}^2}{2} + \frac{x^2}{2}.$$

The energy level sets are the concentric circles and the origin. The phase velocity vector at the phase point (x, y) has components $(y, -x)$. It is perpendicular to the radius vector and equal to it in magnitude. Therefore, the motion of the phase point in the phase plane is a uniform motion around 0: $x = r_0 \cos(\varphi_0 - t)$, $y = r_0 \sin(\varphi_0 - t)$. Each energy level set is a phase curve.

EXAMPLE 2. Suppose that a potential energy is given by the graph in Figure 10. We will draw the energy level sets $\frac{1}{2}y^2 + U(x) = E$. For this, the following facts are helpful.

1. Any equilibrium position of (2) must lie on the x axis of the phase plane. The point $x = \xi$, $y = 0$ is an equilibrium position if ξ is a critical point of the potential energy, i.e., if $(\partial U/\partial x)|_{x=\xi} = 0$.
2. Each level set is a *smooth* curve in a neighborhood of each of its points which is not an equilibrium position (this follows from the implicit function theorem). In particular, if the number E is not a critical value of the potential energy (i.e., is not the value of the potential energy at one of its critical points), then the level set on which the energy is equal to E is a smooth curve.

It follows that in order to study the energy level curve, we should turn our attention to the critical and near-critical values of E. It is convenient here to imagine a little ball rolling in the potential well U.

For example, consider the following argument: "Kinetic energy is nonnegative. This means that potential energy is less than or equal to the total energy. The smaller the potential energy, the greater the velocity." This translates to: "The ball cannot jump out of the potential well, rising

17

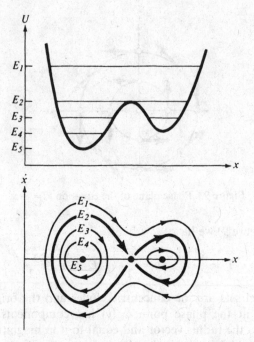

Figure 10   Potential energy and phase curves

higher than the level determined by its initial energy. As it falls into the well, the ball gains velocity." We also notice that the local maximum points of the potential energy are unstable, but the minimum points are stable equilibrium positions.

PROBLEM. Prove this.

PROBLEM. How many phase curves make up the separatrix (figure eight) curve, corresponding to the level E_2?

ANSWER. Three.

PROBLEM. Determine the duration of motion along the separatrix.

ANSWER. It follows from the uniqueness theorem that the time is infinite.

PROBLEM. Show that the time it takes to go from x_1 to x_2 (in one direction) is equal to

$$t_2 - t_1 = \int_{x_1}^{x_2} \frac{dx}{\sqrt{2(E - U(x))}}.$$

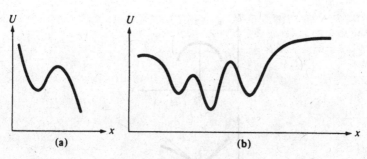

Figure 11   Potential energy

PROBLEM. Draw the phase curves, given the potential energy graphs in Figure 11.

ANSWER. Figure 12.

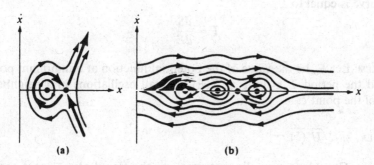

Figure 12   Phase curves

PROBLEM. Draw the phase curves for the "equation of an ideal planar pendulum": $\ddot{x} = -\sin x$.

PROBLEM. Draw the phase curves for the "equation of a pendulum on a rotating axis": $\ddot{x} = -\sin x + M$.

Remark. In these two problems x denotes the angle of displacement of the pendulum. The phase points whose coordinates differ by 2π correspond to the same position of the pendulum. Therefore, in addition to the phase plane, it is natural to look at the phase cylinder $\{x(\bmod 2\pi), y\}$.

PROBLEM. Find the tangent lines to the branches of the critical level corresponding to maximal potential energy $E = U(\xi)$ (Figure 13).

ANSWER. $y = \pm \sqrt{-U''(\xi)}(x - \xi)$.

19

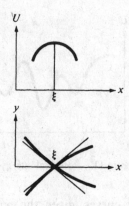

Figure 13   Critical energy level lines

PROBLEM. Let $S(E)$ be the area enclosed by the closed phase curve corresponding to the energy level E. Show that the period of motion along this curve is equal to

$$T = \frac{dS}{dE}.$$

PROBLEM. Let E_0 be the value of the potential function at a minimum point ξ. Find the period $T_0 = \lim_{E \to E_0} T(E)$ of small oscillations in a neighborhood of the point ξ.

ANSWER. $2\pi/\sqrt{U''(\xi)}$.

PROBLEM. Consider a periodic motion along the closed phase curve corresponding to the energy level E. Is it stable in the sense of Liapunov?[12]

ANSWER. No.[13]

D Phase flow

Let M be a point in the phase plane. We look at the solution to system (2) whose initial conditions at $t = 0$ are represented by the point M. We assume that any solution of the system can be extended to the whole time axis. The value of our solution at any value of t depends on M. We denote the resulting phase point (Figure 14) by

$$M(t) = g^t M.$$

In this way we have defined a mapping of the phase plane to itself, $g^t: \mathbb{R}^2 \to \mathbb{R}^2$. By theorems in the theory of ordinary differential equations,

[12] For a definition, see, e.g., p. 155 of *Ordinary Differential Equations* by V. I. Arnold, MIT Press, 1973.

[13] The only exception is the case when the period does not depend on the energy.

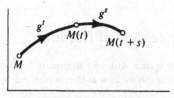

Figure 14   Phase flow

the mapping g^t is a diffeomorphism (a one-to-one differentiable mapping with a differentiable inverse). The diffeomorphisms g^t, $t \in \mathbb{R}$, form a group: $g^{t+s} = g^t \circ g^s$. The mapping g^0 is the identity ($g^0 M = M$), and g^{-t} is the inverse of g^t. The mapping $g: \mathbb{R} \times \mathbb{R}^2 \to \mathbb{R}^2$, defined by $g(t, M) = g^t M$ is differentiable. All these properties together are expressed by saying that the transformations g^t form a *one-parameter group of diffeomorphisms* of the phase plane. This group is also called the *phase flow*, given by system (2) (or Equation (1)).

EXAMPLE. The phase flow given by the equation $\ddot{x} = -x$ is the group g^t of rotations of the phase plane through angle t around the origin.

PROBLEM. Show that the system with potential energy $U = -x^4$ does not define a phase flow.

PROBLEM. Show that if the potential energy is positive, then there is a phase flow.

   Hint. Use the law of conservation of energy to show that a solution can be extended without bound.

PROBLEM. Draw the image of the circle $x^2 + (y-1)^2 < \frac{1}{4}$ under the action of a transformation of the phase flow for the equations (a) of the "inverse pendulum," $\ddot{x} = x$ and (b) of the "nonlinear pendulum," $\ddot{x} = -\sin x$.

ANSWER. Figure 15.

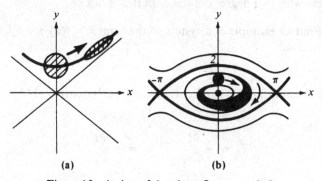

(a)   (b)

Figure 15   Action of the phase flow on a circle

21

5 Systems with two degrees of freedom

Analyzing a general potential system with two degrees of freedom is beyond the capability of modern science. In this paragraph we look at the simplest examples.

A Definitions

By a system with two degrees of freedom we will mean a system defined by the differential equations

(1) $$\ddot{\mathbf{x}} = \mathbf{f}(\mathbf{x}), \qquad \mathbf{x} \in E^2,$$

where $\mathbf{f}$ is a vector field on the plane.

A system is said to be *conservative* if there exists a function $U: E^2 \to \mathbb{R}$ such that $\mathbf{f} = -\partial U/\partial \mathbf{x}$. The equation of motion of a conservative system then has the form[14] $\ddot{\mathbf{x}} = -\partial U/\partial \mathbf{x}$.

B The law of conservation of energy

Theorem. *The total energy of a conservative system is conserved, i.e.,*

$$\frac{dE}{dt} = 0, \quad where \ E = \tfrac{1}{2}\dot{\mathbf{x}}^2 + U(\mathbf{x}), \ \dot{\mathbf{x}}^2 = (\dot{\mathbf{x}}, \dot{\mathbf{x}}).$$

PROOF. $dE/dt = (\dot{\mathbf{x}}, \ddot{\mathbf{x}}) + (\partial U/\partial \mathbf{x}, \dot{\mathbf{x}}) = (\ddot{\mathbf{x}} + (\partial U/\partial \mathbf{x}), \dot{\mathbf{x}}) = 0$ by the equation of motion. $\qquad \square$

Corollary. *If at the initial moment the total energy is equal to E, then all trajectories lie in the region where $U(\mathbf{x}) \leq E$, i.e., a point remains inside the potential well $U(x_1, x_2) \leq E$ for all time.*

Remark. In a system with one degree of freedom it is always possible to introduce the potential energy

$$U(x) = -\int_{x_0}^{x} f(\xi)d\xi.$$

For a system with two degrees of freedom this is not so.

PROBLEM. Find an example of a system of the form $\ddot{\mathbf{x}} = \mathbf{f}(\mathbf{x})$, $\mathbf{x} \in E^2$, which is not conservative.

C Phase space

The equation of motion (1) can be written as the system:

(2)
$$\dot{x}_1 = y_1 \qquad \dot{x}_2 = y_2$$
$$\dot{y}_1 = -\frac{\partial U}{\partial x_1} \qquad \dot{y}_2 = -\frac{\partial U}{\partial x_2}$$

[14] In cartesian coordinates on the plane E^2, $\ddot{x}_1 = -\partial U/\partial x_1$ and $\ddot{x}_2 = -\partial U/\partial x_2$.

The *phase space* of a system with two degrees of freedom is the four-dimensional space with coordinates x_1, x_2, y_1, and y_2.

The system (2) defines the phase velocity vector field in four space as well as[15] the phase flow of the system (a one-parameter group of diffeomorphisms of four-dimensional phase space). The phase curves of (2) are subsets of four-dimensional phase space. All of phase space is partitioned into phase curves. Projecting the phase curves from four space to the x_1, x_2 plane gives the trajectories of our moving point in the x_1, x_2 plane. These trajectories are also called orbits. Orbits can have points of intersection even when the phase curves do not intersect one another. The equation of the law of conservation of energy

$$E = \frac{\dot{\mathbf{x}}^2}{2} + U(\mathbf{x}) = \frac{y_1^2 + y_2^2}{2} + U(x_1, x_2)$$

defines a three-dimensional hypersurface in four space: $E(x_1, x_2, y_1, y_2) = E_0$; this surface, π_{E_0}, remains invariant under the phase flow: $g^t \pi_{E_0} = \pi_{E_0}$. One could say that the phase flow flows along the energy level hypersurfaces. The phase velocity vector field is tangent at every point to π_{E_0}. Therefore, π_{E_0} is entirely composed of phase curves (Figure 16).

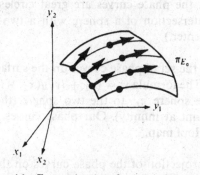

Figure 16 Energy level surface and phase curves

EXAMPLE 1 (*"small oscillations of a spherical pendulum"*). Let $U = \frac{1}{2}(x_1^2 + x_2^2)$. The level sets of the potential energy in the x_1, x_2 plane will be concentric circles (Figure 17).

The equations of motion, $\ddot{x}_1 = -x_1$, $\ddot{x}_2 = -x_2$, are equivalent to the system

$$\dot{x}_1 = y_1 \qquad \dot{x}_2 = y_2$$
$$\dot{y}_1 = -x_1 \qquad \dot{y}_2 = -x_2.$$

This system decomposes into two independent ones; in other words, each of the coordinates x_1 and x_2 changes with time in the same way as in a system with one degree of freedom.

[15] With the usual limitations.

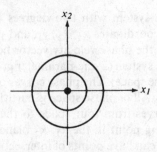

Figure 17   Potential energy level curves for a spherical pendulum

A solution has the form

$$x_1 = c_1 \cos t + c_2 \sin t \qquad x_2 = c_3 \cos t + c_4 \sin t$$
$$y_1 = -c_1 \sin t + c_2 \cos t \qquad y_2 = -c_3 \sin t + c_4 \cos t.$$

It follows from the law of conservation of energy that

$$E = \tfrac{1}{2}(y_1^2 + y_2^2) + \tfrac{1}{2}(x_1^2 + x_2^2) = \text{const},$$

i.e., the level surface π_{E_0} is a sphere in four space.

PROBLEM. Show that the phase curves are great circles of this sphere. (A great circle is the intersection of a sphere with a two-dimensional plane passing through its center.)

PROBLEM. Show that the set of phase curves on the surface π_{E_0} forms a two-dimensional sphere. The formula $w = (x_1 + iy_1)/(x_2 + iy_2)$ gives the "Hopf map" from the three sphere π_{E_0} to the two sphere (the complex w-plane completed by the point at infinity). Our phase curves are the pre-images of points under the Hopf map.

PROBLEM. Find the projection of the phase curves on the x_1, x_2 plane (i.e., draw the orbits of the motion of a point).

EXAMPLE 2 ("*Lissajous figures*"). We look at one more example of a planar motion ("small oscillations with two degrees of freedom"):

$$\ddot{x}_1 = -x_1 \qquad \ddot{x}_2 = -\omega^2 x_2.$$

The potential energy is

$$U = \tfrac{1}{2}x_1^2 + \tfrac{1}{2}\omega^2 x_2^2.$$

From the law of conservation of energy it follows that, if at the initial moment of time the total energy is

$$\tfrac{1}{2}(\dot{x}_1^2 + \dot{x}_2^2) + U(x_1, x_2) = E,$$

then all motions will take place inside the ellipse $U(x_1, x_2) \le E$.

Our system consists of two independent one-dimensional systems. Therefore, the law of conservation of energy is satisfied for each of them separately, i.e., the following quantities are preserved

$$E_1 = \tfrac{1}{2}\dot{x}_1^2 + \tfrac{1}{2}x_1^2 \qquad E_2 = \tfrac{1}{2}\dot{x}_2^2 + \tfrac{1}{2}\omega^2 x_2^2 \qquad (E = E_1 + E_2).$$

Consequently, the variable x_1 is bounded by the region $|x_1| \le A_1$, $A_1 = \sqrt{2E_1(0)}$, and x_2 oscillates within the region $|x_2| \le A_2$. The intersection of these two regions defines a rectangle which contains the orbits (Figure 18).

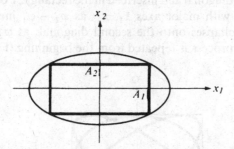

Figure 18   The regions $U \le E$, $U_1 \le E$ and $U_2 \le E$

PROBLEM. Show that this rectangle is inscribed in the ellipse $U \le E$.

The general solution of our equations is $x_1 = A_1 \sin(t + \varphi_1)$, $x_2 = A_2 \sin(\omega t + \varphi_2)$; a moving point independently performs an oscillation with frequency 1 and amplitude A_1 along the horizontal and an oscillation with frequency ω and amplitude A_2 along the vertical.

Consider the following method of describing an orbit in the x_1, x_2 plane. We look at a cylinder with base $2A_1$ and a band of width $2A_2$. We draw on the band a sine wave with period $2\pi A_1/\omega$ and amplitude A_2 and wind the band onto the cylinder (Figure 19). The orthogonal projection of the sinusoid

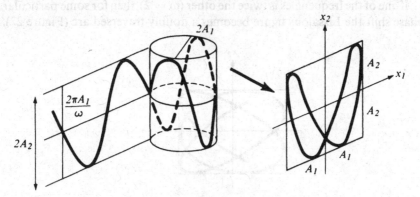

Figure 19   Construction of a Lissajous figure

25

wound around the cylinder onto the x_1, x_2 plane gives the desired orbit, called a *Lissajous figure*.

Lissajous figures can conveniently be seen on an oscilloscope which displays independent harmonic oscillations on the horizontal and vertical axes.

The form of a Lissajous figure very strongly depends on the frequency ω. If $\omega = 1$ (the spherical pendulum of Example 1), then the curve on the cylinder is an ellipse. The projection of this ellipse onto the x_1, x_2 plane depends on the difference $\varphi_2 - \varphi_1$ between the phases. For $\varphi_1 = \varphi_2$ we get a segment of the diagonal of the rectangle; for small $\varphi_2 - \varphi_1$ we get an ellipse close to the diagonal and inscribed in the rectangle. For $\varphi_2 - \varphi_1 = \pi/2$ we get an ellipse with major axes x_1, x_2; as $\varphi_2 - \varphi_1$ increases from $\pi/2$ to π the ellipse collapses onto the second diagonal; as $\varphi_2 - \varphi_1$ increases further the whole process is repeated from the beginning (Figure 20).

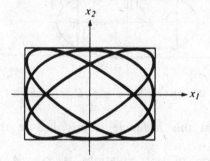

Figure 20   Series of Lissajous figures with $\omega = 1$

Now let the frequencies be only approximately equal: $\omega \approx 1$. The segment of the curve corresponding to $0 \leq t \leq 2\pi$ is very close to an ellipse. The next loop also reminds one of an ellipse, but here the phase shift $\varphi_2 - \varphi_1$ is greater than in the original by $2\pi(\omega - 1)$. Therefore, the Lissajous curve with $\omega \approx 1$ is a distorted ellipse, slowly progressing through all phases from collapsed onto one diagonal to collapsed onto the other (Figure 21).

If one of the frequencies is twice the other ($\omega = 2$), then for some particular phase shift the Lissajous figure becomes a doubly traversed arc (Figure 22).

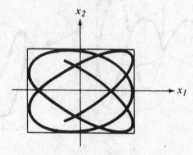

Figure 21   Lissajous figure with $\omega \approx 1$

PROBLEM. Show that this curve is a parabola. By increasing the phase shift $\varphi_2 - \varphi_1$ we get in turn the curves in Fig. 23.

In general, if one of the frequencies is n times bigger than the other ($\omega = n$), then among the graphs of the corresponding Lissajous figures there is the graph of a polynomial of degree n (Figure 24); this polynomial is called a *Chebyshev polynomial*.

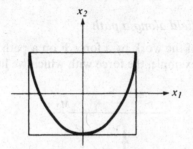

Figure 22   Lissajous figure with $\omega = 2$

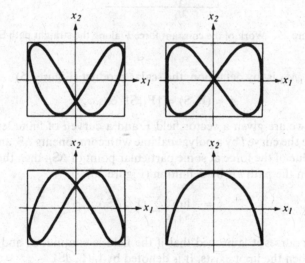

Figure 23   Series of Lissajous figures with $\omega = 2$

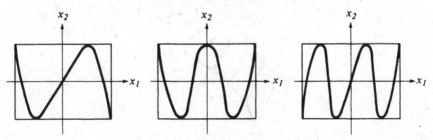

Figure 24   Chebyshev polynomials

27

PROBLEM. Show that if $\omega = m/n$, then the Lissajous figure is a closed algebraic curve; but if ω is irrational, then the Lissajous figure fills the rectangle everywhere densely. What does the corresponding phase trajectory fill out?

6 Conservative force fields

In this section we study the connection between work and potential energy.

A *Work of a force field along a path*

Recall the definition of the work by a force **F** on a path S. The work of the constant force **F** (for example, the force with which we lift up a load) on the

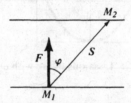

Figure 25   Work of the constant force **F** along the straight path S

path $S = \overrightarrow{M_1 M_2}$ is, by definition, the scalar product (Figure 25)

$$A = (\mathbf{F}, \mathbf{S}) = |\mathbf{F}||\mathbf{S}| \cdot \cos \varphi.$$

Suppose we are given a vector field **F** and a curve l of finite length. We approximate the curve l by a polygonal line with components $\Delta \mathbf{S}_i$ and denote by $\mathbf{F}_i$ the value of the force at some particular point of $\Delta \mathbf{S}_i$; then the work of the field **F** on the path l is by definition (Figure 26)

$$A = \lim_{|\Delta \mathbf{S}_i| \to 0} \sum (\mathbf{F}_i, \Delta \mathbf{S}_i).$$

In analysis courses it is proved that if the field is continuous and the path rectifiable, then the limit exists. It is denoted by $\int_l (\mathbf{F}, d\mathbf{S})$.

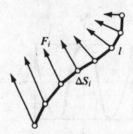

Figure 26   Work of the force field **F** along the path l

B Conditions for a field to be conservative

Theorem. *A vector field* **F** *is conservative if and only if its work along any path* M_1M_2 *depends only on the endpoints of the path, and not on the shape of the path.*

PROOF. Suppose that the work of a field **F** does not depend on the path. Then

$$U(M) = - \int_{M_0}^{M} (\mathbf{F}, d\mathbf{S})$$

is well defined as a function of the point M. It is easy to verify that

$$\mathbf{F} = - \frac{\partial U}{\partial \mathbf{x}},$$

i.e., the field is conservative and U is its potential energy. Of course, the potential energy is defined only up to the additive constant $U(M_0)$, which can be chosen arbitrarily.

Conversely, suppose that the field F is conservative and that U is its potential energy. Then it is easily verified that

$$\int_{M_0}^{M} (\mathbf{F}, d\mathbf{S}) = - U(M) + U(M_0),$$

i.e., the work does not depend on the shape of the path.   □

PROBLEM. Show that the vector field $F_1 = x_2, F_2 = -x_1$ is not conservative (Figure 27).

Figure 27   A non-potential field

PROBLEM. Is the field in the plane minus the origin given by $F_1 = x_2/(x_1^2 + x_2^2)$, $F_2 = -x_1/(x_1^2 + x_2^2)$ conservative? Show that a field is conservative if and only if its work along any closed contour is equal to zero.

C Central fields

Definition. A vector field in the plane E^2 is called *central* with center at 0, if it is invariant with respect to the group of motions[16] of the plane which fix 0.

[16] Including reflections.

PROBLEM. Show that all vectors of a central field lie on rays through 0, and that the magnitude of the vector field at a point depends only on the distance from the point to the center of the field.

It is also useful to look at central fields which are not defined at the point 0.

EXAMPLE. The newtonian field $\mathbf{F} = -k(\mathbf{r}/|r|^3)$ is central, but the field in the problem in Section 6B is not.

Theorem. *Every central field is conservative, and its potential energy depends only on the distance to the center of the field, $U = U(r)$.*

PROOF. According to the previous problem, we may set $\mathbf{F}(\mathbf{r}) = \Phi(r)\mathbf{e}_r$, where $\mathbf{r}$ is the radius vector with respect to 0, r is its length and the unit vector $\mathbf{e}_r = \mathbf{r}/|r|$ its direction. Then

$$\int_{M_1}^{M_2} (\mathbf{F}, d\mathbf{S}) = \int_{r(M_1)}^{r(M_2)} \Phi(r)dr,$$

and this integral is obviously independent of the path. □

PROBLEM. Compute the potential energy of the newtonian field.

Remark. The definitions and theorems of this paragraph can be directly carried over to a euclidean space E^n of any dimension.

7 Angular momentum

We will see later that the invariance of an equation of a mechanical problem with respect to some group of transformations always implies a conservation law. A central field is invariant with respect to the group of rotations. The corresponding first integral is called the angular momentum.

Definition. The *motion of a material point (with unit mass) in a central field on a plane* is defined by the equation

$$\ddot{\mathbf{r}} = \Phi(r)\mathbf{e}_r,$$

where $\mathbf{r}$ is the radius vector beginning at the center of the field 0, r is its length, and $\mathbf{e}_r$ its direction. We will think of our plane as lying in three-dimensional oriented euclidean space.

Definition. The *angular momentum* of a material point of unit mass relative to the point 0 is the vector product

$$\mathbf{M} = [\mathbf{r}, \dot{\mathbf{r}}].$$

The vector $\mathbf{M}$ is perpendicular to our plane and is given by one number: $\mathbf{M} = M\mathbf{n}$, where $\mathbf{n} = [\mathbf{e}_1, \mathbf{e}_2]$ is the normal vector, $\mathbf{e}_1$ and $\mathbf{e}_2$ being an oriented frame in the plane (Figure 28).

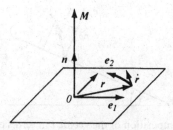

Figure 28   Angular momentum

Remark. In general, the moment of a vector **a** "applied at the point **r**" relative to the point 0 is [**r**, **a**]; for example, in a school statics course one studies the moment of force. [The literal translation of the Russian term for angular momentum is "kinetic moment." (Trans. note)]

A *The law of conservation of angular momentum*

Lemma. *Let* **a** *and* **b** *be two vectors changing with time in the oriented euclidean space* $\mathbb{R}^3$. *Then*

$$\frac{d}{dt}[\mathbf{a}, \mathbf{b}] = [\dot{\mathbf{a}}, \mathbf{b}] + [\mathbf{a}, \dot{\mathbf{b}}].$$

PROOF. This follows from the definition of derivative.   □

Theorem (The law of conservation of angular momentum). *Under motions in a central field, the angular momentum* **M** *relative to the center of the field* 0 *does not change with time.*

PROOF. By definition $\mathbf{M} = [\mathbf{r}, \dot{\mathbf{r}}]$. By the lemma, $\dot{\mathbf{M}} = [\dot{\mathbf{r}}, \dot{\mathbf{r}}] + [\mathbf{r}, \ddot{\mathbf{r}}]$. Since the field is central it is apparent from the equations of motion that the vectors $\ddot{\mathbf{r}}$ and **r** are collinear. Therefore $\dot{\mathbf{M}} = 0$.   □

B *Kepler's law*

The law of conservation of angular momentum was first discovered by Kepler through observation of the motion of Mars. Kepler formulated this law in a slightly different way.

We introduce polar coordinates r, φ on our plane with pole at the center of the field 0. We consider, at the point **r** with coordinates ($|\mathbf{r}| = r$, φ), two unit vectors: $\mathbf{e}_r$, directed along the radius vector so that

$$\mathbf{r} = r\mathbf{e}_r,$$

and $\mathbf{e}_\varphi$, perpendicular to it in the direction of increasing φ. We express the velocity vector $\dot{\mathbf{r}}$ in terms of the basis $\mathbf{e}_r$, $\mathbf{e}_\varphi$ (Figure 29).

Lemma. *We have the relation*

$$\dot{\mathbf{r}} = \dot{r}\mathbf{e}_r + r\dot{\varphi}\mathbf{e}_\varphi.$$

31

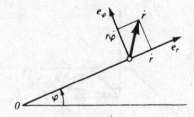

Figure 29   Decomposition of the vector $\dot{\mathbf{r}}$ in terms of the basis $\mathbf{e}_r$, $\mathbf{e}_\varphi$

PROOF. Clearly, the vectors $\mathbf{e}_r$ and $\mathbf{e}_\varphi$ rotate with angular velocity $\dot{\varphi}$, i.e.,

$$\dot{\mathbf{e}}_r = \dot{\varphi}\mathbf{e}_\varphi \qquad \dot{\mathbf{e}}_\varphi = -\dot{\varphi}\mathbf{e}_r.$$

Differentiating the equality $\mathbf{r} = r\mathbf{e}_r$ gives us

$$\dot{\mathbf{r}} = \dot{r}\mathbf{e}_r + r\dot{\mathbf{e}}_r = \dot{r}\mathbf{e}_r + r\dot{\varphi}\mathbf{e}_\varphi. \qquad \square$$

Consequently, the angular momentum is

$$\mathbf{M} = [\mathbf{r}, \dot{\mathbf{r}}] = [\mathbf{r}, \dot{r}\mathbf{e}_r] + [\mathbf{r}, r\dot{\varphi}\mathbf{e}_\varphi] = r\dot{\varphi}[\mathbf{r}, \mathbf{e}_\varphi] = r^2\dot{\varphi}[\mathbf{e}_r, \mathbf{e}_\varphi].$$

Thus, the quantity $M = r^2\dot{\varphi}$ is preserved. This quantity has a simple geometric meaning.

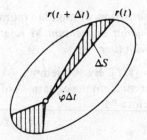

Figure 30   Sectorial velocity

Kepler called the rate of change of the area $S(t)$ swept out by the radius vector the *sectorial velocity* C (Figure 30):

$$C = \frac{dS}{dt}.$$

The law discovered by Kepler through observation of the motion of the planets says: in equal times the radius vector sweeps out equal areas, so that the sectorial velocity is constant, $dS/dt = \text{const}$. This is one formulation of the law of conservation of angular momentum. Since

$$\Delta S = S(t + \Delta t) - S(t) = \tfrac{1}{2}r^2\dot{\varphi}\Delta t + o(\Delta t),$$

this means that the sectorial velocity

$$C = \frac{dS}{dt} = \tfrac{1}{2}r^2\dot\varphi = \tfrac{1}{2}M$$

is half the angular momentum of our point of mass 1, and therefore constant.

EXAMPLE. Some satellites have very elongated orbits. By Kepler's law such a satellite spends most of its time in the distant part of its orbit, where the magnitude of $\dot\varphi$ is small.

8 Investigation of motion in a central field

The law of conservation of angular momentum lets us reduce problems about motion in a central field to problems with one degree of freedom. Thanks to this, motion in a central field can be completely determined.

A Reduction to a one-dimensional problem

We look at the motion of a point (of mass 1) in a central field on the plane:

$$\ddot{\mathbf{r}} = -\frac{\partial U}{\partial \mathbf{r}}, \qquad U = U(r).$$

It is natural to use polar coordinates r, φ.

By the law of conservation of angular momentum the quantity $M = \dot\varphi(t)r^2(t)$ is constant (independent of t).

THEOREM. *For the motion of a material point of unit mass in a central field the distance from the center of the field varies in the same way as r varies in the one-dimensional problem with potential energy*

$$V(r) = U(r) + \frac{M^2}{2r^2}.$$

PROOF. Differentiating the relation shown in Section 7 ($\dot{\mathbf{r}} = \dot r \mathbf{e}_r + r\dot\varphi \mathbf{e}_\varphi$), we find

$$\ddot{\mathbf{r}} = (\ddot r - r\dot\varphi^2)\mathbf{e}_r + (2\dot r \dot\varphi + r\ddot\varphi)\mathbf{e}_\varphi.$$

Since the field is central,

$$\frac{\partial U}{\partial \mathbf{r}} = \frac{\partial U}{\partial r}\mathbf{e}_r.$$

Therefore the equation of motion in polar coordinates takes the form

$$\ddot r - r\dot\varphi^2 = -\frac{\partial U}{\partial r} \qquad 2\dot r\dot\varphi + r\ddot\varphi = 0.$$

33

But, by the law of conservation of angular momentum,

$$\dot{\varphi} = \frac{M}{r^2},$$

where M is a constant independent of t, determined by the initial conditions. Therefore,

$$\ddot{r} = -\frac{\partial U}{\partial r} + r\frac{M^2}{r^4} \quad \text{or} \quad \ddot{r} = -\frac{\partial V}{\partial r}, \quad \text{where } V = U + \frac{M^2}{2r^2}.$$

The quantity $V(r)$ is called the *effective potential energy*.   □

Remark. The total energy in the derived one-dimensional problem

$$E_1 = \frac{\dot{r}^2}{2} + V(r)$$

is the same as the total energy in the original problem

$$E = \frac{\dot{\mathbf{r}}^2}{2} + U(\mathbf{r}),$$

since

$$\frac{\dot{\mathbf{r}}^2}{2} = \frac{\dot{r}^2}{2} + \frac{r^2\dot{\varphi}^2}{2} = \frac{\dot{r}^2}{2} + \frac{M^2}{2r^2}.$$

B *Integration of the equation of motion*

The total energy in the derived one-dimensional problem is conserved. Consequently, the dependence of r on t is defined by the quadrature

$$\dot{r} = \sqrt{2(E - V(r))} \qquad \int dt = \int \frac{dr}{\sqrt{2(E - V(r))}}.$$

Since $\dot{\varphi} = M/r^2$, $d\varphi/dr = (M/r^2)/\sqrt{2(E - V(r))}$, and the equation of the orbit in polar coordinates is found by quadrature,

$$\varphi = \int \frac{M/r^2 \, dr}{\sqrt{2(E - V(r))}}.$$

C *Investigation of the orbit*

We fix the value of the angular momentum at M. The variation of r with time is easy to visualize, if one draws the graph of the effective potential energy $V(r)$ (Figure 31).

Let E be the value of the total energy. All orbits corresponding to the given E and M lie in the region $V(r) \leq E$. On the boundary of this region, $V = E$,

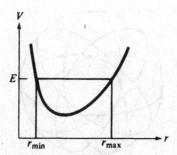

Figure 31   Graph of the effective potential energy

i.e., $\dot{r} = 0$. Therefore, the velocity of the moving point, in general, is not equal to zero since $\dot{\varphi} \neq 0$ for $M \neq 0$.

The inequality $V(r) \leq E$ gives one or several annular regions in the plane:

$$0 \leq r_{min} \leq r \leq r_{max} \leq \infty.$$

If $0 \leq r_{min} < r_{max} < \infty$, then the motion is bounded and takes place inside the ring between the circles of radius r_{min} and r_{max}.

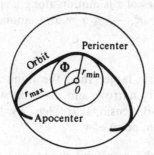

Figure 32   Orbit of a point in a central field

The shape of an orbit is shown in Figure 32. The angle φ varies monotonically while r oscillates periodically between r_{min} and r_{max}. The points where $r = r_{min}$ are called *pericentral*, and where $r = r_{max}$, *apocentral* (if the center is the earth—perigee and apogee; if it is the sun—perihelion and aphelion; if it is the moon—perilune and apolune).

Each of the rays leading from the center to the apocenter or to the pericenter is an axis of symmetry of the orbit.

In general, the orbit is not closed: the angle between the successive pericenters and apocenters is given by the integral

$$\Phi = \int_{r_{min}}^{r_{max}} \frac{M/r^2 \, dr}{\sqrt{2(E - V(r))}}.$$

The angle between two successive pericenters is twice as big.

35

Figure 33   Orbit dense in an annulus

The orbit is closed if the angle Φ is commensurable with 2π, i.e., if $\Phi = 2\pi(m/n)$, where m and n are integers.

It can be shown that if the angle Φ is not commensurable with 2π, then the orbit is everywhere dense in the annulus (Figure 33).

If $r_{min} = r_{max}$, i.e., E is the value of V at a minimum point, then the annulus degenerates to a circle, which is also the orbit.

PROBLEM. For which values of α is motion along a circular orbit in the field with potential energy $U = r^{\alpha}$, $-2 \leq \alpha < \infty$, Liapunov stable?

ANSWER. Only for $\alpha = 2$.

For values of E a little larger than the minimum of V the annulus $r_{min} \leq r \leq r_{max}$ will be very narrow, and the orbit will be close to a circle. In the corresponding one-dimensional problem, r will perform small oscillations close to the minimum point of V.

PROBLEM. Find the angle Φ for an orbit close to the circle of radius r.

Hint. Cf. Section D below.

We now look at the case $r_{max} = \infty$. If $\lim_{r \to \infty} U(r) = \lim_{r \to \infty} V(r) = U_{\infty} < \infty$, then it is possible for orbits to go off to infinity. If the initial energy E is larger than U, then the point goes to infinity with finite velocity $\dot{r}_{\infty} = \sqrt{2(E - U_{\infty})}$. We notice that if $U(r)$ approaches its limit slower than r^{-2}, then the effective potential V will be attracting at infinity (here we assume that the potential U is attracting at infinity).

If, as $r \to 0$, $|U(r)|$ does not grow faster than $M^2/2r^2$, then $r_{min} > 0$ and the orbit never approaches the center. If, however, $U(r) + (M^2/2r^2) \to -\infty$ as $r \to 0$, then it is possible to "fall into the center of the field." Falling into the center of the field is possible even in finite time (for example, in the field $U(r) = -1/r^3$).

PROBLEM. Examine the shape of an orbit in the case when the total energy is equal to the value of the effective energy V at a local maximum point.

D Central fields in which all bounded orbits are closed

It follows from the following sequence of problems that there are only two cases in which all the bounded orbits in a central field are closed, namely,

$$U = ar^2, \qquad a \geq 0$$

and

$$U = -\frac{k}{r}, \qquad k \geq 0.$$

PROBLEM 1. Show that the angle Φ between the pericenter and apocenter is equal to the semiperiod of an oscillation in the one-dimensional system with potential energy $W(x) = U(M/x) + (x^2/2)$.

Hint. The substitution $x = M/r$ gives

$$\Phi = \int_{x_{min}}^{x_{max}} \frac{dx}{\sqrt{2(E - W)}}.$$

PROBLEM 2. Find the angle Φ for an orbit close to the circle of radius r.

ANSWER. $\Phi \approx \Phi_{cir} = \pi(M/r^2\sqrt{V''(r)}) = \pi\sqrt{U'/(3U' + rU'')}$.

PROBLEM 3. For which values of U is the magnitude of Φ_{cir} independent of the radius r?

ANSWER. $U(r) = ar^\alpha$ ($\alpha \geq -2$, $\alpha \neq 0$) and $U(r) = b \log r$.

It follows that $\Phi_{cir} = \pi/\sqrt{\alpha + 2}$ (the logarithmic case corresponds to $\alpha = 0$). For example, for $\alpha = 2$ we have $\Phi_{cir} = \pi/2$, and for $\alpha = -1$ we have $\Phi_{cir} = \pi$.

PROBLEM 4. Let in the situation of problem 3 $U(r) \to \infty$ as $r \to \infty$. Find $\lim_{E \to \infty} \Phi(E, M)$.

ANSWER. $\pi/2$.

Hint. The substitution $x = yx_{max}$ reduces Φ to the form

$$\Phi = \int_{y_{min}}^{1} \frac{dy}{\sqrt{2(W^*(1) - W^*(y))}}, \qquad W^*(y) = \frac{y^2}{2} + \frac{1}{x_{max}^2} U\left(\frac{M}{yx_{max}}\right).$$

As $E \to \infty$ we have $x_{max} \to \infty$ and $y_{min} \to 0$, and the second term in W^* can be discarded.

PROBLEM 5. Let $U(r) = -kr^{-\beta}, 0 < \beta < 2$. Find $\Phi_0 = \lim_{E \to -0} \Phi$.

ANSWER. $\Phi_0 = \int_0^1 dx/\sqrt{x^\beta - x^2} = \pi/(2 - \beta)$. Note that Φ_0 does not depend on M.

PROBLEM 6. Find all central fields in which bounded orbits exist and are all closed.

ANSWER. $U = ar^2$ or $U = -k/r$.

Solution. If all bounded orbits are closed, then, in particular, $\Phi_{cir} = 2\pi(m/n) = $ const. According to Problem 3, $U = ar^\alpha (\alpha \geq -2)$, or $U = b \ln r$ ($\alpha = 0$). In both cases $\Phi_{cir} = \pi/\sqrt{\alpha + 2}$. If $\alpha > 0$, then according to Problem 4, $\lim_{E \to \infty} \Phi(E, M) = \pi/2$. Therefore, $\Phi_{cir} = \pi/2$, $\alpha = 2$. If $\alpha < 0$, then according to Problem 5, $\lim_{E \to -\infty} \Phi(E, M) = \pi/(2 + \alpha)$. Therefore, $\pi/(2 + \alpha) = \pi/\sqrt{2 + \alpha}$, $\alpha = -1$. In the case $\alpha = 0$ we find $\Phi_{cir} = \pi/\sqrt{2}$, which is not commensurable with 2π. Therefore, all bounded orbits can be closed only in fields where $U = ar^2$ or $U = -k/r$. In the field $U = ar^2$, $a > 0$, all the orbits are closed (these are ellipses with center at 0, cf. Example 1, Section 5). In the field $U = -k/r$ all bounded orbits are also closed and also elliptical, as we will now show.

E *Kepler's problem*

This problem concerns motion in a central field with potential $U = -k/r$ and therefore $V(r) = -(k/r) + (M^2/2r^2)$ (Figure 34).

By the general formula

$$\varphi = \int \frac{M/r^2 \, dr}{\sqrt{2(E - V(r))}}.$$

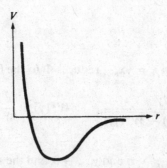

Figure 34   Effective potential of the Kepler problem

Integrating, we get

$$\varphi = \arccos \frac{\dfrac{M}{r} - \dfrac{k}{M}}{\sqrt{2E + \dfrac{k^2}{M^2}}}.$$

To this expression we should have added an arbitrary constant. We will assume it equal to zero; this is equivalent to the choice of an origin of reference for the angle φ at the pericenter. We introduce the following notation:

$$\frac{M^2}{k} = p \qquad \sqrt{1 + \frac{2EM^2}{k^2}} = e.$$

Now we get $\varphi = \arccos((p/r) - 1)/e$, i.e.,

$$r = \frac{p}{1 + e \cos \varphi}.$$

This is the so-called *focal equation* of a conic section. The motion is bounded (Figure 35) for $E < 0$. Then $e < 1$, i.e., the conic section is an ellipse. The number p is called the *parameter* of the ellipse, and e the *eccentricity*. Kepler's first law, which he discovered by observing the motion of Mars, consists of the fact that the planets describe ellipses, with the sun at one focus.

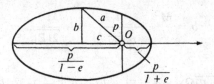

Figure 35   Keplerian ellipse

If we assume that the planets move in a central field of gravity, then Kepler's first law implies Newton's law of gravity: $U = -(k/r)$ (cf. Section 2D above).

The parameter and eccentricity are related with the semi-axes by the formulas

$$2a = \frac{p}{1 - e} + \frac{p}{1 + e} = \frac{2p}{1 - e^2},$$

i.e.,

$$a = \frac{p}{1 - e^2},$$

$e = c/a = \sqrt{a^2 - b^2}/a$, where $c = ae$ is the distance from the center to the focus (cf. Figure 35).

Remark. An ellipse with small eccentricity is very close to a circle.[17] If the distance from the focus to the center is small of first order, then the difference between the semi-axes is of second order: $b = a\sqrt{1 - e^2} \approx a(1 - \frac{1}{2}e^2)$. For example, in the ellipse with major semi-axes of 10 cm and eccentricity 0.1, the difference of the semi-axes is 0.5 mm, and the distance between the focus and the center is 1 cm.

The eccentricities of planets' orbits are very small. Therefore, Kepler originally formulated his first law as follows: the planets move around the sun in circles, but the sun is not at the center.

Kepler's second law, that the sectorial velocity is constant, is true in any central field.

Kepler's third law says that the period of revolution around an elliptical orbit depends only on the size of the major semi-axes.

The squares of the revolution periods of two planets on different elliptical orbits have the same ratio as the cubes of their major semi-axes.[18]

PROOF. We denote by T the period of revolution and by S the area swept out by the radius vector in time T. $2S = MT$, since $M/2$ is the sectorial velocity. But the area of the ellipse, S, is equal to πab, so $T = 2\pi ab/M$. Since

$$a = \frac{M^2/k}{2|E|\frac{M^2}{k^2}} = \frac{k}{2|E|}$$

(from $a = p/(1 - e^2)$), and

$$b = \frac{M^2}{k} \cdot \frac{1}{\sqrt{2|E|}\frac{M}{k}} = \frac{M}{\sqrt{2|E|}},$$

then $T = 2\pi(k/(\sqrt{2|E|})^3)$; but $2|E| = k/a$, so $T = 2\pi a^{3/2}k^{-1/2}$. □

We note that the total energy E depends only on the major semi-axis a of the orbit and is the same for the whole set of elliptical orbits, from a circle of radius a to a line segment of length $2a$.

PROBLEM. At the entry of a satellite into a circular orbit at a distance 300 km from the earth the direction of its velocity deviates from the intended direction by $1°$ towards the earth. How is the perigee changed?

ANSWER. The height of the perigee is less by approximately 110 km.

---

[17] Let a drop of tea fall into a glass of tea close to the center. The waves collect at the symmetric point. The reason is that, by the focal definition of an ellipse, waves radiating from one focus of the ellipse collect at the other.

[18] By planets we mean here points in a central field.

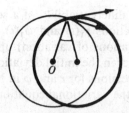

Figure 36   An orbit which is close to circular

Hint. The orbit differs from a circle only to second order, and we can disregard this difference. The radius has the intended value since the initial energy has the intended value. Therefore, we get the true orbit (Figure 36) by twisting the intended orbit through 1°.

PROBLEM. How does the height of the perigee change if the actual velocity is 1 m/sec less than intended?

PROBLEM. The *first cosmic velocity* is the velocity of motion on a circular orbit of radius close to the radius of the earth. Find the magnitude of the first cosmic velocity v_1 and show that $v_2 = \sqrt{2} v_1$ (cf. Section 3B).

ANSWER. 8.1 km/sec.

PROBLEM.[19] During his walk in outer space, the cosmonaut A. Leonov threw the lens cap of his movie camera towards the earth. Describe the motion of the lens cap with respect to the spaceship, taking the velocity of the throw as 10 m/sec.

ANSWER. The lens cap will move relative to the cosmonaut approximately in an ellipse with major axis about 32 km and minor axis about 16 km. The center of the ellipse will be situated 16 km in front of the cosmonaut in his orbit, and the period of circulation around the ellipse will be equal to the period of motion around the orbit.

Hint. We take as our unit of length the radius of the space ship's circular orbit, and we choose a unit of time so that the period of revolution around this orbit is 2π. We must study solutions to Newton's equation

$$\ddot{\mathbf{r}} = -\frac{\mathbf{r}}{r^3},$$

close to the circular solution with $r_0 = 1$, $\varphi_0 = t$. We seek those solutions in the form

$$r = r_0 + r_1 \qquad \varphi = \varphi_0 + \varphi_1 \qquad r_1 \ll 1, \varphi_1 \ll 1.$$

[19] This problem is taken from V. V. Beletskii's delightful book, "Sketches on the Motion of Celestial Bodies," Nauka, 1972.

By the theorem on the differentiability of a solution with respect to its initial conditions, the functions $r_1(t)$ and $\varphi_1(t)$ satisfy a system of linear differential equations (equations of variation) up to small amounts which are of higher than first order in the initial deviation.

By substituting the expressions for r and φ in Newton's equation, we get, after simple computation, the variational equations in the form

$$\ddot{r}_1 = 3r_1 + 2\dot{\varphi}_1 \qquad \ddot{\varphi}_1 = -2\dot{r}_1.$$

After solving these equations for the given initial conditions $(r_1(0) = \varphi_1(0) = \dot{\varphi}_1(0) = 0, \dot{r}_1(0) = -(1/800))$, we get the answer given above.

Disregarding the small quantities of second order gives an effect of under 1/800 of the one obtained (i.e., on the order of 10 meters on one loop). Thus the lens cap describes a 30 km ellipse in an hour-and-a-half, returns to the space ship on the side opposite the earth, and goes past at the distance of a few tens of meters.

Of course, in this calculation we have disregarded the deviation of the orbit from a circle, the effect of forces other than gravity, etc.

9 The motion of a point in three-space

In this paragraph we define the angular momentum relative to an axis and we show that, for motion in an axially symmetric field, it is conserved.

All the results obtained for motion in a plane can be easily carried over to motions in space.

A Conservative fields

We consider a motion in the conservative field

$$\ddot{\mathbf{r}} = -\frac{\partial U}{\partial \mathbf{r}},$$

where $U = U(\mathbf{r})$, $\mathbf{r} \in E^3$.

The law of conservation of energy holds:

$$\frac{dE}{dt} = 0, \quad \text{where } E = \tfrac{1}{2}\dot{\mathbf{r}}^2 + U(\mathbf{r}).$$

B Central fields

For motion in a central field the vector $\mathbf{M} = [\mathbf{r}, \dot{\mathbf{r}}]$ does not change: $d\mathbf{M}/dt = 0$.

Every central field is conservative (this is proved as in the two-dimensional case), and

$$\frac{d\mathbf{M}}{dt} = [\dot{\mathbf{r}}, \dot{\mathbf{r}}] + [\mathbf{r}, \ddot{\mathbf{r}}] = 0,$$

since $\ddot{\mathbf{r}} = -(\partial U/\partial \mathbf{r})$, and the vector $\partial U/\partial \mathbf{r}$ is collinear with $\mathbf{r}$ since the field is central.

Corollary. *For motion in a central field, every orbit is planar.*

PROOF. $(\mathbf{M}, \mathbf{r}) = ([\mathbf{r}, \dot{\mathbf{r}}], \mathbf{r}) = 0$; therefore $\mathbf{r}(t) \perp \mathbf{M}$, and since $\mathbf{M} = \text{const.}$, all orbits lie in the plane perpendicular to $\mathbf{M}$.[20] ☐

Thus the study of orbits in a central field in space reduces to the planar problem examined in the previous paragraph.

PROBLEM. Investigate motion in a central field in n-dimensional euclidean space.

C Axially symmetric fields

Definition. A vector field in E^3 has *axial symmetry* if it is invariant with respect to the group of rotations of space which fix every point of some axis.

PROBLEM. Show that if a field is axially symmetric and conservative, then its potential energy has the form $U = U(r, z)$, where r, φ, and z are cylindrical coordinates.

In particular, it follows from this that the vectors of the field lie in planes through the z axis.

As an example of such a field we can take the gravitational field created by a solid of revolution.

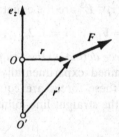

Figure 37   Moment of the vector **F** with respect to an axis

Let z be the axis, oriented by the vector $\mathbf{e}_z$ in three-dimensional euclidean space E^3; **F** a vector in the euclidean linear space $\mathbb{R}^3$; 0 a point on the z axis; $\mathbf{r} = x - 0 \in \mathbb{R}^3$ the radius vector of the point $x \in E^3$ relative to 0 (Figure 37).

Definition. The moment M_z relative to the z axis of the vector **F** applied at the point **r** is the projection onto the z axis of the moment of the vector **F** relative to some point on this axis:

$$M_z = (\mathbf{e}_z, [\mathbf{r}, \mathbf{F}]).$$

[20] The case $\mathbf{M} = 0$ is left to the reader.

43

The number M_z does not depend on the choice of the point 0 on the z axis. In fact, if we look at a point $0'$ on the axis, then by properties of the triple product, $M'_z = (\mathbf{e}_z, [\mathbf{r}', \mathbf{F}]) = ([\mathbf{e}_z, \mathbf{r}'], \mathbf{F}) = ([\mathbf{e}_z, \mathbf{r}], \mathbf{F}) = M_z$.

Remark. M_z depends on the choice of the *direction* of the z axis: if we change $\mathbf{e}_z$ to $-\mathbf{e}_z$, then M_z changes sign.

Theorem. *For a motion in a conservative field with axial symmetry around the z axis, the moment of velocity relative to the z axis is conserved.*

PROOF. $M_z = (\mathbf{e}_z, [\mathbf{r}, \dot{\mathbf{r}}])$. Since $\ddot{\mathbf{r}} = \mathbf{F}$, it follows that $\mathbf{r}$ and $\ddot{\mathbf{r}}$ lie in a plane passing through the z axis, and therefore $[\mathbf{r}, \ddot{\mathbf{r}}]$ is perpendicular to $\mathbf{e}_z$.
Therefore,

$$\dot{M}_z = (\mathbf{e}_z, [\dot{\mathbf{r}}, \dot{\mathbf{r}}]) + (\mathbf{e}_z, [\mathbf{r}, \ddot{\mathbf{r}}]) = 0. \qquad \square$$

Remark. This proof works for any force field in which the force vector $\mathbf{F}$ lies in the plane spanned by $\mathbf{r}$ and $\mathbf{e}_z$.

10 Motions of a system of n points

In this paragraph we prove the laws of conservation of energy, momentum, and angular momentum for systems of material points in E^3.

A Internal and external forces

Newton's equations for the motion of a system of n material points, with masses m_i and radius vectors $\mathbf{r}_i \in E^3$ are the equations

$$m_i \ddot{\mathbf{r}}_i = \mathbf{F}_i, \qquad i = 1, 2, \ldots, n.$$

The vector $\mathbf{F}_i$ is called the *force acting on the i-th point.*

The forces $\mathbf{F}_i$ are determined experimentally. We often observe in a system that for two points these forces are equal in magnitude and act in opposite directions along the straight line joining the points (Figure 38).

Figure 38   Forces of interaction

Such forces are called *forces of interaction* (example: the force of universal gravitation).

If all forces acting on a point of the system are forces of interaction, then the system is said to be *closed*. By definition, the force acting on the i-th point of a closed system is

$$\mathbf{F}_i = \sum_{\substack{j=1 \\ j \neq i}}^{n} \mathbf{F}_{ij}.$$

The vector $\mathbf{F}_{ij}$ is the force with which the j-th point acts on the i-th.

Since the forces $\mathbf{F}_{ij}$ and $\mathbf{F}_{ji}$ are opposite ($\mathbf{F}_{ij} = -\mathbf{F}_{ji}$), we can write them in the form $\mathbf{F}_{ij} = f_{ij}\mathbf{e}_{ij}$, where $f_{ij} = f_{ji}$ is the magnitude of the force and $\mathbf{e}_{ij}$ is the unit vector in the direction from the i-th point to the j-th point.

If the system is not closed, then it is often possible to represent the forces acting on it in the form

$$\mathbf{F}_i = \sum \mathbf{F}_{ij} + \mathbf{F}'_i,$$

where $\mathbf{F}_{ij}$ are forces of interaction and $\mathbf{F}'_i(\mathbf{r}_i)$ is the so-called *external force*.

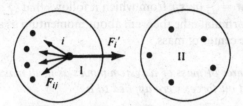

Figure 39   Internal and external forces

EXAMPLE. (Figure 39) We separate a closed system into two parts, I and II. The force $\mathbf{F}_i$ applied to the i-th point of system I is determined by forces of interaction inside system I and forces acting on the i-th point from points of system II, i.e.,

$$\mathbf{F}_i = \sum_{\substack{j \in I \\ j \neq i}} \mathbf{F}_{ij} + \mathbf{F}'_i.$$

$\mathbf{F}'_i$ is the external force with respect to system I.

B The law of conservation of momentum

Definition. The *momentum* of a system is the vector

$$\mathbf{P} = \sum_{i=1}^{n} m_i \dot{\mathbf{r}}_i.$$

Theorem. *The rate of change of momentum of a system is equal to the sum of all external forces acting on points of the system.*

PROOF. $d\mathbf{P}/dt = \sum_{i=1}^{n} m_i \ddot{\mathbf{r}}_i = \sum_{i=1}^{n} \mathbf{F}_i = \sum_{i,j} \mathbf{F}_{ij} + \sum_i \mathbf{F}'_i = \sum_i \mathbf{F}'_i; \sum_{i,j} \mathbf{F}_{ij} = 0$, since for forces of interaction $\mathbf{F}_{ij} = -\mathbf{F}_{ji}$. □

Corollary 1. *The momentum of a closed system is conserved.*

Corollary 2. *If the sum of the exterior forces acting on a system is perpendicular to the x axis, then the projection P_x of the momentum onto the x axis is conserved: $P_x = $ const.*

Definition. The *center of mass* of a system is the point

$$\mathbf{r} = \frac{\sum m_i \mathbf{r}_i}{\sum m_i}.$$

PROBLEM. Show that the center of mass is well defined, i.e., does not depend on the choice of the origin of reference for radius vectors.

The momentum of a system is equal to the momentum of a particle lying at the center of mass of the system and having mass $\sum m_i$.

In fact, $(\sum m_i)\mathbf{r} = \sum (m_i \mathbf{r}_i)$, from which it follows that $(\sum m_i)\dot{\mathbf{r}} = \sum m_i \dot{\mathbf{r}}_i$.

We can now formulate the theorem about momentum as a theorem about the motion of the center of mass.

Theorem. *The center of mass of a system moves as if all masses were concentrated at it and all forces were applied to it.*

PROOF. $(\sum m_i)\dot{\mathbf{r}} = \mathbf{P}$. Therefore, $(\sum m_i)\ddot{\mathbf{r}} = d\mathbf{P}/dt = \sum_i \mathbf{F}_i$. $\qquad\square$

Corollary. *If a system is closed, then its center of mass moves uniformly and linearly.*

C The law of conservation of angular momentum

Definition. The *angular momentum of a material point of mass m relative to the point 0*, is the moment of the momentum vector relative to 0:

$$\mathbf{M} = [\mathbf{r}, m\dot{\mathbf{r}}].$$

The *angular momentum of a system relative to 0* is the sum of the angular momenta of all the points in the system:

$$\mathbf{M} = \sum_{i=1}^{n} [\mathbf{r}_i, m_i \dot{\mathbf{r}}_i].$$

Theorem. *The rate of change of the angular momentum of a system is equal to the sum of the moments of the external forces[21] acting on the points of the system.*

PROOF. $d\mathbf{M}/dt = \sum_{i=1}^{n} [\dot{\mathbf{r}}_i, m_i \dot{\mathbf{r}}_i] + \sum_{i=1}^{n} [\mathbf{r}_i, m_i \ddot{\mathbf{r}}_i]$. The first term is equal to zero, and the second is equal to

$$\sum_{i=1}^{n} [\mathbf{r}_i, \mathbf{F}_i] = \sum_{i=1}^{n} \left[\mathbf{r}_i, \left(\sum_{i \neq j} \mathbf{F}_{ij} + \mathbf{F}_i' \right) \right] = \sum_{i=1}^{n} [\mathbf{r}_i, \mathbf{F}_i'],$$

by Newton's equations.

[21] The moment of force is also called the *torque* [Trans. note].

46

The sum of the moments of two forces of interaction is equal to zero since

$$\mathbf{F}_{ij} = -\mathbf{F}_{ji}, \text{ so } [\mathbf{r}_i, \mathbf{F}_{ij}] + [\mathbf{r}_j, \mathbf{F}_{ji}] = [(\mathbf{r}_i - \mathbf{r}_j), \mathbf{F}_{ij}] = 0.$$

Therefore, the sum of the moments of all forces of interaction is equal to zero:

$$\sum_{i=1}^{n}\left[\mathbf{r}_i, \sum_{i \neq j}\mathbf{F}_{ij}\right] = 0.$$

Therefore, $d\mathbf{M}/dt = \sum_{i=1}^{n}[\mathbf{r}_i, \mathbf{F}'_i]$. □

Corollary 1 (The law of conservation of angular momentum). *If the system is closed, then* $\mathbf{M} = \text{const}$.

We denote the sum of the moments of the external forces by $\mathbf{N} = \sum_{i=1}^{n}[\mathbf{r}_i, \mathbf{F}'_i]$.
Then, by the theorem above, $d\mathbf{M}/dt = \mathbf{N}$, from which we have

Corollary 2. *If the moment of the external forces relative to the z axis is equal to zero, then* M_z *is constant.*

D *The law of conservation of energy*

Definition. The *kinetic energy of a point of mass m* is

$$T = \frac{m\dot{\mathbf{r}}^2}{2}.$$

Definition. The *kinetic energy of a system of mass points* is the sum of the kinetic energies of the points:

$$T = \sum_{i=1}^{n}\frac{m_i\dot{\mathbf{r}}_i^2}{2},$$

where the m_i are the masses of the points and $\dot{\mathbf{r}}_i$ are their velocities.

Theorem. *The increase in the kinetic energy of a system is equal to the sum of the work of all forces acting on the points of the system.*

PROOF.

$$\frac{dT}{dt} = \sum_{i=1}^{n}m_i(\dot{\mathbf{r}}_i, \ddot{\mathbf{r}}_i) = \sum_{i=1}^{r}(\dot{\mathbf{r}}_i, m_i\ddot{\mathbf{r}}_i) = \sum_{i=1}^{n}(\dot{\mathbf{r}}_i, \mathbf{F}_i).$$

Therefore,

$$T(t) - T(t_0) = \int_{t_0}^{t}\frac{dT}{dt}dt = \sum_{i=1}^{n}\int_{t_0}^{t}(\dot{\mathbf{r}}_i, \mathbf{F}_i)dt = \sum_{i=1}^{n}A_i.$$ □

The configuration space of a system of n mass points in E^3 is the direct product of n euclidean spaces: $E^{3n} = E^3 \times \cdots \times E^3$. It has itself the structure of a euclidean space.

Let $\mathbf{r} = (\mathbf{r}_1, \ldots, \mathbf{r}_n)$ be the radius vector of a point in the configuration space, and $\mathbf{F} = (\mathbf{F}_1, \ldots, \mathbf{F}_n)$ the force vector. We can write the theorem above in the form

$$T(t_1) - T(t_0) = \int_{r(t_0)}^{r(t_1)} (\mathbf{F}, d\mathbf{r}) = \int_{t_0}^{t_1} (\dot{\mathbf{r}}, \mathbf{F}) dt.$$

In other words:

The increase in kinetic energy is equal to the work of the "force" $\mathbf{F}$ on the "path" $\mathbf{r}(t)$ in configuration space.

Definition. A system is called conservative if the forces depend only on the location of a point in the system ($\mathbf{F} = \mathbf{F}(\mathbf{r})$), and if the work of $\mathbf{F}$ along any path depends only on the initial and final points of the path:

$$\int_{M_1}^{M_2} (\mathbf{F}, d\mathbf{r}) = \Phi(M_1, M_2).$$

Theorem. *For a system to be conservative it is necessary and sufficient that there exist a potential energy, i.e., a function $U(\mathbf{r})$ such that*

$$\mathbf{F} = -\frac{\partial U}{\partial \mathbf{r}}.$$

PROOF. Cf. Section 6B. □

Theorem. *The total energy of a conservative system ($E = T + U$) is preserved under the motion: $E(t_1) = E(t_0)$.*

PROOF. By what was shown earlier,

$$T(t_1) - T(t_0) = \int_{r(t_0)}^{r(t_1)} (\mathbf{F}, d\mathbf{r}) = U(\mathbf{r}(t_0)) - U(\mathbf{r}(t_1)). \qquad \square$$

Let all the forces acting on the points of a system be divided into forces of interaction and external forces:

$$\mathbf{F}_i = \sum_{i \neq j} \mathbf{F}_{ij} + \mathbf{F}'_i,$$

where $\mathbf{F}_{ij} = -\mathbf{F}_{ji} = f_{ij}\mathbf{e}_{ij}$.

Proposition. *If the forces of interaction depend only on distance, $f_{ij} = f_{ij}(|\mathbf{r}_i - \mathbf{r}_j|)$, then they are conservative.*

PROOF. If a system consists entirely of two points i and j, then, as is easily seen, the potential energy of the interaction is given by the formula

$$U_{ij}(\mathbf{r}) = \int_{r_0}^{r} f_{ij}(\rho)d\rho.$$

We then have

$$-\frac{\partial U_{ij}(|\mathbf{r}_i - \mathbf{r}_j|)}{\partial \mathbf{r}_i} = -f_{ij}\frac{\partial |\mathbf{r}_i - \mathbf{r}_j|}{\partial \mathbf{r}_i} = f_{ij}\mathbf{e}_{ij}.$$

Therefore, the potential energy of the interaction of all the points will be

$$U(\mathbf{r}) = \sum_{i>j} U_{ij}(|\mathbf{r}_i - \mathbf{r}_j|). \qquad \square$$

If the external forces are also conservative, i.e., $\mathbf{F}'_i = -(\partial U'_i/\partial \mathbf{r}_i)$, then the system is conservative, and its total potential energy is

$$U(\mathbf{r}) = \sum_{i>j} U_{ij} + \sum_i U'_i.$$

For such a system the total mechanical energy

$$E = T + U = \sum_i \frac{\dot{\mathbf{r}}_i^2}{2} + \sum_{i>j} U_{ij} + \sum_i U'_i$$

is conserved.

If the system is not conservative, then the total mechanical energy is not generally conserved.

Definition. A decrease in the mechanical energy $E(t_0) - E(t_1)$ is called an increase in the *non-mechanical energy* E':

$$E'(t_1) - E'(t_0) = E(t_0) - E(t_1).$$

Theorem (The law of conservation of energy). *The total energy $H = E + E'$ is conserved.*

This theorem is an obvious corollary of the definition above. Its value lies in the fact that in concrete physical systems, expressions for the size of the non-mechanical energy can be found in terms of other physical quantities (temperature, etc.).

E *Example: The two-body problem*

Suppose that two points with masses m_1 and m_2 interact with potential U, so that the equations of motion have the form

$$m_1\ddot{\mathbf{r}}_1 = -\frac{\partial U}{\partial \mathbf{r}_1} \qquad m_2\ddot{\mathbf{r}}_2 = -\frac{\partial U}{\partial \mathbf{r}_2}, \qquad U = U(|\mathbf{r}_1 - \mathbf{r}_2|).$$

49

Theorem. *The time variation of* $\mathbf{r} = \mathbf{r}_1 - \mathbf{r}_2$ *in the two-body problem is the same as that for the motion of a point of mass* $m = m_1 m_2/(m_1 + m_2)$ *in a field with potential* $U(|\mathbf{r}|)$.

We denote by $\mathbf{r}_0$ the radius vector of the center of mass: $\mathbf{r}_0 = (m_1\mathbf{r}_1 + m_2\mathbf{r}_2)/(m_1 + m_2)$. By the theorem on the conservation of momentum, the point $\mathbf{r}_0$ moves uniformly and linearly.

We now look at the vector $\mathbf{r} = \mathbf{r}_1 - \mathbf{r}_2$. Multiplying the first of the equations of motion by m_2, the second by m_1, and computing, we find that $m_1 m_2 \ddot{\mathbf{r}} = -(m_1 + m_2)(\partial U/\partial \mathbf{r})$, where $U = U(|\mathbf{r}_1 - \mathbf{r}_2|) = U(|\mathbf{r}|)$.

In particular, in the case of a Newtonian attraction, the points describe conic sections with foci at their common center of mass (Figure 40).

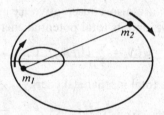

Figure 40   The two body problem

PROBLEM. Determine the major semi-axis of the ellipse which the center of the earth describes around the common center of mass of the earth and the moon. Where is this center of mass, inside the earth or outside? (The mass of the moon is 1/81 times the mass of the earth.)

11 The method of similarity

In some cases it is possible to obtain important information from the form of the equations of motion without solving them, by using the methods of similarity and dimension. The main idea in these methods is to choose a change of scale (of time, length, mass, etc.) under which the equations of motion preserve their form.

A *Example*

Let $\mathbf{r}(t)$ satisfy the equation $m(d^2\mathbf{r}/dt^2) = -(\partial U/\partial \mathbf{r})$. We set $t_1 = \alpha t$ and $m_1 = \alpha^2 m$. Then $\mathbf{r}(t_1)$ satisfies the equation $m_1 \cdot (d^2\mathbf{r}/dt_1^2) = -(\partial U/\partial \mathbf{r})$. In other words:

If the mass of a point is decreased by a factor of 4, then the point can travel the same orbit in the same force field twice as fast.[22]

---

[22] Here we are assuming that U does not depend on m. In the field of gravity, the potential energy U is proportional to m, and therefore the acceleration does not depend on the mass m of the moving point.

B *A problem*

Suppose that the potential energy of a central field is a homogeneous function of degree v:

$$U(\alpha r) = \alpha^v U(r) \quad \text{for any } \alpha > 0.$$

Show that if a curve γ is the orbit of a motion, then the homothetic curve $\alpha\gamma$ is also an orbit (under the appropriate initial conditions). Determine the ratio of the circulation times along these orbits. Deduce from this the isochronicity of the oscillation of a pendulum ($v = 2$) and Kepler's third law ($v = -1$).

PROBLEM. If the radius of a planet is α times the radius of the earth and its mass β times that of the earth, find the ratio of the acceleration of the force of gravity and the first and second cosmic velocities to the corresponding quantities for the earth.

ANSWER. $\gamma = \beta\alpha^{-2}, \delta = \sqrt{\beta/\alpha}$.

For the moon, for example, $\alpha = 1/3.7$ and $\beta = 1/81$. Therefore, the acceleration of gravity is about $1/6$ that of the earth ($\gamma \approx 1/6$), and the cosmic velocities are about $1/5$ those for the earth ($\delta \approx 1/4.7$).

PROBLEM.[23] A desert animal has to cover great distances between sources of water. How does the maximal time the animal can run depend on the size L of the animal?

ANSWER. It is directly proportional to L.

Solution. The store of water is proportional to the volume of the body, i.e., L^3; the evaporation is proportional to the surface area, i.e., L^2. Therefore, the maximal time of a run from one source to another is directly proportional to L.

We notice that the maximal distance an animal can run also grows proportionally to L (cf. the following problem).

PROBLEM.[24] How does the running velocity of an animal on level ground and uphill depend on the size L of the animal?

ANSWER. On level ground $\sim L^0$, uphill $\sim L^{-1}$.

[23] J. M. Smith, *Mathematical Ideas in Biology*. Cambridge University Press, 1968.

[24] *Ibid.*

Solution. The power developed by the animal is proportional to L^2 (the percentage used by muscle is constant at about 25%, the other 75% of the chemical energy is converted to heat; the heat output is proportional to the body surface, i.e., L^2, which means that the effective power is proportional to L^2).

The force of air resistance is directly proportional to the square of the velocity and the area of a cross-section; the power spent on overcoming it is therefore proportional to v^2L^2v. Therefore, $v^3L^2 \sim L^2$, so $v \sim L^0$. In fact, the running velocity on level ground, no smaller for a rabbit than for a horse, in practice does not specifically depend on the size.

The power necessary to run uphill is $mgv \sim L^3v$; since the generated power is $\sim L^2$, we find that $v \sim L^{-1}$. In fact, a dog easily runs up a hill, while a horse slows its pace.

PROBLEM. [24a] How does the height of an animal's jump depend on its size?

ANSWER. $\sim L^0$.

Solution. For a jump of height h one needs energy proportional to L^3h, and the work accomplished by muscular strength F is proportional to FL. The force F is proportional to L^2 (since the strength of bones is proportional to their section). Therefore, $L^3h \sim L^2L$, i.e., the height of a jump does not depend on the size of the animal. In fact, a jerboa and a kangaroo can jump to approximately the same height.

[24a] *Ibid.*

PART II
LAGRANGIAN MECHANICS

Lagrangian mechanics describes motion in a mechanical system by means of the configuration space. The configuration space of a mechanical system has the structure of a differentiable manifold, on which its group of diffeomorphisms acts. The basic ideas and theorems of lagrangian mechanics are invariant under this group,[25] even if formulated in terms of local coordinates.

A lagrangian mechanical system is given by a manifold ("configuration space") and a function on its tangent bundle ("the lagrangian function").

Every one-parameter group of diffeomorphisms of configuration space which fixes the lagrangian function defines a conservation law (i.e., a first integral of the equations of motion).

A newtonian potential system is a particular case of a lagrangian system (the configuration space in this case is euclidean, and the lagrangian function is the difference between the kinetic and potential energies).

The lagrangian point of view allows us to solve completely a series of important mechanical problems, including problems in the theory of small oscillations and in the dynamics of a rigid body.

[25] And even under larger groups of transformations, which also affect time.

Variational principles 3

In this chapter we show that the motions of a newtonian potential system are extremals of a variational principle, "Hamilton's principle of least action."

This fact has many important consequences, including a quick method for writing equations of motion in curvilinear coordinate systems, and a series of qualitative deductions—for example, a theorem on returning to a neighborhood of the initial point.

In this chapter we will use an n-dimensional coordinate space. A vector in such a space is a set of numbers $\mathbf{x} = (x_1, \ldots, x_n)$. Similarly, $\partial f/\partial \mathbf{x}$ means $(\partial f/\partial x_1, \ldots, \partial f/\partial x_n)$, and $(\mathbf{a}, \mathbf{b}) = a_1 b_1 + \cdots + a_n b_n$.

12 Calculus of variations

For what follows, we will need some facts from the calculus of variations. A more detailed exposition can be found in "A Course in the Calculus of Variations" by M. A. Lavrentiev and L. A. Lusternik, M. L., 1938, or G. E. Shilov, "Elementary Functional Analysis," MIT Press, 1974.

The calculus of variations is concerned with the extremals of functions whose domain is an infinite-dimensional space: the space of curves. Such functions are called *functionals*.

An example of a functional is the length of a curve in the euclidean plane: if $\gamma = \{(t, x): x(t) = x, t_0 \leq t \leq t_1\}$, then $\Phi(\gamma) = \int_{t_0}^{t_1} \sqrt{1 + \dot{x}^2}\, dt$.

In general, a functional is any mapping from the space of curves to the real numbers.

We consider an "approximation" γ' to γ, $\gamma' = \{(t, x): x = x(t) + h(t)\}$. We will call it $\gamma' = \gamma + h$. Consider the increment of Φ, $\Phi(\gamma + h) - \Phi(\gamma)$ (Figure 41).

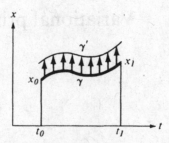

Figure 41   Variation of a curve

A *Variations*

Definition. A functional Φ is called *differentiable*[26] if $\Phi(\gamma + h) - \Phi(\gamma) = F + R$, where F depends linearly on h (i.e., for a fixed γ, $F(h_1 + h_2) = F(h_1) + F(h_2)$ and $F(ch) = cF(h)$), and $R(h, \gamma) = O(h^2)$ in the sense that, for $|h| < \varepsilon$ and $|dh/dt| < \varepsilon$, we have $|R| < C\varepsilon^2$. The linear part of the increment, $F(h)$, is called the *differential*.

It can be shown that if Φ is differentiable, its differential is *uniquely* defined. The differential of a functional is also called its *variation*, and h is called a *variation of the curve*.

EXAMPLE. Let $\gamma = \{(t, x) : x = x(t), t_0 \leq t \leq t_1\}$ be a curve in the (t, x)-plane; $\dot{x} = dx/dt$; $L = L(a, b, c)$ a differentiable function of three variables. We define a functional Φ by

$$\Phi(\gamma) = \int_{t_0}^{t_1} L(x(t), \dot{x}(t), t) dt$$

In case $L = \sqrt{1 + b^2}$, we get the length of γ.

Theorem. *The functional* $\Phi(\gamma) = \int_{t_0}^{t_1} L(x, \dot{x}, t) dt$ *is differentiable, and its derivative is given by the formula*

$$F(h) = \int_{t_0}^{t_1} \left[\frac{\partial L}{\partial x} - \frac{d}{dt} \frac{\partial L}{\partial \dot{x}} \right] h \, dt + \left(\frac{\partial L}{\partial \dot{x}} h \right) \Big|_{t_0}^{t_1}$$

PROOF.

$$\Phi(\gamma + h) - \Phi(\gamma) = \int_{t_0}^{t_1} [L(x + h, \dot{x} + \dot{h}, t) - L(x, \dot{x}, t)] dt$$

$$= \int_{t_0}^{t_1} \left[\frac{\partial L}{\partial x} h + \frac{\partial L}{\partial \dot{x}} \dot{h} \right] dt + O(h^2) = F(h) + R,$$

---

[26] We should specify the class of curves on which Φ is defined and the linear space which contains h. One could assume, for example, that both spaces consist of the infinitely differentiable functions.

where

$$F(h) = \int_{t_0}^{t_1} \left(\frac{\partial L}{\partial x} h + \frac{\partial L}{\partial \dot{x}} \dot{h} \right) dt \quad \text{and} \quad R = O(h^2).$$

Integrating by parts, we find that

$$\int_{t_0}^{t_1} \frac{\partial L}{\partial x} h \, dt = - \int_{t_0}^{t_1} h \frac{d}{dt} \left(\frac{\partial L}{\partial \dot{x}} \right) dt + \left(h \frac{\partial L}{\partial \dot{x}} \right) \Big|_{t_0}^{t_1}. \qquad \square$$

B Extremals

Definition. An *extremal* of a differentiable functional $\Phi(\gamma)$ is a curve γ such that $F(h) = 0$ for all h.

(In exactly the same way that γ is a stationary point of a function if the differential is equal to zero at that point.)

Theorem. *The curve* $\gamma: x = x(t)$ *is an extremal of the functional* $\Phi(\gamma) = \int_{t_0}^{t_1} L(x, \dot{x}, t)dt$ *on the space of curves passing through the points* $x(t_0) = x_0$ *and* $x(t_1) = x_1$, *if and only if*

$$\frac{d}{dt} \left(\frac{\partial L}{\partial \dot{x}} \right) - \frac{\partial L}{\partial x} = 0 \quad \text{along the curve } x(t).$$

Lemma. *If a continuous function* $f(t)$, $t_0 \le t \le t_1$ *satisfies* $\int_{t_0}^{t_1} f(t)h(t)dt = 0$ *for any continuous[27] function* $h(t)$ *with* $h(t_0) = h(t_1) = 0$, *then* $f(t) \equiv 0$.

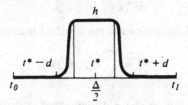

Figure 42   Construction of the function h

PROOF OF THE LEMMA. Let $f(t^*) > 0$ for some t^*, $t_0 < t^* < t_1$. Since f is continuous, $f(t) > c$ in some neighborhood Δ of the point t^*: $t_0 < t^* - d < t < t^* + d < t_1$. Let $h(t)$ be such that $h(t) = 0$ outside Δ, $h(t) > 0$ in Δ, and $h(t) = 1$ in $\Delta/2$ (i.e., for t s.t. $t^* - \frac{1}{2}d < t < t^* + \frac{1}{2}d$). Then, clearly, $\int_{t_0}^{t_1} f(t)h(t) \ge dc > 0$ (Figure 42). This contradiction shows that $f(t^*) = 0$ for all t^*, $t_0 < t^* < t_1$. $\qquad \square$

PROOF OF THE THEOREM. By the preceding theorem,

$$F(h) = - \int_{t_0}^{t_1} \left[\frac{d}{dt} \left(\frac{\partial L}{\partial \dot{x}} \right) - \frac{\partial L}{\partial x} \right] h \, dt + \left(\frac{\partial L}{\partial \dot{x}} h \right) \Big|_{t_0}^{t_1}.$$

[27] Or even for any infinitely differentiable function h.

The term after the integral is equal to zero since $h(t_0) = h(t_1) = 0$. If γ is an extremal, then $F(h) = 0$ for all h with $h(t_0) = h(t_1) = 0$. Therefore,

$$\int_{t_0}^{t_1} f(t)h(t)dt = 0,$$

where

$$f(t) = \frac{d}{dt}\left(\frac{\partial L}{\partial \dot{x}}\right) - \frac{\partial L}{\partial x},$$

for all such h. By the lemma, $f(t) \equiv 0$. Conversely, if $f(t) \equiv 0$, then clearly $F(h) \equiv 0$. $\qquad\qquad\qquad\square$

EXAMPLE. We verify that the extremals of length are straight lines. We have:

$$L = \sqrt{1 + \dot{x}^2} \qquad \frac{\partial L}{\partial x} = 0 \qquad \frac{\partial L}{\partial \dot{x}} = \frac{\dot{x}}{\sqrt{1 + \dot{x}^2}} \qquad \frac{d}{dt}\left(\frac{\dot{x}}{\sqrt{1 + \dot{x}^2}}\right) = 0$$

$$\frac{\dot{x}}{\sqrt{1 + \dot{x}^2}} = c \qquad \dot{x} = c_1 \qquad x = c_1 t + c_2.$$

C The Euler–Lagrange equation

Definition. The equation

$$\frac{d}{dt}\left(\frac{\partial L}{\partial \dot{x}}\right) - \frac{\partial L}{\partial x} = 0$$

is called the *Euler–Lagrange equation* for the functional

$$\Phi = \int_{t_0}^{t_1} L(x, \dot{x}, t)dt.$$

Now let $\mathbf{x}$ be a vector in the n-dimensional coordinate space $\mathbb{R}^n$, $\gamma = \{(t, \mathbf{x}): \mathbf{x} = \mathbf{x}(t), t_0 \leq t \leq t_1\}$ a curve in the $(n + 1)$-dimensional space $\mathbb{R} \times \mathbb{R}^n$, and $L: \mathbb{R}^n \times \mathbb{R}^n \times \mathbb{R} \to \mathbb{R}$ a function of $2n + 1$ variables. As before, we show:

Theorem. *The curve γ is an extremal of the functional $\Phi(\gamma) = \int_{t_0}^{t_1} L(\mathbf{x}, \dot{\mathbf{x}}, t)dt$ on the space of curves joining $(t_0, \mathbf{x}_0)$ and $(t_1, \mathbf{x}_1)$, if and only if the Euler–Lagrange equation is satisfied along γ.*

This is a system of n *second*-order equations, and the solution depends on $2n$ arbitrary constants. The $2n$ conditions $\mathbf{x}(t_0) = \mathbf{x}_0$, $\mathbf{x}(t_1) = \mathbf{x}_1$ are used for finding them.

PROBLEM. Cite examples where there are many extremals connecting two given points, and others where there are none at all.

D *An important remark*

The condition for a curve γ to be an extremal of a functional does not depend on the choice of coordinate system.

For example, the same functional—length of a curve—is given in cartesian and polar coordinates by the different formulas

$$\Phi_{\text{cart}} = \int_{t_0}^{t_1} \sqrt{\dot{x}_1^2 + \dot{x}_2^2}\, dt \qquad \Phi_{\text{pol}} = \int_{t_0}^{t_1} \sqrt{\dot{r}^2 + r^2\dot{\varphi}^2}\, dt.$$

The extremals are the same—straight lines in the plane. The equations of lines in cartesian and polar coordinates are given by different functions: $x_1 = x_1(t)$, $x_2 = x_2(t)$, and $r = f(t)$, $\varphi = \varphi(t)$.

However, both these vector functions satisfy the Euler–Lagrange equation

$$\frac{d}{dt}\frac{\partial L}{\partial \dot{x}} - \frac{\partial L}{\partial x} = 0$$

only, in the first case, when $x_{\text{cart}} = x_1, x_2$ and $L_{\text{cart}} = \sqrt{\dot{x}_1^2 + \dot{x}_2^2}$, and in the second case when $x_{\text{pol}} = r$, φ and $L_{\text{pol}} = \sqrt{\dot{r}^2 + r^2\dot{\varphi}^2}$.

In this way we can easily describe in any coordinates a differential equation for the family of all straight lines.

PROBLEM. Find the differential equation for the family of all straight lines in the plane in polar coordinates.

13 Lagrange's equations

Here we indicate the variational principle whose extremals are solutions of Newton's equations of motion in a potential system.

We compare Newton's equations of dynamics

(1)
$$\frac{d}{dt}(m_i\dot{r}_i) + \frac{\partial U}{\partial r_i} = 0$$

with the Euler–Lagrange equation

$$\frac{d}{dt}\frac{\partial L}{\partial \dot{x}} - \frac{\partial L}{\partial x} = 0.$$

A *Hamilton's principle of least action*

Theorem. *Motions of the mechanical system* (1) *coincide with extremals of the functional*

$$\Phi(\gamma) = \int_{t_0}^{t_1} L\, dt, \quad \text{where } L = T - U$$

is the difference between the kinetic and potential energy.

PROOF. Since $U = U(\mathbf{r})$ and $T = \sum m_i \dot{\mathbf{r}}_i^2/2$, we have $\partial L/\partial \dot{\mathbf{r}}_i = \partial T/\partial \dot{\mathbf{r}}_i = m_i \dot{\mathbf{r}}_i$ and $\partial L/\partial \mathbf{r}_i = -\partial U/\partial \mathbf{r}_i$. □

Corollary. *Let $(q_1, \ldots, q_{3n})$ be any coordinates in the configuration space of a system of n mass points. Then the evolution of $\mathbf{q}$ with time is subject to the Euler–Lagrange equations*

$$\frac{d}{dt}\left(\frac{\partial L}{\partial \dot{\mathbf{q}}}\right) - \frac{\partial L}{\partial \mathbf{q}} = 0, \quad \text{where } L = T - U.$$

PROOF. By the theorem above, a motion is an extremal of the functional $\int L \, dt$. Therefore, in any system of coordinates the Euler–Lagrange equation written in that coordinate system is satisfied. □

Definition. In mechanics we use the following terminology: $L(\mathbf{q}, \dot{\mathbf{q}}, t) = T - U$ is the *Lagrange function* or *lagrangian*, q_i are the *generalized coordinates*, $\dot{q}_i$ are *generalized velocities*, $\partial L/\partial \dot{q}_i = p_i$ are *generalized momenta*, $\partial L/\partial q_i$ are *generalized forces*, $\int_{t_0}^{t_1} L(\mathbf{q}, \dot{\mathbf{q}}, t)dt$ is the *action*, $(d(\partial L/\partial \dot{q}_i)/dt) - (\partial L/\partial q_i) = 0$ are *Lagrange's equations*.

The last theorem is called "Hamilton's form of the principle of least motion" because in many cases the action $\mathbf{q}(t)$ is not only an extremal but is also a *minimum* value of the action functional $\int_{t_0}^{t_1} L \, dt$.

B *The simplest examples*

EXAMPLE 1. For a free mass point in E^3,

$$L = T = \frac{m\dot{\mathbf{r}}^2}{2};$$

in cartesian coordinates $q_i = r_i$ we find

$$L = \frac{m}{2}(\dot{q}_1^2 + \dot{q}_2^2 + \dot{q}_3^2).$$

Here the generalized velocities are the components of the velocity vector, the generalized momenta $p_i = m\dot{q}_i$ are the components of the momentum vector, and Lagrange's equations coincide with Newton's equations $d\mathbf{p}/dt = 0$. The extremals are straight lines. It follows from Hamilton's principle that straight lines are not only shortest (i.e., extremals of the length $\int_{t_0}^{t_1} \sqrt{\dot{q}_1^2 + \dot{q}_2^2 + \dot{q}_3^2} \, dt$) but also extremals of the action $\int_{t_0}^{t_1} (\dot{q}_1^2 + \dot{q}_2^2 + \dot{q}_3^2)dt$.

PROBLEM. Show that this extremum is a minimum.

EXAMPLE 2. We consider planar motion in a central field in polar coordinates $q_1 = r, q_2 = \varphi$. From the relation $\dot{\mathbf{r}} = \dot{r}\mathbf{e}_r + \dot{\varphi}r\mathbf{e}_\varphi$ we find the kinetic energy

$T = \frac{1}{2}m\dot{r}^2 = \frac{1}{2}m(\dot{r}^2 + r^2\dot{\varphi}^2)$ and the lagrangian $L(\mathbf{q}, \dot{\mathbf{q}}) = T(\mathbf{q}, \dot{\mathbf{q}}) - U(\mathbf{q})$, where $U = U(q_1)$.

The generalized momenta will be $\mathbf{p} = \partial L/\partial \dot{\mathbf{q}}$, i.e.,

$$p_1 = m\dot{r} \qquad p_2 = mr^2\dot{\varphi}.$$

The first Lagrange equation $\dot{p}_1 = \partial L/\partial q_1$ takes the form

$$m\ddot{r} = mr\dot{\varphi}^2 - \frac{\partial U}{\partial r}.$$

We already obtained this equation in Section 8.

Since $q_2 = \varphi$ does not enter into L, we have $\partial L/\partial q_2 = 0$. Therefore, the second Lagrange equation will be $\dot{p}_2 = 0$, $p_2 = $ const. This is the law of conservation of angular momentum.

In general, when the field is not central ($U = U(r, \varphi)$), we find $\dot{p}_2 = -\partial U/\partial \varphi$.

This equation can be rewritten in the form $d(\mathbf{M}, \mathbf{e}_z)/dt = N$, where $N = ([\mathbf{r}, \mathbf{F}], \mathbf{e}_z)$ and $\mathbf{F} = -\partial U/\partial \mathbf{r}$. (The rate of change in angular momentum relative to the z axis is equal to the moment of the force $\mathbf{F}$ relative to the z axis.)

In fact, we have $dU = (\partial U/\partial r)dr + (\partial U/\partial \varphi)d\varphi = -(\mathbf{F}, d\mathbf{r}) = -(\mathbf{F}, \mathbf{e}_r)dr - r(\mathbf{F}, \mathbf{e}_\varphi)d\varphi$; therefore, $-\partial U/\partial \varphi = r(\mathbf{F}, \mathbf{e}_\varphi) = r([\mathbf{e}_r, \mathbf{F}], \mathbf{e}_z) = ([\mathbf{r}, \mathbf{F}], \mathbf{e}_z)$.

This example suggests the following generalization of the law of conservation of angular momentum.

Definition. A coordinate q_i is called *cyclic* if it does not enter into the lagrangian: $\partial L/\partial q_i = 0$.

Theorem. *The generalized momentum corresponding to a cyclic coordinate is conserved:* $p_i = $ const.

PROOF. By Lagrange's equation $dp_i/dt = \partial L/\partial q_i = 0$. ☐

14 Legendre transformations

The Legendre transformation is a very useful mathematical tool: it transforms functions on a vector space to functions on the dual space. Legendre transformations are related to projective duality and tangential coordinates in algebraic geometry and the construction of dual Banach spaces in analysis. They are often encountered in physics (for example, in the definition of thermodynamic quantities).

A *Definition*

Let $y = f(x)$ be a convex function, $f''(x) > 0$.

The *Legendre transformation* of the function f is a new function g of a new variable p, which is constructed in the following way (Figure 43). We draw the graph of f in the x, y plane. Let p be a given number. Consider the

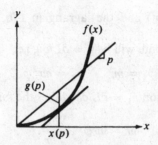

Figure 43   Legendre transformation

straight line $y = px$. We take the point $x = x(p)$ at which the curve is farthest from the straight line in the vertical direction: for each p the function $px - f(x) = F(p, x)$ has a maximum with respect to x at the point $x(p)$. Now we define $g(p) = F(p, x(p))$.

The point $x(p)$ is defined by the extremal condition $\partial F/\partial x = 0$, i.e., $f'(x) = p$. Since f is convex, the point $x(p)$ is unique.[28]

PROBLEM. Show that the domain of g can be a point, a closed interval, or a ray if f is defined on the whole x axis. Prove that if f is defined on a closed interval, then g is defined on the whole p axis.

B Examples

EXAMPLE 1. Let $f(x) = x^2$. Then $F(p, x) = px - x^2$, $x(p) = \frac{1}{2}p$, $g(p) = \frac{1}{4}p^2$.

EXAMPLE 2. Let $f(x) = mx^2/2$. Then $g(p) = p^2/2m$.

EXAMPLE 3. Let $f(x) = x^\alpha/\alpha$. Then $g(p) = p^\beta/\beta$, where $(1/\alpha) + (1/\beta) = 1$ $(\alpha > 1, \beta > 1)$.

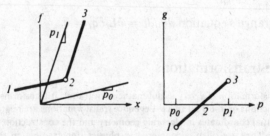

Figure 44   Legendre transformation taking an angle to a line segment

EXAMPLE 4. Let $f(x)$ be a convex polygon. Then $g(p)$ is also a convex polygon, in which the vertices of $f(x)$ correspond to the edges of $g(p)$, and the edges of $f(x)$ to the vertices of $g(p)$. For example, the corner depicted in Figure 44 is transformed to a segment under the Legendre transformation.

[28] If it exists.

C Involutivity

Let us consider a function f which is differentiable as many times as necessary, with $f''(x) > 0$. It is easy to verify that a Legendre transformation takes convex functions to convex functions. Therefore, we can apply it twice.

Theorem. *The Legendre transformation is involutive, i.e., its square is the identity: if under the Legendre transformation f is taken to g, then the Legendre transform of g will again be f.*

PROOF. In order to apply the Legendre transform to g, with variable p, we must by definition look at a new independent variable (which we will call x), construct the function

$$G(x, p) = xp - g(p),$$

and find the point $p(x)$ at which G attains its maximum: $\partial G/\partial p = 0$, i.e., $g'(p) = x$. Then the Legendre transform of $g(p)$ will be the function of x equal to $G(x, p(x))$.

We will show that $G(x, p(x)) = f(x)$. To this end we notice that $G(x, p) = xp - g(p)$ has a simple geometric interpretation: it is the ordinate of the point with abscissa x on the line tangent to the graph of $f(x)$ with slope p

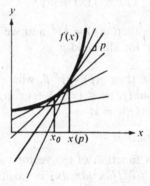

Figure 45   Involutivity of the Legendre transformation

(Figure 45). For fixed p, the function $G(x, p)$ is a linear function of x, with $\partial G/\partial x = p$, and for $x = x(p)$ we have $G(x, p) = xp - g(p) \doteq f(x)$ by the definition of $g(p)$.

Let us now fix $x = x_0$ and vary p. Then the values of $G(x, p)$ will be the ordinates of the points of intersection of the line $x = x_0$ with the line tangent to the graph of $f(x)$ with various slopes p. By the convexity of the graph it follows that all these tangents lie below the curve, and therefore the maximum of $G(x, p)$ for a fixed $x(p_0)$ is equal to $f(x)$ (and is achieved for $p = p(x_0) = f'(x_0)$). □

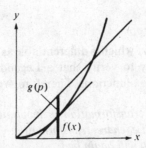

Figure 46   Legendre transformation of a quadratic form

Corollary.[29] *Consider a given family of straight lines* $y = px - g(p)$. *Then its envelope has the equation* $y = f(x)$, *where* f *is the Legendre transform of* g.

D *Young's inequality*

Definition. Two functions, f and g, which are the Legendre transforms of one another are called *dual in the sense of Young*.

By definition of the Legendre transform, $F(x, p) = px - f(x)$ is less than or equal to $g(p)$ for any x and p. From this we have *Young's inequality*:

$$px \le f(x) + g(p).$$

EXAMPLE 1. If $f(x) = \frac{1}{2}x^2$, then $g(p) = \frac{1}{2}p^2$ and we obtain the well-known inequality $px \le \frac{1}{2}x^2 + \frac{1}{2}p^2$ for all x and p.

EXAMPLE 2. If $f(x) = x^\alpha/\alpha$, then $g(p) = p^\beta/\beta$, where $(1/\alpha) + (1/\beta) = 1$, and we obtain *Young's inequality* $px \le (x^\alpha/\alpha) + (p^\beta/\beta)$ for all $x > 0$, $p > 0$, $\alpha > 1$, $\beta > 1$, and $(1/\alpha) + (1/\beta) = 1$.

E *The case of many variables*

Now let $f(\mathbf{x})$ be a convex function of the vector variable $\mathbf{x} = (x_1, \ldots, x_n)$ (i.e., the quadratic form $((\partial^2 f/\partial \mathbf{x}^2)d\mathbf{x}, d\mathbf{x})$ is positive definite). Then the Legendre transform is the function $g(\mathbf{p})$ of the vector variable $\mathbf{p} = (p_1, \ldots, p_n)$, defined as above by the equalities $g(\mathbf{p}) = F(\mathbf{p}, \mathbf{x}(\mathbf{p})) = \max_{\mathbf{x}} F(\mathbf{p}, \mathbf{x})$, where $F(\mathbf{p}, \mathbf{x}) = (\mathbf{p}, \mathbf{x}) - f(\mathbf{x})$ and $\mathbf{p} = \partial f/\partial \mathbf{x}$.

All of the above arguments, including Young's inequality, can be carried over without change to this case.

PROBLEM. Let $f: \mathbb{R}^n \to \mathbb{R}$ be a convex function. Let $\mathbb{R}^{n*}$ denote the dual vector space. Show that the formulas above completely define the mapping $g: \mathbb{R}^{n*} \to \mathbb{R}$ (under the condition that the linear form $df|_{\mathbf{x}}$ ranges over all of $\mathbb{R}^{n*}$ when $\mathbf{x}$ ranges over $\mathbb{R}^n$).

---

[29] One can easily see that this is the theory of "Clairaut's equation."

PROBLEM. Let f be the quadratic form $f(\mathbf{x}) = \sum f_{ij}x_ix_j$. Show that its Legendre transform is again a quadratic form $g(\mathbf{p}) = \sum g_{ij}p_ip_j$, and that the values of both forms at corresponding points coincide (Figure 46):

$$f(\mathbf{x}(\mathbf{p})) = g(\mathbf{p}) \quad \text{and} \quad g(\mathbf{p}(\mathbf{x})) = f(\mathbf{x}).$$

15 Hamilton's equations

By means of a Legendre transformation, a lagrangian system of second-order differential equations is converted into a remarkably symmetrical system of $2n$ first-order equations called a hamiltonian system of equations (or canonical equations).

A Equivalence of Lagrange's and Hamilton's equations

We consider the system of Lagrange's equations $\dot{\mathbf{p}} = \partial L/\partial \mathbf{q}$, where $\mathbf{p} = \partial L/\partial \dot{\mathbf{q}}$, with a given lagrangian function $L: \mathbb{R}^n \times \mathbb{R}^n \times \mathbb{R} \to \mathbb{R}$, which we will assume to be convex[30] with respect to the second argument $\dot{\mathbf{q}}$.

Theorem. *The system of Lagrange's equations is equivalent to the system of $2n$ first-order equations (Hamilton's equations)*

$$\dot{\mathbf{p}} = -\frac{\partial H}{\partial \mathbf{q}}$$

$$\dot{\mathbf{q}} = \frac{\partial H}{\partial \mathbf{p}},$$

where $H(\mathbf{p}, \mathbf{q}, t) = \mathbf{p}\dot{\mathbf{q}} - L(\mathbf{q}, \dot{\mathbf{q}}, t)$ is the Legendre transform of the lagrangian function viewed as a function of $\dot{\mathbf{q}}$.

PROOF. By definition, the Legendre transform of $L(\mathbf{q}, \dot{\mathbf{q}}, t)$ with respect to $\dot{\mathbf{q}}$ is the function $H(\mathbf{p}) = \mathbf{p}\dot{\mathbf{q}} - L(\dot{\mathbf{q}})$, in which $\dot{\mathbf{q}}$ is expressed in terms of $\mathbf{p}$ by the formula $\mathbf{p} = \partial L/\partial \dot{\mathbf{q}}$, and which depends on the parameters $\mathbf{q}$ and t. This function H is called the *hamiltonian*.

The total differential of the hamiltonian

$$dH = \frac{\partial H}{\partial \mathbf{p}} d\mathbf{p} + \frac{\partial H}{\partial \mathbf{q}} d\mathbf{q} + \frac{\partial H}{\partial t} dt$$

is equal to the total differential of $\mathbf{p}\dot{\mathbf{q}} - L$ for $\mathbf{p} = \partial L/\partial \dot{\mathbf{q}}$:

$$dH = \dot{\mathbf{q}} \, d\mathbf{p} - \frac{\partial L}{\partial \mathbf{q}} d\mathbf{q} - \frac{\partial L}{\partial t} dt.$$

Both expressions for dH must be the same. Therefore,

$$\dot{\mathbf{q}} = \frac{\partial H}{\partial \mathbf{p}} \qquad \frac{\partial H}{\partial \mathbf{q}} = -\frac{\partial L}{\partial \mathbf{q}} \qquad \frac{\partial H}{\partial t} = -\frac{\partial L}{\partial t}.$$

[30] In practice this convex function will often be a positive definite quadratic form.

Applying Lagrange's equations $\dot{\mathbf{p}} = \partial L/\partial \mathbf{q}$, we obtain Hamilton's equations.

We have seen that, if $\mathbf{q}(t)$ satisfies Lagrange's equations, then $(\mathbf{p}(t), \mathbf{q}(t))$ satisfies Hamilton's equations. The converse is proved in an analogous manner. Therefore, the systems of Lagrange and Hamilton are equivalent. $\square$

Remark. The theorem just proved applies to all variational problems, not just to the lagrangian equations of mechanics.

B *Hamilton's function and energy*

EXAMPLE. Suppose now that the equations are mechanical, so that the lagrangian has the usual form $L = T - U$, where the kinetic energy T is a quadratic form with respect to $\dot{\mathbf{q}}$:

$$T = \tfrac{1}{2} \sum a_{ij} \dot{q}_i \dot{q}_j, \quad \text{where } a_{ij} = a_{ij}(\mathbf{q}, t) \text{ and } U = U(\mathbf{q}).$$

Theorem. *Under the given assumptions, the hamiltonian H is the total energy $H = T + U$.*

The proof is based on the following lemma on the Legendre transform of a quadratic form.

Lemma. *The values of a quadratic form $f(\mathbf{x})$ and of its Legendre transform $g(\mathbf{p})$ coincide at corresponding points: $f(\mathbf{x}) = g(\mathbf{p})$.*

EXAMPLE. For the form $f(x) = x^2$ this is a well-known property of a tangent to a parabola. For the form $f(x) = \tfrac{1}{2}mx^2$ we have $p = mx$ and $g(p) = p^2/2m = mx^2/2 = f(x)$.

PROOF OF THE LEMMA By Euler's theorem on homogeneous functions $(\partial f/\partial \mathbf{x})\mathbf{x} = 2f$. Therefore, $g(\mathbf{p}(\mathbf{x})) = \mathbf{p}\mathbf{x} - f(\mathbf{x}) = (\partial f/\partial \mathbf{x})\mathbf{x} - f = 2f(\mathbf{x}) - f(\mathbf{x}) = f(\mathbf{x})$. $\square$

PROOF OF THE THEOREM. Reasoning as in the lemma, we find that $H = \mathbf{p}\dot{\mathbf{q}} - L = 2T - (T - U) = T + U$. $\square$

EXAMPLE. For one-dimensional motion

$$\ddot{q} = -\frac{\partial U}{\partial q}.$$

In this case $T = \tfrac{1}{2}\dot{q}^2$, $U = U(q)$, $p = \dot{q}$, $H = \tfrac{1}{2}p^2 + U(q)$ and Hamilton's equations take the form

$$\dot{q} = p$$

$$\dot{p} = -\frac{\partial U}{\partial q}.$$

This example makes it easy to remember which of Hamilton's equations has a minus sign.

Several important corollaries follow from the theorem on the equivalence of the equations of motion to a hamiltonian system. For example, the law of conservation of energy takes the simple form:

Corollary 1. $dH/dt = \partial H/\partial t$. *In particular, for a system whose hamiltonian function does not depend explicitly on time* ($\partial H/\partial t = 0$), *the law of conservation of the hamiltonian function holds*: $H(\mathbf{p}(t), \mathbf{q}(t)) = const$.

PROOF. We consider the variation in H along the trajectory $H(\mathbf{p}(t), \mathbf{q}(t), t)$. Then, by Hamilton's equations,

$$\frac{dH}{dt} = \frac{\partial H}{\partial \mathbf{p}}\left(-\frac{\partial H}{\partial \mathbf{q}}\right) + \frac{\partial H}{\partial \mathbf{q}}\frac{\partial H}{\partial \mathbf{p}} + \frac{\partial H}{\partial t} = \frac{\partial H}{\partial t}. \qquad \square$$

C Cyclic coordinates

When considering central fields, we noticed that a problem could be reduced to a one-dimensional problem by the introduction of polar coordinates. It turns out that, given any symmetry of a problem allowing us to choose a system of coordinates $\mathbf{q}$ in such a way that the hamiltonian function is independent of some of the coordinates, we can find some first integrals and thereby reduce to a problem in a smaller number of coordinates.

Definition. If a coordinate q_1 does not enter into the hamiltonian function $H(p_1, p_2, \ldots, p_n; q_1, \ldots, q_n; t)$, i.e., $\partial H/\partial q_1 = 0$, then it is called *cyclic* (the term comes from the particular case of the angular coordinate in a central field).

Clearly, the coordinate q_1 is cyclic if and only if it does not enter into the lagrangian function ($\partial L/\partial q_1 = 0$). It follows from the hamiltonian form of the equations of motion that:

Corollary 2. *Let* q_1 *be a cyclic coordinate. Then* p_1 *is a first integral. In this case the variation of the remaining coordinates with time is the same as in a system with the* $n - 1$ *independent coordinates* $q_2, \ldots, q_n$ *and with hamiltonian function*

$$H(p_2, \ldots, p_n, q_2, \ldots, q_n, t, c),$$

depending on the parameter $c = p_1$.

PROOF. We set $\mathbf{p}' = (p_2, \ldots, p_n)$ and $\mathbf{q}' = (q_2, \ldots, q_n)$. Then Hamilton's equations take the form

$$\frac{d}{dt}\mathbf{q}' = \frac{\partial H}{\partial \mathbf{p}'} \qquad \frac{d}{dt}q_1 = \frac{\partial H}{\partial p_1}$$

$$\frac{d}{dt}\mathbf{p}' = -\frac{\partial H}{\partial \mathbf{q}'} \qquad \frac{d}{dt}p_1 = 0.$$

The last equation shows that $p_1 = $ const. Therefore, in the system of equations for p' and q', the value of p_1 enters only as a parameter in the hamiltonian function. After this system of $2n - 2$ equations is solved, the equation for q_1 takes the form

$$\frac{d}{dt} q_1 = f(t), \quad \text{where } f(t) = \frac{\partial}{\partial p_1} H(p_1, \mathbf{p}'(t), \mathbf{q}'(t), t)$$

and is easily integrated. □

Almost all the solved problems in mechanics have been solved by means of Corollary 2.

Corollary 3. *Every closed system with two degrees of freedom* $(n = 2)$ *which has a cyclic coordinate is integrable.*

PROOF. In this case the system for p' and q' is one-dimensional and is immediately integrated by means of the integral $H(p', q') = c$. □

16 Liouville's theorem

The phase flow of Hamilton's equations preserves phase volume. It follows, for example, that a hamiltonian system cannot be asymptotically stable.

For simplicity we look at the case in which the hamiltonian function does not depend explicitly on the time: $H = H(\mathbf{p}, \mathbf{q})$.

A *The phase flow*

Definition. The $2n$-dimensional space with coordinates $p_1, \ldots, p_n; q_1, \ldots, q_n$ is called *phase space*.

EXAMPLE. In the case $n = 1$ this is the phase plane of the system $\ddot{x} = -\partial U/\partial x$, which we considered in Section 4.

Just as in this simplest example, the right-hand sides of Hamilton's equations give a vector field: at each point $(\mathbf{p}, \mathbf{q})$ of phase space there is a $2n$-dimensional vector $(-\partial H/\partial \mathbf{q}, \partial H/\partial \mathbf{p})$. We assume that every solution of Hamilton's equations can be extended to the whole time axis.[31]

Definition. The *phase flow* is the one-parameter group of transformations of phase space

$$g^t: (\mathbf{p}(0), \mathbf{q}(0)) \mapsto (\mathbf{p}(t), \mathbf{q}(t)),$$

where $\mathbf{p}(t)$ and $\mathbf{q}(t)$ are solutions of Hamilton's system of equations (Figure 47).

PROBLEM. Show that $\{g^t\}$ is a group.

---

[31] For this it is sufficient, for example, that the level sets of H be compact.

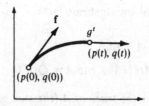

Figure 47 Phase flow

B *Liouville's theorem*

Theorem 1. *The phase flow preserves volume: for any region D we have (Figure 48)*

$$volume \ of \ g^t D = volume \ of \ D.$$

We will prove the following slightly more general proposition also due to Liouville.

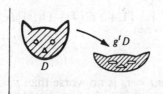

Figure 48 Conservation of volume

Suppose we are given a system of ordinary differential equations $\dot{\mathbf{x}} = \mathbf{f}(\mathbf{x})$, $\mathbf{x} = (x_1, \ldots, x_n)$, whose solution may be extended to the whole time axis. Let $\{g^t\}$ be the corresponding group of transformations:

(1) $$g^t(\mathbf{x}) = \mathbf{x} + \mathbf{f}(\mathbf{x})t + O(t^2), \qquad (t \to 0).$$

Let $D(0)$ be a region in $\mathbf{x}$-space and $v(0)$ its volume;

$$v(t) = \text{volume of } D(t) \qquad D(t) = g^t D(0).$$

Theorem 2. *If div $\mathbf{f} \equiv 0$, then g^t preserves volume: $v(t) = v(0)$.*

C *Proof*

Lemma 1. $(dv/dt)|_{t=0} = \int_{D(0)} \text{div } \mathbf{f} \, dx \quad (dx = dx_1 \cdots dx_n).$

PROOF. For any t, the formula for changing variables in a multiple integral gives

$$v(t) = \int_{D(0)} \det \frac{\partial g^t \mathbf{x}}{\partial \mathbf{x}} \, dx.$$

Calculating $\partial g^t \mathbf{x}/\partial \mathbf{x}$ by formula (1), we find

$$\frac{\partial g^t \mathbf{x}}{\partial \mathbf{x}} = E + \frac{\partial \mathbf{f}}{\partial \mathbf{x}} t + O(t^2) \quad \text{as } t \to 0.$$

We will now use a well-known algebraic fact:

Lemma 2. *For any matrix* $A = (a_{ij})$,

$$\det(E + At) = 1 + t \operatorname{tr} A + O(t^2), \qquad t \to 0,$$

where $\operatorname{tr} A = \sum_{i=1}^{n} a_{ii}$ *is the trace of* A *(the sum of the diagonal elements).*

(The proof of Lemma 2 is obtained by a direct expansion of the determinant: we get 1 and n terms in t; the remaining terms involve t^2, t^3, etc.)

Using this, we have

$$\det \frac{\partial g^t \mathbf{x}}{\partial \mathbf{x}} = 1 + t \operatorname{tr} \frac{\partial \mathbf{f}}{\partial \mathbf{x}} + O(t^2).$$

But $\operatorname{tr} \partial \mathbf{f}/\partial \mathbf{x} = \sum_{i=1}^{n} \partial f_i/\partial x_i = \operatorname{div} \mathbf{f}$. Therefore,

$$v(t) = \int_{D(0)} [1 + t \operatorname{div} \mathbf{f} + O(t^2)] dx,$$

which proves Lemma 1. $\qquad\qquad\square$

PROOF OF THEOREM 2. Since $t = t_0$ is no worse than $t = 0$, Lemma 1 can be written in the form

$$\left. \frac{dv(t)}{dt} \right|_{t=t_0} = \int_{D(t_0)} \operatorname{div} \mathbf{f} \, dx,$$

and if $\operatorname{div} \mathbf{f} \equiv 0$, $dv/dt \equiv 0$. $\qquad\qquad\square$

In particular, for Hamilton's equations we have

$$\operatorname{div} \mathbf{f} = \frac{\partial}{\partial \mathbf{p}} \left(-\frac{\partial H}{\partial \mathbf{q}} \right) + \frac{\partial}{\partial \mathbf{q}} \left(\frac{\partial H}{\partial \mathbf{p}} \right) \equiv 0.$$

This proves Liouville's theorem (Theorem 1). $\qquad\qquad\square$

PROBLEM. Prove Liouville's formula $W = W_0 e^{\int \operatorname{tr} A \, dt}$ for the Wronskian determinant of the linear system $\dot{\mathbf{x}} = A(t)\mathbf{x}$.

Liouville's theorem has many applications.

PROBLEM. Show that in a hamiltonian system it is impossible to have asymptotically stable equilibrium positions and asymptotically stable limit cycles in the phase space.

Liouville's theorem has particularly important applications in statistical mechanics.

Liouville's theorem allows one to apply methods of *ergodic theory*[32] to the study of mechanics. We consider only the simplest example:

D *Poincaré's recurrence theorem*

Let g be a volume-preserving continuous one-to-one mapping which maps a bounded region D of euclidean space onto itself: $gD = D$.

Then in any neighborhood U of any point of D there is a point $x \in U$ which returns to U, i.e., $g^n x \in U$ for some $n > 0$.

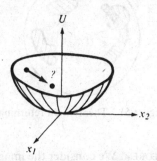

Figure 49   The way a ball will move in an asymmetrical cup is unknown; however Poincaré's theorem predicts that it will return to a neighborhood of the original position.

This theorem applies, for example, to the phase flow g^t of a two-dimensional system whose potential $U(x_1, x_2)$ goes to infinity as $(x_1, x_2) \to \infty$; in this case the invariant bounded region in phase space is given by the condition (Figure 49)

$$D = \{\mathbf{p}, \mathbf{q} : T + U \leq E\}.$$

Poincaré's theorem can be strengthened, showing that almost every moving point returns repeatedly to the vicinity of its initial position. This is one of the few general conclusions which can be drawn about the character of motion. The details of motion are not known at all, even in the case

$$\ddot{\mathbf{x}} = -\frac{\partial U}{\partial \mathbf{x}}, \qquad \text{where } \mathbf{x} = (x_1, x_2).$$

The following prediction is a paradoxical conclusion from the theorems of Poincaré and Liouville: if you open a partition separating a chamber containing gas and a chamber with a vacuum, then after a while the gas molecules will again collect in the first chamber (Figure 50).

The resolution of the paradox lies in the fact that "a while" may be longer than the duration of the solar system's existence.

[32] Cf. for example, the book: Halmos, *Lectures on Ergodic Theory*, 1956 (Mathematical Society of Japan. Publications. No. 3).

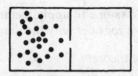

Figure 50   Molecules return to the first chamber.

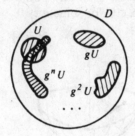

Figure 51   Theorem on returning

PROOF OF POINCARÉ'S THEOREM. We consider the images of the neighborhood
U (Figure 51):

$$U, gU, g^2U, \ldots, g^nU, \ldots$$

All of these have the same volume. If they never intersected, D would have
infinite volume. Therefore, for some $k \geq 0$ and $l \geq 0$, with $k > l$,

$$g^kU \cap g^lU \neq \emptyset.$$

Therefore, $g^{k-l}U \cap U \neq \emptyset$. If y is in this intersection, then $y = g^nx$, with
$x \in U(n = k - l)$. Then $x \in U$ and $g^nx \in U(n = k - l)$. ☐

E Applications of Poincaré's theorem

EXAMPLE 1. Let D be a circle and g rotation through an angle α. If $\alpha = 2\pi(m/n)$, then g^n is the identity, and the theorem is obvious. If α is not commensurable with 2π, then Poincaré's theorem gives

$$\forall \delta > 0, \exists n : |g^nx - x| < \delta \qquad \text{(Figure 52)}.$$

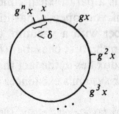

Figure 52   Dense set on the circle

It easily follows that

Theorem. *If* $\alpha \neq 2\pi(m/n)$, *then the set of points* $g^k x$ *is dense*[33] *on the circle* $(k = 1, 2, \ldots)$.

PROBLEM. Show that every orbit of motion in a central field with $U = r^4$ is either closed or densely fills the ring between two circles.

EXAMPLE 2. Let D be the two-dimensional torus and φ_1 and φ_2 angular coordinates on it (longitude and latitude) (Figure 53).

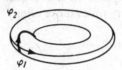

Figure 53   Torus

Consider the system of ordinary differential equations on the torus

$$\dot{\varphi}_1 = \alpha_1 \qquad \dot{\varphi}_2 = \alpha_2.$$

Clearly, div $\mathbf{f} = 0$ and the corresponding motion

$$g^t : (\varphi_1, \varphi_2) \rightarrow (\varphi_1 + \alpha_1 t, \varphi_2 + \alpha_2 t)$$

preserves the volume $d\varphi_1\, d\varphi_2$. From Poincaré's theorem it is easy to deduce

Theorem. *If* α_1/α_2 *is irrational, then the "winding line" on the torus,* $g^t(\varphi_1, \varphi_2)$, *is dense in the torus.*

PROBLEM. Show that if ω is irrational, then the Lissajous figure ($x = \cos t$, $y = \cos \omega t$) is dense in the square $|x| \leq 1, |y| \leq 1$.

EXAMPLE 3. Let D be the n-dimensional torus T^n, i.e., the direct product[34] of n circles:

$$D = \underbrace{S^1 \times S^1 \times \cdots \times S^1}_{n} = T^n.$$

A point on the n-dimensional torus is given by n angular coordinates $\boldsymbol{\varphi} = (\varphi_1, \ldots, \varphi_n)$. Let $\boldsymbol{\alpha} = (\alpha_1, \ldots, \alpha_n)$, and let g^t be the volume-preserving transformation

$$g^t : T^n \rightarrow T^n \qquad \boldsymbol{\varphi} \rightarrow \boldsymbol{\varphi} + \boldsymbol{\alpha} t.$$

[33] A set A is dense in B if there is a point of A in every neighborhood of every point of B.

[34] The direct product of the sets $A, B, \ldots$ is the set of points $(a, b, \ldots)$, with $a \in A, b \in B, \ldots$.

PROBLEM. Under which conditions on α are the following sets dense : (a) the trajectory $\{g^t\varphi\}$; (b) the trajectory $\{g^k\varphi\}$ (t belongs to the group of real numbers $\mathbb{R}$, k to the group of integers $\mathbb{Z}$).

The transformations in Examples 1 to 3 are closely connected to mechanics. But since Poincaré's theorem is abstract, it also has applications unconnected with mechanics.

EXAMPLE 4. Consider the first digits of the numbers 2^n: 1, 2, 4, 8, 1, 3, 6, 1, 2, 5, 1, 2, 4, ...

PROBLEM. Does the digit 7 appear in this sequence? Which digit appears more often, 7 or 8? How many times more often?

Lagrangian mechanics on manifolds

4

In this chapter we introduce the concepts of a differentiable manifold and its tangent bundle. A lagrangian function, given on the tangent bundle, defines a lagrangian "holonomic system" on a manifold. Systems of point masses with holonomic constraints (e.g., a pendulum or a rigid body) are special cases.

17 Holonomic constraints

In this paragraph we define the notion of a system of point masses with holonomic constraints.

A *Example*

Let γ be a smooth curve in the plane. If there is a very strong force field in a neighborhood of γ, directed towards the curve, then a moving point will always be close to γ. In the limit case of an infinite force field, the point must remain on the curve γ. In this case we say that a constraint is put on the system (Figure 54).

To formulate this precisely, we introduce curvilinear coordinates q_1 and q_2 on a neighborhood of γ; q_1 is in the direction of γ and q_2 is distance from the curve.

We consider the system with potential energy

$$U_N = Nq_2^2 + U_0(q_1, q_2),$$

depending on the parameter N (which we will let tend to infinity) (Figure 55).

We consider the initial conditions on γ:

$$q_1(0) = q_1^0 \qquad \dot{q}_1(0) = \dot{q}_1^0 \qquad q_2(0) = 0 \qquad \dot{q}_2(0) = 0.$$

Figure 54   Constraint as an infinitely strong field

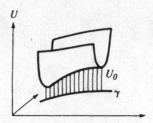

Figure 55   Potential energy U_N

Denote by $q_1 = \varphi(t, N)$ the evolution of the coordinate q_1 under a motion with these initial conditions in the field U_N.

Theorem. *The following limit exists, as $N \to \infty$:*

$$\lim_{N \to \infty} \varphi(t, N) = \psi(t).$$

The limit $q_1 = \psi(t)$ satisfies Lagrange's equation

$$\frac{d}{dt}\left(\frac{\partial L_*}{\partial \dot{q}_1}\right) = \frac{\partial L_*}{\partial q_1},$$

where $L_*(q_1, \dot{q}_1) = T|_{q_2 = \dot{q}_2 = 0} - U_0|_{q_2 = 0}$ *(T is the kinetic energy of motion along γ).*

Thus, as $N \to \infty$, Lagrange's equations for q_1 and q_2 induce Lagrange's equation for $q_1 = \psi(t)$.

We obtain exactly the same result if we replace the plane by the $3n$-dimensional configuration space of n points, consisting of a mechanical system with metric $ds^2 = \sum_{i=1}^{n} m_i \, d\mathbf{r}_i^2$ (the m_i are masses), replace the curve γ by a submanifold of the $3n$-dimensional space, replace q_1 by some coordinates $\mathbf{q}_1$ on γ, and replace q_2 by some coordinates $\mathbf{q}_2$ in the directions perpendicular to γ. If the potential energy has the form

$$U = U_0(\mathbf{q}_1, \mathbf{q}_2) + N\mathbf{q}_2^2,$$

then as $N \to \infty$, a motion on γ is defined by Lagrange's equations with the lagrangian function

$$L_* = T|_{\mathbf{q}_2 = \dot{\mathbf{q}}_2 = 0} - U_0|_{\mathbf{q}_2 = 0}.$$

B Definition of a system with constraints

We will not prove the theorem above,[35] but neither will we use it. We need it only to justify the following.

Definition. Let γ be an m-dimensional surface in the $3n$-dimensional configuration space of the points $r_1, \ldots, r_n$ with masses $m_1, \ldots, m_n$. Let $q = (q_1, \ldots, q_m)$ be some coordinates on $\gamma: r_i = r_i(q)$. The system described by the equations

$$\frac{d}{dt}\frac{\partial L}{\partial \dot{q}} = \frac{\partial L}{\partial q} \qquad L = \tfrac{1}{2}\sum m_i \dot{r}_i^2 + U(q)$$

is called a system of n points with $3n - m$ ideal *holonomic constraints*. The surface γ is called the *configuration space of the system with constraints*.

If the surface γ is given by $k = 3n - m$ functionally independent equations $f_1(r) = 0, \ldots, f_k(r) = 0$, then we say that the system is constrained by the relations $f_1 = 0, \ldots, f_k = 0$.

Holonomic constraints also could have been defined as the limiting case of a system with a large potential energy. The meaning of these constraints in mechanics lies in the experimentally determined fact that many mechanical systems belong to this class more or less exactly.

From now on, for convenience, we will call ideal holonomic constraints simply constraints. Other constraints will not be considered in this book.

18 Differentiable manifolds

The configuration space of a system with constraints is a differentiable manifold. In this paragraph we give the elementary facts about differentiable manifolds.

A Definition of a differentiable manifold

A set M is given the structure of a differentiable manifold if M is provided with a finite or countable collection of *charts*, so that every point is represented in at least one chart.

A chart is an open set U in the euclidean coordinate space $q = (q_1, \ldots, q_n)$, together with a one-to-one mapping φ of U onto some subset of M, $\varphi: U \to \varphi U \subset M$.

We assume that if points p and p' in two charts U and U' have the same image in M, then p and p' have neighborhoods $V \subset U$ and $V' \subset U'$ with the same image in M (Figure 56). In this way we get a mapping $\varphi'^{-1}\varphi: V \to V'$.

This is a mapping of the region V of the euclidean space q onto the region V' of the euclidean space q', and it is given by n functions of n variables,

---

[35] The proof is based on the fact that, due to the conservation of energy, a moving point cannot move further from γ than $cN^{-1/2}$, which approaches zero as $N \to \infty$.

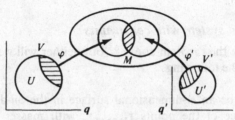

Figure 56   Compatible charts

$\mathbf{q}' = \mathbf{q}'(\mathbf{q})$, $(\mathbf{q} = \mathbf{q}(\mathbf{q}'))$. The charts U and U' are called *compatible* if these functions are differentiable.[36]

An *atlas* is a union of compatible charts. Two atlases are *equivalent* if their union is also an atlas.

A differentiable manifold is a class of equivalent atlases. We will consider only *connected* manifolds.[37] Then the number n will be the same for all charts; it is called the *dimension* of the manifold.

A *neighborhood* of a point on a manifold is the image under a mapping $\varphi: U \to M$ of a neighborhood of the representation of this point in a chart U. We will assume that every two different points have non-intersecting neighborhoods.

B Examples

EXAMPLE 1. Euclidean space $\mathbb{R}^n$ is a manifold, with an atlas consisting of one chart.

EXAMPLE 2. The sphere $S^2 = \{(x, y, z): x^2 + y^2 + z^2 = 1\}$ has the structure of a manifold, with atlas, for example, consisting of two charts (U_i, φ_i, $i = 1, 2$) in stereographic projection (Figure 57). An analogous construction applies to the n-sphere

$$S^n = \{(x_1, \ldots, x_{n+1}): \sum x_i^2 = 1\}.$$

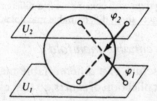

Figure 57   Atlas of a sphere

EXAMPLE 3. Consider a planar pendulum. Its configuration space—the circle S^1—is a manifold. The usual atlas is furnished by the angular coordinates $\varphi: \mathbb{R}^1 \to S^1$, $U_1 = (-\pi, \pi)$, $U_2 = (0, 2\pi)$ (Figure 58).

EXAMPLE 4. The configuration space of the "spherical" mathematical pendulum is the two-dimensional sphere S^2 (Figure 58).

---

[36] By differentiable here we mean r times continuously differentiable; the exact value of r ($1 \le r \le \infty$) is immaterial (we may take $r = \infty$, for example).

[37] A manifold is connected if it cannot be divided into two disjoint open subsets.

Figure 58   Planar, spherical and double planar pendulums

EXAMPLE 5. The configuration space of a "planar double pendulum" is the direct product of two circles, i.e., the two-torus $T^2 = S^1 \times S^1$ (Figure 58).

EXAMPLE 6. The configuration space of a spherical double pendulum is the direct product of two spheres, $S^2 \times S^2$.

EXAMPLE 7. A rigid line segment in the (q_1, q_2)-plane has for its configuration space the manifold $\mathbb{R}^2 \times S^1$, with coordinates q_1, q_2, q_3 (Figure 59). It is covered by two charts.

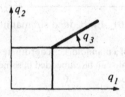

Figure 59   Configuration space of a segment in the plane

EXAMPLE 8. A rigid right triangle OAB moves around the vertex O. The position of the triangle is given by three numbers: the direction $OA \in S^2$ is given by two numbers, and if OA is given, one can rotate $OB \in S^1$ around the axis OA (Figure 60).

Connected with the position of the triangle OAB is an orthogonal right-handed frame, $\mathbf{e}_1 = OA/|OA|, \mathbf{e}_2 = OB/|OB|, \mathbf{e}_3 = [\mathbf{e}_1, \mathbf{e}_2]$. The correspondence is one-to-one; therefore the position of the triangle is given by an orthogonal three-by-three matrix with determinant 1.

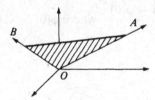

Figure 60   Configuration space of a triangle

The set of all three-by-three matrices is the nine-dimensional space $\mathbb{R}^9$. Six orthogonality conditions select out two three-dimensional connected manifolds of matrices with determinant $+1$ and -1. The rotations of three-space (determinant $+1$) form a group, which we call SO(3).

Therefore, the *configuration space of the triangle OAB is* SO(3).

PROBLEM. Show that SO(3) is homeomorphic to three-dimensional real projective space.

Definition. The dimension of the configuration space is called the *number of degrees of freedom.*

EXAMPLE 9. Consider a system of k rods in a closed chain with hinged joints.

PROBLEM. How many degrees of freedom does this system have?

EXAMPLE 10. *Embedded manifolds.* We say that M is an embedded k-dimensional sub-manifold of euclidean space $\mathbb{R}^n$ (Figure 61) if in a neighborhood U of every point $\mathbf{x} \in M$ there are $n - k$ functions $f_1: U \to \mathbb{R}, f_2: U \to \mathbb{R}, \ldots, f_{n-k}: U \to \mathbb{R}$ such that the intersection of U with M is given by the equations $f_1 = 0, \ldots, f_{n-k} = 0$, and the vectors $\operatorname{grad} f_1, \ldots, \operatorname{grad} f_{n-k}$ at $\mathbf{x}$ are linearly independent.

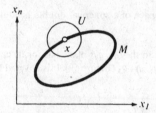

Figure 61   Embedded submanifold

It is easy to give M the structure of a manifold, i.e., coordinates in a neighborhood of $\mathbf{x}$ (how?).

It can be shown that every manifold can be embedded in some euclidean space. In Example 8, SO(3) is a subset of $\mathbb{R}^9$.

PROBLEM. Show that SO(3) is embedded in $\mathbb{R}^9$, and at the same time, that SO(3) is a manifold.

C *Tangent space*

If M is a k-dimensional manifold embedded in E^n, then at every point $\mathbf{x}$ we have a k-dimensional tangent space $TM_{\mathbf{x}}$. Namely, $TM_{\mathbf{x}}$ is the orthogonal complement to $\{\operatorname{grad} f_1, \ldots, \operatorname{grad} f_{n-k}\}$ (Figure 62). The vectors of the tangent space $TM_{\mathbf{x}}$ based at $\mathbf{x}$ are called tangent vectors to M at $\mathbf{x}$. We can also define these vectors directly as velocity vectors of curves in M:

$$\dot{\mathbf{x}} = \lim_{t \to 0} \frac{\varphi(t) - \varphi(0)}{t} \quad \text{where } \varphi(0) = \mathbf{x}, \varphi(t) \in M.$$

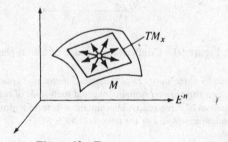

Figure 62   Tangent space

The definition of tangent vectors can also be given in intrinsic terms, independent of the embedding of M into E^n.

We will call two curves $\mathbf{x} = \varphi(t)$ and $\mathbf{x} = \psi(t)$ *equivalent* if $\varphi(0) = \psi(0) = \mathbf{x}$ and $\lim_{t \to 0} (\varphi(t) - \psi(t))/t = 0$ in some chart. Then this tangent relationship is true in any chart (prove this!).

Definition. A *tangent vector* to a manifold M at the point $\mathbf{x}$ is an equivalence class of curves $\varphi(t)$, with $\varphi(0) = \mathbf{x}$.

It is easy to define the operations of multiplication of a tangent vector by a number and addition of tangent vectors. The set of tangent vectors to M at $\mathbf{x}$ forms a *vector space $TM_\mathbf{x}$*. This space is also called the *tangent space* to M at $\mathbf{x}$.

For embedded manifolds the definition above agrees with the previous definition. Its advantage lies in the fact that it also holds for abstract manifolds, not embedded anywhere.

Definition. Let U be a chart of an atlas for M with coordinates $q_1, \ldots, q_n$. Then the *components* of the tangent vector to the curve $\mathbf{q} = \varphi(t)$ are the numbers $\xi_1, \ldots, \xi_n$, where $\xi_i = (d\varphi_i/dt)|_{t=0}$.

D The tangent bundle

The union of the tangent spaces to M at the various points, $\bigcup_{\mathbf{x} \in M} TM_\mathbf{x}$, has a natural differentiable manifold structure, the dimension of which is twice the dimension of M.

This manifold is called the *tangent bundle* of M and is denoted by TM. A point of TM is a vector ξ, tangent to M at some point $\mathbf{x}$. Local coordinates on TM are constructed as follows. Let $q_1, \ldots, q_n$ be local coordinates on M, and $\xi_1, \ldots, \xi_n$ components of a tangent vector in this coordinate system. Then the $2n$ numbers $(q_1, \ldots, q_n, \xi_1, \ldots, \xi_n)$ give a local coordinate system on TM. One sometimes writes dq_i for ξ_i.

The mapping $p: TM \to M$ which takes a tangent vector ξ to the point $\mathbf{x} \in M$ at which the vector is tangent to M ($\xi \in TM_\mathbf{x}$), is called the *natural projection*. The inverse image of a point $\mathbf{x} \in M$ under the natural projection, $p^{-1}(\mathbf{x})$, is the tangent space $TM_\mathbf{x}$. This space is called the *fiber of the tangent bundle over the point* $\mathbf{x}$.

E Riemannian manifolds

If M is a manifold embedded in euclidean space, then the metric on euclidean space allows us to measure the lengths of curves, angles between vectors, volumes, etc. All of these quantities are expressed by means of the lengths of tangent vectors, that is, by the positive-definite quadratic form given on every tangent space $TM_\mathbf{x}$ (Figure 63):

$$TM_\mathbf{x} \to \mathbb{R} \qquad \xi \to \langle \xi, \xi \rangle.$$

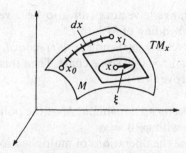

Figure 63   Riemannian metric

For example, the length of a curve on a manifold is expressed using this form as $l(\gamma) = \int_{x_0}^{x_1} \sqrt{\langle d\mathbf{x}, d\mathbf{x}\rangle}$, or, if the curve is given parametrically, $\gamma: [t_0, t_1] \to M, t \to \mathbf{x}(t) \in M$, then

$$l(\gamma) = \int_{t_0}^{t_1} \sqrt{\langle \dot{\mathbf{x}}, \dot{\mathbf{x}}\rangle} dt.$$

Definition. A differentiable manifold with a fixed positive-definite quadratic form $\langle \xi, \xi \rangle$ on every tangent space TM_x is called a *Riemannian manifold*. The quadratic form is called the *Riemannian metric*.

Remark. Let U be a chart of an atlas for M with coordinates $q_1, \ldots, q_n$. Then a Riemannian metric is given by the formula

$$ds^2 = \sum_{i, j = 1}^{n} a_{ij}(q)dq_i \, dq_j \qquad a_{ij} = a_{ji},$$

where dq_i are the coordinates of a tangent vector.

The functions $a_{ij}(q)$ are assumed to be differentiable as many times as necessary.

F *The derivative map*

Let $f: M \to N$ be a mapping of a manifold M to a manifold N. f is called *differentiable* if in local coordinates on M and N it is given by differentiable functions.

Definition. The *derivative* of a differentiable mapping $f: M \to N$ at a point $\mathbf{x} \in M$ is the linear map of the tangent spaces

$$f_{*\mathbf{x}}: TM_{\mathbf{x}} \to TN_{f(\mathbf{x})},$$

which is given in the following way (Figure 64):

Let $\mathbf{v} \in TM_{\mathbf{x}}$. Consider a curve $\varphi: \mathbb{R} \to M$ with $\varphi(0) = \mathbf{x}$, and velocity vector $(d\varphi/dt)|_{t=0} = \mathbf{v}$. Then $f_{*\mathbf{x}}\mathbf{v}$ is the velocity vector of the curve $f \circ \varphi: \mathbb{R} \to N$,

$$f_{*\mathbf{x}}\mathbf{v} = \frac{d}{dt}\bigg|_{t=0} f(\varphi(t)).$$

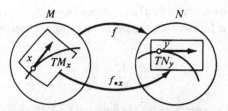

Figure 64   Derivative of a mapping

PROBLEM. Show that the vector $f_{*\mathbf{x}}\mathbf{v}$ does not depend on the curve φ, but only on the vector $\mathbf{v}$.

PROBLEM. Show that the map $f_{*\mathbf{x}}: TM_\mathbf{x} \to TN_{f(\mathbf{x})}$ is linear.

PROBLEM. Let $\mathbf{x} = (x_1, \ldots, x_m)$ be coordinates in a neighborhood of $\mathbf{x} \in M$, and $\mathbf{y} = (y_1, \ldots, y_n)$ be coordinates in a neighborhood of $\mathbf{y} \in N$. Let $\boldsymbol{\xi}$ be the set of components of the vector $\mathbf{v}$, and $\boldsymbol{\eta}$ the set of components of the vector $f_{*\mathbf{x}}\mathbf{v}$. Show that

$$\boldsymbol{\eta} = \frac{\partial \mathbf{y}}{\partial \mathbf{x}}\boldsymbol{\xi}, \quad \text{i.e.,} \quad \eta_i = \sum_j \frac{\partial y_i}{\partial x_j}\xi_j.$$

Taking the union of the mappings $f_{*\mathbf{x}}$ for all $\mathbf{x}$, we get a mapping of the whole tangent bundle

$$f_*: TM \to TN \qquad f_*\mathbf{v} = f_{*\mathbf{x}}\mathbf{v} \quad \text{for } \mathbf{v} \in TM_\mathbf{x}.$$

PROBLEM. Show that f_* is a differentiable map.

PROBLEM. Let $f: M \to N, g: N \to K$, and $h = g \cdot f: M \to K$. Show that $h_* = g_* \cdot f_*$.

19 Lagrangian dynamical systems

In this paragraph we define lagrangian dynamical systems on manifolds. Systems with holonomic constraints are a particular case.

A *Definition of a lagrangian system*

Let M be a differentiable manifold, TM its tangent bundle, and $L: TM \to \mathbb{R}$ a differentiable function. A map $\gamma: \mathbb{R} \to M$ is called a *motion in the lagrangian system with configuration manifold M and lagrangian function L* if γ is an extremal of the functional

$$\Phi(\gamma) = \int_{t_0}^{t_1} L(\dot\gamma)dt,$$

where $\dot\gamma$ is the velocity vector $\dot\gamma(t) \in TM_{\gamma(t)}$.

EXAMPLE. Let M be a region in a coordinate space with coordinates $\mathbf{q} = (q_1, \ldots, q_n)$. The lagrangian function $L: TM \to \mathbb{R}$ may be written in the form of a function $L(\mathbf{q}, \dot{\mathbf{q}})$ of the $2n$ coordinates. As we showed in Section 12, the evolution of coordinates of a point moving with time satisfies Lagrange's equations.

Theorem. *The evolution of the local coordinates* $\mathbf{q} = (q_1, \ldots, q_n)$ *of a point* $\gamma(t)$ *under motion in a lagrangian system on a manifold satisfies the Lagrange equations*

$$\frac{d}{dt}\frac{\partial L}{\partial \dot{\mathbf{q}}} = \frac{\partial L}{\partial \mathbf{q}},$$

where $L(\mathbf{q}, \dot{\mathbf{q}})$ *is the expression for the function* $L : TM \to \mathbb{R}$ *in the coordinates* $\mathbf{q}$ *and* $\dot{\mathbf{q}}$ *on* TM.

We often encounter the following special case.

B *Natural systems*

Let M be a Riemannian manifold. The quadratic form on each tangent space,

$$T = \tfrac{1}{2}\langle \mathbf{v}, \mathbf{v} \rangle \qquad \mathbf{v} \in TM_{\mathbf{x}},$$

is called the *kinetic energy*. A differentiable function $U : M \to \mathbb{R}$ is called a *potential energy*.

Definition. A lagrangian system on a Riemannian manifold is called *natural* if the lagrangian function is equal to the difference between kinetic and potential energies: $L = T - U$.

EXAMPLE. Consider two mass points m_1 and m_2 joined by a line segment of length l in the (x, y)-plane. Then a configuration space of three dimensions

$$M = \mathbb{R}^2 \times S^1 \subset \mathbb{R}^2 \times \mathbb{R}^2$$

is defined in the four-dimensional configuration space $\mathbb{R}^2 \times \mathbb{R}^2$ of two free points (x_1, y_1) and (x_2, y_2) by the condition $\sqrt{(x_1 - x_2)^2 + (y_1 - y_2)^2} = l$ (Figure 65).

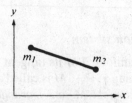

Figure 65   Segment in the plane

There is a quadratic form on the tangent space to the four-dimensional space (x_1, x_2, y_1, y_2):

$$m_1(\dot{x}_1^2 + \dot{y}_1^2) + m_2(\dot{x}_2^2 + \dot{y}_2^2).$$

Our three-dimensional manifold, as it is embedded in the four-dimensional one, is provided with a Riemannian metric. The holonomic system thus obtained is called in mechanics a line segment of fixed length in the (x, y)-plane. The kinetic energy is given by the formula

$$T = m_1 \frac{\dot{x}_1^2 + \dot{y}_1^2}{2} + m_2 \frac{\dot{x}_2^2 + \dot{y}_2^2}{2}.$$

C Systems with holonomic constraints

In Section 17 we defined the notion of a system of point masses with holonomic constraints. We will now show that such a system is natural.

Consider the configuration manifold M of a system with constraints as embedded in the $3n$-dimensional configuration space of a system of free points. The metric on the $3n$-dimensional space is given by the quadratic form $\sum_{i=1}^{n} m_i \dot{\mathbf{r}}_i^2$. The embedded Riemannian manifold M with potential energy U coincides with the system defined in Section 17 or with the limiting case of the system with potential $U + N\mathbf{q}_2^2$, $N \to \infty$, which grows rapidly outside of M.

D Procedure for solving problems with constraints

1. Determine the configuration manifold and introduce coordinates $q_1, \ldots, q_k$ (in a neighborhood of each of its points).
2. Express the kinetic energy $T = \sum \frac{1}{2} m_i \dot{\mathbf{r}}_i^2$ as a quadratic form in the generalized velocities

$$T = \tfrac{1}{2} \sum a_{ij}(\mathbf{q}) \dot{q}_i \dot{q}_j.$$

3. Construct the lagrangian function $L = T - U(\mathbf{q})$ and solve Lagrange's equations.

EXAMPLE. We consider the motion of a point mass of mass 1 on a surface of revolution in three-dimensional space. It can be shown that the orbits are geodesics on the surface. In cylindrical coordinates r, φ, z the surface is given (locally) in the form $r = r(z)$ or $z = z(r)$. The kinetic energy has the form (Figure 66)

$$T = \tfrac{1}{2}(\dot{x}^2 + \dot{y}^2 + \dot{z}^2) = \tfrac{1}{2}[(1 + r_z'^2)\dot{z}^2 + r^2(z)\dot{\varphi}^2]$$

in coordinates φ and z, and

$$T = \tfrac{1}{2}(\dot{x}^2 + \dot{y}^2 + \dot{z}^2) = \tfrac{1}{2}[(1 + z_r'^2)\dot{r}^2 + r^2\dot{\varphi}^2]$$

in coordinates r and φ. (We have used the identity $\dot{x}^2 + \dot{y}^2 = \dot{r}^2 + r^2\dot{\varphi}^2$.)

The lagrangian function L is equal to T. In both coordinate systems φ is a cyclic coordinate. The corresponding momentum is preserved: $p_\varphi = r^2\dot{\varphi}$ is nothing other than the z-component of

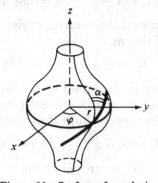

Figure 66 Surface of revolution

angular momentum. Since the system has two degrees of freedom, knowing the cyclic coordinate φ is sufficient for integrating the problem completely (cf. Corollary 3, Section 15).

We can obtain more easily a clear picture of the orbits by reasoning slightly differently. Denote by α the angle of the orbit with a meridian. We have $r\dot{\varphi} = |v| \sin \alpha$, where $|v|$ is the magnitude of the velocity vector (Figure 66).

By the law of conservation of energy, $H = L = T$ is preserved. Therefore, $|v| = $ const, so the conservation law for p_φ takes the form

$$r \sin \alpha = \text{const}$$

("Clairaut's theorem").

This relationship shows that the motion takes place in the region $|\sin \alpha| \le 1$, i.e., $r \ge r_0 \sin \alpha_0$. Furthermore, the inclination of the orbit from the meridian increases as the radius r decreases. When the radius reaches the smallest possible value, $r = r_0 \sin \alpha_0$, the orbit is reflected and returns to the region with larger r (Figure 67).

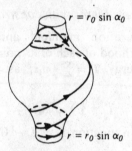

Figure 67   Geodesics on a surface of revolution

PROBLEM. Show that the geodesics on a convex surface of revolution are divided into three classes: meridians, closed curves, and geodesics dense in a ring $r \ge c$.

PROBLEM. Study the behavior of geodesics on the surface of a torus $((r - R)^2 + z^2 = \rho^2)$.

E *Non-autonomous systems*

A *lagrangian non-autonomous system* differs from the autonomous systems, which we have been studying until now, by the additional dependence of the lagrangian function on time:

$$L: TM \times \mathbb{R} \to \mathbb{R} \qquad L = L(\mathbf{q}, \dot{\mathbf{q}}, t).$$

In particular, both the kinetic and potential energies can depend on time in a non-autonomous natural system:

$$T: TM \times \mathbb{R} \to \mathbb{R} \qquad U: M \times \mathbb{R} \to \mathbb{R} \qquad T = T(\mathbf{q}, \dot{\mathbf{q}}, t) \qquad U = U(\mathbf{q}, t).$$

A system of n mass points, constrained by holonomic constraints dependent on time, is defined with the help of a time-dependent submanifold of the configuration space of a free system. Such a manifold is given by a mapping

$$i: M \times \mathbb{R} \to E^{3n} \qquad i(\mathbf{q}, t) = \mathbf{x},$$

which, for any fixed $t \in \mathbb{R}$, defines an embedding $M \to E^{3n}$. The formula of section D remains true for non-autonomous systems.

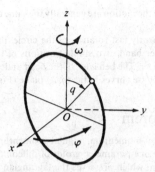

Figure 68   Bead on a rotating circle

EXAMPLE. Consider the motion of a bead along a vertical circle of radius r (Figure 68) which rotates with angular velocity ω around the vertical axis passing through the center O of the circle. The manifold M is the circle. Let q be the angular coordinate on the circle, measured from the highest point.

Let x, y, and z be cartesian coordinates in E^3 with origin O and vertical axis z. Let φ be the angle of the plane of the circle with the plane xOz. By hypothesis, $\varphi = \omega t$. The mapping $i: M \times \mathbb{R} \to E^3$ is given by the formula

$$i(q, t) = (r \sin q \cos \omega t, r \sin q \sin \omega t, r \cos q).$$

From this formula (or, more simply, from an "infinitesimal right triangle") we find that

$$T = \frac{m}{2}(\omega^2 r^2 \sin^2 q + r^2 \dot{q}^2) \qquad U = mgr \cos q.$$

In this case the lagrangian function $L = T - U$ turns out to be independent of t, although the constraint does depend on time. Furthermore, the lagrangian function turns out to be the same as in the one-dimensional system with kinetic energy

$$T_0 = \frac{M}{2} \dot{q}^2 \qquad M = mr^2,$$

and with potential energy

$$V = A \cos q - B \sin^2 q, \qquad A = mgr, \ B = \frac{m}{2} \omega^2 r^2.$$

The form of the phase portrait depends on the ratio between A and B. For $2B < A$ (i.e., for a rotation of the circle slow enough that $\omega^2 r < g$), the lowest position of the bead ($q = \pi$) is

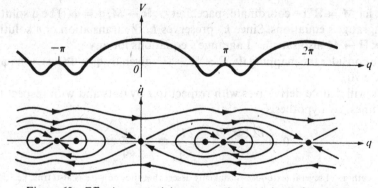

Figure 69   Effective potential energy and phase plane of the bead

stable and the characteristics of the motion are generally the same as in the case of a mathematical pendulum ($\omega = 0$).

For $2B > A$, i.e., for sufficiently fast rotation of the circle, the lowest position of the bead becomes unstable; on the other hand, two stable positions of the bead appear on the circle, where $\cos q = -A/2B = -g/\omega^2 r$. The behavior of the bead under all possible initial conditions is clear from the shape of the phase curves in the $(q, \dot{q})$-plane (Figure 69).

20 E. Noether's theorem

Various laws of conservation (of momentum, angular momentum, etc.) are particular cases of one general theorem: to every one-parameter group of diffeomorphisms of the configuration manifold of a lagrangian system which preserves the lagrangian function, there corresponds a first integral of the equations of motion.

A Formulation of the theorem

Let M be a smooth manifold, $L: TM \to \mathbb{R}$ a smooth function on its tangent bundle TM. Let $h: M \to M$ be a smooth map.

Definition. A lagrangian system (M, L) *admits the mapping h* if for any tangent vector $\mathbf{v} \in TM$,

$$L(h_* \mathbf{v}) = L(\mathbf{v}).$$

EXAMPLE. Let $M = \{(x_1, x_2, x_3)\}$, $L = (m/2)(\dot{x}_1^2 + \dot{x}_2^2 + \dot{x}_3^2) - U(x_2, x_3)$. The system admits the translation $h: (x_1, x_2, x_3) \to (x_1 + s, x_2, x_3)$ along the x_1 axis and does not admit, generally speaking, translations along the x_2 axis.

Noether's theorem. *If the system (M, L) admits the one-parameter group of diffeomorphisms $h^s: M \to M$, $s \in \mathbb{R}$, then the lagrangian system of equations corresponding to L has a first integral $I: TM \to \mathbb{R}$.*

In local coordinates q on M the integral I is written in the form

$$I(\mathbf{q}, \dot{\mathbf{q}}) = \frac{\partial L}{\partial \dot{\mathbf{q}}} \frac{dh^s(\mathbf{q})}{ds}\bigg|_{s=0}.$$

B Proof

First, let $M = \mathbb{R}^n$ be coordinate space. Let $\boldsymbol{\varphi}: \mathbb{R} \to M$, $\mathbf{q} = \boldsymbol{\varphi}(t)$ be a solution to Lagrange's equations. Since h_*^s preserves L, the translation of a solution, $h^s \circ \boldsymbol{\varphi}: \mathbb{R} \to M$ also satisfies Lagrange's equations for any s.[38]

We consider the mapping $\boldsymbol{\Phi}: \mathbb{R} \times \mathbb{R} \to \mathbb{R}^n$, given by $\mathbf{q} = \boldsymbol{\Phi}(s, t) = h^s(\boldsymbol{\varphi}(t))$ (Figure 70).

We will denote derivatives with respect to t by dots and with respect to s by primes. By hypothesis

$$(1) \qquad 0 = \frac{\partial L(\boldsymbol{\Phi}, \dot{\boldsymbol{\Phi}})}{\partial s} = \frac{\partial L}{\partial \mathbf{q}} \cdot \boldsymbol{\Phi}' + \frac{\partial L}{\partial \dot{\mathbf{q}}} \dot{\boldsymbol{\Phi}}',$$

---

[38] The authors of several textbooks mistakenly assert that the converse is also true, i.e., that if h^s takes solutions to solutions, then h_*^s preserves L.

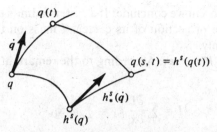

Figure 70   Noether's theorem

where the partial derivatives of L are taken at the point $\mathbf{q} = \boldsymbol{\Phi}(s, t)$, $\dot{\mathbf{q}} = \dot{\boldsymbol{\Phi}}(s, t)$.

As we stated above, the mapping $\boldsymbol{\Phi}|_{s=\text{const}}: \mathbb{R} \to \mathbb{R}^n$ for any fixed s satisfies Lagrange's equation

$$\frac{\partial}{\partial t}\left[\frac{\partial L}{\partial \dot{\mathbf{q}}}(\boldsymbol{\Phi}(s, t), \dot{\boldsymbol{\Phi}}(s, t)) \right] = \frac{\partial L}{\partial \mathbf{q}}(\boldsymbol{\Phi}(s, t), \dot{\boldsymbol{\Phi}}(s, t)).$$

We introduce the notation $\mathbf{F}(s, t) = (\partial L/\partial \dot{\mathbf{q}})(\boldsymbol{\Phi}(s, t), \dot{\boldsymbol{\Phi}}(s, t))$ and substitute $\partial \mathbf{F}/\partial t$ for $\partial L/\partial \mathbf{q}$ in (1).

Writing $\dot{\mathbf{q}}'$ as $d\mathbf{q}'/dt$, we get

$$0 = \left(\frac{d}{dt}\frac{\partial L}{\partial \dot{\mathbf{q}}} \right)\mathbf{q}' + \frac{\partial L}{\partial \dot{\mathbf{q}}}\left(\frac{d}{dt}\mathbf{q}' \right) = \frac{d}{dt}\left(\frac{\partial L}{\partial \dot{\mathbf{q}}}\mathbf{q}' \right) = \frac{dI}{dt}. \qquad \square$$

Remark. The first integral $I = (\partial L/\partial \dot{\mathbf{q}})\mathbf{q}'$ is defined above using local coordinates $\mathbf{q}$. It turns out that the *value of $I(\mathbf{v})$ does not depend on the choice of coordinate system $\mathbf{q}$.*

In fact, I is the rate of change of $L(\mathbf{v})$ when the vector $\mathbf{v} \in TM_{\mathbf{x}}$ varies inside $TM_{\mathbf{x}}$ with velocity $(d/ds)|_{s=0}h^s\mathbf{x}$. Therefore, $I(\mathbf{v})$ is well defined as a function of the tangent vector $\mathbf{v} \in TM_{\mathbf{x}}$. Noether's theorem is proved in the same way when M is a manifold.

C Examples

EXAMPLE 1. Consider a system of point masses with masses m_i:

$$L = \sum m_i \frac{\dot{\mathbf{x}}_i^2}{2} - U(\mathbf{x}) \qquad \mathbf{x}_i = x_{i1}\mathbf{e}_1 + x_{i2}\mathbf{e}_2 + x_{i3}\mathbf{e}_3,$$

constrained by the conditions $f_j(\mathbf{x}) = 0$. We assume that the system admits translations along the $\mathbf{e}_1$ axis:

$$h^s: \mathbf{x}_i \to \mathbf{x}_i + s\mathbf{e}_1 \quad \text{for all } i.$$

In other words, the constraints admit motions of the system as a whole along the $\mathbf{e}_1$ axis, and the potential energy does not change under these.

By Noether's theorem we conclude: If a system admits translations along the $\mathbf{e}_1$ axis, then the projection of its center of mass on the $\mathbf{e}_1$ axis moves linearly and uniformly.

In fact, $(d/ds)|_{s=0} h^s \mathbf{x}_i = \mathbf{e}_1$. According to the remark at the end of B, the quantity

$$I = \sum \frac{\partial L}{\partial \dot{\mathbf{x}}_i} \mathbf{e}_1 = \sum m_i \dot{x}_{i1}$$

is preserved, i.e., the first component P_1 of the momentum vector is preserved. We showed this earlier for a system without constraints.

EXAMPLE 2. If a system admits rotations around the $\mathbf{e}_1$ axis, then the angular momentum with respect to this axis,

$$M_1 = \sum_i ([\mathbf{x}_i, m_i \dot{\mathbf{x}}_i], \mathbf{e}_1)$$

is conserved.

It is easy to verify that if h^s is rotation around the $\mathbf{e}_1$ axis by the angle s, then $(d/ds)|_{s=0} h^s \mathbf{x}_i = [\mathbf{e}_1, \mathbf{x}_i]$, from which it follows that

$$I = \sum_i \frac{\partial L}{\partial \dot{\mathbf{x}}_i} [\mathbf{e}_1, \mathbf{x}_i] = \sum_i (m_i \dot{\mathbf{x}}_i, [\mathbf{e}_1, \mathbf{x}_i]) = \sum_i ([\mathbf{x}_i, m_i \dot{\mathbf{x}}_i], \mathbf{e}_1).$$

PROBLEM 1. Suppose that a particle moves in the field of the uniform helical line $x = \cos \varphi$, $y = \sin \varphi, z = c\varphi$. Find the law of conservation corresponding to this helical symmetry.

ANSWER. In any system which admits helical motions leaving our helical line fixed, the quantity $I = cP_3 + M_3$ is conserved.

PROBLEM 2. Suppose that a rigid body is moving under its own inertia. Show that its center of mass moves linearly and uniformly. If the center of mass is at rest, then the angular momentum with respect to it is conserved.

PROBLEM 3. What quantity is conserved under the motion of a heavy rigid body if it is fixed at some point O? What if, in addition, the body is symmetric with respect to an axis passing through O?

PROBLEM 4. Extend Noether's theorem to non-autonomous lagrangian systems.

Hint. Let $M_1 = M \times \mathbb{R}$ be the extended configuration space (the direct product of the configuration manifold M with the time axis $\mathbb{R}$).

Define a function $L_1 : TM_1 \to \mathbb{R}$ by

$$L \frac{dt}{d\tau};$$

i.e., in local coordinates $\mathbf{q}, t$ on M_1 we define it by the formula

$$L_1\left(\mathbf{q}, t, \frac{d\mathbf{q}}{d\tau}, \frac{dt}{d\tau}\right) = L\left(\mathbf{q}, \frac{d\mathbf{q}/d\tau}{dt/d\tau}, t\right) \frac{dt}{d\tau}.$$

We apply Noether's theorem to the lagrangian system (M_1, L_1).

If L_1 admits the transformations $h^s: M_1 \to M_1$, we obtain a first integral $I_1: TM_1 \to \mathbb{R}$. Since $\int L\, dt = \int L_1\, d\tau$, this reduces to a first integral $I: TM \times \mathbb{R} \to \mathbb{R}$ of the original system. If, in local coordinates $(\mathbf{q}, t)$ on M_1, we have $I_1 = I_1(\mathbf{q}, t, d\mathbf{q}/d\tau, dt/d\tau)$, then $I(\mathbf{q}, \dot{\mathbf{q}}, t) = I_1(\mathbf{q}, t, \dot{\mathbf{q}}, 1)$.

In particular, if L does not depend on time, L_1 admits translations along time, $h^s(\mathbf{q}, t) = (\mathbf{q}, t + s)$. The corresponding first integral I is the energy integral.

21 D'Alembert's principle

We give here a new definition of a system of point masses with holonomic constraints and prove its equivalence to the definition given in Section 17.

A Example

Consider the holonomic system (M, L), where M is a surface in three-dimensional space $\{\mathbf{x}\}$:

$$L = \tfrac{1}{2}m\dot{\mathbf{x}}^2 - U(\mathbf{x}).$$

In mechanical terms, "the mass point $\mathbf{x}$ of mass m must remain on the smooth surface M."

Consider a motion of the point, $\mathbf{x}(t)$. If Newton's equations $m\ddot{\mathbf{x}} + (\partial U/\partial \mathbf{x}) = 0$ were satisfied, then in the absence of external forces ($U = 0$) the trajectory would be a straight line and could not lie on the surface M.

From the point of view of Newton, this indicates the presence of a new force "forcing the point to stay on the surface."

Definition. The quantity

$$\mathbf{R} = m\ddot{\mathbf{x}} + \frac{\partial U}{\partial \mathbf{x}}$$

is called the *constraint force* (Figure 71).

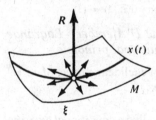

Figure 71   Constraint force

If we take the constraint force $\mathbf{R}(t)$ into account, Newton's equations are obviously satisfied:

$$m\ddot{\mathbf{x}} = -\frac{\partial U}{\partial \mathbf{x}} + \mathbf{R}.$$

The physical meaning of the constraint force becomes clear if we consider our system with constraints as the limit of systems with potential energy $U + NU_1$ as $N \to \infty$, where $U_1(\mathbf{x}) = \rho^2(\mathbf{x}, M)$. For large N the constraint potential NU_1 produces a rapidly changing force

$\mathbf{F} = -N\,\partial U_1/\partial\mathbf{x}$; when we pass to the limit ($N \to \infty$) the average value of the force $\mathbf{F}$ under oscillations of $\mathbf{x}$ near M is $\mathbf{R}$. The force $\mathbf{F}$ is perpendicular to M. Therefore, the constraint force $\mathbf{R}$ is perpendicular to M: $(\mathbf{R}, \xi) = 0$ for every tangent vector ξ.

B Formulation of the D'Alembert–Lagrange principle

In mechanics, tangent vectors to the configuration manifold are called *virtual variations*. The D'Alembert–Lagrange principle states:

$$\left(m\ddot{\mathbf{x}} + \frac{\partial U}{\partial\mathbf{x}}, \xi\right) = 0$$

for any virtual variation ξ, or stated differently, the *work of the constraint force on any virtual variation is zero.*

For a system of points $\mathbf{x}_i$ with masses m_i the constraint forces $\mathbf{R}_i$ are defined by $\mathbf{R}_i = m_i\ddot{\mathbf{x}}_i + (\partial U/\partial\mathbf{x}_i)$, and D'Alembert's principle has the form $\sum (\mathbf{R}_i, \xi_i) = 0$, *or* $\sum ((m_i\ddot{\mathbf{x}}_i + (\partial U/\partial\mathbf{x}_i), \xi_i) = 0$, i.e., the sum of the works of the constraint forces on any virtual variation $\{\xi_i\} \in TM_{\mathbf{x}}$ is zero.

Constraints with the property described above are called *ideal*.

If we define a system with holonomic constraints as a limit as $N \to \infty$, then the D'Alembert–Lagrange principle becomes a theorem: its proof is sketched above for the simplest case.

It is possible, however, to *define* an ideal holonomic constraint using the D'Alembert–Lagrange principle. In this way we have three definitions of holonomic systems with constraints:

1. The limit of systems with potential energies $U + NU_1$ as $N \to \infty$.
2. A holonomic system (M, L), where M is a smooth submanifold of the configuration space of a system without constraints and L is the lagrangian.
3. A system which complies with the D'Alembert–Lagrange principle.

All three definitions are mathematically equivalent.

The proof of the implications (1) $\Rightarrow$ (2) and (1) $\Rightarrow$ (3) is sketched above and will not be given in further detail. We will now show that (2) $\Leftrightarrow$ (3).

C The equivalence of the D'Alembert–Lagrange principle and the variational principle

Let M be a submanifold of euclidean space, $M \subset \mathbb{R}^N$, and $\mathbf{x}: \mathbb{R} \to M$ a curve, with $\mathbf{x}(t_0) = \mathbf{x}_0$, $\mathbf{x}(t_1) = \mathbf{x}_1$.

Definition. The curve $\mathbf{x}$ is called a *conditional extremal* of the action functional

$$\Phi = \int_{t_0}^{t_1} \left\{\frac{\dot{\mathbf{x}}^2}{2} - U(\mathbf{x})\right\}dt,$$

if the differential $\delta\Phi$ is equal to zero *under the condition* that the variation consists of nearby curves[39] joining $\mathbf{x}_0$ to $\mathbf{x}_1$ in M.

[39] Strictly speaking, in order to define a variation $\delta\Phi$, one must define on the set of curves near $\mathbf{x}$ on M the structure of a region in a vector space. This can be done using coordinates on M; however, the property of being a conditional extremal does not depend on the choice of a co-ordinate system.

We will write

(1) $$\delta_M \Phi = 0.$$

Clearly, Equation (1) is equivalent to the Lagrange equations

$$\frac{d}{dt}\frac{\partial L}{\partial \dot{\mathbf{q}}} = \frac{\partial L}{\partial \mathbf{q}} \qquad L = \frac{\dot{\mathbf{x}}^2}{2} - U(\mathbf{x}) \qquad \mathbf{x} = \mathbf{x}(\mathbf{q}),$$

in some local coordinate system $\mathbf{q}$ on M.

Theorem. *A curve* $\mathbf{x}: \mathbb{R} \to M \subset \mathbb{R}^N$ *is a conditional extremal of the action (i.e., satisfies Equation (1)) if and only if it satisfies D'Alembert's equation*

(2) $$\left(\ddot{\mathbf{x}} + \frac{\partial U}{\partial \mathbf{x}}, \, \boldsymbol{\xi} \right) = 0, \qquad \forall \boldsymbol{\xi} \in TM_{\mathbf{x}}.$$

Lemma. *Let* $\mathbf{f}: \{t: t_0 \le t \le t_1\} \to \mathbb{R}^N$ *be a continuous vector field. If, for every continuous tangent vector field* $\boldsymbol{\xi}$, *tangent to* M *along* $\mathbf{x}$ *(i.e.,* $\boldsymbol{\xi}(t) \in TM_{\mathbf{x}(t)}$, *with* $\boldsymbol{\xi}(t) = 0$ *for* $t = t_0, t_1$), *we have*

$$\int_{t_0}^{t_1} \mathbf{f}(t)\boldsymbol{\xi}(t)dt = 0,$$

then the field $\mathbf{f}(t)$ *is perpendicular to* M *at every point* $\mathbf{x}(t)$ *(i.e.,* $(\mathbf{f}(t), \mathbf{h}) = 0$ *for every vector* $\mathbf{h} \in TM_{\mathbf{x}(t)}$) *(Figure 72).*

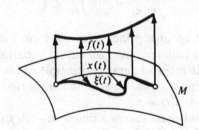

Figure 72   Lemma about the normal field

The proof of the lemma repeats the argument which we used to derive the Euler–Lagrange equations in Section 12.

PROOF OF THE THEOREM. We compare the value of Φ on the two curves $\mathbf{x}(t)$ and $\mathbf{x}(t) + \boldsymbol{\xi}(t)$, where $\boldsymbol{\xi}(t_0) = \boldsymbol{\xi}(t_1) = 0$. Integrating by parts, we obtain

$$\delta \Phi = \int_{t_0}^{t_1} \left(\dot{\mathbf{x}}\dot{\boldsymbol{\xi}} - \frac{\partial U}{\partial \mathbf{x}}\boldsymbol{\xi} \right)dt = -\int_{t_0}^{t_1} \left(\ddot{\mathbf{x}} + \frac{\partial U}{\partial \mathbf{x}} \right)\boldsymbol{\xi}\, dt.$$

It is obvious from this formula[40] that Equation (1), $\delta_M \Phi = 0$, is equivalent to the collection of equations

(3)
$$\int_{t_0}^{t_1} \left(\ddot{\mathbf{x}} + \frac{\partial U}{\partial \mathbf{x}} \right) \boldsymbol{\xi} \, dt = 0.$$

for all tangent vector fields $\boldsymbol{\xi}(t) \in TM_{\mathbf{x}(t)}$ with $\boldsymbol{\xi}(t_0) = \boldsymbol{\xi}(t_1) = 0$. By the lemma (where we must set $\mathbf{f} = \ddot{\mathbf{x}} + (\partial U/\partial \mathbf{x})$) the collection of equations (3) is equivalent to the D'Alembert-Lagrange equation (2).   □

D Remarks

Remark 1. We derive the *D'Alembert–Lagrange principle for a system of n points* $\mathbf{x}_i \in \mathbb{R}^3$, $i = 1, \ldots, n$, with masses m_i, with holonomic constraints, from the above theorem.

In the coordinates $\bar{\mathbf{x}} = \{\bar{\mathbf{x}}_i = \sqrt{m_i}\mathbf{x}_i\}$, the kinetic energy takes the form $T = \frac{1}{2} \sum m_i \dot{\mathbf{x}}_i^2 = \frac{1}{2}\dot{\bar{\mathbf{x}}}^2$.

By the theorem, the extremals of the principle of least action satisfy the condition

$$\left(\ddot{\bar{\mathbf{x}}} + \frac{\partial U}{\partial \bar{\mathbf{x}}}, \boldsymbol{\xi} \right) = 0$$

(the D'Alembert-Lagrange principle for points in $\mathbb{R}^{3n}$: the $3n$-dimensional reaction force is orthogonal to the manifold M in the metric T). Returning to the coordinates $\mathbf{x}_i$, we get

$$0 = \left(\sqrt{m_i}\ddot{\mathbf{x}}_i + \frac{\partial U}{\partial \sqrt{m_i}\mathbf{x}_i}, \sqrt{m_i}\boldsymbol{\xi}_i \right) = \sum_i \left(m_i\ddot{\mathbf{x}}_i + \frac{\partial U}{\partial \mathbf{x}_i}, \boldsymbol{\xi}_i \right),$$

i.e., the D'Alembert-Lagrange principle in the form indicated earlier: the sum of the work of the reaction forces on virtual variations is zero.

Remark 2. The D'Alembert-Lagrange principle can be given in a slightly different form if we turn to statics. An *equilibrium position* is a point $\mathbf{x}_0$ which is the orbit of a motion: $\mathbf{x}(t) = \mathbf{x}_0$.

Suppose that a point mass moves along a smooth surface M under the influence of the force $\mathbf{f} = -\partial U/\partial \mathbf{x}$.

Theorem. *The point* $\mathbf{x}_0$ *in M is an equilibrium position if and only if the force is orthogonal to the surface at* $\mathbf{x}_0$: $(\mathbf{f}(\mathbf{x}_0), \boldsymbol{\xi}) = 0$ *for all* $\boldsymbol{\xi} \in TM_{\mathbf{x}_0}$.

This follows from the D'Alembert-Lagrange equations in view of the fact that $\ddot{\mathbf{x}} = 0$.

Definition. $-m\ddot{\mathbf{x}}$ is called the *force of inertia*.

---

[40] The distance of the points $\mathbf{x}(t) + \boldsymbol{\xi}(t)$ from M is small of second-order compared with $\xi(t)$.

Now the D'Alembert–Lagrange principle takes the form:

Theorem. *If the forces of inertia are added to the acting forces,* **x** *becomes an equilibrium position.*

PROOF. D'Alembert's equation

$$(-m\ddot{\mathbf{x}} + \mathbf{f}, \boldsymbol{\xi}) = 0$$

expresses the fact, as in the preceding theorem, that **x** is an equilibrium position of a system with forces $-m\ddot{\mathbf{x}} + \mathbf{f}$. □

Entirely analogous statements are true for systems of points: If $\mathbf{x} = \{\mathbf{x}_i\}$ are equilibrium positions, then the sum of the work of the forces acting on the virtual variations is equal to zero. If the forces of inertia $-m_i\ddot{\mathbf{x}}_i(t)$ are added to the acting forces, then the position $\mathbf{x}(t)$ becomes an equilibrium position.

Now a problem about motions can be reduced to a problem about equilibrium under actions of other forces.

Remark 3. Up to now we have not considered cases when the constraints depend on time. All that was said above carries over to such constraints without any changes.

EXAMPLE. Consider a bead sliding along a rod which is tilted at an angle α to the vertical axis and is rotating uniformly with angular velocity ω around

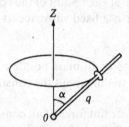

Figure 73   Bead on a rotating rod

this axis (its weight is negligible). For our coordinate q we take the distance from the point 0 (Figure 73). The kinetic energy and lagrangian are:

$$L = T = \tfrac{1}{2}mv^2 = \tfrac{1}{2}m\dot{q}^2 + \tfrac{1}{2}m\omega^2 r^2,$$

$$r = q \sin \alpha.$$

Lagrange's equation: $m\ddot{q} = m\omega^2 q \sin^2 \alpha$.

The constraint force at each moment is orthogonal to virtual variations (i.e., to the direction of the rod), but is not at all orthogonal to the actual trajectory.

Remark 4. It is easy to derive conservation laws from the D'Alembert–Lagrange equations. For example, if translation along the x_1 axis $\boldsymbol{\xi}_i = \mathbf{e}_1$ is among the virtual variations, then the sum of the work of the constraint forces on this variation is equal to zero:

$$\sum(\mathbf{R}_i, \mathbf{e}_1) = (\sum \mathbf{R}_i, \mathbf{e}_1) = 0.$$

If we now consider constraint forces as external forces, then we notice that the sum of the first components of the external forces is equal to zero. This means that the first component, P_1, of the momentum vector is preserved.

We obtained this same result earlier from Noether's theorem.

Remark 5. We emphasize once again that the holonomic character of some particular physical constraint or another (to a given degree of exactness) is a question of experiment. From the mathematical point of view, the holonomic character of a constraint is a postulate of physical origin; it can be introduced in various equivalent forms, for example, in the form of the principle of least action (1) or the D'Alembert–Lagrange principle (2), but, when defining the constraints, the term always refers to experimental facts which go beyond Newton's equations.

Remark 6. Our terminology differs somewhat from that used in mechanics textbooks, where the D'Alembert–Lagrange principle is extended to a wider class of systems ("non-holonomic systems with ideal constraints"). In this book we will not consider non-holonomic systems. We remark only that one example of a non-holonomic system is a sphere rolling on a plane without slipping. In the tangent space at each point of the configuration manifold of a non-holonomic system there is a fixed subspace to which the velocity vector must belong.

Remark 7. If a system consists of mass points connected by rods, hinges, etc., then the need may arise to talk about the constraint force of some particular constraint.

We defined the total "constraint force of all constraints" $\mathbf{R}_i$ for every mass point m_i. The concept of a constraint force for an individual constraint is *impossible* to define, as may be already seen from the simple example of a beam resting on three columns. If we try to define constraint forces of the columns, $\mathbf{R}_1$, $\mathbf{R}_2$, $\mathbf{R}_3$ by passing to a limit (considering the columns as very rigid springs), then we may become convinced that the result depends on the distribution of rigidity.

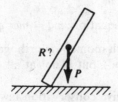

Figure 74 Constraint force on a rod

Problems for students are selected so that this difficulty does not arise.

PROBLEM. A rod of weight P, tilted at an angle of 60° to the plane of a table, begins to fall with initial velocity zero (Figure 74). Find the constraint force of the table at the initial moment, considering the table as (a) absolutely smooth and (b) absolutely rough. (In the first case, the holonomic constraint holds the end of the rod on the plane of the table, and in the second case, at a given point.)

5 Oscillations

Because linear equations are easy to solve and study, the theory of linear oscillations is the most highly developed area of mechanics. In many non-linear problems, linearization produces a satisfactory approximate solution. Even when this is not the case, the study of the linear part of a problem is often a first step, to be followed by the study of the relation between motions in a nonlinear system and in its linear model.

22 Linearization

We give here the definition of small oscillations.

A Equilibrium positions

Definition. A point $\mathbf{x}_0$ is called an *equilibrium position* of the system

$$(1) \qquad \frac{d\mathbf{x}}{dt} = \mathbf{f}(\mathbf{x}), \qquad \mathbf{x} \in \mathbb{R}^n$$

if $\mathbf{x}(t) \equiv \mathbf{x}_0$ is a solution of this system. In other words, $\mathbf{f}(\mathbf{x}_0) = 0$, i.e., the vector field $\mathbf{f}(\mathbf{x})$ is zero at $\mathbf{x}_0$.

EXAMPLE. Consider the natural dynamical system with lagrangian function $L(\mathbf{q}, \dot{\mathbf{q}}) = T - U$, where $T = \frac{1}{2} \sum a_{ij}(\mathbf{q})\dot{q}_i \dot{q}_j \geq 0$ and $U = U(\mathbf{q})$:

$$(2) \qquad \frac{d}{dt} \frac{\partial L}{\partial \dot{\mathbf{q}}} = \frac{\partial L}{\partial \mathbf{q}}, \qquad \mathbf{q} = (q_1, \ldots, q_n).$$

Lagrange's equations can be written in the form of a system of $2n$ first-order equations of form (1). We will try to find an equilibrium position:

Theorem. *The point* $\mathbf{q} = \mathbf{q}_0$, $\dot{\mathbf{q}} = \dot{\mathbf{q}}_0$ *will be an equilibrium position if and only if* $\dot{\mathbf{q}}_0 = 0$ *and* $\mathbf{q}_0$ *is a critical point of the potential energy, i.e.,*

(3)
$$\left. \frac{\partial U}{\partial \mathbf{q}} \right|_{\mathbf{q}_0} = 0.$$

PROOF. We write down Lagrange's equations

$$\frac{d}{dt} \frac{\partial T}{\partial \dot{\mathbf{q}}} = \frac{\partial T}{\partial \mathbf{q}} - \frac{\partial U}{\partial \mathbf{q}}.$$

From (2) it is clear that, for $\dot{\mathbf{q}} = 0$, we will have $\partial T / \partial \mathbf{q} = 0$ and $\partial T / \partial \dot{\mathbf{q}} = 0$. Therefore, $\mathbf{q} = \mathbf{q}_0$ is a solution in case (3) holds and only in that case. $\square$

B Stability of equilibrium positions

We will now investigate motions with initial conditions close to an equilibrium position.

Theorem. *If the point* $\mathbf{q}_0$ *is a strict local minimum of the potential energy* U, *then the equilibrium* $\mathbf{q} = \mathbf{q}_0$ *is stable in the sense of Liapunov.*

PROOF. Let $U(\mathbf{q}_0) = h$. For sufficiently small $\varepsilon > 0$, the connected component of the set $\{\mathbf{q}: U(\mathbf{q}) \leq h + \varepsilon\}$ containing $\mathbf{q}_0$ will be an arbitrarily small neighborhood of $\mathbf{q}_0$ (Figure 75). Furthermore, the connected component of the corresponding region in phase space $\mathbf{p}$, $\mathbf{q}$, $\{\mathbf{p}, \mathbf{q}: E(\mathbf{p}, \mathbf{q}) \leq h + \varepsilon\}$, (where $\mathbf{p} = \partial T / \partial \dot{\mathbf{q}}$ is the momentum and $E = T + U$ is the total energy) will be an arbitrarily small neighborhood of the point $\mathbf{p} = 0$, $\mathbf{q} = \mathbf{q}_0$.

But the region $\{\mathbf{p}, \mathbf{q}: E \leq h + \varepsilon\}$ is invariant with respect to the phase flow by the law of conservation of energy. Therefore, for initial conditions $\mathbf{p}(0)$, $\mathbf{q}(0)$ close enough to $(0, \mathbf{q}_0)$, every phase trajectory $(\mathbf{p}(t), \mathbf{q}(t))$ is close to $(0, \mathbf{q}_0)$. $\square$

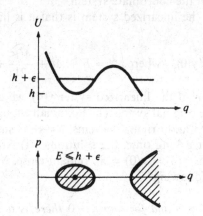

Figure 75 Stable equilibrium position

PROBLEM. Can an equilibrium position $q = q_0$, $p = 0$ be asymptotically stable?

PROBLEM. Show that in an *analytic* system with one degree of freedom an equilibrium position q_0 which is not a strict local minimum of the potential energy is not stable in the sense of Liapunov. Produce an example of an infinitely differentiable system where this is not true.

Remark. It seems likely that in an analytic system with n degrees of freedom, an equilibrium position which is not a minimum point is unstable; but this has never been proved for $n > 2$.

C Linearization of a differential equation

We now turn to the general system (1). In studying solutions of (1) which are close to an equilibrium position x_0, we often use a *linearization*. Assume that $x_0 = 0$ (the general case is reduced to this one by a translation of the co-ordinate system). Then the first term of the Taylor series for f is linear:

$$f(x) = Ax + R_2(x), \qquad A = \frac{\partial f}{\partial x}\bigg|_0 \text{ and } R_2 = O(x^2),$$

where the linear operator A is given in coordinates $x_1, \ldots, x_n$ by the matrix a_{ij}:

$$A(x)_i = \sum a_{ij} x_j; \qquad a_{ij} = \frac{\partial f_i}{\partial x_j}.$$

Definition. The passage from system (1) to the system

$$(4) \qquad\qquad \frac{dy}{dt} = Ay \qquad (x \in \mathbb{R}^n, y \in TR_0^n)$$

is called the *linearization* of (1).

PROBLEM. Show that linearization is a well-defined operation: the operator A does not depend on the coordinate system.

The advantage of the linearized system is that it is linear and therefore easily solved:

$$y(t) = e^{At} y(0), \quad \text{where } e^{At} = E + At + \frac{A^2 t^2}{2!} + \cdots.$$

Knowing the solution of the linearized system (4), we can say something about solutions of the original system (1). For small enough x, the difference between the linearized and original systems, $R_2(x)$, is small in comparison with x. Therefore, for a long time, the solutions $y(t)$, $x(t)$ of both systems with initial conditions $y(0) = x(0) = x_0$ remain close. More explicitly, we can easily prove the following:

Theorem. *For any $T > 0$ and for any $\varepsilon > 0$ there is a $\delta > 0$ such that if $|x(0)| < \delta$, then $|x(t) - y(t)| < \varepsilon\delta$ for all t in the interval $0 < t < T$.*

D Linearization of a lagrangian system

We return again to the lagrangian system (2) and try to linearize it in a neighborhood of the equilibrium position $\mathbf{q} = \mathbf{q}_0$. In order to simplify the formulas, we choose a coordinate system so that $\mathbf{q}_0 = 0$.

Theorem. *In order to linearize the lagrangian system (2) in a neighborhood of the equilibrium position $\mathbf{q} = 0$, it is sufficient to replace the kinetic energy $T = \frac{1}{2}a_{ij}(\mathbf{q})\dot{q}_i\dot{q}_j$ by its value at $\mathbf{q} = 0$,*

$$T_2 = \tfrac{1}{2}\sum a_{ij}\dot{q}_i\dot{q}_j, \qquad a_{ij} = a_{ij}(0),$$

and replace the potential energy $U(\mathbf{q})$ by its quadratic part

$$U_2 = \tfrac{1}{2}\sum b_{ij}q_iq_j, \qquad b_{ij} = \left.\frac{\partial^2 U}{\partial q_i\,\partial q_j}\right|_{q=0}.$$

PROOF. We reduce the lagrangian system to the form (1) by using the canonical variables $\mathbf{p}$ and $\mathbf{q}$:

$$\dot{\mathbf{p}} = -\frac{\partial H}{\partial \mathbf{q}} \qquad \dot{\mathbf{q}} = \frac{\partial H}{\partial \mathbf{p}}, \qquad H(\mathbf{p}, \mathbf{q}) = T + U.$$

Since $\mathbf{p} = \mathbf{q} = 0$ is an equilibrium position, the expansions of the right-hand sides in Taylor series at zero begin with terms that are linear in $\mathbf{p}$ and $\mathbf{q}$. Since the right-hand sides are partial derivatives, these *linear* terms are determined by the *quadratic* terms H_2 of the expansion for $H(\mathbf{p}, \mathbf{q})$. But H_2 is precisely the hamiltonian function of the system with lagrangian $L_2 = T_2 - U_2$, since, clearly, $H_2 = T_2(\mathbf{p}) + U_2(\mathbf{q})$. Therefore, the linearized equations of motion are the equations of motion for the system described in the theorem with $L_2 = T_2 - U_2$. ☐

EXAMPLE. We consider the system with one degree of freedom:

$$T = \tfrac{1}{2}a(q)\dot{q}^2, \qquad U = U(q).$$

Let $q = q_0$ be a stable equilibrium position: $(\partial U/\partial q)|_{q=q_0} = 0, (\partial^2 U/\partial q^2)|_{q=q_0} > 0$ (Figure 76).

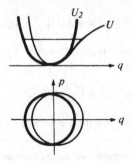

Figure 76 Linearization

As we know from the phase portrait, for initial conditions close to $q = q_0$, $p = 0$, the solution is periodic with period τ depending, generally speaking, on the initial conditions. The above two theorems imply

Corollary. *The period τ of oscillations close to the equilibrium position q_0 approaches the limit $\tau_0 = 2\pi/\omega_0$, (where $\omega_0^2 = b/a$, $b = (\partial^2 U/\partial q^2)|_{q=q_0}$, and $a = a(q_0)$) as the amplitudes of the oscillations decrease.*

PROOF. For the linearized system, $T_2 = \frac{1}{2}a\dot{q}^2$ and $U_2 = \frac{1}{2}bq^2$ (taking $q_0 = 0$). The solutions to Lagrange's equation $\ddot{q} = -\omega_0^2 q$ have period $\tau_0 = 2\pi/\omega_0$:

$$q = c_1 \cos \omega_0 t + c_2 \sin \omega_0 t$$

for any initial amplitude. □

E Small oscillations

Definition. Motions in a linearized system $(L_2 = T_2 - U_2)$ are called small oscillations[41] near an equilibrium $\mathbf{q} = \mathbf{q}_0$. In a one-dimensional problem the numbers τ_0 and ω_0 are called the *period* and the *frequency* of small oscillations.

PROBLEM. Find the period of small oscillations of a bead of mass 1 on a wire $y = U(x)$ in a gravitational field with $g = 1$, near an equilibrium position $x = x_0$ (Figure 77).

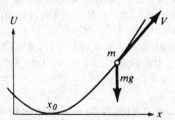

Figure 77   Bead on a wire

Solution. We have

$$U = mgy = U(x)$$

$$T = \frac{1}{2}mv^2 = \frac{1}{2}\left[1 + \left(\frac{\partial U}{\partial x}\right)^2\right]\dot{x}^2.$$

Let x_0 be a stable equilibrium position: $(\partial U/\partial x)|_{x_0} = 0$; $(\partial^2 U/\partial x^2)|_{x_0} > 0$. Then the frequency of small oscillations, ω, is defined by the formula

$$\omega^2 = \left(\frac{\partial^2 U}{\partial x^2}\right)\bigg|_{x_0},$$

since, for the linearized system, $T_2 = \frac{1}{2}\dot{q}^2$ and $U_2 = \frac{1}{2}\omega^2 q^2$ $(q = x - x_0)$.

---

[41] If the equilibrium position is unstable, we will talk about "unstable small oscillations" even though these motions may not have an oscillatory character.

PROBLEM. Show that not only a small oscillation, but any motion of the bead is equivalent to a motion in some one-dimensional system with lagrangian function $L = \frac{1}{2}\dot{q}^2 - V(q)$.

Hint. Take length along the wire for q.

23 Small oscillations

We show here that a lagrangian system undergoing small oscillations decomposes into a direct product of systems with one degree of freedom.

A A problem about pairs of forms

We will consider in more detail the problem of small oscillations. In other words, we consider a system whose kinetic and potential energies are quadratic forms

(1) $$T = \frac{1}{2}(A\dot{q}, \dot{q}) \qquad U = \frac{1}{2}(Bq, q) \qquad q \in \mathbb{R}^n, \dot{q} \in \mathbb{R}^n.$$

The kinetic energy is a positive-definite form.

In order to integrate Lagrange's equations, we will make a special choice of coordinates.

As we know from linear algebra, a pair of quadratic forms (Aq, q), (Bq, q), the first of which is positive-definite, can be reduced to principal axes by a linear change of coordinates:[42]

$$Q = Cq \qquad Q = (Q_1, \ldots, Q_n).$$

In addition, the coordinates Q can be chosen so that the form (Aq, q) decomposes into the sum of squares (Q, Q). Let Q be such coordinates; then, since $\dot{Q} = C\dot{q}$, we have

(2) $$T = \frac{1}{2}\sum_{i=1}^{n}\dot{Q}_i^2 \qquad U = \frac{1}{2}\sum_{i=1}^{n}\lambda_i Q_i^2.$$

The numbers λ_i are called *the eigenvalues of the form B with respect to A*.

PROBLEM. Show that the eigenvalues of B with respect to A satisfy the characteristic equation

(3) $$\det|B - \lambda A| = 0,$$

all the roots of which are, therefore, real (the matrices A and B are symmetric and $A > 0$).

B Characteristic oscillations

In the coordinates Q the lagrangian system decomposes into n independent equations

(4) $$\ddot{Q}_i = -\lambda_i Q_i.$$

---

[42] If one wants to, one can introduce a euclidean structure by taking the first form as the scalar product, and then reducing the second form to the principal axes by a transformation which is orthogonal with respect to this euclidean structure.

Therefore we have proved:

Theorem. *A system performing small oscillations is the direct product of n one-dimensional systems performing small oscillations.*

For the one-dimensional systems, there are three possible cases:
Case 1: $\lambda = \omega^2 > 0$; the solution is $Q = C_1 \cos \omega t + C_2 \sin \omega t$ (oscillation)
Case 2: $\lambda = 0$; the solution is $Q = C_1 + C_2 t$ (neutral equilibrium)
Case 3: $\lambda = -k^2 < 0$; the solution is $Q = C_1 \cosh kt + C_2 \sinh kt$
(instability)

Corollary. *Suppose one of the eigenvalues of* (3) *is positive:* $\lambda = \omega^2 > 0$. *Then system* (1) *can perform a small oscillation of the form*

(5) $$\mathbf{q}(t) = (C_1 \cos \omega t + C_2 \sin \omega t)\boldsymbol{\xi},$$

where $\boldsymbol{\xi}$ *is an eigenvector corresponding to* λ *(Figure 78):*

$$B\boldsymbol{\xi} = \lambda A \boldsymbol{\xi}.$$

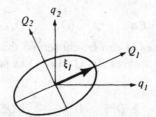

Figure 78   Characteristic oscillation

This oscillation is the product of the one-dimensional motion $Q_i = C_1 \cos \omega_i t + C_2 \sin \omega_i t$ and the trivial motion $Q_j = 0$ $(j \neq i)$.

Definition. The periodic motion (5) is called a *characteristic oscillation* of system (1), and the number ω is called the *characteristic frequency*.

Remark. Characteristic oscillations are also called principal oscillations or normal modes. A nonpositive λ also has eigenvectors; we will also call the corresponding motions "characteristic oscillations," although they are not periodic; the corresponding "characteristic frequencies" are imaginary.

PROBLEM. Show that the number of independent real characteristic oscillations is equal to the dimension of the largest positive-definite subspace for the potential energy $\frac{1}{2}(B\mathbf{q}, \mathbf{q})$.

Now the result may be formulated as follows:

Theorem. *The system* (1) *has n characteristic oscillations, the directions of which are pairwise orthogonal with respect to the scalar product given by the kinetic energy A.*

PROOF. The coordinate system **Q** is orthogonal with respect to the scalar product $(A\mathbf{q}, \mathbf{q})$ by (2).　　　　　　　　　　　　　　　　　　□

C *Decomposition into characteristic oscillations*

It follows from the above theorem that:

Corollary. *Every small oscillation is a sum of characteristic oscillations.*

A sum of characteristic oscillations is generally not periodic (remember the Lissajous figures!).

To decompose a motion into a sum of characteristic oscillations, it is sufficient to project the initial conditions $\mathbf{q}, \dot{\mathbf{q}}$ onto the characteristic directions ξ_i and solve the corresponding one-dimensional problems (4).

Therefore, the Lagrange equations for system (1) can be solved in the following way. We first look for characteristic oscillations of the form $\mathbf{q} = e^{i\omega t}\xi$. Substituting these into Lagrange's equations

$$\frac{d}{dt}A\dot{\mathbf{q}} = -B\mathbf{q},$$

we find

$$(B - \omega^2 A)\xi = 0.$$

From the characteristic equation (3) we find n eigenvalues $\lambda_k = \omega_k^2$. To these there correspond n pairwise orthogonal eigenvectors ξ_k. A general solution in the case $\lambda \neq 0$ has the form

$$\mathbf{q}(t) = \text{Re} \sum_{k=1}^{n} C_k \, e^{i\omega_k t}\xi_k.$$

Remark. This result is also true when some of the λ are multiple eigenvalues.

Thus, in a lagrangian system, as opposed to a general system of linear differential equations, resonance terms of the form $t \sin \omega t$, etc. do not arise, even in the case of multiple eigenvalues.

D *Examples*

EXAMPLE 1. Consider the system of two identical mathematical pendulums of length $l_1 = l_2 = 1$ and mass $m_1 = m_2 = 1$ in a gravitational field with $g = 1$. Suppose that the pendulums are connected by a weightless spring whose length is equal to the distance between the points of suspension (Figure 79). Denote by q_1 and q_2 the angles of inclination of the pendulums. Then

105

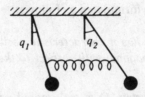

Figure 79   Identical connected pendulums

for small oscillations, $T = \frac{1}{2}(\dot{q}_1^2 + \dot{q}_2^2)$ and $U = \frac{1}{2}(q_1^2 + q_2^2 + \alpha(q_1 - q_2)^2)$, where $\frac{1}{2}\alpha(q_1 - q_2)^2$ is the potential energy of the elasticity of the spring. Set

$$Q_1 = \frac{q_1 + q_2}{\sqrt{2}} \quad \text{and} \quad Q_2 = \frac{q_1 - q_2}{\sqrt{2}}.$$

Then

$$q_1 = \frac{Q_1 + Q_2}{\sqrt{2}} \quad \text{and} \quad q_2 = \frac{Q_1 - Q_2}{\sqrt{2}},$$

and both forms are reduced to principal axes:

$$T = \frac{1}{2}(\dot{Q}_1^2 + \dot{Q}_2^2) \qquad U = \frac{1}{2}(\omega_1^2 Q_1^2 + \omega_2^2 Q_2^2).$$

where $\omega_1 = 1$ and $\omega_2 = \sqrt{1 + 2\alpha}$ (Figure 80). So the two characteristic oscillations are as follows (Figure 81):

1. $Q_2 = 0$, i.e., $q_1 = q_2$; both pendulums move in phase with the original frequency 1, and the spring has no effect;
2. $Q_1 = 0$, i.e., $q_1 = -q_2$: the pendulums move in opposite phase with increased frequency $\omega_2 > 1$ due to the action of the spring.

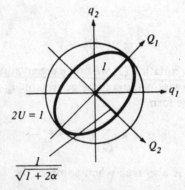

Figure 80   Configuration space of the connected pendulums

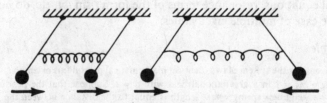

Figure 81   Characteristic oscillations of the connected pendulums

Now let the spring be very weak: $\alpha \ll 1$. Then an interesting effect called *exchange of energy* occurs.

EXAMPLE 2. Suppose that the pendulums are at rest at the initial moment, and one of them is given velocity $\dot{q}_1 = v$. We will show that after some time T the first pendulum will be almost stationary, and all the energy will have gone to the second.

It follows from the initial conditions that $Q_1(0) = Q_2(0) = 0$. Therefore, $Q_1 = c_1 \sin t$, and $Q_2 = c_2 \sin \omega t$ with $\omega = \sqrt{1 + 2\alpha} \approx 1 + \alpha$ ($\alpha \ll 1$). But $\dot{Q}_1(0) = \dot{Q}_2(0) = v/\sqrt{2}$. Therefore, $c_1 = v/\sqrt{2}$ and $c_2 = v/\omega\sqrt{2}$, and our solution has the form

$$q_1 = \frac{v}{2}\left(\sin t + \frac{1}{\omega}\sin \omega t\right) \qquad q_2 = \frac{v}{2}\left(\sin t - \frac{1}{\omega}\sin \omega t\right)$$

or, disregarding the term $v(1 - (1/\omega))\sin \omega t$, which is small since α is,

$$q_1 \approx \frac{v}{2}(\sin t + \sin \omega t) = v\cos \varepsilon t \sin \omega' t,$$

$$q_2 \approx \frac{v}{2}(\sin t - \sin \omega t) = -v\cos \omega' t \sin \varepsilon t,$$

$$\varepsilon = \frac{\omega - 1}{2} \approx \frac{\alpha}{2} \qquad \omega' = \frac{\omega + 1}{2} \approx 1.$$

The quantity $\varepsilon \approx \alpha/2$ is small, since α is; therefore q_1 undergoes an oscillation of frequency $\omega' \approx 1$ with slowly changing amplitude $v\cos \varepsilon t$ (Figure 82).

After time $T = \pi/2\varepsilon \approx \pi/\alpha$, essentially only the second pendulum will be oscillating; after $2T$, again only the first, etc. ("beats") (Figure 83).

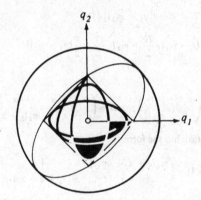

Figure 82 Beats: trajectories in the configuration space

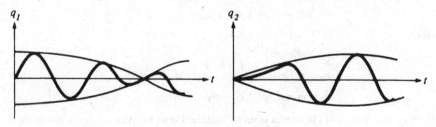

Figure 83 Beats

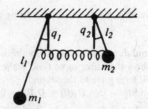

Figure 84   Connected pendulums

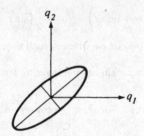

Figure 85   Potential energy of strongly connected pendulums

EXAMPLE 3. We investigate the characteristic oscillations of two different pendulums ($m_1 \neq m_2$, $l_1 \neq l_2, g = 1$), connected by a spring with energy $\frac{1}{2}\alpha(q_1 - q_2)^2$ (Figure 84). How do the characteristic frequencies behave as $\alpha \to 0$ or as $\alpha \to \infty$?

We have

$$T = \tfrac{1}{2}(m_1 l_1^2 \dot{q}_1^2 + m_2 l_2^2 \dot{q}_2^2)$$

$$U = m_1 l_1 \frac{q_1^2}{2} + m_2 l_2 \frac{q_2^2}{2} + \frac{\alpha}{2}(q_1 - q_2)^2.$$

Therefore (Figure 85),

$$A = \begin{pmatrix} m_1 l_1^2 & 0 \\ 0 & m_2 l_2^2 \end{pmatrix} \qquad B = \begin{pmatrix} m_1 l_1 + \alpha & -\alpha \\ -\alpha & m_2 l_2 + \alpha \end{pmatrix}$$

and the characteristic equation has the form

$$\det(B - \lambda A) = \begin{pmatrix} m_1 l_1 + \alpha - \lambda m_1 l_1^2 & -\alpha \\ -\alpha & m_2 l_2 + \alpha - \lambda m_2 l_2^2 \end{pmatrix} = 0$$

or

$$a\lambda^2 - (b_0 + b_1\alpha)\lambda + (c_0 + c_1\alpha) = 0,$$

where

$$a = m_1 m_2 l_1^2 l_2^2$$

$$b_0 = m_1 l_1 m_2 l_2 (l_1 + l_2) \qquad b_1 = m_1 l_1^2 + m_2 l_2^2$$

$$c_0 = m_1 m_2 l_1 l_2 \qquad c_1 = m_1 l_1 + m_2 l_2.$$

This is the equation of a hyperbola in the (α, λ)-plane (Figure 86). As $\alpha \to 0$ (weak spring) the frequencies approach the frequencies of free pendulums ($\omega_{1,2}^2 = l_{1,2}^{-1}$); as $\alpha \to \infty$, one of the

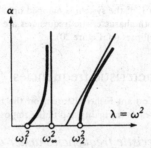

Figure 86   Dependence of characteristic frequencies on the stiffness of the spring

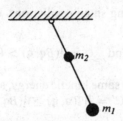

Figure 87   Limiting case of pendulums connected by an infinitely stiff spring

frequencies tends to ∞, while the other approaches the characteristic frequency ω_∞ of a pendulum with two masses on one rod (Figure 87):

$$\omega_\infty^2 = \frac{m_1 l_1 + m_2 l_2}{m_1 l_1^2 + m_2 l_2^2}.$$

PROBLEM. Investigate the characteristic oscillations of a planar double pendulum (Figure 88).

PROBLEM. Find the shape of the trajectories of the small oscillations of a point mass on the plane, sitting inside an equilateral triangle and connected by identical springs to the vertices (Figure 89).

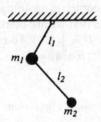

Figure 88   Double pendulum

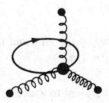

Figure 89   System with an infinite set of characteristic oscillations

109

Solution. Under rotation by 120° the system is mapped onto itself. Consequently, all directions are characteristic, and both characteristic frequencies are the same: $U = \frac{1}{2}\omega^2(x^2 + y^2)$. Therefore, the trajectories are ellipses (cf. Figure 20).

24 Behavior of characteristic frequencies

We prove here the Rayleigh-Courant-Fisher theorem on the behavior of characteristic frequencies of a system under increases in rigidity and under imposed constraints.

A *Behavior of characteristic frequencies under a change in rigidity*

Consider a system performing small oscillations, with kinetic and potential energies

$$T = \tfrac{1}{2}(A\dot{\mathbf{q}}, \dot{\mathbf{q}}) > 0 \quad \text{and} \quad U = \tfrac{1}{2}(B\mathbf{q}, \mathbf{q}) > 0 \quad \text{for all } \mathbf{q}, \dot{\mathbf{q}} \neq 0.$$

Definition. A system with the same kinetic energy, and a new potential energy U', is called *more rigid* if $U' = \frac{1}{2}(B'\mathbf{q}, \mathbf{q}) \geq \frac{1}{2}(B\mathbf{q}, \mathbf{q}) = U$ for all $\mathbf{q}$.

We wish to understand how the characteristic frequencies change under an increase in the rigidity of a system.

PROBLEM. Discuss the one-dimensional case.

Theorem 1. *Under an increase in rigidity, all the characteristic frequencies are increased, i.e., if $\omega_1 \leq \omega_2 \leq \cdots \leq \omega_n$ are the characteristic frequencies of the less rigid system, and $\omega'_1 \leq \omega'_2 \leq \cdots \leq \omega'_n$ are the characteristic frequencies of the more rigid system, then $\omega_1 \leq \omega'_1; \omega_2 \leq \omega'_2; \ldots; \omega_n \leq \omega'_n$.*

This theorem has a simple geometric meaning. Without loss of generality we may assume that $A = E$, i.e., that we are considering the euclidean structure given by the kinetic energy $T = \frac{1}{2}(\dot{\mathbf{q}}, \dot{\mathbf{q}})$. To each system we associate the ellipsoids $E: (B\mathbf{q}, \mathbf{q}) = 1$ and $E': (B'\mathbf{q}, \mathbf{q}) = 1$.

It is clear that

Lemma 1. *If the system U' is more rigid than U, then the corresponding ellipsoid E' lies inside E.*

It is also clear that

Lemma 2. *The major semi-axes of the ellipsoid are the inverses of the characteristic frequencies $\omega_i: \omega_i = 1/a_i$.*

Therefore, Theorem 1 is equivalent to the following geometric proposition (Figure 90).

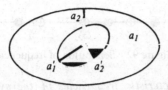

Figure 90 The semi-axes of the inside ellipse are smaller.

Theorem 2. *If the ellipsoid E with semi-axes $a_1 \geq a_2 \geq \cdots \geq a_n$ contains the ellipsoid E' with semi-axes $a'_1 \geq a'_2 \geq \cdots \geq a'_n$, both ellipses having the same center, then the semi-axes of the inside ellipsoid are smaller:*

$$a_1 \geq a'_1, a_2 \geq a'_2, \ldots, a_n \geq a'_n.$$

EXAMPLE. Under an increase in the rigidity α of the spring connecting the pendulums of Example 3, Section 23, the potential energy grows, and by Theorem 1, the characteristic frequencies grow: $d\omega_i/d\alpha > 0$.

Now consider the case when the rigidity of the spring approaches infinity, $\alpha \to \infty$. Then in the limit the pendulums are rigidly connected and we get a system with one degree of freedom; the limiting characteristic frequency ω_∞ satisfies $\omega_1 < \omega_\infty < \omega_2$.

B Behavior of characteristic frequencies under the imposition of a constraint

We return to a general system with n degrees of freedom, and let $T = \frac{1}{2}(\dot{\mathbf{q}}, \dot{\mathbf{q}})$ and $U = \frac{1}{2}(B\mathbf{q}, \mathbf{q})$ ($\mathbf{q} \in \mathbb{R}^n$) be the kinetic and potential energies of a system performing small oscillations.

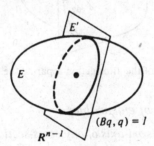

Figure 91 Linear constraint

Let $\mathbb{R}^{n-1} \subset \mathbb{R}^n$ be an $(n-1)$-dimensional subspace in $\mathbb{R}^n$ (Figure 91). Consider the system with $n-1$ degrees of freedom ($\mathbf{q} \in \mathbb{R}^{n-1}$) whose kinetic and potential energies are the restrictions of T and U to $\mathbb{R}^{n-1}$. We say that this system is obtained from the original by *imposition of a linear constraint*.

Let $\omega_1 \leq \omega_2 \leq \cdots \leq \omega_n$ be the n characteristic frequencies of the original system, and

$$\omega'_1 \leq \omega'_2 \leq \cdots \leq \omega'_{n-1}$$

the $(n-1)$ characteristic frequencies of the system with a constraint.

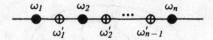

Figure 92   Separation of frequencies

Theorem 3. *The characteristic frequencies of the system with a constraint separate the characteristic frequencies of the original system (Figure 92):*

$$\omega_1 \le \omega_1' \le \omega_2 \le \omega_2' \le \cdots \le \omega_{n-1} \le \omega_{n-1}' \le \omega_n.$$

By Lemma 2 this theorem is equivalent to the following geometric proposition.

Theorem 4. *Consider the cross-section of the n-dimensional ellipsoid $E = \{\mathbf{q}: (B\mathbf{q}, \mathbf{q}) = 1\}$ with semi-axes $a_1 \ge a_2 \ge \cdots \ge a_n$ by a hyperplane $\mathbb{R}^{n-1}$ through its center. Then the semi-axes of this $(n-1)$-dimensional ellipsoid—the cross-section E'—separate the semi-axes of the ellipsoid E' (Figure 93):*

$$a_1 \ge a_1' \ge a_2 \ge a_2' \ge \cdots \ge a_{n-1} \ge a_{n-1}' \ge a_n.$$

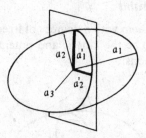

Figure 93   The semi-axes of the intersection separate the semi-axes of the ellipsoid

C *Extremal properties of eigenvalues*

Theorem 5. *The smallest semi-axis of any cross-section of the ellipsoid E with semi-axes $a_1 \ge a_2 \ge \cdots \ge a_n$ by a subspace $\mathbb{R}^k$ is less than or equal to a_k:*

$$a_k = \max_{\{\mathbb{R}^k\}} \ \min_{\mathbf{x} \in \mathbb{R}^k \cap E} \|\mathbf{x}\|$$

(the upper bound is attained on the subspace spanned by the semi-axes $a_1 \ge a_2 \ge \cdots \ge a_k$).

PROOF.[43] Consider the subspace $\mathbb{R}^{n-k+1}$ spanned by the axes $a_k \ge a_{k+1} \ge \cdots \ge a_n$. Its dimension is $n - k + 1$. Therefore, it intersects $\mathbb{R}^k$. Let $\mathbf{x}$ be a point of the intersection lying on the ellipsoid. Then $\|\mathbf{x}\| \le a_k$, since $\mathbf{x} \in \mathbb{R}^{n-k+1}$.

---

[43] It is useful to think of the case $n = 3, k = 2$.

Since $l \leq \|\mathbf{x}\|$, where l is the length of the smallest semi-axis of the ellipsoid $E \cap \mathbb{R}^k$, l must be no larger than a_k. $\qquad\square$

PROOF OF THEOREM 2. The smallest semi-axis of every k-dimensional section of the inner ellipsoid $\mathbb{R}^k \cap E'$ is less than or equal to the smallest semi-axis of $\mathbb{R}^k \cap E$. By Theorem 5,

$$a'_k = \max_{\{\mathbb{R}^k\}} \min_{\mathbf{x} \in \mathbb{R}^k \cap E'} \|\mathbf{x}\| \leq \max_{\{\mathbb{R}^k\}} \min_{\mathbf{x} \in \mathbb{R}^k \cap E} \|\mathbf{x}\| = a_k. \qquad\square$$

PROOF OF THEOREM 4. The inequality $a'_k \leq a_k$ follows from Theorem 5, since in the calculation of a_k the maximum is taken over a larger set. To prove the inequality $a'_k \geq a_{k+1}$, we intersect $\mathbb{R}^{n-1}$ with any $k+1$-dimensional subspace $\mathbb{R}^{k+1}$. The intersection has dimension greater than or equal to k. The smallest semi-axis of the ellipsoid $E' \cap \mathbb{R}^{k+1}$ is greater than or equal to the smallest semi-axis of $E \cap \mathbb{R}^{k+1}$. By Theorem 5,

$$a'_k = \max_{\{\mathbb{R}^k \subset \mathbb{R}^{n-1}\}} \min_{\mathbf{x} \in \mathbb{R}^k \cap E'} \|\mathbf{x}\| \geq \max_{\{\mathbb{R}^{k+1} \subset \mathbb{R}^n\}} \min_{\mathbf{x} \in \mathbb{R}^{k+1} \cap E'} \|\mathbf{x}\|$$

$$\geq \max_{\{\mathbb{R}^{k+1} \subset \mathbb{R}^n\}} \min_{\mathbf{x} \in \mathbb{R}^{k+1} \cap E} \|\mathbf{x}\| = a_{k+1}. \qquad\square$$

Theorems 1 and 3 follow directly from those just proven.

PROBLEM. Show that if we increase the kinetic energy of a system without decreasing the potential energy (for example, we increase the mass on a given spring), then every characteristic frequency decreases.

PROBLEM. Show that under the orthogonal projection of an ellipsoid lying in one subspace of euclidean space onto another subspace, all the semi-axes are decreased.

PROBLEM. Suppose that a quadratic form $A(\varepsilon)$ on euclidean space $\mathbb{R}^n$ is a continuously differentiable function of the parameter ε. Show that every characteristic frequency depends differentiably on ε, and find the derivatives.

ANSWER. Let $\lambda_1, \ldots, \lambda_k$ be the eigenvalues of $A(0)$. To every eigenvalue λ_i of multiplicity ν_i there corresponds a subspace $\mathbb{R}^{\nu_i}$. The derivatives of the eigenvalues of $A(\varepsilon)$ at 0 are equal to the eigenvalues of the restricted form $B = (dA/d\varepsilon)|_{\varepsilon=0}$ on $\mathbb{R}^{\nu_i}$.

In particular, if all the eigenvalues of $A(0)$ are simple, then their derivatives are equal to the diagonal elements of the matrix B in the characteristic basis for $A(0)$.

It follows from this problem that when a form is increased, its eigenvalues grow. In this way we obtain new proofs of Theorems 1 and 2.

PROBLEM. How does the pitch of a bell change when a crack appears in the bell?

25 Parametric resonance

If the parameters of a system vary periodically with time, then an equilibrium position can be unstable, even if it is stable for each fixed value of the parameter. This instability is what makes it possible to swing on a swing.

A *Dynamical systems whose parameters vary periodically with time*

EXAMPLE 1. A *swing*: the length of the equivalent mathematical pendulum $l(t)$ varies periodically with time: $l(t + T) = l(t)$ (Figure 94).

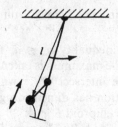

Figure 94   Swing

EXAMPLE 2. A pendulum in a periodically varying gravitational field (for example, the moon) is described by *Hill's equation*:

$$(1) \qquad \ddot{q} = -\omega^2(t)q \qquad \omega(t + T) = \omega(t)$$

EXAMPLE 3. A pendulum suspended from a point which periodically oscillates vertically is also described by an equation of the form (1).

For systems with periodically varying parameters the right-hand side of the equations of motion are periodic functions of t. The equations of motion can be written in the form of a system of first-order ordinary differential equations

$$(2) \qquad \dot{\mathbf{x}} = \mathbf{f}(\mathbf{x}, t) \qquad \mathbf{f}(\mathbf{x}, t + T) = \mathbf{f}(\mathbf{x}, t), \qquad \mathbf{x} \in \mathbb{R}^n$$

with periodic right-hand sides. For example, Equation (1) can be written as the system

$$(3) \qquad \left. \begin{aligned} \dot{x}_1 &= x_2 \\ \dot{x}_2 &= -\omega^2 x_1 \end{aligned} \right\} \omega(t + T) = \omega(t).$$

B *The mapping at a period*

Recall the general properties of the system (2). We denote by $g^t: \mathbb{R}^n \to \mathbb{R}^n$ the mapping taking $\mathbf{x} \in \mathbb{R}^n$ to the value at time t, $g^t\mathbf{x} = \boldsymbol{\varphi}(t)$, of the solution $\boldsymbol{\varphi}$ of system (2) with initial conditions $\boldsymbol{\varphi}(0) = \mathbf{x}$ (Figure 95).

The mappings g^t do not form a group: in general,

$$g^{t+s} \neq g^t g^s \neq g^s g^t.$$

PROBLEM. Show that $\{g^t\}$ is a group if and only if the right-hand sides $\mathbf{f}$ do not depend on t.

PROBLEM. Show that, if T is the period of $\mathbf{f}$, then $g^{T+s} = g^s \cdot g^T$ and, in particular, $g^{nT} = (g^T)^n$, so that the mappings g^{nT} (n an integer) form a group.

114

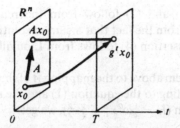

Figure 95   Mapping at a period

The mapping $g^T: \mathbb{R}^n \to \mathbb{R}^n$ plays an important role in what is to come; we will call it the *mapping at a period* and will denote it by

$$A: \mathbb{R}^n \to \mathbb{R}^n \qquad A\mathbf{x}(0) = \mathbf{x}(T).$$

EXAMPLE. For the systems

$$\begin{cases} \dot{x}_1 = x_2 \\ \dot{x}_2 = -x_1 \end{cases} \qquad \begin{cases} \dot{x}_1 = x_1 \\ \dot{x}_2 = -x_2, \end{cases}$$

which can be considered periodic with any period T, the mapping A is a rotation or a hyperbolic rotation (Figure 96).

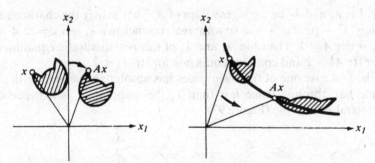

Figure 96   Rotation and hyperbolic rotation

Theorem.

1. *The point $\mathbf{x}_0$ is a fixed point of the mapping A ($A\mathbf{x}_0 = \mathbf{x}_0$) if and only if the solution with initial conditions $\mathbf{x}(0) = \mathbf{x}_0$ is periodic with period T.*
2. *The periodic solution $\mathbf{x}(t)$ is Liapunov stable (asymptotically stable) if and only if the fixed point $\mathbf{x}_0$ of the mapping A is Liapunov stable (asymptotically stable).[44]*
3. *If the system (2) is linear, i.e., $f(\mathbf{x}, t) = f(t)\mathbf{x}$ is a linear function of $\mathbf{x}$, then A is linear.*
4. *If the system (2) is hamiltonian, then A preserves volume: $\det A_* = 1$.*

[44] A fixed point $\mathbf{x}_0$ of the mapping A is Liapunov stable (respectively, asymptotically stable) if $\forall \varepsilon > 0$, $\exists \delta > 0$ such that if $|\mathbf{x} - \mathbf{x}_0| < \delta$, then $|A^n\mathbf{x} - A^n\mathbf{x}_0| < \varepsilon$ for all $0 < n < \infty$ (respectively, $A^n\mathbf{x} - A^n\mathbf{x}_0 \to 0$ as $n \to \infty$).

PROOF. Assertions (1) and (2) follow from the relationship $g^{T+s} = g^s A$. Assertion (3) follows from the fact that a sum of solutions of a linear system is again a solution. Assertion (4) follows from Liouville's theorem. ☐

We apply the theorem above to the mapping A of the phase plane $\{(x_1, x_2)\}$ onto itself, corresponding to the equation (1) and the system (3). Since (3) is linear and hamiltonian ($H = \frac{1}{2}\omega^2 x_1^2 + \frac{1}{2}x_2^2$), we get:

Corollary. *The mapping A is linear, and preserves area* (det $A = 1$). *The trivial solution of Equation* (1) *is stable if and only if the mapping A is stable.*

PROBLEM. Show that a rotation of the plane is a stable mapping, and a hyperbolic rotation is unstable.

C Linear mappings of the plane to itself which preserve area

Theorem. *Let A be the matrix of a linear mapping of the plane to itself which preserves area* (det $A = 1$). *Then the mapping A is stable if $|\text{tr } A| < 2$, and unstable if $|\text{tr } A| > 2$* (tr $A = a_{11} + a_{22}$).

PROOF. Let λ_1 and λ_2 be the eigenvalues of A. They satisfy the characteristic equation $\lambda^2 - (\text{tr } A)\lambda + 1 = 0$ with real coefficients $\lambda_1 + \lambda_2 = \text{tr } A$ and $\lambda_1 \cdot \lambda_2 = \det A = 1$. The roots λ_1 and λ_2 of this real quadratic equation are real for $|\text{tr } A| > 2$ and complex conjugate for $|\text{tr } A| < 2$.

In the first case one of the eigenvalues has absolute value greater than 1, and one has absolute value less than 1; the mapping A is a hyperbolic rotation and is unstable (Figure 97).

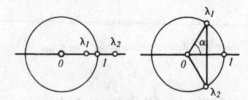

Figure 97   Eigenvalues of the mapping A

In the second case the eigenvalues lie on the unit circle (Figure 97):

$$1 = \lambda_1 \cdot \lambda_2 = \lambda_1 \cdot \bar{\lambda}_1 = |\lambda_1|^2.$$

The mapping A is equivalent to a rotation through angle α (where $\lambda_{1,2} = e^{\pm i\alpha}$), i.e., it may be reduced to a rotation by means of an appropriate choice of coordinates on the plane. Therefore, it is stable. ☐

In this way, every question about the stability of the trivial solution of an equation of the form (1) is reduced to computation of the trace of the matrix

A. Unfortunately, the calculation of this trace can be done explicitly only in special cases. It is always possible to find the trace approximately by numerically integrating the equation on the interval $0 \le t \le T$. In the important case when $\omega(t)$ is close to a constant, some simple general arguments can help.

D Strong stability

Definition. The trivial solution of a hamiltonian linear system is *strongly stable* if it is stable, and if the trivial solution of every sufficiently close linear hamiltonian system is also stable.[45]

The two theorems above imply:

Corollary. *If $|\mathrm{tr}\ A| < 2$, then the trivial solution is strongly stable.*

PROOF. If $|\mathrm{tr}\ A| < 2$, then a mapping A' corresponding to a sufficiently close system will also have $|\mathrm{tr}\ A'| < 2$. ☐

Let us apply this to a system with almost constant (only slightly varying) coefficients. Consider, for example, the equation

(4) $$\ddot{x} = -\omega^2(1 + \varepsilon a(t))x, \qquad \varepsilon \ll 1$$

where $a(t + 2\pi) = a(t)$, e.g., $a(t) = \cos t$ (Figure 98) (a pendulum whose frequency oscillates near ω with small amplitude and period 2π).[46]

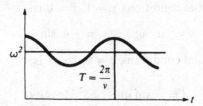

Figure 98   Instantaneous frequency as a function of time

We will represent each system of the form (4) by a point in the plane of parameters ε, $\omega > 0$. Clearly, the stable systems with $|\mathrm{tr}\ A| < 2$ form an open set in the (ω, ε)-plane; so do the unstable systems with $|\mathrm{tr}\ A| > 2$ (Figure 99).

The boundary of stability is given by the equation $|\mathrm{tr}\ A| = 2$.

Theorem. *All points on the ω-axis except the integers and half-integers $\omega = k/2, k = 0, 1, 2, \ldots$ correspond to strongly stable systems (4).*

[45] The distance between two linear systems with periodic coefficients, $\dot{x} = B_1(t)x$, $\dot{x} = B_2(t)x$, is defined as the maximum over t of the distance between the operators $B_1(t)$ and $B_2(t)$.

[46] In the case $a(t) = \cos t$, Equation (4) is called *Mathieu's equation.*

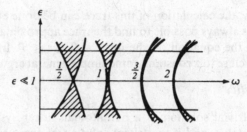

Figure 99   Zones of parametric resonance

Thus, the set of unstable systems can approach the ω-axis only at the points $\omega = k/2$. In other words, swinging a swing by small periodic changes of the length is possible only in the case when one period of the change in length is close to a whole number of half-periods of characteristic oscillations —a result well known experimentally.

The proof of the theorem above is based on the fact that for $\varepsilon = 0$, Equation (4) has constant coefficients and is clearly solvable.

PROBLEM. Calculate the matrix of the transformation A after period $T = 2\pi$ in the basis $x, \dot{x}$ for system (4) with $\varepsilon = 0$.

Solution. The general solution is:

$$x = c_1 \cos \omega t + c_2 \sin \omega t.$$

The solution with initial conditions $x = 1, \dot{x} = 0$ is:

$$x = \cos \omega t \qquad \dot{x} = -\omega \sin \omega t.$$

The solution with initial conditions $x = 0, \dot{x} = 1$ is:

$$x = \frac{1}{\omega} \sin \omega t \qquad \dot{x} = \cos \omega t.$$

ANSWER.

$$A = \begin{bmatrix} \cos 2\pi\omega & \dfrac{1}{\omega} \sin 2\pi\omega \\ -\omega \sin 2\pi\omega & \cos 2\pi\omega \end{bmatrix}.$$

Therefore, $|\operatorname{tr} A| = |2 \cos 2\omega\pi| < 2$ if $\omega \neq k/2, k = 0, 1, \ldots$, and the theorem follows from the preceding corollary.

A more careful analysis[47] shows that in general (and for $a(t) = \cos t$) the region of instability (shaded in Figure 99) in fact approaches the ω-axis near the points $\omega = k/2, k = 1, 2, \ldots$.

[47] Cf., for example, the problem analyzed below.

Thus, for $\omega \approx k/2$, $k = 1, 2, \ldots$, the lowest equilibrium position of the idealized swing (4) is unstable and it swings under an arbitrarily small periodic change of length. This phenomenon is called *parametric resonance*. A characteristic property of parametric resonance is that it is strongest when the frequency of the variation of the parameter ν (in Equation (4), $\nu = 1$) is twice the characteristic frequency ω.

Remark. Theoretically, parametric resonance can be observed for the infinite collection of cases $\omega/\nu \approx k/2$, $k = 1, 2, \ldots$. In practice, it is usually observed only when k is small ($k = 1, 2$, and more rarely, 3). The reason is that:

1. For large k the region of instability approaches the ω-axis in a very narrow "tongue" and the resonance frequencies ω must satisfy very rigid bounds ($\sim \varepsilon \theta^k$, where $\theta \in (0, 1)$ depends on the width of the analyticity band for the function $a(t)$ in (4)).
2. The instability itself is weak for large k, since $|\operatorname{tr} A| - 2$ is small and the eigenvalues are close to 1 for large k.
3. If there is an arbitrarily small amount of friction, then there is a minimal value ε_k of the amplitude in order for parametric resonance to begin (for ε less than this the oscillation dies out). As k grows, ε_k grows quickly (Figure 100).

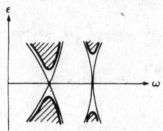

Figure 100   Influence of friction on parametric resonance

We also notice that for Equation (4) the size of x grows without bound in the unstable case. In real systems, oscillations attain only finite amplitudes, since for large x the linear equation (4) itself loses influence, and we must consider the nonlinear effects.

PROBLEM. Find the shape of the region of stability in the ε, ω-plane for the system described by the equations

$$\ddot{x} = -f^2(t)x \qquad f(t) = \begin{cases} \omega + \varepsilon & 0 < t < \pi \\ \omega - \varepsilon & \pi < t < 2\pi \end{cases} \qquad \varepsilon \ll 1$$

$$f(t + 2\pi) = f(t).$$

Solution. It follows from the solution of the preceding problem that $A = A_2 A_1$, where

$$A_k = \begin{bmatrix} c_k & \dfrac{1}{\omega} s_k \\ -\omega_k s_k & c_k \end{bmatrix}$$

$c_k = \cos \pi \omega_k$, $s_k = \sin \pi \omega_k$, $\omega_{1,2} = \omega \pm \varepsilon$.

Therefore, the boundary of the zone of stability has the equation

$$(5) \qquad |\operatorname{tr} A| = \left| 2c_1 c_2 - \left(\frac{\omega_1}{\omega_2} + \frac{\omega_2}{\omega_1} \right) s_1 s_2 \right| = 2.$$

Since $\varepsilon \ll 1$, we have $\omega_1/\omega_2 = (\omega + \varepsilon)/(\omega - \varepsilon) \approx 1$. We introduce the notation

$$\frac{\omega_1}{\omega_2} + \frac{\omega_2}{\omega_1} = 2(1 + \Delta).$$

Then, as is easily computed, $\Delta = (2\varepsilon^2/\omega^2) + O(\varepsilon^4) \ll 1$. Using the relations $2c_1 c_2 = \cos 2\pi\varepsilon + \cos 2\pi\omega$ and $2s_1 s_2 = \cos 2\pi\varepsilon - \cos 2\pi\omega$, we rewrite Equation (5) in the form

$$-\Delta \cos 2\pi\varepsilon + (2 + \Delta)\cos 2\pi\omega = \pm 2$$

or

$$(6a) \qquad \cos 2\pi\omega = \frac{2 + \Delta \cos 2\pi\varepsilon}{2 + \Delta}$$

$$(6b) \qquad \cos 2\pi\omega = \frac{-2 + \Delta \cos 2\pi\varepsilon}{2 + \Delta}$$

In the first case $\cos 2\pi\omega \approx 1$. Therefore, we set

$$\omega = k + a, |a| \ll 1 \qquad \cos 2\pi\omega = \cos 2\pi a = 1 - 2\pi^2 a^2 + O(a^4).$$

We rewrite Equation (6a) in the form

$$\cos 2\pi\omega = 1 - \frac{\Delta}{2 + \Delta}(1 - \cos 2\pi\varepsilon)$$

or $2\pi^2 a^2 + O(a^4) = \Delta \pi^2 \varepsilon^2 + O(\varepsilon^4)$.

Substituting in the value $\Delta = (2\varepsilon^2/\omega^2) + O(\varepsilon^4)$, we find

$$a = \pm \frac{\varepsilon^2}{\omega^2} + o(\varepsilon^2), \quad \text{i.e.,} \quad \omega = k \pm \frac{\varepsilon^2}{k^2} + o(\varepsilon^2).$$

Equation (6b) is solved analogously; for the result we get

$$\omega = k + \frac{1}{2} \pm \frac{\varepsilon}{\pi(k + \frac{1}{2})} + o(\varepsilon).$$

Therefore the answer has the form depicted in Figure 101.

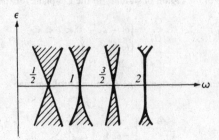

Figure 101   Zones of parametric resonance for $f = \omega \pm \varepsilon$.

E Stability of an inverted pendulum with vertically oscillating point of suspension

PROBLEM. Can the topmost, usually unstable, equilibrium position of a pendulum become stable if the point of suspension oscillates in the vertical direction (Figure 102)?

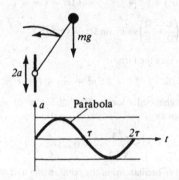

Figure 102   Inverted pendulum with oscillating point of suspension

Let the length of the pendulum be l, the amplitude of the oscillation of the point of suspension be $a \ll l$, the period of oscillation of the point of suspension 2τ, and, moreover, in the course of every half-period let the acceleration of the point of suspension be *constant* and equal to $\pm c$ (then $c = 8a/\tau^2$). It turns out that for fast enough oscillations of the point of suspension ($\tau \ll 1$) the topmost equilibrium becomes stable.

Solution. The equation of motion can be written in the form $\ddot{x} = (\omega^2 \pm d^2)x$ (the sign changes after time τ), where $\omega^2 = g/l$ and $d^2 = c/l$. If the oscillation of the suspension is fast enough, then $d^2 > \omega^2$ ($d^2 = 8a/l\tau^2$).

As in the previous problem, $A = A_2 A_1$, where

$$A_1 = \begin{bmatrix} \operatorname{ch} k\tau & \dfrac{1}{k}\operatorname{sh} k\tau \\[2mm] k\operatorname{sh} k\tau & \operatorname{ch} k\tau \end{bmatrix} \qquad A_2 = \begin{bmatrix} \cos \Omega\tau & \dfrac{1}{\Omega}\sin \Omega\tau \\[2mm] -\Omega \sin \Omega\tau & \cos \Omega\tau \end{bmatrix}$$

$$k^2 = d^2 + \omega^2, \qquad\qquad \Omega^2 = d^2 - \omega^2.$$

The stability condition $|\operatorname{tr} A| < 2$ therefore has the form

(7)
$$\left| 2 \operatorname{ch} k\tau \cos \Omega\tau + \left(\frac{k}{\Omega} - \frac{\Omega}{k}\right)\operatorname{sh} k\tau \sin \Omega\tau \right| < 2$$

We will show that this condition is fulfilled for sufficiently fast oscillations of the point of suspension, i.e., when $c \gg g$. We introduce the dimensionless variables ε, μ:

$$\frac{a}{l} = \varepsilon^2 \ll 1 \qquad \frac{g}{c} = \mu^2 \ll 1.$$

121

Then

$$k\tau = 2\sqrt{2}\,\varepsilon\sqrt{1 + \mu^2} \qquad \Omega\tau = 2\sqrt{2}\,\varepsilon\sqrt{1 - \mu^2}$$

$$\frac{k}{\Omega} - \frac{\Omega}{k} = \sqrt{\frac{1 + \mu^2}{1 - \mu^2}} - \sqrt{\frac{1 - \mu^2}{1 + \mu^2}} = 2\mu^2 + O(\mu^4).$$

Therefore, for small ε and μ we have the following expansion with error $o(\varepsilon^4 + \mu^4)$:

$$\operatorname{ch} k\tau = 1 + 4\varepsilon^2(1 + \mu^2) + \tfrac{8}{3}c^4 + \cdots \qquad \cos \Omega\tau = 1 - 4\varepsilon^2(1 - \mu^2) + \tfrac{8}{3}\varepsilon^4 + \cdots$$

$$\left(\frac{k}{\Omega} - \frac{\Omega}{k}\right)\operatorname{sh} k\tau \sin \Omega\tau = 16\varepsilon^2\mu^2 + \cdots$$

so the stability condition (7) takes the form

$$2(1 - 16\varepsilon^4 + \tfrac{16}{3}\varepsilon^4 + 8\varepsilon^2\mu^2 + \cdots) + 16\varepsilon^2\mu^2 < 2,$$

i.e., disregarding the small higher-order terms, $\tfrac{4}{3}16\varepsilon^4 \geq 32\mu^2\varepsilon^2$ or $\mu \leq \varepsilon\sqrt{2/3}$, or $g/c \leq 2a/3l$. This condition can be rewritten as

$$N \geq \sqrt{\frac{3}{64}}\,\omega\frac{l}{a} \approx 0.22\omega\frac{l}{a},$$

where $N = 1/2\tau$ is the number of oscillations of the point in one unit of time. For example, if the length of the pendulum l is 20 cm, and the amplitude of the oscillation of the point of suspension a is 1 cm, then

$$N \geq 0.22\sqrt{\frac{980}{20}} \cdot 20 \approx 31 \text{ (oscillations per second).}$$

For example, the topmost position is stable if the frequency of oscillation of the point of suspension is greater than 40 per second.

Rigid bodies

6

In this chapter we study in detail some very special mechanical problems. These problems are traditionally included in a course on classical mechanics, first because they were solved by Euler and Lagrange, and also because we live in three-dimensional euclidean space, so that most of the mechanical systems with a finite number of degrees of freedom which we are likely to encounter consist of rigid bodies.

26 Motion in a moving coordinate system

In this paragraph we define angular velocity.

A Moving coordinate systems

We look at a lagrangian system described in coordinates $\mathbf{q}, t$ by the lagrangian function $L(\mathbf{q}, \dot{\mathbf{q}}, t)$. It will often be useful to shift to a moving coordinate system $\mathbf{Q} = \mathbf{Q}(\mathbf{q}, t)$.

To write the equations of motion in a moving system, it is sufficient to express the lagrangian function in the new coordinates.

Theorem. *If the trajectory γ: $\mathbf{q} = \boldsymbol{\varphi}(t)$ of Lagrange's equations $d(\partial L/\partial \dot{\mathbf{q}})/dt = \partial L/\partial \mathbf{q}$ is written as γ: $\mathbf{Q} = \boldsymbol{\Phi}(t)$ in the local coordinates $\mathbf{Q}, t$ (where $\mathbf{Q} = \mathbf{Q}(\mathbf{q}, t)$), then the function $\boldsymbol{\Phi}(t)$ satisfies Lagrange's equations $d(\partial L'/\partial \dot{\mathbf{Q}})/dt = \partial L'/\partial \mathbf{Q}$, where $L'(\mathbf{Q}, \dot{\mathbf{Q}}, t) = L(\mathbf{q}, \dot{\mathbf{q}}, t)$.*

PROOF. The trajectory γ is an extremal: $\delta \int_\gamma L(\mathbf{q}, \dot{\mathbf{q}}, t) dt = 0$. Therefore, $\delta \int_\gamma L'(\mathbf{Q}, \dot{\mathbf{Q}}, t) dt = 0$ and $\boldsymbol{\Phi}(t)$ satisfies Lagrange's equations. $\square$

B *Motions, rotations, and translational motions*

We consider, in particular, the important case where **q** is the cartesian radius vector of a point relative to an inertial coordinate system k (which we will call *stationary*), and **Q** is the cartesian radius vector of the same point relative to a *moving* coordinate system K.

Definition. Let k and K be oriented euclidean spaces. A *motion* of K relative to k is a mapping smoothly depending on t:

$$D_t \colon K \to k,$$

which preserves the metric and the orientation (Figure 103).

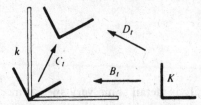

Figure 103   The motion D_t decomposed as the product of a rotation B_t and translation C_t

Definition. A motion D_t is called a *rotation* if it takes the origin of K to the origin of k, i.e., if D_t is a linear operator.

Theorem. *Every motion D_t can be uniquely written as the composition of a rotation $B_t \colon K \to k$ and a translation $C_t \colon k \to k$:*

$$D_t = C_t B_t,$$

where $C_t \mathbf{q} = \mathbf{q} + \mathbf{r}(t)$, $(\mathbf{q}, \mathbf{r} \in k)$.

PROOF. We set $\mathbf{r}(t) = D_t 0$, $B_t = C_t^{-1} D_t$. Then $B_t 0 = 0$. ☐

Definition. A motion D_t is called *translational* if the mapping $B_t \colon K \to k$ corresponding to it does not depend on t: $B_t = B_0 = B$, $D_t \mathbf{Q} = B\mathbf{Q} + \mathbf{r}(t)$.

We will call k a stationary coordinate system, K a moving one, and $\mathbf{q}(t) \in k$ the radius-vector of a point moving relative to the stationary system; if

(1) $$\mathbf{q}(t) = D_t \mathbf{Q}(t) = B_t \mathbf{Q}(t) + \mathbf{r}(t)$$

(Figure 104), $\mathbf{Q}(t)$ is called the radius vector of the point relative to the moving system.

Warning. The vector $B_t \mathbf{Q}(t) \in k$ should not be confused with $\mathbf{Q}(t) \in K$—they lie in different spaces!

124

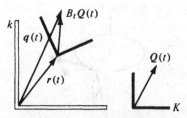

Figure 104   Radius vector of a point with respect to stationary (**q**) and moving (**Q**) coordinate systems

C Addition of velocities

We will now express the "absolute velocity" $\dot{\mathbf{q}}$ in terms of the relative motion $\mathbf{Q}(t)$ and the motion of the coordinate system, D_t. By differentiating with respect to t in formula (1) we find a formula for the addition of velocities

$$(2) \qquad \dot{\mathbf{q}} = \dot{B}\mathbf{Q} + B\dot{\mathbf{Q}} + \dot{\mathbf{r}}.$$

In order to clarify the meaning of the three terms in (2), we consider the following special cases.

The case of translational motion $(\dot{B} = 0)$

In this case Equation (2) gives $\dot{\mathbf{q}} = B\dot{\mathbf{Q}} + \dot{\mathbf{r}}$. In other words, we have shown

Theorem. *If the moving system K has a translational motion relative to k, then the absolute velocity is equal to the sum of the relative velocity and the velocity of the motion of the system K:*

$$(3) \qquad \mathbf{v} = \mathbf{v}' + \mathbf{v}_0,$$

where

$\mathbf{v} = \dot{\mathbf{q}} \in k$ *is the absolute velocity,*

$\mathbf{v}' = B\dot{\mathbf{Q}} \in k$ *is the relative velocity (distinct from* $\dot{\mathbf{Q}} \in K$ *!)*

$\mathbf{v}_0 = \dot{\mathbf{r}} \in k$ *is the velocity of motion of the moving coordinate system.*

D Angular velocity

In the case of a rotation of K the relationship between the relative and absolute velocities is not so simple. We first consider the case when our point is at rest in K (i.e., $\dot{\mathbf{Q}} = 0$) and the coordinate system K rotates (i.e., $\mathbf{r} = 0$). In this case the motion of the point $\mathbf{q}(t)$ is called a *transferred rotation*.

EXAMPLE. *Rotation with fixed angular velocity* $\boldsymbol{\omega} \in k$. Let $U(t): k \rightarrow k$ be the rotation of the space k around the $\boldsymbol{\omega}$-axis through the angle $|\boldsymbol{\omega}|t$. Then $B(t) = U(t)B(0)$ is called a *uniform rotation of K with angular velocity* $\boldsymbol{\omega}$.

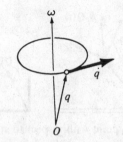

Figure 105   Angular velocity

Clearly, the velocity of the transferred motion of the point q in this case is given by the formula (Figure 105)

$$\dot{\mathbf{q}} = [\boldsymbol{\omega}, \mathbf{q}].$$

We now turn to the general case of a rotation of K ($\mathbf{r} = 0, \dot{\mathbf{Q}} = 0$).

Theorem. *At every moment of time t, there is a vector $\boldsymbol{\omega}(t) \in k$ such that the transferred velocity is expressed by the formula*

$$(4) \qquad\qquad \dot{\mathbf{q}} = [\boldsymbol{\omega}, \mathbf{q}], \qquad \forall \mathbf{q} \in k.$$

The vector $\boldsymbol{\omega}$ is called the *instantaneous angular velocity*; clearly, it is defined uniquely by Equation (4).

Corollary. *Suppose that a rigid body K rotates around a stationary point 0 of the space k. Then at every moment of time there exists an instantaneous axis of rotation—the straight line in the body passing through 0 such that the velocity of its points at the given moment of time is equal to zero. The velocity of the remaining points is perpendicular to this straight line and is proportional to the distance from it.*

The instantaneous axis of rotation in k is given by its vector $\boldsymbol{\omega}$; in K the corresponding vector is denoted by $\boldsymbol{\Omega} = B^{-1}\boldsymbol{\omega} \in K$; $\boldsymbol{\Omega}$ is called *the vector of angular velocity in the body.*

EXAMPLE. The angular velocity of the earth is directed from the center to the North Pole; its length is equal to $2\pi/3600 \cdot 24 \text{ sec}^{-1} \approx 7.3 \cdot 10^{-5} \text{ sec}^{-1}$.

PROOF OF THE THEOREM. By (2) we have

$$\dot{\mathbf{q}} = \dot{B}\mathbf{Q}.$$

Therefore, if we express $\mathbf{Q}$ in terms of $\mathbf{q}$, we get $\dot{\mathbf{q}} = \dot{B}B^{-1}\mathbf{q} = A\mathbf{q}$, where $A = \dot{B}B^{-1}: k \to k$ is a linear operator on k.

Lemma 1. *The operator A is skew-symmetric: $A^t + A = 0$.*

PROOF. Since $B: K \to k$ is an orthogonal operator from one euclidean space to another, its transpose is its inverse: $B^t = B^{-1}: k \to K$. By differentiating the relationship $BB^t = E$ with respect to t, we get

$$\dot{B}B^t + B\dot{B}^t = 0 \qquad \dot{B}B^{-1} + (\dot{B}B^{-1})^t = 0. \qquad \square$$

Lemma 2. *Every skew-symmetric operator A on a three-dimensional oriented euclidean space is the operator of vector multiplication by a fixed vector:*

$$A\mathbf{q} = [\boldsymbol{\omega}, \mathbf{q}] \quad \text{for all } \mathbf{q} \in \mathbb{R}^3.$$

PROOF. The skew-symmetric operators from $\mathbb{R}^3$ to $\mathbb{R}^3$ form a linear space. Its dimension is 3, since a skew-symmetric 3×3 matrix is determined by its three elements below the diagonal.

The operator of vector multiplication by $\boldsymbol{\omega}$ is linear and skew-symmetric. The operators of vector multiplication by all possible vectors $\boldsymbol{\omega}$ in three-space form a linear subspace of the space of all skew-symmetric operators.

The dimension of this subspace is equal to 3. Therefore, the subspace of vector multiplications is the space of all skew-symmetric operators. $\square$

CONCLUSION OF THE PROOF OF THE THEOREM. By Lemmas 1 and 2,

$$\dot{\mathbf{q}} = A\mathbf{q} = [\boldsymbol{\omega}, \mathbf{q}]. \qquad \square$$

In cartesian coordinates the operator A is given by an antisymmetric matrix; we denote its elements by $\pm\omega_{1,2,3}$:

$$A = \begin{pmatrix} 0 & -\omega_3 & \omega_2 \\ \omega_3 & 0 & -\omega_1 \\ -\omega_2 & \omega_1 & 0 \end{pmatrix}.$$

In this notation the vector $\boldsymbol{\omega} = \omega_1\mathbf{e}_1 + \omega_2\mathbf{e}_2 + \omega_3\mathbf{e}_3$ will be an eigenvector with eigenvalue 0. By applying A to the vector $\mathbf{q} = q_1\mathbf{e}_1 + q_2\mathbf{e}_2 + q_3\mathbf{e}_3$, we obtain by a direct calculation

$$A\mathbf{q} = [\boldsymbol{\omega}, \mathbf{q}].$$

E Transferred velocity

The case of purely rotational motion

Suppose now that the system K rotates ($\mathbf{r} = 0$), and that a point in K is moving ($\dot{\mathbf{Q}} \neq 0$). From (2) we find (Figure 106)

$$\dot{\mathbf{q}} = \dot{B}\mathbf{Q} + B\dot{\mathbf{Q}} = [\boldsymbol{\omega}, \mathbf{q}] + \mathbf{v}'.$$

In other words, we have shown

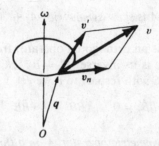

Figure 106   Addition of velocities

Theorem. *If a moving system K rotates relative to $0 \in k$, then the absolute velocity is equal to the sum of the relative velocity and the transferred velocity:*

$$\mathbf{v} = \mathbf{v}' + \mathbf{v}_n,$$

where

(5)
$\mathbf{v} = \dot{\mathbf{q}} \in k$ *is the absolute velocity*

$\mathbf{v}' = B\dot{\mathbf{Q}} \in k$ *is the relative velocity*

$\mathbf{v}_n = \dot{B}\mathbf{Q} = [\boldsymbol{\omega}, \mathbf{q}] \in k$ *is the transferred velocity of rotation.*

Finally, the general case can be reduced to the two cases above, if we consider an auxiliary system K_1 which moves by translation with respect to k and with respect to which K moves by rotating around $0 \in K_1$. From formula (2) one can see that

$$\mathbf{v} = \mathbf{v}' + \mathbf{v}_n + \mathbf{v}_0,$$

where

$\mathbf{v} = \dot{\mathbf{q}} \in k$ is the absolute velocity,

$\mathbf{v}' = B\dot{\mathbf{Q}} \in k$ is the relative velocity,

$\mathbf{v}_n = \dot{B}\mathbf{Q} = [\boldsymbol{\omega}, \mathbf{q} - \mathbf{r}] \in k$ is the transferred velocity of rotation,

and

$\mathbf{v}_0 = \dot{\mathbf{r}} \in k$ is the velocity of motion of the moving coordinate system.

PROBLEM. Show that the angular velocity of a rigid body does not depend on the choice of origin of the moving system K in the body.

PROBLEM. Show that the most general movement of a rigid body is a helical movement, i.e., the composition of a rotation through angle φ around some axis and a translation by h along it.

PROBLEM. A watch lies on a table. Find the angular velocity of the hands of the watch: (a) relative to the earth, (b) relative to an inertial coordinate system.

128

Hint. If we are given three coordinate systems k, K_1, and K_2, then the angular velocity of K_2 relative to k is equal to the sum of the angular velocities of K_1 relative to k and of K_2 relative to K_1, since

$$(E + A_1 t + \cdots)(E + A_2 t + \cdots) = E + (A_1 + A_2)t + \cdots.$$

27 Inertial forces and the Coriolis force

The equations of motion in a non-inertial coordinate system differ from the equations of motion in an inertial system by additional terms called inertial forces. This allows us to detect experimentally the non-inertial nature of a system (for example, the rotation of the earth around its axis).

A Coordinate systems moving by translation

Theorem. *In a coordinate system K which moves by translation relative to an inertial system k, the motion of a mechanical system takes place as if the coordinate system were inertial, but on every point of mass m an additional "inertial force" acted:* $\mathbf{F} = -m\ddot{\mathbf{r}}$, *where $\ddot{\mathbf{r}}$ is the acceleration of the system K.*

Proof. If $\mathbf{Q} = \mathbf{q} - \mathbf{r}(t)$, then $m\ddot{\mathbf{Q}} = m\ddot{\mathbf{q}} - m\ddot{\mathbf{r}}$. The effect of the translation of the coordinate system is reduced in this way to the appearance of an additional homogeneous force field $-m\mathbf{W}$, where $\mathbf{W}$ is the acceleration of the origin. $\qquad\square$

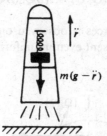

$m(g - \ddot{r})$

Figure 107   Overload

EXAMPLE 1. At the moment of takeoff, a rocket has acceleration $\ddot{\mathbf{r}}$ directed upward (Figure 107). Thus, the coordinate system K connected to the rocket is not inertial, and an observer inside can detect the existence of a force field $m\mathbf{W}$ and measure the inertial force, for example, by means of weighted springs. In this case the inertial force is called *overload.**

EXAMPLE 2. When jumping from a loft, a person has acceleration $\mathbf{g}$, directed downwards. Thus, the sum of the inertial force and the force of gravity is equal to zero; weighted springs show that the weight of any object is equal to zero, so such a state is called *weightlessness.* In exactly the same way, weightlessness is observed in the free ballistic flight of a satellite since the force of inertia is opposite to the gravitational force of the earth.

EXAMPLE 3. If the point of suspension of a pendulum moves with acceleration $\mathbf{W}(t)$, then the pendulum moves as if the force of gravity $\mathbf{g}$ were variable and equal to $\mathbf{g} - \mathbf{W}(t)$.

* Translator's note. The word *overload* is the literal translation of the Russian term *peregruzka*. There does not seem to be an English term for this particular kind of inertial force.

B Rotating coordinate systems

Let $B_t: K \rightarrow k$ be a rotation of the coordinate system K relative to the stationary coordinate system k. We will denote by $\mathbf{Q}(t) \in K$ the radius vector of a moving point in the moving coordinate system, and by $\mathbf{q}(t) = B_t\mathbf{Q}(t) \in k$ the radius vector in the stationary system. The vector of angular velocity in the moving coordinate system is denoted, as in Section 26, by $\mathbf{\Omega}$. We assume that the motion of the point $\mathbf{q}$ in k is subject to Newton's equation $m\ddot{\mathbf{q}} = \mathbf{f}(\mathbf{q}, \dot{\mathbf{q}})$.

Theorem. *Motion in a rotating coordinate system takes place as if three additional inertial forces acted on every moving point $\mathbf{Q}$ of mass m:*

1. *the inertial force of rotation:* $m[\dot{\mathbf{\Omega}}, \mathbf{Q}]$,
2. *the Coriolis force:* $2m[\mathbf{\Omega}, \dot{\mathbf{Q}}]$, *and*
3. *the centrifugal force:* $m[\mathbf{\Omega}, [\mathbf{\Omega}, \mathbf{Q}]]$.

Thus

$$m\ddot{\mathbf{Q}} = \mathbf{F} - m[\dot{\mathbf{\Omega}}, \mathbf{Q}] - 2m[\mathbf{\Omega}, \dot{\mathbf{Q}}] - m[\mathbf{\Omega}, [\mathbf{\Omega}, \mathbf{Q}]],$$

where

$$B\mathbf{F}(\mathbf{Q}, \dot{\mathbf{Q}}) = \mathbf{f}(B\mathbf{Q}, (B\mathbf{Q})^{\cdot}).$$

The first of the inertial forces is observed only in nonuniform rotation. The second and third are present even in uniform rotation.

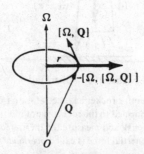

Figure 108 Centrifugal force of inertia

The centrifugal force (Figure 108) is always directed outward from the instantaneous axis of rotation $\mathbf{\Omega}$; it has magnitude $|\mathbf{\Omega}|^2 r$, where r is the distance to this axis. This force does not depend on the velocity of the relative motion, and acts even on a body at rest in the coordinate system K.

The Coriolis force depends on the velocity $\dot{\mathbf{Q}}$. In the northern hemisphere of the earth it deflects every body moving along the earth to the right, and every falling body eastward.

PROOF OF THE THEOREM. We notice that for any vector $X \in K$ we have $\dot{B}X = B[\Omega, X]$. In fact, by Section 26, $\dot{B}X = [\omega, x] = [B\Omega, BX]$. This is equal to $B[\Omega, X]$ since the operator B preserves the metric and orientation, and therefore the vector product.

Since $q = BQ$ we see that $\dot{q} = \dot{B}Q + B\dot{Q} = B(\dot{Q} + [\Omega, Q])$. Differentiating once more, we obtain

$$\ddot{q} = \dot{B}(\dot{Q} + [\Omega, Q]) + B(\ddot{Q} + [\dot{\Omega}, Q] + [\Omega, \dot{Q}])$$
$$= B([\Omega, (\dot{Q} + [\Omega, Q])] + \ddot{Q} + [\dot{\Omega}, Q] + [\Omega, \dot{Q}])$$
$$= B(\ddot{Q} + 2[\Omega, \dot{Q}] + [\Omega, [\Omega, Q]] + [\dot{\Omega}, Q]). \qquad \square$$

(We again used the relationship $\dot{B}X = B[\Omega, X]$; this time $X = \dot{Q} + [\Omega, Q]$.)

We will consider in more detail the effect of the earth's rotation on laboratory experiments. Since the earth rotates practically uniformly, we can take $\dot{\Omega} = 0$. The centrifugal force has its largest value at the equator, where it attains $\Omega^2 \rho/g \approx (7.3 \times 10^{-5})^2 \cdot 6.4 \times 10^6/9.8 \approx 3/1000$ the weight. Within the limits of a laboratory it changes little, so to observe it one must travel some distance. Thus, within the limits of a laboratory the rotation of the earth appears only in the form of the Coriolis force: in the coordinate system Q associated to the earth, we have, with good accuracy,

$$\frac{d}{dt} m\dot{Q} = mg + 2m[\dot{Q}, \Omega]$$

(the centrifugal force is taken into account in g).

EXAMPLE 1. A stone is thrown (without initial velocity) into a 250 m deep mine shaft at the latitude of Leningrad. How far does it deviate from the vertical?

We solve the equation

$$\ddot{Q} = g + 2[\dot{Q}, \Omega]$$

by the following approach, taking $\Omega \ll 1$. We set (Figure 109)

$$Q = Q_1 + Q_2,$$

where $\dot{Q}_2(0) = Q_2(0) = 0$ and $Q_1 = Q_1(0) + gt^2/2$. For Q_2, we then get

$$\ddot{Q}_2 = 2[gt, \Omega] + O(\Omega^2) \qquad Q_2 \approx \frac{t^3}{3}[g, \Omega] \approx \frac{2t}{3}[h, \Omega] \qquad h = \frac{gt^2}{2}.$$

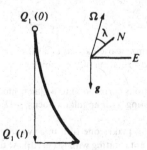

Figure 109 Displacement of a falling stone by Coriolis force

131

From this it is apparent that the stone lands about

$$\frac{2t}{3}|\mathbf{h}||\mathbf{\Omega}|\cos \lambda \approx \frac{2 \cdot 7}{3} \cdot 250 \cdot 7 \cdot 10^{-5} \cdot \tfrac{1}{2}m \approx 4 \text{ cm}$$

to the east.

PROBLEM. By how much would the Coriolis force displace a missile fired vertically upwards at Leningrad from falling back onto its launching pad, if the missile rose 1 kilometer?

EXAMPLE 2 (*The Foucault pendulum*). Consider small oscillations of an ideal pendulum, taking into account the Coriolis force. Let $\mathbf{e}_x$, $\mathbf{e}_y$, and $\mathbf{e}_z$ be the axes of a coordinate system associated to the earth, with $\mathbf{e}_z$ directed upwards, and $\mathbf{e}_x$ and $\mathbf{e}_y$ in the horizontal plane (Figure 110). In

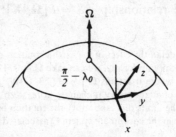

Figure 110   Coordinate system for studying the motion of a Foucault pendulum

the approximation of small oscillations, $\dot{z} = 0$ (in comparison with $\dot{x}$ and $\dot{y}$); therefore, the horizontal component of the Coriolis force will be $2m\dot{y}\Omega_z\mathbf{e}_x - 2m\dot{x}\Omega_z\mathbf{e}_y$. From this we get the equations of motion

$$\begin{cases} \ddot{x} = -\omega^2 x + 2\dot{y}\Omega_z, \\ \ddot{y} = -\omega^2 y - 2\dot{x}\Omega_z, \end{cases} \quad (\Omega_z = |\mathbf{\Omega}|\sin \lambda_0, \text{ where } \lambda_0 \text{ is the latitude})$$

If we set $x + iy = w$, then $\dot{w} = \dot{x} + i\dot{y}$, $\ddot{w} = \ddot{x} + i\ddot{y}$, and the two equations reduce to one complex equation

$$\ddot{w} + i2\Omega_z\dot{w} + \omega^2 w = 0.$$

We solve it: $w = e^{\lambda t}$, $\lambda^2 + 2i\Omega_z\lambda + \omega^2 = 0$, $\lambda = -i\Omega_z \pm i\sqrt{\Omega_z^2 + \omega^2}$. But $\Omega_z^2 \ll \omega^2$. Therefore, $\sqrt{\Omega_z^2 + \omega^2} = \omega + O(\Omega_z^2)$, from which it follows, by disregarding Ω_z^2, that

$$\lambda \approx -i\Omega_z \pm i\omega$$

or, to the same accuracy,

$$w = e^{-i\Omega_z t}(c_1 e^{i\omega t} + c_2 e^{-i\omega t}).$$

For $\Omega_z = 0$ we get the usual harmonic oscillations of a spherical pendulum. We see that the effect of the Coriolis force reduces to a rotation of the whole picture with angular velocity $-\Omega_z$, where $|\Omega_z| = |\mathbf{\Omega}|\sin \lambda_0$.

In particular, if the initial conditions correspond to a planar motion ($y(0) = \dot{y}(0) = 0$), then the plane of oscillation will be rotating with angular velocity $-\Omega_z$ with respect to the earth's coordinate system (Figure 111).

At a pole, the plane of oscillation makes one turn in a twenty-four-hour day (and is fixed with respect to a coordinate system not rotating with the earth). At the latitude of Moscow ($56°$) the plane of oscillation turns 0.83 of a rotation in a twenty-four-hour day, i.e., $12.5°$ in an hour.

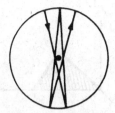

Figure 111   Trajectory of a Foucault pendulum

PROBLEM. A river flows with velocity 3 km/hr. For what radius of curvature of a river bend is the Coriolis force from the earth's rotation greater than the centrifugal force determined by the flow of the river?

ANSWER. The radius of curvature must be least on the order of 10 km for a river of medium width.

The solution of this problem explains why a large river in the northern hemisphere (for example, the Volga in the middle of its course), undermines the base of its right bank, while a river like the Moscow River, with its abrupt bends of small radius, undermines either the left or right (whichever is outward from the bend) bank.

28 Rigid bodies

In this paragraph we define a rigid body and its inertia tensor, inertia ellipsoid, moments of inertia, and axes of inertia.

A The configuration manifold of a rigid body

Definition. A *rigid body* is a system of point masses, constrained by holonomic relations expressed by the fact that the distance between points is constant:

$$(1) \qquad |\mathbf{x}_i - \mathbf{x}_j| = r_{ij} = \text{const.}$$

Theorem. *The configuration manifold of a rigid body is a six-dimensional manifold, namely, $\mathbb{R}^3 \times SO(3)$ (the direct product of a three-dimensional space $\mathbb{R}^3$ and the group $SO(3)$ of its rotations), as long as there are three points in the body not in a straight line.*

PROOF. Let $\mathbf{x}_1$, $\mathbf{x}_2$, and $\mathbf{x}_3$ be three points of the body which do not lie in a straight line. Consider the right-handed orthonormal frame whose first vector is in the direction of $\mathbf{x}_2 - \mathbf{x}_1$, and whose second is on the $\mathbf{x}_3$ side in the $\mathbf{x}_1\mathbf{x}_2\mathbf{x}_3$-plane (Figure 112). It follows from the conditions $|\mathbf{x}_i - \mathbf{x}_j| = r_{ij}$ ($i = 1, 2, 3$), that the positions of all the points of the body are uniquely determined by the positions of $\mathbf{x}_1$, $\mathbf{x}_2$, and $\mathbf{x}_3$, which are given by the position of the frame. Finally, the space of frames in $\mathbb{R}^3$ is $\mathbb{R}^3 \times SO(3)$, since every frame is obtained from a fixed one by a rotation and a translation.[48] $\square$

---

[48] Strictly speaking, the configuration space of a rigid body is $\mathbb{R}^3 \times O(3)$, and $\mathbb{R}^3 \times SO(3)$ is only one of the two connected components of this manifold, corresponding to the orientation of the body.

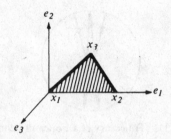

Figure 112   Configuration manifold of a rigid body

PROBLEM. Find the configuration space of a rigid body, all of whose points lie on a line.

ANSWER. $\mathbb{R}^3 \times S^2$.

Definition. A rigid body with a fixed point O is a system of point masses constrained by the condition $\mathbf{x}_1 = O$ in addition to conditions (1).

Clearly, its configuration manifold is the three-dimensional rotation group SO(3).

B *Conservation laws*

Consider the problem of the motion of a free rigid body under its own inertia, outside of any force field. For an (approximate) example we can use the rolling of a spaceship.

The system admits all translational displacements: they do not change the lagrangian function. By Noether's theorem there exist three first integrals: the three components of the vector of momentum. Therefore, we have shown

Theorem. *Under the free motion of a rigid body, its center of mass moves uniformly and linearly.*

Now we can look at an inertial coordinate system in which the center of inertia is stationary. Then we have

Corollary. *A free rigid body rotates about its center of mass as if the center of mass were fixed at a stationary point O.*

In this way, the problem is reduced to the problem, with three degrees of freedom, of the motion of a rigid body around a fixed point O. We will study this problem in more detail (not necessarily assuming that O is the center of mass of the body).

The lagrangian function admits all rotations around O. By Noether's theorem there exist three corresponding first integrals: the three components of the vector of angular momentum. The total energy of the system, $E = T$, is also conserved (here it is equal to the kinetic energy). Therefore, we have shown

Theorem. *In the problem of the motion of a rigid body around a stationary point O, in the absence of outside forces, there are four first integrals: M_x, M_y, M_z, and E.*

From this theorem we can get qualitative conclusions about the motion without any calculation.

The position and velocity of the body are determined by a point in the six-dimensional manifold $TSO(3)$—the tangent bundle of the configuration manifold $SO(3)$. The first integrals M_x, M_y, M_z, and E are four functions on $TSO(3)$. One can verify that in the general case (if the body does not have any particular symmetry) these four functions are independent. Therefore, the four equations

$$M_x = C_1 \qquad M_y = C_2 \qquad M_z = C_3 \qquad E = C_4 > 0$$

define a two-dimensional submanifold V_c in the six-dimensional manifold $TSO(3)$.

This manifold is invariant: if the initial conditions of motion give a point on V_c, then for all time of the motion, the point in $TSO(3)$ corresponding to the position and velocity of the body remains in V_c.

Therefore, V_c admits a tangent vector field (namely, the field of velocities of the motion on $TSO(3)$); for $C_4 > 0$ this field cannot have singular points. Furthermore, it is easy to verify that V_c is compact (using E) and orientable (since $TSO(3)$ is orientable).[49]

In topology it is proved that the only connected orientable compact two-dimensional manifolds are the spheres with n handles, $n \geq 0$ (Figure 113). Of these, only the torus ($n = 1$) admits a tangent vector field without singular points. Therefore, the invariant manifold V_c is a two-dimensional torus (or several tori).

We will see later that one can choose angular coordinates φ_1, φ_2, (mod 2π) on this torus such that a motion represented by a point of V_c is given by the equations $\dot{\varphi}_1 = \omega_1(c)$, $\dot{\varphi}_2 = \omega_2(c)$.

---

[49] The following assertions are easy to prove:

1. Let $f_1, \ldots, f_k \colon M \to \mathbb{R}$ be functions on an oriented manifold M. Consider the set V given by the equations $f_1 = c_1, \ldots, f_k = c_k$. Assume that the gradients of $f_1, \ldots, f_k$ are linearly independent at each point. Then V is orientable.
2. The direct product of orientable manifolds is orientable.
3. The tangent bundle $TSO(3)$ is the direct product $\mathbb{R}^3 \times SO(3)$. A manifold whose tangent bundle is a direct product is called *parallelizable*. The group $SO(3)$ (like every Lie group) is parallelizable.
4. A parallelizable manifold is orientable.

It follows from assertions 1–4 that $SO(3)$, $TSO(3)$, and V_c are orientable.

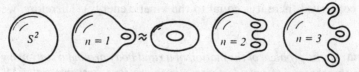

Figure 113   Two-dimensional compact connected orientable manifolds

In other words, a rotation of a rigid body is represented by the super-position of two periodic motions with (usually) different periods: if the frequencies ω_1 and ω_2 are non-commensurable, then the body never returns to its original state of motion. The magnitudes of the frequencies ω_1 and ω_2 depend on the initial conditions C.

C The inertia operator[50]

We now go on to the quantitative theory and introduce the following notation. Let k be a stationary coordinate system and K a coordinate system rotating together with the body around the point O: in K the body is at rest.

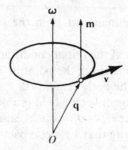

Figure 114   Radius vector and vectors of velocity, angular velocity and angular momentum of a point of the body in space

Every vector in K is carried over to k by an operator B. Corresponding vectors in K and k will be denoted by the same letter; capital for K and lower case for k. So, for example (Figure 114),

$\mathbf{q} \in k$ is the radius vector of a point in space;

$\mathbf{Q} \in K$ is its radius vector in the body, $\mathbf{q} = B\mathbf{Q}$;

$\mathbf{v} = \dot{\mathbf{q}} \in k$ is the velocity vector of a point in space;

$\mathbf{V} \in K$ is the same vector in the body, $\mathbf{v} = B\mathbf{V}$;

$\boldsymbol{\omega} \in k$ is the angular velocity in space;

$\boldsymbol{\Omega} \in K$ is the angular velocity in the body, $\boldsymbol{\omega} = B\boldsymbol{\Omega}$;

$\mathbf{m} \in k$ is the angular momentum in space;

$\mathbf{M} \in K$ is the angular momentum in the body, $\mathbf{m} = B\mathbf{M}$.

Since the operator $B: K \to k$ preserves the metric and orientation, it preserves the scalar and vector products.

[50] Often called the inertia *tensor* (translator's note).

136

By definition of angular velocity (Section 26),

$$\mathbf{v} = [\boldsymbol{\omega}, \mathbf{q}].$$

By definition of the angular momentum of a point of mass m with respect to O,

$$\mathbf{m} = [\mathbf{q}, m\mathbf{v}] = m[\mathbf{q}, [\boldsymbol{\omega}, \mathbf{q}]].$$

Therefore,

$$\mathbf{M} = m[\mathbf{Q}, [\boldsymbol{\Omega}, \mathbf{Q}]].$$

Hence, there is a linear operator transforming $\boldsymbol{\Omega}$ to $\mathbf{M}$:

$$A: K \to K \qquad A\boldsymbol{\Omega} = \mathbf{M}.$$

This operator still depends on a point of the body $(\mathbf{Q})$ and its mass (m).

Lemma. *The operator A is symmetric.*

PROOF. In view of the relation $([a, b], c) = ([c, a], b)$ we have, for any $\mathbf{X}$ and $\mathbf{Y}$ in K,

$$(A\mathbf{X}, \mathbf{Y}) = m([\mathbf{Q}, [\mathbf{X}, \mathbf{Q}]], \mathbf{Y}) = m([\mathbf{Y}, \mathbf{Q}], [\mathbf{X}, \mathbf{Q}]),$$

and the last expression is symmetric in $\mathbf{X}$ and $\mathbf{Y}$. $\qquad\square$

By substituting the vector of angular velocity $\boldsymbol{\Omega}$ for $\mathbf{X}$ and $\mathbf{Y}$ and noticing that $[\boldsymbol{\Omega}, \mathbf{Q}]^2 = \mathbf{V}^2 = \mathbf{v}^2$, we obtain

Corollary. *The kinetic energy of a point of a body is a quadratic form with respect to the vector of angular velocity $\boldsymbol{\Omega}$, namely:*

$$T = \tfrac{1}{2}(A\boldsymbol{\Omega}, \boldsymbol{\Omega}) = \tfrac{1}{2}(\mathbf{M}, \boldsymbol{\Omega}).$$

The symmetric operator A is called the *inertia operator* (or tensor) *of the point $\mathbf{Q}$.*

If a body consists of many points $\mathbf{Q}_i$ with masses m_i, then by summing we obtain

Theorem. *The angular momentum $\mathbf{M}$ of a rigid body with respect to a stationary point O depends linearly on the angular velocity $\boldsymbol{\Omega}$, i.e., there exists a linear operator $A: K \to K$, $A\boldsymbol{\Omega} = \mathbf{M}$. The operator A is symmetric.*

The kinetic energy of a body is a quadratic form with respect to the angular velocity $\boldsymbol{\Omega}$,

$$T = \tfrac{1}{2}(A\boldsymbol{\Omega}, \boldsymbol{\Omega}) = \tfrac{1}{2}(\mathbf{M}, \boldsymbol{\Omega}).$$

PROOF. By definition, the angular momentum of a body is equal to the sum of the angular momenta of its points:

$$\mathbf{M} = \sum_i \mathbf{M}_i = \sum_i A_i\boldsymbol{\Omega} = A\boldsymbol{\Omega}, \qquad \text{where } A = \sum_i A_i.$$

Since by the lemma the inertia operator A_i of every point is symmetric, the operator A is also symmetric. For kinetic energy we obtain, by definition,

$$T = \sum_i T_i = \sum_i \tfrac{1}{2}(\mathbf{M}_i, \boldsymbol{\Omega}) = \tfrac{1}{2}(\mathbf{M}, \boldsymbol{\Omega}) = \tfrac{1}{2}(A\boldsymbol{\Omega}, \boldsymbol{\Omega}). \qquad \square$$

D Principal axes

Like every symmetric operator, A has three mutually orthogonal characteristic directions. Let $\mathbf{e}_1$, $\mathbf{e}_2$, and $\mathbf{e}_3 \in K$ be their unit vectors and I_1, I_2, and I_3 their eigenvalues. In the basis $\mathbf{e}_i$, the inertia operator and the kinetic energy have a particularly simple form:

$$M_i = I_i \Omega_i$$

$$T = \tfrac{1}{2}(I_1 \Omega_1^2 + I_2 \Omega_2^2 + I_3 \Omega_3^2).$$

The axes $\mathbf{e}_i$ are called the principal axes of the body at the point O.

Finally, if the numbers I_1, I_2, and I_3 are not all different, then the axes $\mathbf{e}_i$ are not uniquely defined. We will further clarify the meaning of the eigenvalues I_1, I_2, and I_3.

Theorem. *For a rotation of a rigid body fixed at a point O, with angular velocity*
$\boldsymbol{\Omega} = \Omega \mathbf{e}$ ($\Omega = |\boldsymbol{\Omega}|$) *around the* $\mathbf{e}$ *axis, the kinetic energy is equal to*

$$T = \tfrac{1}{2} I_{\mathbf{e}} \Omega^2, \quad \text{where } I_{\mathbf{e}} = \sum_i m_i r_i^2$$

and r_i is the distance of the i-th point to the $\mathbf{e}$ axis (Figure 115).

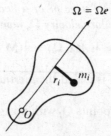

Figure 115   Kinetic energy of a body rotating around an axis

PROOF. By definition $T = \tfrac{1}{2} \sum m_i v_i^2$; but $|\mathbf{v}_i| = \Omega r_i$, so $T = \tfrac{1}{2}(\sum m_i r_i^2)\Omega^2$.
$\qquad \square$

The number $I_{\mathbf{e}}$ depends on the direction $\mathbf{e}$ of the axis of rotation $\boldsymbol{\Omega}$ in the body.

Definition. $I_{\mathbf{e}}$ is called the *moment of inertia of the body with respect to the* $\mathbf{e}$ *axis*:

$$I_{\mathbf{e}} = \sum_i m_i r_i^2.$$

By comparing the two expressions for T we obtain:

Corollary. *The eigenvalues I_i of the inertia operator A are the moments of inertia of the body with respect to the principal axes $\mathbf{e}_i$.*

E The inertia ellipsoid

In order to study the dependence of the moment of inertia $I_\mathbf{e}$ upon the direction of the axis $\mathbf{e}$ in a body, we consider the vectors $\mathbf{e}/\sqrt{I_\mathbf{e}}$, where the unit vector $\mathbf{e}$ runs over the unit sphere.

Theorem. *The vectors $\mathbf{e}/\sqrt{I_\mathbf{e}}$ form an ellipsoid in K.*

PROOF. If $\mathbf{\Omega} = \mathbf{e}/\sqrt{I_\mathbf{e}}$, then the quadratic form $T = \frac{1}{2}(A\mathbf{\Omega}, \mathbf{\Omega})$ is equal to $\frac{1}{2}$. Therefore, $\{\mathbf{\Omega}\}$ is the level set of a positive-definite quadratic form, i.e., an ellipsoid. $\qquad\square$

One could say that this ellipsoid consists of those angular velocity vectors $\mathbf{\Omega}$ whose kinetic energy is equal to $\frac{1}{2}$.

Definition. The ellipsoid $\{\mathbf{\Omega}: (A\mathbf{\Omega}, \mathbf{\Omega}) = 1\}$ is called the *inertia ellipsoid of the body* at the point 0 (Figure 116).

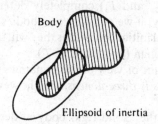

Body

Ellipsoid of inertia

Figure 116   Ellipsoid of inertia

In terms of the principal axes $\mathbf{e}_i$, the equation of the inertia ellipsoid has the form

$$I_1\Omega_1^2 + I_2\Omega_2^2 + I_3\Omega_3^2 = 1.$$

Therefore *the principal axes of the inertia ellipsoid are directed along the principal axes of the inertia tensor, and their lengths are inversely proportional to* $\sqrt{I_i}$.

Remark. If a body is stretched out along some axis, then the moment of inertia with respect to this axis is small, and consequently, the inertia ellipsoid is also stretched out along this axis; thus, the inertia ellipsoid may resemble the shape of the body.

If a body has an axis of symmetry of order k passing through O (so that it coincides with itself after rotation by $2\pi/k$ around the axis), then the inertia ellipsoid also has the same symmetry with respect to this axis. But a triaxial

139

ellipsoid does not have axes of symmetry of order $k > 2$. Therefore, every axis of symmetry of a body of order $k > 2$ is an axis of rotation of the inertia ellipsoid and, therefore, a principal axis.

EXAMPLE. The inertia ellipsoid of three points of mass m at the vertices of an equilateral triangle with center O is an ellipsoid of revolution around an axis normal to the plane of the triangle (Figure 117).

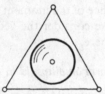

Figure 117   Ellipsoid of inertia of an equilateral triangle

If there are several such axes, then the inertia ellipsoid is a sphere, and any axis is principal.

PROBLEM. Draw the line through the center of a cube such that the sum of the squares of its distances from the vertices of the cube is: (a) largest, (b) smallest.

We now remark that the inertia ellipsoid (or the inertia operator or the moments of inertia I_1, I_2, and I_3) completely determines the rotational characteristics of our body: if we consider two bodies with identical inertia ellipsoids, then for identical initial conditions they will move identically (since they have the same lagrangian function $L = T$).

Therefore, from the point of view of the dynamics of rotation around 0, *the space of all rigid bodies is three-dimensional*, however many points compose the body.

We can even consider the "solid rigid body of density $\rho(\mathbf{Q})$," having in mind the limit as $\Delta\mathbf{Q} \to 0$ of the sequence of bodies with a finite number of points $\mathbf{Q}_i$ with masses $\rho(\mathbf{Q}_i)\Delta\mathbf{Q}_i$ (Figure 118) or, what amounts to the same thing, any body with moments of inertia

$$I_e = \iiint \rho(\mathbf{Q})r^2(\mathbf{Q})d\mathbf{Q},$$

where r is the distance from $\mathbf{Q}$ to the e axis.

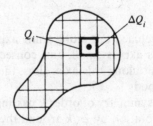

Figure 118   Continuous solid rigid body

140

EXAMPLE. Find the principal axes and moments of inertia of the uniform planar plate $|x| \leq a$, $|y| \leq b, z = 0$ with respect to O.

Solution. Since the plate has three planes of symmetry, the inertia ellipsoid has the same planes of symmetry and, therefore, principal axes x, y, and z. Furthermore,

$$I_y = \int_{-a}^{a} \int_{-b}^{b} x^2 \rho \, dx \, dy = \frac{ma^2}{3}.$$

In the same way

$$I_x = \frac{mb^2}{3}.$$

Clearly, $I_z = I_x + I_y$.

PROBLEM. Show that the moments of inertia of any body satisfy the triangle inequalities

$$I_1 \leq I_2 + I_3 \qquad I_2 \leq I_3 + I_1 \quad \text{and} \quad I_3 \leq I_1 + I_2.$$

and that equality holds only for a planar body.

PROBLEM. Find the axes and moments of inertia of a homogeneous ellipsoid of mass m with semiaxes a, b, and c relative to the center O.

Hint. First look at the sphere.

PROBLEM. Prove Steiner's theorem: The moments of inertia of any rigid body relative to two parallel axes, one of which passes through the center of mass, are related by the equation

$$I = I_0 + mr^2,$$

where m is the mass of the body, r is the distance between the axes, and I_0 is the moment of inertia relative to the axis passing through the center of mass.

Thus the moment of inertia relative to an axis passing through the center of mass is less than the moment of inertia relative to any parallel axis.

PROBLEM. Find the principal axes and moments of inertia of a uniform tetrahedron relative to its vertices.

PROBLEM. Draw the angular momentum vector $\mathbf{M}$ for a body with a given inertia ellipsoid rotating with a given angular velocity $\boldsymbol{\Omega}$.

ANSWER. $\mathbf{M}$ is in the direction normal to the inertia ellipsoid at a point on the $\boldsymbol{\Omega}$ axis (Figure 119).

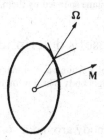

Figure 119 Angular velocity, ellipsoid of inertia and angular momentum

Figure 120   Behavior of moments of inertia as the body becomes smaller

PROBLEM. A piece is cut off a rigid body fixed at the stationary point O. How are the principal moments of inertia changed? (Figure 120).

ANSWER. All three principal moments are decreased.
   Hint. Cf. Section 24.

PROBLEM. A small mass ε is added to a rigid body with moments of inertia $I_1 > I_2 > I_3$ at the point $\mathbf{Q} = x_1\mathbf{e}_1 + x_2\mathbf{e}_2 + x_3\mathbf{e}_3$. Find the change in I_1 and $\mathbf{e}_1$ with error $O(\varepsilon^2)$.
   Solution. The center of mass is displaced by a distance of order ε. Therefore, the moments of inertia of the old body with respect to the parallel axes passing through the old and new centers of mass differ in magnitude of order ε^2. At the same time, the addition of mass changes the moment of inertia relative to any fixed axis by order ε. Therefore, we can disregard the displacement of the center of mass for calculations with error $O(\varepsilon^2)$.
   Thus, after addition of a small mass the kinetic energy takes the form

$$T = T_0 + \tfrac{1}{2}\varepsilon[\mathbf{\Omega}, \mathbf{Q}]^2 + O(\varepsilon^2),$$

where $T_0 = \tfrac{1}{2}(I_1\Omega_1^2 + I_2\Omega_2^2 + I_3\Omega_3^2)$ is the kinetic energy of the original body. We look for the eigenvalue $I_1(\varepsilon)$ and eigenvector $\mathbf{e}_1(\varepsilon)$ of the inertia operator in the form of a Taylor series in ε. By equating coefficients of ε in the relation $A(\varepsilon)\mathbf{e}_1(\varepsilon) = I_1(\varepsilon)\mathbf{e}_1(\varepsilon)$, we find that, within error $O(\varepsilon^2)$:

$$I_1(\varepsilon) \approx I_1 + \varepsilon(x_2^2 + x_3^2) \quad \text{and} \quad \mathbf{e}_1(\varepsilon) \approx \mathbf{e}_1 + \varepsilon\left(\frac{x_1 x_2}{I_2 - I_1}\mathbf{e}_2 + \frac{x_1 x_3}{I_3 - I_1}\mathbf{e}_3\right).$$

From the formula for $I_1(\varepsilon)$ it is clear that the change in the principal moments of inertia (to the first approximation in ε) is as if neither the center of mass nor the principal axes changed. The formula for $\mathbf{e}_1(\varepsilon)$ demonstrates how the directions of the principal axes change: the largest principal axis of the inertia ellipsoid approaches the added point, and the smallest recedes from it. Furthermore, the addition of a small mass on one of the principal planes of the inertia ellipsoid rotates the two axes lying in this plane and does not change the direction of the third axis. The appearance of the differences of moments of inertia in the denominator is connected with the fact that the major axes of an ellipsoid of revolution are not defined. If the inertia ellipsoid is nearly an ellipsoid of revolution (i.e., $I_1 \approx I_2$) then the addition of a small mass could strongly turn the axes $\mathbf{e}_1$ and $\mathbf{e}_2$ in the plane spanned by them.

29 Euler's equations. Poinsot's description of the motion

Here we study the motion of a rigid body around a stationary point in the absence of outside forces and the similar motion of a free rigid body. The motion turns out to have two frequencies.

A Euler's equations

Consider the motion of a rigid body around a stationary point O. Let $\mathbf{M}$ be the angular momentum vector of the body relative to O in the body, $\mathbf{\Omega}$ the angular velocity vector in the body, and A the inertia operator ($A\Omega = \mathbf{M}$); the vectors Ω and $\mathbf{M}$ belong to the moving coordinate system K (Section 26). The angular momentum vector of the body relative to O in space, $\mathbf{m} = B\mathbf{M}$, is preserved under the motion (Section 28B).

Therefore, the vector $\mathbf{M}$ in the body ($\mathbf{M} \in K$) must move so that $\mathbf{m} = B_t\mathbf{M}(t)$ does not change when t changes.

Theorem

(1)
$$\frac{d\mathbf{M}}{dt} = [\mathbf{M}, \Omega].$$

PROOF. We apply formula (5), Section 26 for the velocity of the motion of the "point" $\mathbf{M}(t) \in K$ with respect to the stationary space k. We get

$$\dot{\mathbf{m}} = B\dot{\mathbf{M}} + [\omega, \mathbf{m}] = B(\dot{\mathbf{M}} + [\Omega, \mathbf{M}]).$$

But since the angular momentum $\mathbf{m}$ with respect to the space is preserved ($\dot{\mathbf{m}} = 0$), $\dot{\mathbf{M}} + [\Omega, \mathbf{M}] = 0$. $\qquad\square$

Relation (1) is called the *Euler equations*. Since $\mathbf{M} = A\Omega$, (1) can be viewed as a differential equation for $\mathbf{M}$ (or for Ω). If

$$\Omega = \Omega_1\mathbf{e}_1 + \Omega_2\mathbf{e}_2 + \Omega_3\mathbf{e}_3 \quad \text{and} \quad \mathbf{M} = M_1\mathbf{e}_1 + M_2\mathbf{e}_2 + M_3\mathbf{e}_3$$

are the decompositions of Ω and $\mathbf{M}$ with respect to the principal axes at O, then $M_i = I_i\Omega_i$ and (1) becomes the system of three equations

(2)
$$\frac{dM_1}{dt} = a_1 M_2 M_3, \qquad \frac{dM_2}{dt} = a_2 M_3 M_1, \qquad \frac{dM_3}{dt} = a_3 M_1 M_2,$$

where $a_1 = (I_2 - I_3)/I_2 I_3$, $a_2 = (I_3 - I_1)/I_3 I_1$, and $a_3 = (I_1 - I_2)/I_1 I_2$, or, in the form of a system of three equations for the three components of the angular velocity,

$$I_1 \frac{d\Omega_1}{dt} = (I_2 - I_3)\Omega_2\Omega_3,$$

$$I_2 \frac{d\Omega_2}{dt} = (I_3 - I_1)\Omega_3\Omega_1,$$

$$I_3 \frac{d\Omega_3}{dt} = (I_1 - I_2)\Omega_1\Omega_2.$$

Remark. Suppose that outside forces act on the body, the sum of whose moments with respect to O is equal to $\mathbf{n}$ in the stationary coordinate system and $\mathbf{N}$ in the moving system ($\mathbf{n} = B\mathbf{N}$). Then

$$\dot{\mathbf{m}} = \mathbf{n}$$

and the Euler equations take the form

$$\frac{d\mathbf{M}}{dt} = [\mathbf{M}, \Omega] + \mathbf{N}.$$

B Solutions of the Euler equations

Lemma. *The Euler equations* (2) *have two quadratic first integrals*

$$2E = \frac{M_1^2}{I_1} + \frac{M_2^2}{I_2} + \frac{M_3^2}{I_3} \quad \text{and} \quad M^2 = M_1^2 + M_2^2 + M_3^2.$$

PROOF. E is preserved by the law of conservation of energy, and M^2 by the law of conservation of angular momentum **m**, since $\mathbf{m}^2 = \mathbf{M}^2 = M^2$.   □

Thus, **M** lies in the intersection of an ellipsoid and a sphere. In order to study the structure of the curves of intersection we will fix the ellipsoid $E > 0$ and change the radius M of the sphere (Figure 121).

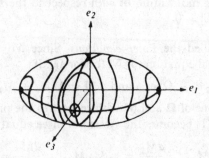

Figure 121   Trajectories of Euler's equation on an energy level surface

We assume that $I_1 > I_2 > I_3$. The semiaxes of the ellipsoid will be $\sqrt{2EI_1} > \sqrt{2EI_2} > \sqrt{2EI_3}$. If the radius M of the sphere is less than the smallest semiaxes or larger than the largest ($M < \sqrt{2EI_3}$ or $M > \sqrt{2EI_1}$), then the intersection is empty, and no actual motion corresponds to such values of E and M. If the radius of the sphere is equal to the smallest semi-axes, then the intersection consists of two points. Increasing the radius, so that $\sqrt{2EI_3} < M < \sqrt{2EI_2}$, we get two curves around the ends of the small-est semiaxes. In exactly the same way, if the radius of the sphere is equal to the largest semiaxes we get their ends, and if it is a little smaller we get two closed curves close to the ends of the largest semiaxes. Finally, if $M = \sqrt{2EI_2}$, the intersection consists of two circles.

Each of the six ends of the semiaxes of the ellipsoid is a separate trajectory of the Euler equations (2)—a stationary position of the vector **M**. It corre-sponds to a fixed value of the vector of angular velocity directed along one of the principal axes $\mathbf{e}_i$; during such a motion, **Ω** remains collinear with **M**. Therefore, the vector of angular velocity retains its position **ω** in space collinear with **m**: the body simply rotates with fixed angular velocity around the principal axis of inertia $\mathbf{e}_i$, which is stationary in space.

Definition. A motion of a body, under which its angular velocity remains constant (ω = const, Ω = const) is called a *stationary rotation*.

We have proved:

Theorem. *A rigid body fixed at a point O admits a stationary rotation around any of the three principal axes* e_1, e_2, *and* e_3.

If, as we assumed, $I_1 > I_2 > I_3$, then the right-hand side of the Euler equations does not become 0 anywhere else, i.e., there are no other stationary rotations.

We will now investigate the stability (in the sense of Liapunov) of solutions to the Euler equations.

Theorem. *The stationary solutions* $\mathbf{M} = M_1 e_1$ *and* $\mathbf{M} = M_3 e_3$ *of the Euler equations corresponding to the largest and smallest principal axes are stable, while the solution corresponding to the middle axis* ($\mathbf{M} = M_2 e_2$) *is unstable.*

PROOF. For a small deviation of the initial condition from $M_1 e_1$ or $M_3 e_3$, the trajectory will be a small closed curve, while for a small deviation from $M_2 e_2$ it will be a large one. ☐

PROBLEM. Are stationary *rotations of the body* around the largest and smallest principal axes Liapunov stable?

ANSWER. No.

C *Poinsot's description of the motion*

It is easy to visualize the motion of the angular momentum and angular velocity vectors in a body ($\mathbf{M}$ and Ω)—they are periodic if $M \neq \sqrt{2EI_i}$.

In order to see how a body rotates *in space*, we look at its inertia ellipsoid.

$$E = \{\Omega : (A\Omega, \Omega) = 1\} \subset K,$$

where $A : \Omega \to \mathbf{M}$ is the symmetric operator of inertia of the body fixed at O.

At every moment of time the ellipsoid E occupies a position $B_t E$ in the stationary space k.

Theorem (Poinsot). *The inertia ellipsoid rolls without slipping along a stationary plane perpendicular to the angular momentum vector* $\mathbf{m}$ *(Figure 122).*

PROOF. Consider a plane π perpendicular to the momentum vector $\mathbf{m}$ and tangent to the inertia ellipsoid $B_t E$. There are two such planes, and at the point of tangency the normal to the ellipsoid is parallel to $\mathbf{m}$.

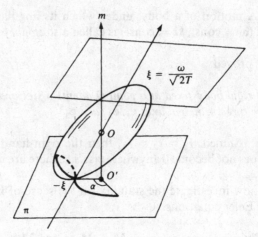

Figure 122   Rolling of the ellipsoid of inertia on the invariable plane

But the inertia ellipsoid E has normal $\mathbf{grad}(A\Omega, \Omega) = 2A\Omega = 2\mathbf{M}$ at the point Ω. Therefore, at the points $\pm\xi = \omega/\sqrt{2T}$ of the ω axis, the normal to $B_t E$ is collinear with $\mathbf{m}$.

So the plane π is tangent to $B_t E$ at the points $\pm\xi$ on the instantaneous axis of rotation. But the scalar product of ξ with the stationary vector $\mathbf{m}$ is equal to $\pm(1/\sqrt{2T})(\mathbf{m}, \omega) = \pm\sqrt{2T}$, and is therefore constant. So the distance of the plane π from O does not change, i.e., π is stationary.

Since the point of tangency lies on the instantaneous axis of rotation, its velocity is equal to zero. This implies that the ellipsoid $B_t E$ rolls without slipping along π.   □

Translator's remark: The plane π is sometimes called the *invariable plane.*

Corollary. *Under initial conditions close to a stationary rotation around the large (or small) axis of inertia, the angular velocity always remains close to its initial position, not only in the body (Ω) but also in space (ω).*

We now consider the trajectory of the point of tangency in the stationary plane π. When the point of tangency makes an entire revolution on the ellipsoid, the initial conditions are repeated except that the body has turned through some angle α around the $\mathbf{m}$ axis. The second revolution will be exactly like the first; if $\alpha = 2\pi(p/q)$, the motion is completely periodic; if the angle is not commensurable with 2π, the body will never return to its initial state.

In this case the trajectory of the point of tangency is dense in an annulus with center O' in the plane (Figure 123).

PROBLEM. Show that the connected components of the invariant two-dimensional manifold V_c (Section 28B) in the six-dimensional space $TSO(3)$

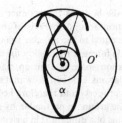

Figure 123   Trajectory of the point of contact on the invariable plane

are tori, and that one can choose coordinates φ_1 and φ_2 mod 2π on them so that $\dot{\varphi}_1 = \omega_1(c)$ and $\dot{\varphi}_2 = \omega_2(c)$.

Hint. Take the phase of the periodic variation of M as φ_1.

We now look at the important special case when the inertia ellipsoid is an ellipsoid of revolution:

$$I_2 = I_3 \neq I_1.$$

In this case the axis of the ellipsoid $B_t\mathbf{e}_1$, the instantaneous axis of rotation $\boldsymbol{\omega}$, and the vector $\mathbf{m}$ always lie in one plane. The angles between them and the length of the vector $\boldsymbol{\omega}$ are preserved; the axes of rotation ($\boldsymbol{\omega}$) and symmetry ($B_t\mathbf{e}_1$) sweep out cones around the angular momentum vector $\mathbf{m}$ with the same angular velocity (Figure 124). This motion around $\mathbf{m}$ is called *precession*.

PROBLEM. Find the angular velocity of precession.

ANSWER. Decompose the angular velocity vector $\boldsymbol{\omega}$ into components in the directions of the angular momentum vector $\mathbf{m}$ and the axis of the body $B_t\mathbf{e}_1$. The first component gives the angular velocity of precession, $\omega_{pr} = M/I_2$.

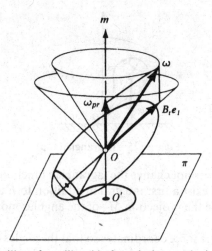

Figure 124   Rolling of an ellipsoid of revolution on the invariable plane

Hint. Represent the motion of the body as the product of a rotation around the axis of momentum and a subsequent rotation around the axis of the body. The sum of the angular velocity vectors of these rotations is equal to the angular velocity vector of the product.

Remark. In the absence of outside forces, a rigid body fixed at a point O is represented by a lagrangian system whose configuration space is a group, namely SO(3), and the lagrangian function is invariant under left translations. One can show that a significant part of Euler's theory of rigid body motion uses only this property and therefore holds for an arbitrary left-invariant lagrangian system on an arbitrary Lie group. In particular, by applying this theory to the group of volume-preserving diffeomorphisms of a domain D in a riemannian manifold, one can obtain the basic theorems of the hydrodynamics of an ideal fluid. (See Appendix 2.)

30 Lagrange's top

We consider here the motion of an axially symmetric rigid body fixed at a stationary point in a uniform force field. This motion is composed of three periodic processes: rotation, precession, and nutation.

A *Euler angles*

Consider a rigid body fixed at a stationary point O and subject to the action of the gravitational force $m\mathbf{g}$. The problem of the motion of such a "heavy rigid body" has not yet been solved in the general case and in some sense is unsolvable.

In this problem with three degrees of freedom, only two first integrals are known: the total energy $E = T + U$, and the projection M_z of the angular momentum on the vertical. There is an important special case in which the problem can be completely solved—the case of a *symmetric top*. A symmetric or lagrangian top is a rigid body fixed at a stationary point O whose inertia ellipsoid at O is an ellipsoid of revolution and whose center of gravity lies on the axis of symmetry $\mathbf{e}_3$ (Figure 125). In this case, a rotation

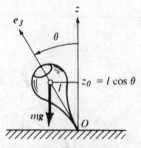

Figure 125   Lagrangian top

around the $\mathbf{e}_3$ axis does not change the lagrangian function, and by Noether's theorem there must exist a first integral in addition to E and M_z (as we will see, it turns out to be the projection M_3 of the angular momentum vector on the $\mathbf{e}_3$ axis).

If we can introduce three coordinates so that the angles of rotation around the z axis and around the axis of the top are among them, then these coordinates will be cyclic, and the problem with three degrees of freedom will reduce to a problem with one degree of freedom (for the third coordinate).

Such a choice of coordinates on the configuration space SO(3) is possible; these coordinates φ, ψ, θ are called the *Euler angles* and form a local coordinate system in SO(3) similar to geographical coordinates on the sphere: they exclude the poles and are multiple-valued on one meridian.

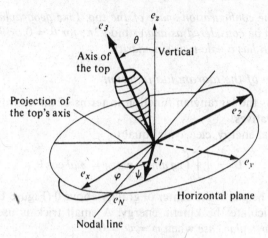

Figure 126   Euler angles

We introduce the following notation (Figure 126):

e_x, e_y, and e_z are the unit vectors of a right-handed cartesian stationary coordinate system at the stationary point O;

e_1, e_2, and e_3 are the unit vectors of a right moving coordinate system connected to the body, directed along the principal axes at O;

$I_1 = I_2 \neq I_3$ are the moments of inertia of the body at O;

e_N is the unit vector of the axis $[e_z , e_3]$, called the "line of nodes" (all vectors are in the "stationary space" k).

In order to carry the stationary frame (e_x, e_y, e_z) into the moving frame (e_1, e_2, e_3), we must perform three rotations:

1. Through an angle φ around the e_z axis. Under this rotation, e_z remains fixed, and e_x goes to e_N.
2. Through an angle θ around the e_N axis. Under this rotation, e_z goes to e_3, and e_N remains fixed.
3. Through an angle ψ around the e_3 axis. Under this rotation, e_N goes to e_1, and e_3 stays fixed.

After all three rotations, e_x has gone to e_1, and e_z to e_3; therefore, e_y goes to e_2.

149

The angles φ, ψ, and θ are called the *Euler angles*. It is easy to prove:

Theorem. *To every triple of numbers φ, θ, ψ the construction above associates a rotation of three-dimensional space, $B(\varphi, \theta, \psi) \in SO(3)$, taking the frame $(\mathbf{e}_x, \mathbf{e}_y, \mathbf{e}_z)$ into the frame $(\mathbf{e}_1, \mathbf{e}_2, \mathbf{e}_3)$. In addition, the mapping $(\varphi, \theta, \psi) \to B(\varphi, \theta, \psi)$ gives local coordinates*

$$0 < \varphi < 2\pi \qquad 0 < \psi < 2\pi \qquad 0 < \theta < \pi$$

on $SO(3)$, the configuration space of the top. Like geographical longitude, φ and ψ can be considered as angles mod 2π; for $\theta = 0$ or $\theta = \pi$ the map $(\varphi, \theta, \psi) \to B$ has a pole-type singularity.

B Calculation of the lagrangian function

We will express the lagrangian function in terms of the coordinates φ, θ, ψ and their derivatives.

The potential energy, clearly, is equal to

$$U = \iiint zg\, dm = mgz_0 = mgl \cos \theta,$$

where z_0 is the height of the center of gravity above 0 (Figure 125).

We now calculate the kinetic energy. A small trick is useful here: we consider the *particular case* when $\varphi = \psi = 0$.

Lemma. *The angular velocity of a top is expressed in terms of the derivatives of the Euler angles by the formula*

$$\boldsymbol{\omega} = \dot{\theta}\mathbf{e}_1 + (\dot{\varphi} \sin \theta)\mathbf{e}_2 + (\dot{\psi} + \dot{\varphi} \cos \theta)\mathbf{e}_3,$$

if $\varphi = \psi = 0$.

PROOF. We look at the velocity of a point of the top occupying the position $\mathbf{r}$ at time t. After time dt this point takes the position (within $(dt)^2$)

$$B(\varphi + d\varphi, \theta + d\theta, \psi + d\psi)B^{-1}(\varphi, \theta, \psi)\mathbf{r},$$

where $d\varphi = \dot{\varphi}\, dt$, $d\theta = \dot{\theta}\, dt$ and $d\psi = \dot{\psi}\, dt$.

Consequently, to the same accuracy the displacement vector is the sum of the three terms

$$B(\varphi + d\varphi, \theta, \psi)B^{-1}(\varphi, \theta, \psi)\mathbf{r} - \mathbf{r} = [\boldsymbol{\omega}_\varphi, \mathbf{r}]dt,$$
$$B(\varphi, \theta + d\theta, \psi)B^{-1}(\varphi, \theta, \psi)\mathbf{r} - \mathbf{r} = [\boldsymbol{\omega}_\theta, \mathbf{r}]dt,$$
$$B(\varphi, \theta, \psi + d\psi)B^{-1}(\varphi, \theta, \psi)\mathbf{r} - \mathbf{r} = [\boldsymbol{\omega}_\psi, \mathbf{r}]dt$$

(the angular velocities $\boldsymbol{\omega}_\varphi$, $\boldsymbol{\omega}_\theta$, and $\boldsymbol{\omega}_\psi$ are defined by these formulas).

Therefore, the velocity of the point $\mathbf{r}$ is $\mathbf{v} = [\boldsymbol{\omega}_\phi + \boldsymbol{\omega}_\theta + \boldsymbol{\omega}_\psi, \mathbf{r}]$, so the angular velocity of the body is

$$\boldsymbol{\omega} = \boldsymbol{\omega}_\varphi + \boldsymbol{\omega}_\theta + \boldsymbol{\omega}_\psi,$$

where the terms are defined by the formulas above.

It remains to decompose the vectors $\boldsymbol{\omega}_\varphi$, $\boldsymbol{\omega}_\theta$, and $\boldsymbol{\omega}_\psi$ with respect to $\mathbf{e}_1$, $\mathbf{e}_2$, and $\mathbf{e}_3$. We have not yet used the fact that $\varphi = \psi = 0$. If $\varphi = \psi = 0$, then

$$B(\varphi + d\varphi, \theta, \psi)B^{-1}(\varphi, \theta, \psi)$$

is simply a rotation around the axis $\mathbf{e}_z$ through an angle $d\varphi$, so

$$\boldsymbol{\omega}_\varphi = \dot{\varphi}\mathbf{e}_z.$$

Furthermore, $B(\varphi, \theta + d\theta, \psi)B^{-1}(\varphi, \theta, \psi)$ is simply a rotation around the axis $\mathbf{e}_N = \mathbf{e}_x = \mathbf{e}_1$ through an angle $d\theta$ in the case $\varphi = \psi = 0$, so

$$\boldsymbol{\omega}_\theta = \dot{\theta}\mathbf{e}_1.$$

Finally, $B(\varphi, \theta, \psi + d\psi)B^{-1}(\varphi, \theta, \psi)$ is a rotation through an angle $d\psi$ around the axis $\mathbf{e}_3$, so

$$\boldsymbol{\omega}_\psi = \dot{\psi}\mathbf{e}_3.$$

In short, for $\varphi = \psi = 0$ we have

$$\boldsymbol{\omega} = \dot{\varphi}\mathbf{e}_z + \dot{\theta}\mathbf{e}_1 + \dot{\psi}\mathbf{e}_3.$$

But, clearly, for $\varphi = \psi = 0$

$$\mathbf{e}_z = \mathbf{e}_3 \cos\theta + \mathbf{e}_2 \sin\theta.$$

So the components of the angular velocity along the principal axes $\mathbf{e}_1$, $\mathbf{e}_2$, and $\mathbf{e}_3$ are

$$\omega_1 = \dot{\theta} \qquad \omega_2 = \dot{\varphi}\sin\theta \qquad \omega_3 = \dot{\psi} + \dot{\varphi}\cos\theta. \qquad \square$$

Since $T = \frac{1}{2}(I_1\omega_1^2 + I_2\omega_2^2 + I_3\omega_3^2)$, the kinetic energy for $\varphi = \psi = 0$ is given by the formula

$$T = \frac{I_1}{2}(\dot{\theta}^2 + \dot{\varphi}^2 \sin^2\theta) + \frac{I_3}{2}(\dot{\psi} + \dot{\varphi}\cos\theta)^2.$$

But the kinetic energy cannot depend on φ and ψ: these are cyclic coordinates, and by a choice of origin of reference for φ and ψ which does not change T we can always make $\varphi = 0$ and $\psi = 0$. Thus the formula we got for the kinetic energy is true for all φ and ψ.

In this way we obtain the lagrangian function

$$L = \frac{I_1}{2}(\dot{\theta}^2 + \dot{\varphi}^2 \sin^2\theta) + \frac{I_3}{2}(\dot{\psi} + \dot{\varphi}\cos\theta)^2 - mgl\cos\theta.$$

C Investigation of the motion

To the cyclic coordinates φ and ψ there correspond the first integrals

$$\frac{\partial L}{\partial \dot{\varphi}} = M_z = \dot{\varphi}(I_1 \sin^2\theta + I_3 \cos^2\theta) + \dot{\psi}I_3 \cos\theta$$

$$\frac{\partial L}{\partial \dot{\psi}} = M_3 = \dot{\varphi}I_3 \cos\theta + \dot{\psi}I_3.$$

Theorem. *The inclination θ of the axis of the top to the vertical changes with time in the same way as in the one-dimensional system with energy*

$$E' = \frac{I_1}{2}\dot{\theta}^2 + U_{\text{eff}}(\theta),$$

where the effective potential energy is given by the formula

$$U_{\text{eff}} = \frac{(M_z - M_3\cos\theta)^2}{2I_1\sin^2\theta} + mgl\cos\theta.$$

PROOF. Following the general theory, we express $\dot{\phi}$ and $\dot{\psi}$ in terms of M_3 and M_z. We get the total energy of the system as

$$E = \frac{I_1}{2}\dot{\theta}^2 + \frac{M_3^2}{2I_3} + mgl\cos\theta + \frac{(M_z - M_3\cos\theta)^2}{2I_1\sin^2\theta}$$

and

$$\dot{\phi} = \frac{M_z - M_3\cos\theta}{I_1\sin^2\theta}.$$

The number $M_3^2/2I_3 = E - E'$, independent of θ, does not affect the equation for θ. □

In order to study the one-dimensional system above it is convenient to make the substitution $\cos\theta = u\,(-1 \le u \le 1)$.

We also write

$$\frac{M_z}{I_1} = a \qquad \frac{M_3}{I_1} = b \qquad \frac{2E'}{I_1} = \alpha \qquad \frac{2mgl}{I_1} = \beta > 0.$$

Then we can rewrite the law of conservation of energy E' as

$$\dot{u}^2 = f(u),$$

where $f(u) = (\alpha - \beta u)(1 - u^2) - (a - bu)^2$, and the law of variation of the azimuth φ as

$$\dot{\phi} = \frac{a - bu}{1 - u^2}.$$

We notice that $f(u)$ is a polynomial of degree 3, $f(+\infty) = +\infty$, and $f(\pm 1) = -(a \mp b)^2 < 0$ if $a \ne \pm b$. On the other hand, actual motions correspond to constants a, b, α, and β for which $f(u) \ge 0$ for some $-1 \le u \le 1$. Thus $f(u)$ has exactly two real roots u_1 and u_2 on the interval $-1 \le u \le 1$ (and one for $u > 1$, Figure 127). Therefore, the inclination θ of the axis of the top changes periodically between two limit values θ_1 and θ_2 (Figure 128). This periodic change in inclination is called *nutation*.

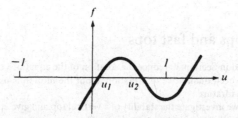

Figure 127 Graph of the function $f(u)$

We now consider the motion of the azimuth of the axis of the top. The point of intersection of the axis with the unit sphere moves in the ring between the parallels θ_1 and θ_2. The variation of the azimuth of the axis is determined by the equation

$$\dot{\varphi} = \frac{a - bu}{1 - u^2}.$$

If the root u' of the equation $a = bu$ lies outside of (u_1, u_2), then the angle φ varies monotonically and the axis traces a curve like a sinusoid on the unit sphere (Figure 128(a)). If the root u' of the equation $a = bu$ lies inside (u_1, u_2), then the rate of change of φ is in opposite directions on the parallels θ_1 and θ_2, and the axis traces a looping curve in the sphere (Figure 128(b)).

If the root u' of $a = bu$ lies on the boundary (e.g., $u' = u_2$), then the axis traces a curve with cusps (Figure 128(c)).

The last case, although exceptional, is observed every time we release the axis of a top launched at inclination θ_2 without initial velocity; the top first falls, but then rises again.

The azimuthal motion of the top is called *precession*. The complete motion of the top consists of rotation around its own axis, nutation, and precession. Each of the three motions has its own frequency. If the frequencies are incommensurable, the top never returns to its initial position, although it approaches it arbitrarily closely.

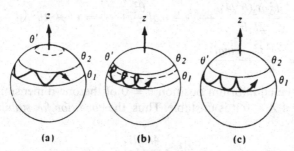

Figure 128 Path of the top's axis on the unit sphere

31 Sleeping tops and fast tops

The formulas obtained in Section 30 reduce the solution of the equations of motion of a top to elliptic integrals. However, qualitative information about the motion is usually easy to obtain without turning to quadrature.

In this paragraph we investigate the stability of a vertical top and give approximate formulas for the motion of a rapidly spinning top.

A Sleeping tops

We consider first the particular solution of the equations of motion in which the axis of the top is always vertical ($\theta = 0$) and the angular velocity is constant (a "sleeping" top). In this case, clearly, $M_z = M_3 = I_3 \omega_3$ (Figure 129).

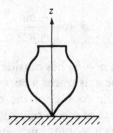

Figure 129   Sleeping top

PROBLEM. Show that a stationary rotation around the vertical axis is always Liapunov unstable.

We will look at the motion of the *axis of the top*, and not of the top itself. Will the axis of the top stably remain close to the vertical, i.e., will θ remain small? Expressing the effective potential energy of the system

$$U_{\text{eff}} = \frac{(M_z - M_3 \cos \theta)^2}{2I_1 \sin^2 \theta} + mgl \cos \theta$$

as a power series in θ, we find

$$U_{\text{eff}} = \frac{I_3^2 \omega_3^2 (\theta^4/4)}{2I_1 \theta^2} + \cdots - mgl \frac{\theta^2}{2} + \cdots = C + A\theta^2 + \cdots,$$

$$A = \frac{\omega_3^2 I_3^2}{8I_1} - \frac{mgl}{2}.$$

If $A > 0$, the equilibrium position $\theta = 0$ of the one-dimensional system is stable, and if $A < 0$ it is unstable. Thus, the *condition for stability has the form*

$$\omega_3^2 > \frac{4mglI_1}{I_3^2}.$$

When friction reduces the velocity of a sleeping top to below this limit, the top wakes up.

PROBLEM. Show that, for $\omega_3^2 > 4mgl\,I_1/I_3^2$, the axis of a sleeping top is stable with respect to perturbations which change the values of M_z and M_3, as well as θ.

B Fast tops

A top is called *fast* if the kinetic energy of its rotation is large in comparison with its potential energy:

$$\tfrac{1}{2}I_3\omega_3^2 \gg mgl.$$

It is clear from a similarity argument that multiplying the angular velocity by N is exactly equivalent to dividing the weight by N^2.

Theorem. *If, while the initial position of a top is preserved, the angular velocity is multiplied by N, then the trajectory of the top will be exactly the same as if the angular velocity remained as it was and the acceleration of gravity g were divided by N^2. In the case of large angular velocity the trajectory clearly goes N times faster.*[51]

In this way we can study the case $g \to 0$ and apply the results to study the case $\omega \to \infty$.

To begin, we consider the case $g = 0$, i.e., the motion of a symmetric top in the absence of gravity. We compare two descriptions of this motion: Lagrange's (Section 30C) and Poinsot's (Section 29C).

We first consider Lagrange's equation for the variation of the angle of inclination θ of the top's axis.

Lemma. *In the absence of gravity, the angle θ_0 satisfying $M_z = M_3 \cos\theta_0$ is a stable equilibrium position of the equation of motion of the top's axis. The frequency of small oscillations of θ near this equilibrium position is equal to*

$$\omega_{\text{nut}} = \frac{I_3\omega_3}{I_1}.$$

PROOF. In the absence of gravity the effective potential energy reduces to

$$U_{\text{eff}} = \frac{(M_z - M_3\cos\theta)^2}{2I_1\sin^2\theta}.$$

This nonnegative function has the minimum value of zero for the angle $\theta = \theta_0$ determined by the condition $M_z = M_3\cos\theta_0$ (Figure 130). Thus, the angle of inclination θ_0 of the top's axis

---

[51] Denote by $\varphi_g(t, \xi)$ the position of the top at time t with initial condition $\xi \in TSO(3)$ and gravitational acceleration g. Then the theorem says that

$$\varphi_g(t, N\xi) = \varphi_{N^{-2}g}(Nt, \xi).$$

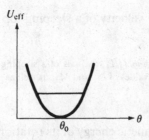

Figure 130   Effective potential energy of a top

to the vertical is stably stationary: for small deviations of the initial angle θ from θ_0, there will be periodic oscillations of θ near θ_0 (nutation). The frequency of these oscillations is easily determined by the following general formula: the frequency ω of small oscillations in a one-dimensional system with energy

$$E = \frac{a\dot{x}^2}{2} + U(x), \qquad U(x_0) = \min U(x)$$

is given (Section 22D) by the formula

$$\omega^2 = \frac{U''(x_o)}{a}.$$

The energy of the one-dimensional system describing oscillations of the inclination of the top's axis is

$$\frac{I_1}{2}\dot{\theta}^2 + U_{\text{eff}}.$$

For $\theta = \theta_0 + x$ we find $M_z - M_3 \cos\theta = M_3(\cos\theta_0 - \cos(\theta_0 + x)) = M_3 x \sin\theta_0 + O(x^2)$

$$U_{\text{eff}} = \frac{M_3^2 \cdot x^2 \cdot \sin^2\theta_0}{2I_1 \sin^2\theta_0} + o(x^2) = \frac{I_3^2\omega_3^2}{2I_1}x^2 + \cdots,$$

from which we obtain the expression for the frequency of nutation

$$\omega_{\text{nut}} = \frac{I_3\omega_3}{I_1}. \qquad\qquad \square$$

From the formula $\dot{\phi} = (M_z - M_3 \cos\theta)/I_1 \sin^2\theta$ it is clear that, for $\theta = \theta_0$, the azimuth of the axis does not change with time: the axis is stationary. The azimuthal motion of the axis under small deviations of θ from θ_0 could also be studied with the help of this formula, but we will deal with it differently.

The motion of a top in the absence of gravity can be considered in Poinsot's description. Then the axis of the top rotates uniformly around the angular momentum vector, preserving its position in space. Thus, the axis of the top describes a circle on the sphere whose center corresponds to the angular momentum vector (Figure 131).

Remark. Now the motion of the top's axis, which according to Lagrange was called *nutation*, is called *precession* in Poinsot's description of motion.

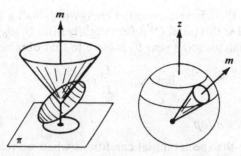

Figure 131 Comparison of the descriptions of the motion of a top according to Lagrange and Poinsot

This means that the formula obtained above for the frequency of a small nutation, $\omega_{\text{nut}} = I_3\omega_3/I_1$, agrees with the formula for the frequency of precession $\omega = M/I_1$ in Poinsot's description: when the amplitude of nutation approaches zero, $I_3\omega_3 \to M$.

C A top in a weak field

We go now to the case when the force of gravity is not absent, but is very small (the values of M_z and M_3 are fixed). In this case a term $mgl\cos\theta$, small together with its derivatives, is added to the effective potential energy. We will show that this term slightly changes the frequency of nutation.

Lemma. *Suppose that the function $f(x)$ has a minimum at $x = 0$ and Taylor expansion $f(x) = Ax^2/2 + \ldots, A > 0$. Suppose that the function $h(x)$ has Taylor expansion $h(x) = B + Cx + \cdots$. Then, for sufficiently small ε, the function $f_\varepsilon(x) = f(x) + \varepsilon h(x)$ has a minimum at the point (Figure 132)*

$$x_\varepsilon = -\frac{C\varepsilon}{A} + O(\varepsilon^2),$$

which is close to zero. In addition, $f_\varepsilon''(x_\varepsilon) = A + O(\varepsilon)$.

PROOF. We have $f_\varepsilon'(x) = Ax + C\varepsilon + O(x^2) + O(\varepsilon x)$, and the result is obtained by applying the implicit function theorem to $f_\varepsilon'(x)$. □

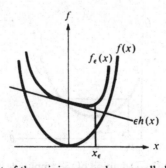

Figure 132 Displacement of the minimum under a small change of the function

By the lemma, the effective potential energy for small g has a minimum θ_g close to θ_0, and at this point U'' differs slightly from $U''(\theta_0)$. Therefore, the frequency of a small nutation near θ_0 is close to that obtained for $g = 0$:

$$\lim_{g \to 0} \omega_{\text{nut}} = \frac{I_3}{I_1}\,\omega_3.$$

D A rapidly thrown top

We now consider the special initial conditions when we release the axis of the top without an initial push from a position with inclination θ_0 to the vertical.

Theorem. *If the axis of the top is stationary at the initial moment ($\dot{\varphi} = \dot{\theta} = 0$) and the top is rotating rapidly around its axis ($\omega_3 \to \infty$), which is inclined from the vertical with angle $\theta_0(M_z = M_3 \cos \theta_0)$, then asymptotically, as $\omega_3 \to \infty$,*

1. *the nutation frequency is proportional to the angular velocity;*
2. *the amplitude of nutation is inversely proportional to the square of the angular velocity;*
3. *the frequency of precession is inversely proportional to the angular velocity;*
4. *the following asymptotic formulas hold (as $\omega_3 \to \infty$):*

$$\omega_{\text{nut}} \sim \frac{I_3}{I_1}\,\omega_3 \qquad a_{\text{nut}} \sim \frac{I_1 mgl}{I_3^2 \omega_3^2}\sin \theta_0 \qquad \omega_{\text{prec}} \sim \frac{mgl}{I_3 \omega_3}$$

(here $f(\omega_3) \sim g(\omega_3)$ if $\lim_{\omega_3 \to \infty}(f/g) = 1$).

For the proof, we look at the case when the initial angular velocity is fixed, but $g \to 0$. Then by interpreting the formulas with the aid of a similarity argument (cf. Section B), we obtain the theorem.

We already know from Section 30C that under our initial conditions the axis of the top traces a curve with cusps on the sphere.

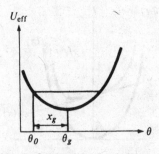

Figure 133   Definition of the amplitude of nutation

We apply the lemma to locate the minimum point θ_g of the effective potential energy. We set (Figure 133)

$$\theta = \theta_0 + x \qquad \cos\theta = \cos\theta_0 - x\sin\theta_0 + \cdots.$$

Then we obtain, as above, the Taylor expansion in x at θ_0

$$U_{\text{eff}}|_{g=0} = \frac{I_3^2\omega_3^2}{2I_1}x^2 + \cdots, \qquad mgl\cos\theta = mgl\cos\theta_0 - xmgl\sin\theta_0 + \cdots.$$

Applying the lemma to $f = U_{\text{eff}}|_{g=0}$, $g = \varepsilon$, $h = ml\cos(\theta_0 + x)$, we find that the minimum of the effective potential energy U_{eff} is attained at angle of inclination

$$\theta_g = \theta_0 + x_g \qquad x_g = \frac{I_1 ml \sin\theta_0}{I_3^2\omega_3^2}g + O(g^2).$$

Thus the inclination θ of the top's axis will oscillate near θ_g (Figure 134). But, at the initial moment,

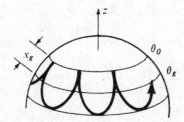

Figure 134   Motion of a top's axis

$\theta = \theta_0$ and $\dot\theta = 0$. This means that θ_0 corresponds to the highest position of the axis of the top. Thus, for small g, the amplitude of nutation is asymptotically equal to

$$a_{\text{nut}} \sim x_g \sim \frac{I_1 ml \sin\theta_0}{I_3^2\omega_3^2}g \qquad (g \to 0).$$

We now find the precessional motion of the axis. From the general formula

$$\dot\varphi = \frac{M_z - M_3\cos\theta}{I_1\sin^2\theta}$$

for $M_z = M_3\cos\theta_0$ and $\theta = \theta_0 + x$, we find that $M_z - M_3\cos\theta = M_3 x\sin\theta_0 + \cdots$; so

$$\dot\varphi = \frac{M_3}{I_1\sin\theta_0}x + \cdots.$$

But x oscillates harmonically between 0 and $2x_g$ (up to $O(g^2)$). Therefore, the average value of the velocity of precession over the period of nutation is asymptotically equal to

$$\bar{\dot\varphi} \sim \frac{M_3}{I_1\sin\theta_0}x_g \sim \frac{mgl}{I_3\omega_3} \qquad (g \to 0).$$

PROBLEM. Show that

$$\lim_{g\to 0}\lim_{t\to\infty}\frac{\varphi(t) - \varphi(0)}{tmgl/I_3\omega_3} = 1.$$

PART III
HAMILTONIAN MECHANICS

Hamiltonian mechanics is geometry in phase space. Phase space has the structure of a symplectic manifold. The group of symplectic diffeomorphisms acts on phase space. The basic concepts and theorems of hamiltonian mechanics (even when formulated in terms of local symplectic coordinates) are invariant under this group (and under the larger group of transformations which also transform time).

A hamiltonian mechanical system is given by an even-dimensional manifold (the "phase space"), a symplectic structure on it (the "Poincaré integral invariant") and a function on it (the "hamiltonian function"). Every one-parameter group of symplectic diffeomorphisms of the phase space preserving the hamiltonian function is associated to a first integral of the equations of motion.

Lagrangian mechanics is contained in hamiltonian mechanics as a special case (the phase space in this case is the cotangent bundle of the configuration space, and the hamiltonian function is the Legendre transform of the lagrangian function).

The hamiltonian point of view allows us to solve completely a series of mechanical problems which do not yield solutions by other means (for example, the problem of attraction by two stationary centers and the problem of geodesics on the triaxial ellipsoid). The hamiltonian point of view has even greater value for the approximate methods of perturbation theory (celestial mechanics), for understanding the general character of motion in complicated mechanical systems (ergodic theory, statistical mechanics) and in connection with other areas of mathematical physics (optics, quantum mechanics, etc.).

Differential forms

7

Exterior differential forms arise when concepts such as the work of a field along a path and the flux of a fluid through a surface are generalized to higher dimensions.

Hamiltonian mechanics cannot be understood without differential forms. The information we need about differential forms involves exterior multiplication, exterior differentiation, integration, and Stokes' formula.

32 Exterior forms

Here we define exterior algebraic forms

A 1-forms

Let $\mathbb{R}^n$ be an n-dimensional real vector space.[52] We will denote vectors in this space by $\xi, \eta, \ldots$.

Definition. A form of degree 1 (or a 1-form) is a linear function $\omega: \mathbb{R}^n \to \mathbb{R}$, i.e.,

$$\omega(\lambda_1 \xi_1 + \lambda_2 \xi_2) = \lambda_1 \omega(\xi_1) + \lambda_2 \omega(\xi_2), \qquad \lambda_1, \lambda_2 \in \mathbb{R} \text{ and } \xi_1, \xi_2 \in \mathbb{R}^n.$$

We recall the basic facts about 1-forms from linear algebra. The set of all 1-forms becomes a real vector space if we define the sum of two forms by

$$(\omega_1 + \omega_2)(\xi) = \omega_1(\xi) + \omega_2(\xi),$$

and scalar multiplication by

$$(\lambda \omega)(\xi) = \lambda \omega(\xi).$$

[52] It is essential to note that we do not fix any special euclidean structure on $\mathbb{R}^n$. In some *examples* we use such a structure; in these cases this will be specifically stated ("euclidean $\mathbb{R}^n$").

163

The space of 1-forms on $\mathbb{R}^n$ is itself n-dimensional, and is also called the dual space $(\mathbb{R}^n)^*$.

Suppose that we have chosen a linear coordinate system $x_1, \ldots, x_n$ on $\mathbb{R}^n$. Each coordinate x_i is itself a 1-form. These n 1-forms are linearly independent. Therefore, every 1-form ω has the form

$$\omega = a_1 x_1 + \cdots + a_n x_n, \qquad a_i \in \mathbb{R}.$$

The value of ω on a vector ξ is equal to

$$\omega(\xi) = a_1 x_1(\xi) + \cdots + a_n x_n(\xi),$$

where $x_1(\xi), \ldots, x_n(\xi)$ are the components of ξ in the chosen coordinate system.

EXAMPLE. If a uniform force field $\mathbf{F}$ is given on euclidean $\mathbb{R}^3$, its work A on the displacement ξ is a 1-form acting on ξ (Figure 135).

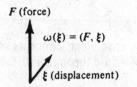

$$\omega(\xi) = (F, \xi)$$

Figure 135   The work of a force is a 1-form acting on the displacement.

B 2-forms

Definition. An exterior form of degree 2 (or a 2-form) is a function on pairs of vectors $\omega^2: \mathbb{R}^n \times \mathbb{R}^n \to \mathbb{R}$, which is bilinear and skew symmetric:

$$\omega^2(\lambda_1 \xi_1 + \lambda_2 \xi_2, \xi_3) = \lambda_1 \omega^2(\xi_1, \xi_3) + \lambda_2 \omega^2(\xi_2, \xi_3)$$

$$\omega^2(\xi_1, \xi_2) = -\omega^2(\xi_2, \xi_1),$$

$$\forall \lambda_1, \lambda_2 \in \mathbb{R}, \xi_1, \xi_2, \xi_3 \in \mathbb{R}^n.$$

EXAMPLE 1. Let $S(\xi_1, \xi_2)$ be the oriented area of the parallelogram constructed on the vectors ξ_1 and ξ_2 of the oriented euclidean plane $\mathbb{R}^2$, i.e.,

$$S(\xi_1, \xi_2) = \begin{vmatrix} \xi_{11} & \xi_{12} \\ \xi_{21} & \xi_{22} \end{vmatrix}, \quad \text{where } \xi_1 = \xi_{11} e_1 + \xi_{12} e_2, \xi_2 = \xi_{21} e_1 + \xi_{22} e_2,$$

with e_1, e_2 a basis giving the orientation on $\mathbb{R}^2$.

It is easy to see that $S(\xi_1, \xi_2)$ is a 2-form (Figure 136).

EXAMPLE 2. Let $\mathbf{v}$ be a uniform velocity vector field for a fluid in three-dimensional oriented euclidean space (Figure 137). Then the flux of the fluid over the area of the parallelogram ξ_1, ξ_2 is a bilinear skew symmetric function of ξ_1 and ξ_2, i.e., a 2-form defined by the triple scalar product

$$\omega^2(\xi_1 \xi_2) = (\mathbf{v}, \xi_1, \xi_2).$$

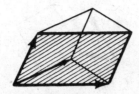

Figure 136   Oriented area is a 2-form.

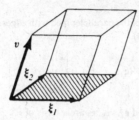

Figure 137   Flux of a fluid through a surface is a 2-form.

EXAMPLE 3. The oriented area of the projection of the parallelogram with sides ξ_1 and ξ_2 on the x_1, x_2-plane in euclidean $\mathbb{R}^3$ is a 2-form.

PROBLEM 1. Show that for every 2-form ω^2 on $\mathbb{R}^n$ we have

$$\omega^2(\xi, \xi) = 0, \qquad \forall \xi \in \mathbb{R}^n.$$

Solution. By skew symmetry, $\omega^2(\xi, \xi) = -\omega^2(\xi, \xi)$.

The set of all 2-forms on $\mathbb{R}^n$ becomes a real vector space if we define the addition of forms by the formula

$$(\omega_1 + \omega_2)(\xi_1, \xi_2) = \omega_1(\xi_1, \xi_2) + \omega_2(\xi_1, \xi_2)$$

and multiplication by scalars by the formula

$$(\lambda\omega)(\xi_1, \xi_2) = \lambda\omega(\xi_1, \xi_2).$$

PROBLEM 2. Show that this space is finite-dimensional, and find its dimension.
ANSWER. $n(n-1)/2$; a basis is shown below.

C *k*-forms

Definition. An exterior form of degree k, or a k-form, is a function of k vectors which is k-linear and antisymmetric:

$$\omega(\lambda_1\xi_1' + \lambda_2\xi_1'', \xi_2, \ldots, \xi_k) = \lambda_1\omega(\xi_1', \xi_2, \ldots, \xi_k) + \lambda_2\omega(\xi_1'', \xi_2, \ldots, \xi_k)$$

$$\omega(\xi_{i_1}, \ldots, \xi_{i_k}) = (-1)^\nu \omega(\xi_1, \ldots, \xi_k),$$

where

$$\nu = \begin{cases} 0 & \text{if the permutation } i_1, \ldots, i_k \text{ is even}; \\ 1 & \text{if the permutation } i_1, \ldots, i_k \text{ is odd.} \end{cases}$$

165

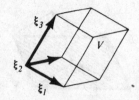

Figure 138   Oriented volume is a 3-form.

EXAMPLE 1. The oriented volume of the parallelepiped with edges $\xi_1, \ldots, \xi_n$ in oriented euclidean space $\mathbb{R}^n$ is an n-form (Figure 138).

$$V(\xi_1, \ldots, \xi_n) = \begin{vmatrix} \xi_{11} & \cdots & \xi_{1n} \\ \vdots & & \vdots \\ \xi_{n1} & \cdots & \xi_{nn} \end{vmatrix}.$$

where $\xi_i = \xi_{i1}e_1 + \cdots + \xi_{in}e_n$ and $e_1, \ldots, e_n$ are a basis of $\mathbb{R}^n$.

EXAMPLE 2. Let $\mathbb{R}^k$ be an oriented k-plane in n-dimensional euclidean space $\mathbb{R}^n$. Then the k-dimensional oriented volume of the projection of the parallelepiped with edges $\xi_1, \xi_2, \ldots,$ $\xi_k \in \mathbb{R}^n$ onto $\mathbb{R}^k$ is a k-form on $\mathbb{R}^n$.

The set of all k-forms in $\mathbb{R}^n$ form a real vector space if we introduce operations of addition

$$(\omega_1 + \omega_2)(\xi) = \omega_1(\xi) + \omega_2(\xi), \qquad \xi = \{\xi_1, \ldots, \xi_k\}, \xi_j \in \mathbb{R}^n,$$

and multiplication by scalars

$$(\lambda\omega)(\xi) = \lambda\omega(\xi).$$

PROBLEM 3. Show that this vector space is finite-dimensional and find its dimension.
ANSWER. $\binom{n}{k}$: a basis is shown below.

D The exterior product of two 1-forms

We now introduce one more operation: exterior multiplication of forms. If ω^k is a k-form and ω^l is an l-form on $\mathbb{R}^n$, then their exterior product $\omega^k \wedge \omega^l$ will be a $k + l$-form. We first define the exterior product of 1-forms, which associates to every pair of 1-forms ω_1, ω_2 on $\mathbb{R}^n$ a 2-form $\omega_1 \wedge \omega_2$ on $\mathbb{R}^n$.

Let ξ be a vector in $\mathbb{R}^n$. Given two 1-forms ω_1 and ω_2, we can define a mapping of $\mathbb{R}^n$ to the plane $\mathbb{R} \times \mathbb{R}$ by associating to $\xi \in \mathbb{R}^n$ the vector $\omega(\xi)$ with components $\omega_1(\xi)$ and $\omega_2(\xi)$ in the plane with coordinates ω_1, ω_2 (Figure 139).

Definition. The value of the exterior product $\omega_1 \wedge \omega_2$ on the pair of vectors $\xi_1, \xi_2 \in \mathbb{R}^n$ is the oriented area of the image of the parallelogram with sides $\omega(\xi_1)$ and $\omega(\xi_2)$ on the ω_1, ω_2-plane:

$$(\omega_1 \wedge \omega_2)(\xi_1, \xi_2) = \begin{vmatrix} \omega_1(\xi_1) & \omega_2(\xi_1) \\ \omega_1(\xi_2) & \omega_2(\xi_2) \end{vmatrix}.$$

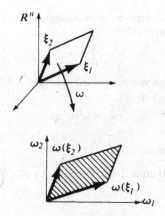

Figure 139   Definition of the exterior product of two 1-forms

PROBLEM 4. Show that $\omega_1 \wedge \omega_2$ really is a 2-form.

PROBLEM 5. Show that the mapping

$$(\omega_1, \omega_2) \to \omega_1 \wedge \omega_2$$

is bilinear and skew symmetric:

$$\omega_1 \wedge \omega_2 = -\omega_2 \wedge \omega_1,$$

$$(\lambda'\omega_1' + \lambda''\omega_1'') \wedge \omega_2 = \lambda'\omega_1' \wedge \omega_2 + \lambda''\omega_1'' \wedge \omega_2.$$

Hint. The determinant is bilinear and skew-symmetric not only with respect to rows, but also with respect to columns.

Now suppose we have chosen a system of linear coordinates on $\mathbb{R}^n$, i.e., we are given n independent 1-forms $x_1, \ldots, x_n$. We will call these forms *basic*.

The exterior products of the basic forms are the 2-forms $x_i \wedge x_j$. By skew-symmetry, $x_i \wedge x_i = 0$ and $x_i \wedge x_j = -x_j \wedge x_i$. The geometric meaning of the form $x_i \wedge x_j$ is very simple: its value on the pair of vectors ξ_1, ξ_2 is equal to the oriented area of the image of the parallelogram ξ_1, ξ_2 on the coordinate plane x_i, x_j under the projection parallel to the remaining coordinate directions.

PROBLEM 6. Show that the $\binom{n}{2} = n(n-1)/2$ forms $x_i \wedge x_j$ $(i < j)$ are linearly independent.

In particular, in three-dimensional euclidean space (x_1, x_2, x_3), the area of the projection on the (x_1, x_2)-plane is $x_1 \wedge x_2$, on the (x_2, x_3)-plane it is $x_2 \wedge x_3$, and on the (x_3, x_1)-plane it is $x_3 \wedge x_1$.

PROBLEM 7. Show that every 2-form in the three-dimensional space (x_1, x_2, x_3) is of the form

$$Px_2 \wedge x_3 + Qx_3 \wedge x_1 + Rx_1 \wedge x_2.$$

167

PROBLEM 8. Show that every 2-form on the n-dimensional space with coordinates $x_1, \ldots, x_n$ can be uniquely represented in the form

$$\omega^2 = \sum_{i<j} a_{ij} x_i \wedge x_j.$$

Hint. Let e_i be the i-th basis vector, i.e., $x_i(e_i) = 1$, $x_j(e_i) = 0$ for $i \neq j$. Look at the value of the form ω^2 on the pair e_i, e_j. Then

$$a_{ij} = \omega^2(e_i, e_j).$$

E Exterior monomials

Suppose that we are given k 1-forms $\omega_1, \ldots, \omega_k$. We define their exterior product $\omega_1 \wedge \cdots \wedge \omega_k$.

Definition. Set

$$(\omega_1 \wedge \cdots \wedge \omega_k)(\xi_1, \ldots, \xi_k) = \begin{vmatrix} \omega_1(\xi_1) & \cdots & \omega_k(\xi_1) \\ \vdots & & \vdots \\ \omega_1(\xi_k) & \cdots & \omega_k(\xi_k) \end{vmatrix}.$$

In other words, the value of a product of 1-forms on the parallelepiped $\xi_1, \ldots, \xi_k$ is equal to the oriented volume of the image of the parallelepiped in the oriented euclidean coordinate space $\mathbb{R}^k$ under the mapping $\xi \to (\omega_1(\xi), \ldots, \omega_k(\xi))$.

PROBLEM 9. Show that $\omega_1 \wedge \cdots \wedge \omega_k$ is a k-form.

PROBLEM 10. Show that the operation of exterior product of 1-forms gives a multi-linear skew-symmetric mapping

$$(\omega_1, \ldots, \omega_k) \to \omega_1 \wedge \ldots \wedge \omega_k.$$

In other words,

$$(\lambda' \omega_1' + \lambda'' \omega_1'') \wedge \omega_2 \wedge \cdots \wedge \omega_k = \lambda' \omega_1' \wedge \omega_2 \wedge \cdots \wedge \omega_k + \lambda'' \omega_1'' \wedge \omega_2 \wedge \cdots \wedge \omega_k$$

and

$$\omega_{i_1} \wedge \cdots \wedge \omega_{i_k} = (-1)^\nu \omega_1 \wedge \cdots \wedge \omega_k,$$

where

$$\nu = \begin{cases} 0 & \text{if the permutation } i_1, \ldots, i_k \text{ is even,} \\ 1 & \text{if the permutation } i_1, \ldots, i_k \text{ is odd.} \end{cases}$$

Now consider a coordinate system on $\mathbb{R}^n$ given by the basic forms $x_1, \ldots, x_n$. The exterior product of k basic forms

$$x_{i_1} \wedge \cdots \wedge x_{i_k}, \qquad 1 \le i_m \le n,$$

is the oriented volume of the image of a k-parallelepiped on the k-plane $(x_{i_1}, \ldots, x_{i_k})$ under the projection parallel to the remaining coordinate directions.

PROBLEM 11. Show that, if two of the indices $i_1, \ldots, i_k$ are the same, then the form $x_{i_1} \wedge \cdots \wedge x_{i_k}$ is zero.

PROBLEM 12. Show that the forms

$$x_{i_1} \wedge \cdots \wedge x_{i_k}, \quad \text{where } 1 \le i_1 < i_2 < \cdots < i_k \le n,$$

are linearly independent.

The number of such forms is clearly $\binom{n}{k}$. We will call them *basic k-forms*.

PROBLEM 13. Show that every k-form on $\mathbb{R}^n$ can be uniquely represented as a linear combination of basic forms:

$$\omega^k = \sum_{1 \le i_1 < \cdots < i_k \le n} a_{i_1, \ldots, i_k} x_{i_1} \wedge \cdots \wedge x_{i_k}.$$

Hint. $a_{i_1, \ldots, i_k} = \omega^k(\mathbf{e}_{i_1}, \ldots, \mathbf{e}_{i_k})$.

It follows as a result of this problem that the dimension of the vector space of k-forms on $\mathbb{R}^n$ is equal to $\binom{n}{k}$. In particular, for $k = n$, $\binom{n}{k} = 1$, from which follows

Corollary. *Every n-form on $\mathbb{R}^n$ is either the oriented volume of a parallelepiped with some choice of unit volume, or zero*:

$$\omega^n = a \cdot x_1 \wedge \cdots \wedge x_n.$$

PROBLEM 14. Show that every k-form on $\mathbb{R}^n$ with $k > n$ is zero.

We now consider the product of a k-form ω^k and an l-form ω^l. First, suppose that we are given two monomials

$$\omega^k = \omega_1 \wedge \cdots \wedge \omega_k \quad \text{and} \quad \omega^l = \omega_{k+1} \wedge \cdots \wedge \omega_{k+l},$$

where $\omega_1, \ldots, \omega_{k+l}$ are 1-forms. We define their product $\omega^k \wedge \omega^l$ to be the monomial

$$(\omega_1 \wedge \cdots \wedge \omega_k) \wedge (\omega_{k+1} \wedge \cdots \wedge \omega_{k+l})$$
$$= \omega_1 \wedge \cdots \wedge \omega_k \wedge \omega_{k+1} \wedge \cdots \wedge \omega_{k+l}.$$

PROBLEM 15. Show that the product of monomials is associative:

$$(\omega^k \wedge \omega^l) \wedge \omega^m = \omega^k \wedge (\omega^l \wedge \omega^m)$$

and skew-commutative:

$$\omega^k \wedge \omega^l = (-1)^{kl} \omega^l \wedge \omega^k.$$

Hint. In order to move each of the l factors of ω^l forward, we need k inversions with the k factors of ω^k.

Remark. It is useful to remember that skew-commutativity means commutativity only if one of the degrees k and l is even, and anti-commutativity if both degrees k and l are odd.

33 Exterior multiplication

We define here the operation of exterior multiplication of forms and show that it is skew-commutative, distributive, and associative.

A Definition of exterior multiplication

We now define the exterior multiplication of an arbitrary k-form ω^k by an arbitrary l-form ω^l. The result $\omega^k \wedge \omega^l$ will be a $k + l$-form. The operation of multiplication turns out to be:

1. skew-commutative: $\omega^k \wedge \omega^l = (-1)^{kl}\omega^l \wedge \omega^k$;
2. distributive: $(\lambda_1\omega_1^k + \lambda_2\omega_2^k) \wedge \omega^l = \lambda_1\omega_1^k \wedge \omega^l + \lambda_2\omega_2^k \wedge \omega^l$;
3. associative: $(\omega^k \wedge \omega^l) \wedge \omega^m = \omega^k \wedge (\omega^l \wedge \omega^m)$.

Definition. The exterior product $\omega^k \wedge \omega^l$ of a k-form ω^k on $\mathbb{R}^n$ with an l-form ω^l on $\mathbb{R}^n$ is the $k + l$-form on $\mathbb{R}^n$ whose value on the $k + l$ vectors $\xi_1, \ldots, \xi_k, \xi_{k+1}, \ldots, \xi_{k+l} \in \mathbb{R}^n$ is equal to

$$(1) \quad (\omega^k \wedge \omega^l)(\xi_1, \ldots, \xi_{k+l}) = \sum (-1)^\nu \omega^k(\xi_{i_1}, \ldots, \xi_{i_k})\omega^l(\xi_{j_1}, \ldots, \xi_{j_l}),$$

where $i_1 < \cdots < i_k$ and $j_1 < \cdots < j_l$; $(i_1, \ldots, i_k, j_1, \ldots, j_l)$ is a permutation of the numbers $(1, 2, \ldots, k + l)$; and

$$\nu = \begin{cases} 1 & \text{if this permutation is odd}; \\ 0 & \text{if this permutation is even}. \end{cases}$$

In other words, every partition of the $k + l$ vectors $\xi_1, \ldots, \xi_{k+l}$ into two groups (of k and of l vectors) gives one term in our sum (1). This term is equal to the product of the value of the k-form ω^k on the k vectors of the first group with the value of the l-form ω^l on the l vectors of the second group, with sign $+$ or $-$ depending on how the vectors are ordered in the groups. If they are ordered in such a way that the k vectors of the first group and the l vectors of the second group written in succession form an even permutation of the vectors $\xi_1, \xi_2, \ldots, \xi_{k+l}$, then we take the sign to be $+$, and if they form an odd permutation we take the sign to be $-$.

EXAMPLE. If $k = l = 1$, then there are just two partitions: ξ_1, ξ_2 and ξ_2, ξ_1. Therefore,

$$(\omega_1 \wedge \omega_2)(\xi_1, \xi_2) = \omega_1(\xi_1)\omega_2(\xi_2) - \omega_2(\xi_1)\omega_1(\xi_2),$$

which agrees with the definition of multiplication of 1-forms in Section 32.

PROBLEM 1. Show that the definition above actually defines a $k + l$-form (i.e., that the value of $(\omega^k \wedge \omega^l)(\xi_1, \ldots, \xi_{k+l})$ depends linearly and skew-symmetrically on the vectors ξ).

B *Properties of the exterior product*

Theorem. *The exterior multiplication of forms defined above is skew-commutative, distributive, and associative. For monomials it coincides with the multiplication defined in Section 32.*

The proof of skew-commutativity is based on the simplest properties of even and odd permutations (cf. the problem at the end of Section 32) and will be left to the reader.

Distributivity follows from the fact that every term in (1) is linear with respect to ω^k and ω^l.

The proof of associativity requires a little more combinatorics. Since the corresponding arguments are customarily carried out in algebra courses for the proof of Laplace's theorem on the expansion of a determinant by column minors, we may use this theorem.[53]

We begin with the following observation: if associativity is proved for the terms of a sum, then it is also true for the sum, i.e.,

$$\left.\begin{array}{l}(\omega_1' \wedge \omega_2) \wedge \omega_3 = \omega_1' \wedge (\omega_2 \wedge \omega_3)\\(\omega_1'' \wedge \omega_2) \wedge \omega_3 = \omega_1'' \wedge (\omega_2 \wedge \omega_3)\end{array}\right\}\ \text{implies}$$
$$((\omega_1' + \omega_1'') \wedge \omega_2) \wedge \omega_3 = (\omega_1' + \omega_1'') \wedge (\omega_2 \wedge \omega_3).$$

For, by distributivity, which has already been proved, we have

$$((\omega_1' + \omega_1'') \wedge \omega_2) \wedge \omega_3 = ((\omega_1' \wedge \omega_2) \wedge \omega_3) + ((\omega_1'' \wedge \omega_2) \wedge \omega_3),$$

$$(\omega_1' + \omega_1'') \wedge (\omega_2 \wedge \omega_3) = (\omega_1' \wedge (\omega_2 \wedge \omega_3)) + (\omega_1'' \wedge (\omega_2 \wedge \omega_3)).$$

We already know from Section 32 (Problem 13) that every form on $\mathbb{R}^n$ is a sum of monomials; therefore, it is enough to show associativity for multiplication of monomials.

Since we have not yet proved the equivalence of the definition in Section 32 of multiplication of k 1-forms with the general definition (1), we will temporarily denote the multiplication of k 1-forms by the symbol $\overline{\wedge}$, so that our monomials have the form

$$\omega^k = \omega_1 \overline{\wedge} \cdots \overline{\wedge} \omega_k \quad \text{and} \quad \omega^l = \omega_{k+1} \overline{\wedge} \cdots \overline{\wedge} \omega_{k+l},$$

where $\omega_1, \ldots, \omega_{k+l}$ are 1-forms.

---

[53] A direct proof of associativity (also containing a proof of Laplace's theorem) consists of checking the signs in the identity

$$((\omega^k \wedge \omega^l) \wedge \omega^m)(\xi_1, \ldots, \xi_{k+l+m}) = \sum \pm \omega^k(\xi_{i_1}, \ldots, \xi_{i_k})\omega^l(\xi_{j_1}, \ldots, \xi_{j_l})\omega^m(\xi_{h_1}, \ldots \xi_{h_m}),$$

where $i_1 < \cdots < i_k, j_1 < \cdots < j_l, h_1 < \cdots < h_m$; $(i_1, \ldots, h_m)$ is a permutation of the numbers $(1, \ldots, k + l + m)$.

Lemma. *The exterior product of two monomials is a monomial:*

$$(\omega_1 \; \overline{\wedge} \; \cdots \; \overline{\wedge} \; \omega_k) \wedge (\omega_{k+1} \; \overline{\wedge} \; \cdots \; \overline{\wedge} \; \omega_{k+l})$$
$$= \omega_1 \; \overline{\wedge} \; \cdots \; \overline{\wedge} \; \omega_k \; \overline{\wedge} \; \omega_{k+1} \; \overline{\wedge} \; \cdots \; \overline{\wedge} \; \omega_{k+l}.$$

PROOF. We calculate the values of the left and right sides on $k + l$ vectors $\xi_1, \ldots, \xi_{k+l}$. The value of the left side, by formula (1), is equal to the sum of the products

$$\sum \pm \det_{1 \le i \le k} |\omega_i(\xi_{i_m})| \cdot \det_{k < i \le k+l} |\omega_i(\xi_{j_m})|$$

of the minors of the first k columns of the determinant of order $k + l$ and the remaining minors. Laplace's theorem on the expansion by minors of the first k columns asserts exactly that this sum, with the same rule of sign choice as in Definition (1), is equal to the determinant $\det |\omega_i(\xi_j)|$. $\qquad\square$

It follows from the lemma that the operations $\overline{\wedge}$ and $\wedge$ coincide: we get, in turn,

$$\omega_1 \; \overline{\wedge} \; \omega_2 = \omega_1 \wedge \omega_2,$$

$$\omega_1 \; \overline{\wedge} \; \omega_2 \; \overline{\wedge} \; \omega_3 = (\omega_1 \; \overline{\wedge} \; \omega_2) \wedge \omega_3 = (\omega_1 \wedge \omega_2) \wedge \omega_3,$$

$$\omega_1 \; \overline{\wedge} \; \omega_2 \; \overline{\wedge} \; \cdots \; \overline{\wedge} \; \omega_k = (\cdots ((\omega_1 \wedge \omega_2) \wedge \omega_3) \wedge \cdots \wedge \omega_k).$$

The associativity of $\wedge$-multiplication of monomials therefore follows from the obvious associativity of $\overline{\wedge}$-multiplication of 1-forms. Thus, in view of the observation made above, associativity is proved in the general case.

PROBLEM 2. Show that the exterior square of a 1-form, or, in general, of a form of odd order, is equal to zero: $\omega^k \wedge \omega^k = 0$ if k is odd.

EXAMPLE 1. Consider a coordinate system $p_1, \ldots, p_n, q_1, \ldots, q_n$ on $\mathbb{R}^{2n}$ and the 2-form $\omega^2 = \sum_{i=1}^{n} p_i \wedge q_i$.

[Geometrically, this form signifies the sum of the oriented areas of the projection of a parallelogram on the n two-dimensional coordinate planes $(p_1, q_1), \ldots, (p_n, q_n)$. Later, we will see that the 2-form ω^2 has a special meaning for hamiltonian mechanics. It can be shown that every nondegenerate[54] 2-form on $\mathbb{R}^{2n}$ has the form ω^2 in some coordinate system $(p_1, \ldots, q_n)$.]

PROBLEM 3. Find the exterior square of the 2-form ω^2.

ANSWER.
$$\omega^2 \wedge \omega^2 = -2 \sum_{i > j} p_i \wedge p_j \wedge q_i \wedge q_j.$$

PROBLEM 4. Find the exterior k-th power of ω^2.

ANSWER.
$$\underbrace{\omega^2 \wedge \omega^2 \wedge \cdots \wedge \omega^2}_{k} = \pm k! \sum_{i_1 < \cdots < i_k} p_{i_1} \wedge \cdots \wedge p_{i_k} \wedge q_{i_1} \wedge \cdots \wedge q_{i_k}.$$

[54] A bilinear form ω^2 is nondegenerate if $\forall \xi \ne 0, \exists \eta : \omega^2(\xi, \eta) \ne 0$. See Section 41B.

In particular,

$$\underbrace{\omega^2 \wedge \cdots \wedge \omega^2}_{n} = \pm n! \, p_1 \wedge \cdots \wedge p_n \wedge q_1 \wedge \cdots \wedge q_n$$

is, up to a factor, the volume of a $2n$-dimensional parallelepiped in $\mathbb{R}^{2n}$.

EXAMPLE 2. Consider the oriented euclidean space $\mathbb{R}^3$. Every vector $\mathbf{A} \in \mathbb{R}^3$ determines a 1-form $\omega_{\mathbf{A}}^1$, by $\omega_{\mathbf{A}}^1(\xi) = (\mathbf{A}, \xi)$ (scalar product) and a 2-form $\omega_{\mathbf{A}}^2$ by

$$\omega_{\mathbf{A}}^2(\xi_1, \xi_2) = (\mathbf{A}, \xi_1, \xi_2) \qquad \text{(triple scalar product)}.$$

PROBLEM 5. Show that the maps $\mathbf{A} \to \omega_{\mathbf{A}}^1$ and $\mathbf{A} \to \omega_{\mathbf{A}}^2$ establish isomorphisms of the linear space $\mathbb{R}^3$ of vectors $\mathbf{A}$ with the linear spaces of 1-forms on $\mathbb{R}^3$ and 2-forms on $\mathbb{R}^3$. If we choose an orthonormal oriented coordinate system (x_1, x_2, x_3) on $\mathbb{R}^3$, then

$$\omega_{\mathbf{A}}^1 = A_1 x_1 + A_2 x_2 + A_3 x_3$$

and

$$\omega_{\mathbf{A}}^2 = A_1 x_2 \wedge x_3 + A_2 x_3 \wedge x_1 + A_3 x_1 \wedge x_2.$$

Remark. Thus the isomorphisms do not depend on the choice of the orthonormal oriented coordinate system (x_1, x_2, x_3). But they do depend on the choice of the euclidean structure on $\mathbb{R}^3$, and the isomorphism $\mathbf{A} \to \omega_{\mathbf{A}}^2$ also depends on the orientation (coming implicitly in the definition of triple scalar product).

PROBLEM 6. Show that, under the isomorphisms established above, the exterior product of 1-forms becomes the vector product in $\mathbb{R}^3$, i.e., that

$$\omega_{\mathbf{A}}^1 \wedge \omega_{\mathbf{B}}^1 = \omega_{[\mathbf{A},\mathbf{B}]}^1 \quad \text{for any } \mathbf{A}, \mathbf{B} \in \mathbb{R}^3.$$

In this way the exterior product of 1-forms can be considered as an extension of the vector product in $\mathbb{R}^3$ to higher dimensions. However, in the n-dimensional case, the product is not a vector in the same space: the space of 2-forms on $\mathbb{R}^n$ is isomorphic to $\mathbb{R}^n$ only for $n = 3$.

PROBLEM 7. Show that, under the isomorphisms established above, the exterior product of a 1-form and a 2-form becomes the scalar product of vectors in $\mathbb{R}^3$:

$$\omega_{\mathbf{A}}^1 \wedge \omega_{\mathbf{B}}^2 = (\mathbf{A}, \mathbf{B}) x_1 \wedge x_2 \wedge x_3.$$

C Behavior under mappings

Let $f: \mathbb{R}^m \to \mathbb{R}^n$ be a linear map, and ω^k an exterior k-form on $\mathbb{R}^n$. Then there is a k-form $f^*\omega^k$ on $\mathbb{R}^m$, whose value on the k vectors $\xi_1, \ldots, \xi_k \in \mathbb{R}^m$ is equal to the value of ω^k on their images:

$$(f^*\omega^k)(\xi_1, \ldots, \xi_k) = \omega^k(f\xi_1, \ldots, f\xi_k).$$

PROBLEM 8. Verify that $f^*\omega^k$ is an exterior form.

PROBLEM 9. Verify that f^* is a linear operator from the space of k-forms on $\mathbb{R}^n$ to the space of k-forms on $\mathbb{R}^m$ (the star *superscript* means that f^* acts in the opposite direction from f).

PROBLEM 10. Let $f: \mathbb{R}^m \to \mathbb{R}^n$ and $g: \mathbb{R}^n \to \mathbb{R}^p$. Verify that $(g \circ f)^* = f^* \circ g^*$.

PROBLEM 11. Verify that f^* preserves exterior multiplication: $f^*(\omega^k \wedge \omega^l) = (f^*\omega^k) \wedge (f^*\omega^l)$.

34 Differential forms

We give here the definition of differential forms on differentiable manifolds.

A *Differential 1-forms*

The simplest example of a differential form is the differential of a function.

EXAMPLE. Consider the function $y = f(x) = x^2$. Its differential $df = 2x\,dx$ depends on the point x and on the "increment of the argument," i.e., on the tangent vector ξ to the x axis. We fix the point x. Then the differential of the function at x, $df|_x$, depends *linearly on* ξ. So, if $x = 1$ and the coordinate of the tangent vector ξ is equal to 1, then $df = 2$, and if the coordinate of ξ is equal to 10, then $df = 20$ (Figure 140).

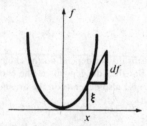

Figure 140   Differential of a function

Let $f: M \to \mathbb{R}$ be a differentiable function on the manifold M (we can imagine a "function of many variables" $f: \mathbb{R}^n \to \mathbb{R}$). The differential $df|_x$ of f at $\mathbf{x}$ is a linear map

$$df_{\mathbf{x}}: TM_{\mathbf{x}} \to \mathbb{R}$$

of the tangent space to M at $\mathbf{x}$ into the real line. We recall from Section 18F the definition of this map:

Let $\xi \in TM_{\mathbf{x}}$ be the velocity vector of the curve $\mathbf{x}(t): \mathbb{R} \to M$; $\mathbf{x}(0) = \mathbf{x}$ and $\dot{\mathbf{x}}(0) = \xi$. Then, by definition,

$$df_{\mathbf{x}}(\xi) = \frac{d}{dt}\bigg|_{t=0} f(\mathbf{x}(t)).$$

PROBLEM 1. Let ξ be the velocity vector of the plane curve $x(t) = \cos t$, $y(t) = \sin t$ at $t = 0$. Calculate the values of the differentials dx and dy of the functions x and y on the vector ξ (Figure 141).

ANSWER. $dx|_{(1,0)}(\xi) = 0,\ dy|_{(1,0)}(\xi) = 1$

Note that the differential of a function f at a point $\mathbf{x} \in M$ is a 1-form $df_{\mathbf{x}}$ on the tangent space $TM_{\mathbf{x}}$.

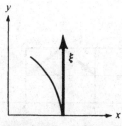

Figure 141   Problem 1

The differential df of f on the manifold M is a smooth map of the tangent bundle TM to the line

$$df: TM \to \mathbb{R} \qquad \left(TM = \bigcup_{\mathbf{x}} TM_{\mathbf{x}} \right).$$

This map is differentiable and is linear on each tangent space $TM_{\mathbf{x}} \subset TM$.

Definition. *A differential form of degree 1 (or a 1-form) on a manifold M is a smooth map*

$$\omega: TM \to \mathbb{R}$$

of the tangent bundle of M to the line, linear on each tangent space $TM_{\mathbf{x}}$.

One could say that a *differential 1-form on M is an algebraic 1-form on $TM_{\mathbf{x}}$ which is "differentiable with respect to $\mathbf{x}$."*

PROBLEM 2. Show that every differential 1-form on the line is the differential of some function.

PROBLEM 3. Find differential 1-forms on the circle and the plane which are not the differential of any function.

B The general form of a differential 1-form on $\mathbb{R}^n$

We take as our manifold M a vector space with coordinates $x_1, \ldots, x_n$. Recall that the components $\xi_1, \ldots, \xi_n$ of a tangent vector $\xi \in T\mathbb{R}^n_{\mathbf{x}}$ are the values of the differentials $dx_1, \ldots, dx_n$ on the vector ξ. These n 1-forms on $T\mathbb{R}^n_{\mathbf{x}}$ are linearly independent. Thus the 1-forms $dx_1, \ldots, dx_n$ form a basis for the n-dimensional space of 1-forms on $T\mathbb{R}^n_{\mathbf{x}}$, and every 1-form on $T\mathbb{R}^n_{\mathbf{x}}$ can be uniquely written in the form $a_1 \, dx_1 + \cdots + a_n \, dx_n$, where the a_i are real coefficients. Now let ω be an arbitrary differential 1-form on $\mathbb{R}^n$. At every point $\mathbf{x}$ it can be expanded uniquely in the basis $dx_1, \ldots, dx_n$. From this we get:

Theorem. *Every differential 1-form on the space $\mathbb{R}^n$ with a given coordinate system $x_1, \ldots, x_n$ can be written uniquely in the form*

$$\omega = a_1(x)dx_1 + \cdots + a_n(x)dx_n,$$

where the coefficients $a_i(x)$ are smooth functions.

175

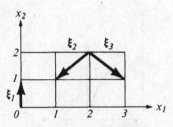

Figure 142   Problem 4

PROBLEM 4. Calculate the value of the forms $\omega_1 = dx_1$, $\omega_2 = x_1 dx_2$, and $\omega_3 = dr^2 (r^2 = x_1^2 + x_2^2)$ on the vectors ξ_1, ξ_2, and ξ_3 (Figure 142).

ANSWER.

	ξ_1	ξ_2	ξ_3
ω_1	0	-1	1
ω_2	0	-2	-2
ω_3	0	-8	0

PROBLEM 5. Let $x_1, \ldots, x_n$ be functions on a manifold M forming a local coordinate system in some region. Show that every 1-form on this region can be uniquely written in the form $\omega = a_1(x) \, dx_1 + \cdots + a_n(x) \, dx_n$.

C Differential k-forms

Definition. *A differential k-form $\omega^k|_x$ at a point **x** of a manifold M is an exterior k-form on the tangent space TM_x to M at **x**, i.e., a k-linear skew-symmetric function of k vectors $\xi_1, \ldots, \xi_k$ tangent to M at **x**.*

If such a form $\omega^k|_x$ is given at every point **x** of the manifold M and if it is differentiable, then we say that we are given a *k-form ω^k on the manifold M.*

PROBLEM 6. Put a natural differentiable manifold structure on the set whose elements are k-tuples of vectors tangent to M at some point **x**.

A differential k-form is a smooth map from the manifold of Problem 6 to the line.

PROBLEM 7. Show that the k-forms on M form a vector space (infinite-dimensional if k does not exceed the dimension of M).

Differential forms can be multiplied by functions as well as by numbers. Therefore, the set of C^∞ differential k-forms has a natural structure as a module over the ring of infinitely differentiable real functions on M.

D *The general form of a differential k-form on* $\mathbb{R}^n$

Take as the manifold M the vector space $\mathbb{R}^n$ with fixed coordinate functions $x_1, \ldots, x_n: \mathbb{R}^n \to \mathbb{R}$. Fix a point $\mathbf{x}$. We saw above that the n 1-forms $dx_1, \ldots, dx_n$ form a basis of the space of 1-forms on the tangent space $T\mathbb{R}_{\mathbf{x}}^n$.

Consider exterior products of the basic forms:

$$dx_{i_1} \wedge \cdots \wedge dx_{i_k}, \qquad i_1 < \cdots < i_k.$$

In Section 32 we saw that these $\binom{n}{k}$ k-forms form a basis of the space of exterior k-forms on $T\mathbb{R}_{\mathbf{x}}^n$. Therefore, every exterior k-form on $T\mathbb{R}_{\mathbf{x}}^n$ can be written uniquely in the form

$$\sum_{i_1 < \cdots < i_k} a_{i_1, \ldots, i_k} \, dx_{i_1} \wedge \cdots \wedge dx_{i_k}.$$

Now let ω be an arbitrary differential k-form on $\mathbb{R}^n$. At every point $\mathbf{x}$ it can be uniquely expressed in terms of the basis above. From this follows:

Theorem. *Every differential k-form on the space $\mathbb{R}^n$ with a given coordinate system $x_1, \ldots, x_n$ can be written uniquely in the form*

$$\omega^k = \sum_{i_1 < \cdots < i_k} a_{i_1, \ldots, i_k}(\mathbf{x}) dx_{i_1} \wedge \cdots \wedge dx_{i_k},$$

where the $a_{i_1, \ldots, i_k}(\mathbf{x})$ are smooth functions on $\mathbb{R}^n$.

PROBLEM 8. Calculate the value of the forms $\omega_1 = dx_1 \wedge dx_2$, $\omega_2 = x_1 \, dx_1 \wedge dx_2 - x_2 \, dx_2 \wedge dx_1$, and $\omega_3 = r \, dr \wedge d\varphi$ (where $x_1 = r \cos \varphi$ and $x_2 = r \sin \varphi$) on the pairs of vectors (ξ_1, η_1), (ξ_2, η_2), and (ξ_3, η_3) (Figure 143).

ANSWER.

	(ξ_1, η_1)	(ξ_2, η_2)	(ξ_3, η_3)
ω_1	1	1	-1
ω_2	2	1	-3
ω_3	1	1	-1

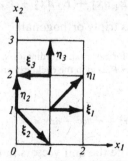

Figure 143   Problem 8

PROBLEM 9. Calculate the value of the forms $\omega_1 = dx_2 \wedge dx_3$, $\omega_2 = x_1\,dx_3 \wedge dx_2$, and $\omega_3 = dx_3 \wedge dr^2$ $(r^2 = x_1^2 + x_2^2 + x_3^2)$, on the pair of vectors $\boldsymbol{\xi} = (1, 1, 1)$, $\boldsymbol{\eta} = (1, 2, 3)$ at the point $\mathbf{x} = (2, 0, 0)$.

ANSWER. $\omega_1 = 1$, $\omega_2 = -2$, $\omega_3 = -8$.

PROBLEM 10. Let $x_1, \ldots, x_n : M \to \mathbb{R}$ be functions on a manifold which form a local coordinate system on some region. Show that every differential form on this region can be written uniquely in the form

$$\omega^k = \sum_{i_1 < \cdots < i_k} a_{i_1, \ldots, i_k}(\mathbf{x})\, dx_{i_1} \wedge \cdots \wedge dx_{i_k}.$$

EXAMPLE. Change of variables in a form. Suppose that we are given two coordinate systems on $\mathbb{R}^3$: x_1, x_2, x_3 and y_1, y_2, y_3. Let ω be a 2-form on $\mathbb{R}^3$. Then, by the theorem above, ω can be written in the system of x-coordinates as $\omega = X_1\,dx_2 \wedge dx_3 + X_2\,dx_3 \wedge dx_1 + X_3\,dx_1 \wedge dx_2$, where X_1, X_2, and X_3 are functions of x_1, x_2, and x_3, and in the system of y-coordinates as $\omega = Y_1\,dy_2 \wedge dy_3 + Y_2\,dy_3 \wedge dy_1 + Y_3\,dy_1 \wedge dy_2$, where Y_1, Y_2, and Y_3 are functions of y_1, y_2, and y_3.

PROBLEM 11. Given the form written in the x-coordinates (i.e., the X_i) and the change of variables formulas $\mathbf{x} = \mathbf{x}(\mathbf{y})$, write the form in y-coordinates, i.e., find Y.

Solution. We have $dx_i = (\partial x_i/\partial y_1)\,dy_1 + (\partial x_i/\partial y_2)\,dy_2 + (\partial x_i/\partial y_3)\,dy_3$. Therefore,

$$dx_2 \wedge dx_3 = \left(\frac{\partial x_2}{\partial y_1}\,dy_1 + \frac{\partial x_2}{\partial y_2}\,dy_2 + \frac{\partial x_2}{\partial y_3}\,dy_3\right) \wedge \left(\frac{\partial x_3}{\partial y_1}\,dy_1 + \frac{\partial x_3}{\partial y_2}\,dy_2 + \frac{\partial x_3}{\partial y_3}\,dy_3\right),$$

from which we get

$$Y_3 = X_1 \left|\frac{D(x_2, x_3)}{D(y_1, y_2)}\right| + X_2 \left|\frac{D(x_3, x_1)}{D(y_1, y_2)}\right| + X_3 \left|\frac{D(x_1, x_2)}{D(y_1, y_2)}\right|, \text{ etc.}$$

E *Appendix. Differential forms in three-dimensional spaces*

Let M be a three-dimensional oriented riemannian manifold (in all future examples M will be euclidean three-space $\mathbb{R}^3$). Let x_1, x_2, and x_3 *be local coordinates, and let the square of the length element have the form*

$$ds^2 = E_1\,dx_1^2 + E_2\,dx_2^2 + E_3\,dx_3^2$$

(i.e., the coordinate system is triply orthogonal).

PROBLEM 12. Find E_1, E_2, and E_3 for cartesian coordinates x, y, z, for cylindrical coordinates r, φ, z and for spherical coordinates R, φ, θ in the euclidean space $\mathbb{R}^3$ (Figure 144).

ANSWER.

$$ds^2 = dx^2 + dy^2 + dz^2 = dr^2 + r^2\,d\varphi^2 + dz^2 = dR^2 + R^2\cos^2\theta\,d\varphi^2 + R^2\,d\theta^2.$$

We let $\mathbf{e}_1$, $\mathbf{e}_2$, and $\mathbf{e}_3$ denote the unit vectors in the coordinate directions. These three vectors form a basis of the tangent space.

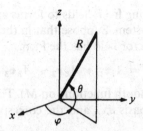

Figure 144 Problem 12

PROBLEM 13. Find the values of the forms dx_1, dx_2, and dx_3 on the vectors $\mathbf{e}_1, \mathbf{e}_2$, and $\mathbf{e}_3$.

ANSWER. $dx_i(\mathbf{e}_i) = 1/\sqrt{E_i}$, the rest are zero. In particular, for cartesian coordinates $dx(\mathbf{e}_x) = dy(\mathbf{e}_y) = dz(\mathbf{e}_z) = 1$; for cylindrical coordinates $dr(\mathbf{e}_r) = dz(\mathbf{e}_z) = 1$ and $d\varphi(\mathbf{e}_\varphi) = 1/r$ (Figure 145), for spherical coordinates $dR(\mathbf{e}_R) = 1$, $d\varphi(\mathbf{e}_\varphi) = 1/R \cos \theta$ and $d\theta(\mathbf{e}_\theta) = 1/R$.

The metric and orientation on the manifold M furnish the tangent space to M at every point with the structure of an oriented euclidean three-dimensional space. In terms of this structure, we can talk about scalar, vector, and triple scalar products.

PROBLEM 14. Calculate $[\mathbf{e}_1, \mathbf{e}_2]$, $(\mathbf{e}_R, \mathbf{e}_\theta)$, and $(\mathbf{e}_z, \mathbf{e}_x, \mathbf{e}_y)$.

ANSWER. $\mathbf{e}_3, 0, 1$.

In an oriented euclidean three-space every vector $\mathbf{A}$ corresponds to a 1-form ω_A^1 and a 2-form ω_A^2, defined by the conditions

$$\omega_A^1(\xi) = (\mathbf{A}, \xi) \qquad \omega_A^2(\xi, \eta) = (\mathbf{A}, \xi, \eta), \qquad \xi, \eta \in \mathbb{R}^3.$$

The correspondence between vector fields and forms does not depend on the system of coordinates, but only on the euclidean structure and orientation. Therefore, every vector field $\mathbf{A}$ on our manifold M corresponds to a differential 1-form ω_A^1 on M and a differential 2-form ω_A^2 on M.

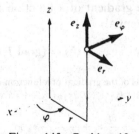

Figure 145 Problem 13

The formulas for changing from fields to forms and back have a different form in each coordinate system. Suppose that in the coordinates $x_1, x_2,$ and x_3 described above, the vector field has the form

$$\mathbf{A} = A_1\mathbf{e}_1 + A_2\mathbf{e}_2 + A_3\mathbf{e}_3$$

(the components A_i are smooth functions on M). The corresponding 1-form $\omega_\mathbf{A}^1$ decomposes over the basis dx_i, and the corresponding 2-form over the basis $dx_i \wedge dx_j$.

PROBLEM 15. Given the components of the vector field $\mathbf{A}$, find the decompositions of the 1-form $\omega_\mathbf{A}^1$ and the 2-form $\omega_\mathbf{A}^2$.

Solution. We have $\omega_\mathbf{A}^1(\mathbf{e}_1) = (\mathbf{A}, \mathbf{e}_1) = A_1$. Also, $(a_1\,dx_1 + a_2\,dx_2 + a_3\,dx_3)(\mathbf{e}_1) = a_1\,dx_1(\mathbf{e}_1) = a_1/\sqrt{E_1}$. From this we get that $a_1 = A_1\sqrt{E_1}$, so that

$$\omega_\mathbf{A}^1 = A_1\sqrt{E_1}\,dx_1 + A_2\sqrt{E_2}\,dx_2 + A_3\sqrt{E_3}\,dx_3.$$

In the same way, we have $\omega_\mathbf{A}^2(\mathbf{e}_2, \mathbf{e}_3) = (\mathbf{A}, \mathbf{e}_2, \mathbf{e}_3) = A_1$. Also,

$$(\alpha_1\,dx_2 \wedge dx_3 + \alpha_2\,dx_3 \wedge dx_1 + \alpha_3\,dx_1 \wedge dx_2)(\mathbf{e}_2, \mathbf{e}_3) = \alpha_1\,\frac{1}{\sqrt{E_2 E_3}}.$$

Hence, $\alpha_1 = A_1\sqrt{E_2 E_3}$, i.e.,

$$\omega_\mathbf{A}^2 = A_1\sqrt{E_2 E_3}\,dx_2 \wedge dx_3 + A_2\sqrt{E_3 E_1}\,dx_3 \wedge dx_1 + A_3\sqrt{E_1 E_2}\,dx_1 \wedge dx_2.$$

In particular, in cartesian, cylindrical, and spherical coordinates on $\mathbb{R}^3$ the vector field

$$\mathbf{A} = A_x\mathbf{e}_x + A_y\mathbf{e}_y + A_z\mathbf{e}_z = A_r\mathbf{e}_r + A_\varphi\mathbf{e}_\varphi + A_z\mathbf{e}_z = A_R\mathbf{e}_R + A_\varphi\mathbf{e}_\varphi + A_u\mathbf{e}_u$$

corresponds to the 1-form

$$\omega_\mathbf{A}^1 = A_x\,dx + A_y\,dy + A_z\,dz = A_r\,dr + rA_\varphi\,d\varphi + A_z\,dz = A_R\,dR + R\cos\theta A_\varphi\,d\varphi + RA_\theta\,d\theta$$

and the 2-form

$$\begin{aligned}\omega_\mathbf{A}^2 &= A_x\,dy \wedge dz + A_y\,dz \wedge dx + A_z\,dx \wedge dy \\ &= rA_r\,d\varphi \wedge dz + A\,dz \wedge dr + rA_z\,dr \wedge d\varphi \\ &= R^2\cos\theta A_R\,d\varphi \wedge d\theta + RA_\varphi\,d\theta \wedge dR + R\cos\theta A_z\,dR \wedge d\varphi.\end{aligned}$$

An example of a vector field on a manifold M is the gradient of a function $f: M \to \mathbb{R}$. Recall that the gradient of a function is the vector field **grad** f corresponding to the differential:

$$\omega_{\mathbf{grad}\,f}^1 = df, \quad \text{i.e.,} \quad df(\xi) = (\mathbf{grad}\,f, \xi) \qquad \forall\xi.$$

PROBLEM 16. Find the components of the gradient of a function in the basis $\mathbf{e}_1, \mathbf{e}_2, \mathbf{e}_3$.

Solution. We have $df = (\partial f/\partial x_1)\,dx_1 + (\partial f/\partial x_2)\,dx_2 + (\partial f/\partial x_3)\,dx_3$. By the problem above

$$\mathbf{grad}\,f = \frac{1}{\sqrt{E_1}}\frac{\partial f}{\partial x_1}\mathbf{e}_1 + \frac{1}{\sqrt{E_2}}\frac{\partial f}{\partial x_2}\mathbf{e}_2 + \frac{1}{\sqrt{E_3}}\frac{\partial f}{\partial x_3}\mathbf{e}_3.$$

In particular, in cartesian, cylindrical, and spherical coordinates

$$\mathbf{grad}\, f = \frac{\partial f}{\partial x}\,\mathbf{e}_x + \frac{\partial f}{\partial y}\,\mathbf{e}_y + \frac{\partial f}{\partial z}\,\mathbf{e}_z = \frac{\partial f}{\partial r}\,\mathbf{e}_r + \frac{1}{r}\frac{\partial f}{\partial \varphi}\,\mathbf{e}_\varphi + \frac{\partial f}{\partial z}\,\mathbf{e}_z$$

$$= \frac{\partial f}{\partial R}\,\mathbf{e}_R + \frac{1}{R\cos\theta}\frac{\partial f}{\partial \varphi}\,\mathbf{e}_\varphi + \frac{1}{R}\frac{\partial f}{\partial \theta}\,\mathbf{e}_\theta.$$

35 Integration of differential forms

We define here the concepts of a chain, the boundary of a chain, and the integration of a form over a chain.

The integral of a differential form is a higher-dimensional generalization of such ideas as the flux of a fluid across a surface or the work of a force along a path.

A *The integral of a 1-form along a path*

We begin by integrating a 1-form ω^1 on a manifold M. Let

$$\gamma: [0 \leq t \leq 1] \to M$$

be a smooth map (the "path of integration"). The integral of the form ω^1 on the path γ is defined as a limit of Riemann sums. Every Riemann sum consists of the values of the form ω^1 on some tangent vectors $\boldsymbol{\xi}_i$ (Figure 146):

$$\int_\gamma \omega^1 = \lim_{\Delta \to 0} \sum_{i=1}^{n} \omega^1(\boldsymbol{\xi}_i).$$

The tangent vectors $\boldsymbol{\xi}_i$ are constructed in the following way. The interval $0 \leq t \leq 1$ is divided into parts $\Delta_i: t_i \leq t \leq t_{i+1}$ by the points t_i. The interval Δ_i can be looked at as a tangent vector Δ_i to the t axis at the point t_i. Its image in the tangent space to M at the point $\gamma(t_i)$ is

$$\boldsymbol{\xi}_i = d\gamma|_{t_i}(\Delta_i) \in TM_{\gamma(t_i)}.$$

The sum has a limit as the largest of the intervals Δ_i tends to zero. It is called the integral of the 1-form ω^1 along the path γ.

The definition of the integral of a k-form along a k-dimensional surface follows an analogous pattern. The surface of integration is partitioned into

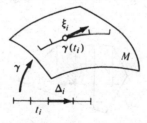

Figure 146   Integrating a 1-form along a path

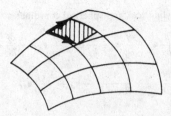

Figure 147   Integrating a 2-form over a surface

small curvilinear k-dimensional parallelepipeds (Figure 147); these parallelepipeds are replaced by parallelepipeds in the tangent space. The sum of the
values of the form on the parallelepipeds in the tangent space approaches
the integral as the partition is refined. We will first consider a particular case.

B *The integral of a k-form on oriented euclidean space* $\mathbb{R}^k$

Let $x_1, \ldots, x_k$ be an oriented coordinate system on $\mathbb{R}^k$. Then every k-form
on $\mathbb{R}^k$ is proportional to the form $dx_1 \wedge \cdots \wedge dx_k$, i.e., it has the form
$\omega^k = \varphi(x)dx_1 \wedge \cdots \wedge dx_k$, where $\varphi(x)$ is a smooth function.

Let D be a bounded convex polyhedron in $\mathbb{R}^k$ (Figure 148). By definition,
the integral of the form ω^k on D is the integral of the function φ:

$$\int_D \omega^k = \int_D \varphi(x)dx_1, \ldots, dx_k,$$

where the integral on the right is understood to be the usual limit of Riemann
sums.

Such a definition follows the pattern outlined above, since in this case the
tangent space to the manifold is identified with the manifold.

PROBLEM 1. Show that $\int_D \omega^k$ depends linearly on ω^k.

PROBLEM 2. Show that if we divide D into two distinct polyhedra D_1 and D_2, then

$$\int_D \omega^k = \int_{D_1} \omega^k + \int_{D_2} \omega^k.$$

In the general case (a k-form on an n-dimensional space) it is not so easy
to identify the elements of the partition with tangent parallelepipeds; we will
consider this case below.

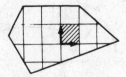

Figure 148   Integrating a k-form in k-dimensional space

C The behavior of differential forms under maps

Let $f: M \to N$ be a differentiable map of a smooth manifold M to a smooth manifold N, and let ω be a differential k-form on N (Figure 149). Then, a well-defined k-form arises also on M: it is denoted by $f^*\omega$ and is defined by the relation

$$(f^*\omega)(\xi_1, \ldots, \xi_k) = \omega(f_*\xi_1, \ldots, f_*\xi_k)$$

for any tangent vectors $\xi_1, \ldots, \xi_k \in TM_x$. Here f_* is the differential of the map f. In other words, the value of the form $f^*\omega$ on the vectors $\xi_1, \ldots, \xi_k$ is equal to the value of ω on the images of these vectors.

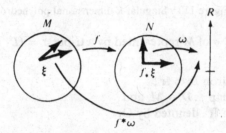

Figure 149   A form on N induces a form on M.

EXAMPLE. If $y = f(x_1, x_2) = x_1^2 + x_2^2$ and $\omega = dy$, then

$$f^*\omega = 2x_1 \, dx_1 + 2x_2 \, dx_2.$$

PROBLEM 3. Show that $f^*\omega$ is a k-form on M.

PROBLEM 4. Show that the map f^* preserves operations on forms:

$$f^*(\lambda_1\omega_1 + \lambda_2\omega_2) = \lambda_1 f^*(\omega_1) + \lambda_2 f^*(\omega_2),$$
$$f^*(\omega_1 \wedge \omega_2) = (f^*\omega_1) \wedge (f^*\omega_2).$$

PROBLEM 5. Let $g: L \to M$ be a differentiable map. Show that $(fg)^* = g^*f^*$.

PROBLEM 6. Let D_1 and D_2 be two compact, convex polyhedra in the oriented k-dimensional space $\mathbb{R}^k$ and $f: D_1 \to D_2$ a differentiable map which is an orientation-preserving diffeomorphism[55] of the interior of D_1 onto the interior of D_2. Then, for any differential k-form ω^k on D_2,

$$\int_{D_1} f^*\omega^k = \int_{D_2} \omega^k.$$

Hint. This is the change of variables theorem for a multiple integral:

$$\int_{D_1} \frac{\partial(y_1, \ldots, y_n)}{\partial(x_1, \ldots, x_n)} \varphi(y(x)) dx_1 \cdots dx_n = \int_{D_2} \varphi(y) dy_1 \cdots dy_n.$$

[55] i.e., one-to-one with a differentiable inverse.

D Integration of a k-form on an n-dimensional manifold

Let ω be a differential k-form on an n-dimensional manifold M. Let D be a bounded convex k-dimensional polyhedron in k-dimensional euclidean space $\mathbb{R}^k$ (Figure 150). The role of "path of integration" will be played by a

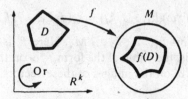

Figure 150   Singular k-dimensional polyhedron

k-dimensional cell[56] σ of M represented by a triple $\sigma = (D, f, \text{Or})$ consisting of

1. a convex polyhedron $D \subset \mathbb{R}^k$,
2. a differentiable map $f: D \to M$, and
3. an orientation on $\mathbb{R}^k$, denoted by Or.

Definition. The integral of the k-form ω over the k-dimensional cell σ is the integral of the corresponding form over the polyhedron D

$$\int_\sigma \omega = \int_D f^* \omega.$$

PROBLEM 7. Show that the integral depends linearly on the form:

$$\int_\sigma \lambda_1 \omega_1 + \lambda_2 \omega_2 = \lambda_1 \int_\sigma \omega_1 + \lambda_2 \int_\sigma \omega_2.$$

The k-dimensional cell which differs from σ only by the choice of orientation is called the *negative of* σ and is denoted by $-\sigma$ or $-1 \cdot \sigma$ (Figure 151).

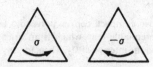

Figure 151   Problem 8

PROBLEM 8. Show that, under a change of orientation, the integral changes sign:

$$\int_{-\sigma} \omega = -\int_\sigma \omega.$$

---

[56] The cell σ is usually called a *singular k-dimensional polyhedron*.

E Chains

The set $f(D)$ is not necessarily a smooth submanifold of M. It could have "self-intersections" or "folds" and could even be reduced to a point. However, even in the one-dimensional case, it is clear that it is inconvenient to restrict ourselves to contours of integration consisting of one piece: it is useful to be able to consider contours consisting of several pieces which can be traversed in either direction, perhaps more than once. The analogous concept in higher dimensions is called a *chain*.

Definition. *A chain of dimension k on a manifold M consists of a finite collection of k-dimensional oriented cells $\sigma_1, \ldots, \sigma_r$ in M and integers $m_1, \ldots, m_r$, called multiplicities (the multiplicities can be positive, negative, or zero). A chain is denoted by*

$$c_k = m_1\sigma_1 + \cdots + m_r\sigma_r.$$

We introduce the natural identifications

$$m_1\sigma + m_2\sigma = (m_1 + m_2)\sigma$$

$$m_1\sigma_1 + m_2\sigma_2 = m_2\sigma_2 + m_1\sigma_1 \qquad 0\sigma = 0 \qquad c_k + 0 = c_k.$$

PROBLEM 9. Show that the set of all k-chains on M forms a commutative group if we define the addition of chains by the formula

$$(m_1\sigma_1 + \cdots + m_r\sigma_r) + (m_1'\sigma_1' + \cdots + m_{r'}'\sigma_{r'}') = m_1\sigma_1 + \cdots + m_r\sigma_r + m_1'\sigma_1' + \cdots + m_{r'}'\sigma_{r'}'.$$

F Example: the boundary of a polyhedron

Let D be a convex oriented k-dimensional polyhedron in k-dimensional euclidean space $\mathbb{R}^k$. The *boundary* of D is the $(k-1)$-chain ∂D on $\mathbb{R}^k$ defined in the following way (Figure 152).

The cells σ_i of the chain ∂D are the $(k-1)$-dimensional faces D_i of the polyhedron D, together with maps $f_i : D_i \to \mathbb{R}^k$ embedding the faces in $\mathbb{R}^k$ and orientations Or_i defined below; the multiplicities are equal to 1:

$$\partial D = \sum \sigma_i \qquad \sigma_i = (D_i, f_i, \mathrm{Or}_i).$$

Rule of orientation of the boundary. Let $e_1, \ldots, e_k$ be an oriented frame in $\mathbb{R}^k$. Let D_i be one of the faces of D. We choose an interior point of D_i and there

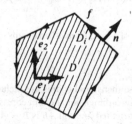

Figure 152 Oriented boundary

construct a vector **n** outwardly normal to the polyhedron D. An orienting frame for the face D_i will be a frame $\mathbf{f}_1, \ldots, \mathbf{f}_{k-1}$ on D_i such that the frame $(\mathbf{n}, \mathbf{f}_1, \ldots, \mathbf{f}_{k-1})$ is oriented *correctly* (i.e., the same way as the frame $\mathbf{e}_1, \ldots, \mathbf{e}_k$).

The *boundary of a chain* is defined in an analogous way. Let $\sigma = (D, f, \text{Or})$ be a k-dimensional cell in the manifold M. Its boundary $\partial\sigma$ is the $(k-1)$ chain: $\partial\sigma = \sum \sigma_i$ consisting of the cells $\sigma_i = (D_i, f_i, \text{Or}_i)$, where the D_i are the $(k-1)$-dimensional faces of D, Or_i are orientations chosen by the rule above, and f_i are the restrictions of the mapping $f: D \to M$ to the face D_i.

The boundary ∂c_k of the k-dimensional chain c_k in M is the sum of the boundaries of the cells of c_k with multiplicities (Figure 153):

$$\partial c_k = \partial(m_1\sigma_1 + \cdots + m_r\sigma_r) = m_1\,\partial\sigma_1 + \cdots + m_r\,\partial\sigma_r.$$

Obviously, ∂c_k is a $(k-1)$-chain on M.[5,7]

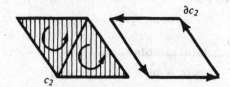

Figure 153   Boundary of a chain

PROBLEM 10. Show that the boundary of the boundary of any chain is zero: $\partial\partial c_k = 0$.

Hint. By the linearity of ∂ it is enough to show that $\partial\partial D = 0$ for a convex polyhedron D. It remains to verify that every $(k-2)$-dimensional face of D appears in $\partial\partial D$ twice, with opposite signs. It is enough to prove this for $k = 2$ (planar cross-sections).

G The integral of a form over a chain

Let ω^k be a k-form on M, and c_k a k-chain on M, $c_k = \sum m_i\sigma_i$. The *integral of the form ω^k over the chain c_k* is the sum of the integrals on the cells, counting multiplicities:

$$\int_{c_k} \omega^k = \sum m_i \int_{\sigma_i} \omega^k.$$

PROBLEM 11. Show that the integral depends linearly on the form:

$$\int_{c_k} \omega_1^k + \omega_2^k = \int_{c_k} \omega_1^k + \int_{c_k} \omega_2^k.$$

PROBLEM 12. Show that integration of a fixed form ω^k on chains c_k defines a homomorphism from the group of chains to the line.

[5,7] We are taking $k > 1$ here. One-dimensional chains are included in the general scheme if we make the following definitions: a zero-dimensional chain consists of a collection of points with multiplicities; the boundary of an oriented interval $\vec{AB}$ is $B - A$ (the point B with multiplicity 1 and A with multiplicity -1); the boundary of a point is empty.

EXAMPLE 1. Let M be the plane $\{(p, q)\}$, ω^1 the form pdq, and c_1 the chain consisting of one cell σ with multiplicity 1:

$$[0 \leq t \leq 2\pi] \xrightarrow{f} (p = \cos t, q = \sin t).$$

Then $\int_{c_1} pdq = \pi$. In general, if a chain c_1 represents the boundary of a region G (Figure 154), then $\int_{c_1} pdq$ is equal to the area of G with sign $+$ or $-$ depending on whether the pair of vectors (outward normal, oriented boundary vector) has the same or opposite orientation as the pair (p axis, q axis).

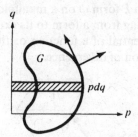

Figure 154   The integral of the form $p\, dq$ over the boundary of a region is equal to the area of the region.

EXAMPLE 2. Let M be the oriented three-dimensional euclidean space $\mathbb{R}^3$. Then every 1-form on M corresponds to some vector field $\mathbf{A}$ $(\omega^1 = \omega_{\mathbf{A}}^1)$, where

$$\omega_{\mathbf{A}}^1(\xi) = (\mathbf{A}, \xi).$$

The integral of $\omega_{\mathbf{A}}^1$ on a chain c_1 representing a curve l is called the *circulation of the field* $\mathbf{A}$ *over the curve* l:

$$\int_{c_1} \omega_{\mathbf{A}}^1 = \int_l (\mathbf{A}, dl).$$

Every 2-form on M also corresponds to some field $\mathbf{A}$ $(\omega^2 = \omega_{\mathbf{A}}^2$, where $\omega_{\mathbf{A}}^2(\xi, \eta) = (\mathbf{A}, \xi, \eta))$.

The integral of the form $\omega_{\mathbf{A}}^2$ on a chain c_2 representing an oriented surface S is called *the flux of the field* $\mathbf{A}$ *through the surface* S:

$$\int_{c_2} \omega_{\mathbf{A}}^2 = \int_S (\mathbf{A}, d\mathbf{n}).$$

PROBLEM 13. Find the flux of the field $\mathbf{A} = (1/R^2)\mathbf{e}_R$ over the surface of the sphere $x^2 + y^2 + z^2 = 1$, oriented by the vectors $\mathbf{e}_x, \mathbf{e}_y$ at the point $z = 1$. Find the flux of the same field over the surface of the ellipsoid $(x^2/a^2) + (y^2/b^2) + z^2 = 1$ oriented the same way.

   Hint. Cf. Section 36H.

PROBLEM 14. Suppose that, in the $2n$-dimensional space $\mathbb{R}^n = \{(p_1, \ldots, p_n; q_1, \ldots, q_n)\}$, we are given a 2-chain c_2 representing a two-dimensional oriented surface S with boundary l. Find

$$\int_{c_2} dp_1 \wedge dq_1 + \cdots + dp_n \wedge dq_n \text{ and } \int_l p_1 dq_1 + \cdots + p_n dq_n.$$

ANSWER. The sum of the oriented areas of the projection of S on the two-dimensional coordinate planes p_i, q_i.

36 Exterior differentiation

We define here exterior differentiation of k-forms and prove Stokes' theorem: the integral of the derivative of a form over a chain is equal to the integral of the form itself over the boundary of the chain.

A *Example: the divergence of a vector field*

The exterior derivative of a k-form ω on a manifold M is a $(k + 1)$-form $d\omega$ on the same manifold. Going from a form to its exterior derivative is analogous to forming the differential of a function or the divergence of a vector field. We recall the definition of divergence.

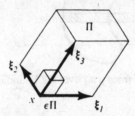

Figure 155   Definition of divergence of a vector field

Let $\mathbf{A}$ be a vector field on the oriented euclidean three-space $\mathbb{R}^3$, and let S be the boundary of a parallelepiped Π with edges ξ_1, ξ_2, and ξ_3 at the vertex x (Figure 155). Consider the ("outward") flux of the field $\mathbf{A}$ through the surface S:

$$F(\Pi) = \int_S (\mathbf{A}, d\mathbf{n}).$$

If the parallelepiped Π is very small, the flux F is approximately proportional to the product of the volume of the parallelepiped, $V = (\xi_1, \xi_2, \xi_3)$, and the "source density" at the point x. This is the limit

$$\lim_{\varepsilon \to 0} \frac{F(\varepsilon\Pi)}{\varepsilon^3 V}$$

where $\varepsilon\Pi$ is the parallelepiped with edges $\varepsilon\xi_1, \varepsilon\xi_2, \varepsilon\xi_3$. This limit does not depend on the choice of the parallelepiped Π but only on the point x, and is called the *divergence*, div $\mathbf{A}$, of the field $\mathbf{A}$ at x.

To go to higher-dimensional cases, we note that the "flux of $\mathbf{A}$ through a surface element" is the 2-form which we called ω_A^2. The divergence, then, is the density in the expression for the 3-form

$$\omega^3 = \text{div } \mathbf{A} \, dx \wedge dy \wedge dz,$$

$$\omega^3(\xi_1, \xi_2, \xi_3) = \text{div } \mathbf{A} \cdot V(\xi_1, \xi_2, \xi_3),$$

characterizing the "sources in an elementary parallelepiped."

188

The exterior derivative $d\omega^k$ of a k-form ω^k on an n-dimensional manifold M may be defined as the principal multilinear part of the integral of ω^k over the boundaries of $(k + 1)$-dimensional parallelepipeds.

B Definition of the exterior derivative

We define the value of the form $d\omega$ on $k + 1$ vectors $\boldsymbol{\xi}_1, \ldots, \boldsymbol{\xi}_{k+1}$ tangent to M at $\mathbf{x}$. To do this, we choose some coordinate system in a neighborhood of $\mathbf{x}$ on M, i.e., a differentiable map f of a neighborhood of the point 0 in euclidean space $\mathbb{R}^n$ to a neighborhood of $\mathbf{x}$ in M (Figure 156).

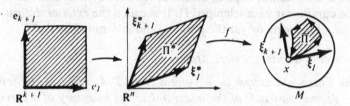

Figure 156   The curvilinear parallelepiped Π.

The pre-images of the vectors $\boldsymbol{\xi}_1, \ldots, \boldsymbol{\xi}_{k+1} \in TM_{\mathbf{x}}$ under the differential of f lie in the tangent space to $\mathbb{R}^n$ at 0. This tangent space can be naturally identified with $\mathbb{R}^n$, so we may consider the pre-images to be vectors

$$\boldsymbol{\xi}_1^*, \ldots, \boldsymbol{\xi}_{k+1}^* \in \mathbb{R}^n.$$

We take the parallelepiped Π^* in $\mathbb{R}^n$ spanned by these vectors (strictly speaking, we must look at the standard oriented cube in $\mathbb{R}^{k+1}$ and its linear map onto Π^*, taking the edges $\mathbf{e}_1, \ldots, \mathbf{e}_{k+1}$ to $\boldsymbol{\xi}_1^*, \ldots, \boldsymbol{\xi}_{k+1}^*$, as a $(k + 1)$-dimensional cell in $\mathbb{R}^n$). The map f takes the parallelepiped Π^* to a $(k + 1)$-dimensional cell on M (a "curvilinear parallelepiped"). The boundary of the cell Π is a k-chain, $\partial\Pi$. Consider the integral of the form ω^k on the boundary $\partial\Pi$ of Π:

$$F(\boldsymbol{\xi}_1, \ldots, \boldsymbol{\xi}_{k+1}) = \int_{\partial\Pi} \omega^k.$$

EXAMPLE. We will call a smooth function $\varphi: M \to R$ a 0-form on M. The integral of the 0-form φ on the 0-chain $c_0 = \sum m_i A_i$ (where the m_i are integers and the A_i points of M) is

$$\int_{c_0} \varphi = \sum m_i \varphi(A_i).$$

Then the definition above gives the "increment" $F(\boldsymbol{\xi}_1) = \varphi(x_1) - \varphi(x)$ (Figure 157) of the function φ, and the principal linear part of $F(\boldsymbol{\xi}_1)$ at 0 is simply the differential of φ.

PROBLEM 1. Show that the function $F(\boldsymbol{\xi}_1, \ldots, \boldsymbol{\xi}_{k+1})$ is skew-symmetric with respect to $\boldsymbol{\xi}$.

It turns out that the principal $(k + 1)$-linear part of the "increment" $F(\boldsymbol{\xi}_1, \ldots, \boldsymbol{\xi}_{k+1})$ is an exterior $(k + 1)$-form on the tangent space $TM_{\mathbf{x}}$ to M

Figure 157 The integral over the boundary of a one-dimensional parallelepiped is the change in the function.

at **x**. This form does not depend on the coordinate system that was used to define the curvilinear parallelepiped Π. It is called the *exterior derivative*, or *differential*, of the form ω^k (at the point **x**) and is denoted by $d\omega^k$.

C A theorem on exterior derivatives

Theorem. *There is a unique $(k + 1)$-form Ω on TM_x which is the principal $(k + 1)$-linear part at 0 of the integral over the boundary of a curvilinear parallelepiped, $F(\xi_1, \ldots, \xi_{k+1})$; i.e.,*

(1) $$F(\varepsilon\xi_1, \ldots, \varepsilon\xi_{k+1}) = \varepsilon^{k+1}\Omega(\xi_1, \ldots, \xi_{k+1}) + o(\varepsilon^{k+1}) \qquad (\varepsilon \to 0).$$

The form Ω does not depend on the choice of coordinates involved in the definition of F. If, in the local coordinate system $x_1, \ldots, x_n$ on M, the form ω^k is written as

$$\omega^k = \sum a_{i_1, \ldots, i_k} \, dx_{i_1} \wedge \cdots \wedge dx_{i_k},$$

then Ω is written as

(2) $$\Omega = d\omega^k = \sum da_{i_1, \ldots, i_k} \wedge dx_{i_1} \wedge \cdots \wedge dx_{i_k}.$$

We will carry out the proof of this theorem for the case of a form $\omega^1 = a(x_1, x_2)dx_1$ on the x_1, x_2 plane. The proof in the general case is entirely analogous, but the calculations are somewhat longer.

We calculate $F(\xi, \eta)$, i.e., the integral of ω^1 on the boundary of the parallelogram Π with sides ξ and η and vertex at 0 (Figure 158). The chain $\partial \Pi$ is

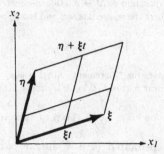

Figure 158 Theorem on exterior derivatives

given by the mappings of the interval $0 \leq t \leq 1$ to the plane $t \to \xi t$, $t \to \xi + \eta t$, $t \to \eta t$, and $t \to \eta + \xi t$ with multiplicities $1, 1, -1$, and -1. Therefore,

$$\int_{\partial \Pi} \omega^1 = \int_0^1 [a(\xi t) - a(\xi t + \eta)]\xi_1 - [a(\eta t) - a(\eta t + \xi)]\eta_1 \, dt$$

where $\xi_1 = dx_1(\xi)$, $\eta_1 = dx_1(\eta)$, $\xi_2 = dx_2(\xi)$, and $\eta_2 = dx_2(\eta)$ are the components of the vectors ξ and η. But

$$a(\xi t + \eta) - a(\xi t) = \frac{\partial a}{\partial x_1}\eta_1 + \frac{\partial a}{\partial x_2}\eta_2 + O(\xi^2, \eta^2)$$

(the derivatives are taken at $x_1 = x_2 = 0$). In the same way

$$a(\eta t + \xi) - a(\eta t) = \frac{\partial a}{\partial x_1}\xi_1 + \frac{\partial a}{\partial x_2}\xi_2 + O(\xi^2, \eta^2).$$

By using these expressions in the integral, we find that

$$F(\xi, \eta) = \int_{\partial \Pi} \omega^1 = \frac{\partial a}{\partial x_2}(\xi_2 \eta_1 - \xi_1 \eta_2) + o(\xi^2, \eta^2).$$

The principal bilinear part of F, as promised in (1), turns out to be the value of the exterior 2-form

$$\Omega = \frac{\partial a}{\partial x_2} dx_2 \wedge dx_1$$

on the pair of vectors ξ, η. Thus the form obtained is given by formula (2), since

$$da \wedge dx_1 = \frac{\partial a}{\partial x_1} dx_1 \wedge dx_1 + \frac{\partial a}{\partial x_2} dx_2 \wedge dx_1 = \frac{\partial a}{\partial x_2} dx_2 \wedge dx_1.$$

Finally, if the coordinate system x_1, x_2 is changed to another (Figure 159), the parallelogram Π is changed to a nearby curvilinear parallelogram Π', so that the difference in the values of the integrals, $\int_{\partial \Pi} \omega^1 - \int_{\partial \Pi'} \omega^1$ will be small of more than second order (prove it!). $\square$

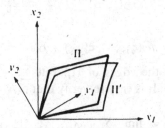

Figure 159 Independence of the exterior derivative from the coordinate system

PROBLEM 2. Carry out the proof of the theorem in the general case.

PROBLEM 3. Prove the formulas for differentiating a sum and a product:

$$d(\omega_1 + \omega_2) = d\omega_1 + d\omega_2.$$

and

$$d(\omega^k \wedge \omega^l) = d\omega^k \wedge \omega^l + (-1)^k \omega^k \wedge d\omega^l.$$

PROBLEM 4. Show that the differential of a differential is equal to zero: $dd = 0$.

PROBLEM 5. Let $f: M \to N$ be a smooth map and ω a k-form on N. Show that $f^*(d\omega) = d(f^*\omega)$.

D Stokes' formula

One of the most important corollaries of the theorem on exterior derivatives is the Newton-Leibniz-Gauss-Green-Ostrogradskii-Stokes-Poincaré formula:

$$(3) \qquad \int_{\partial c} \omega = \int_c d\omega,$$

where c is any $(k + 1)$-chain on a manifold M and ω is any k-form on M.

To prove this formula it is sufficient to prove it for the case when the chain consists of one cell σ. We assume first that this cell σ is given by an oriented parallelepiped $\Pi \subset \mathbb{R}^{k+1}$ (Figure 160).

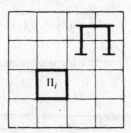

Figure 160   Proof of Stokes' formula for a parallelepiped

We partition Π into N^{k+1} small equal parallelepipeds Π_i similar to Π. Then, clearly,

$$\int_{\partial \Pi} \omega = \sum_{i=1}^{N^{k+1}} F_i, \quad \text{where } F_i = \int_{\partial \Pi_i} \omega.$$

By formula (1) we have

$$F_i = d\omega(\xi_1^i, \ldots, \xi_{k+1}^i) + o(N^{-(k+1)}),$$

where $\xi_1^i, \ldots, \xi_{k+1}^i$ are the edges of Π_i. But $\sum_{i=1}^{N^{k+1}} d\omega(\xi_1^i, \ldots, \xi_{k+1}^i)$ is a Riemann sum for $\int_\Pi d\omega$. It is easy to verify that $o(N^{-(k+1)})$ is uniform, so

$$\lim_{N \to \infty} \sum_{i=1}^{N^{k+1}} F_i = \lim_{N \to \infty} \sum_{i=1}^{N^{k+1}} d\omega(\xi_1^i, \ldots, \xi_{k+1}^i) = \int_\Pi d\omega.$$

Finally, we obtain

$$\int_{\partial \Pi} \omega = \sum F_i = \lim_{N \to \infty} \sum F_i = \int_{\Pi} d\omega.$$

Formula (3) follows automatically from this for any chain whose polyhedra are parallelepipeds.

To prove formula (3) for any convex polyhedron D, it is enough to prove it for a simplex,[58] since D can always be partitioned into simplices (Figure 161):

$$D = \sum D_i \qquad \partial D = \sum \partial D_i.$$

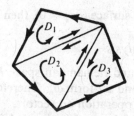

Figure 161   Division of a convex polyhedron into simplices

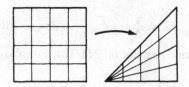

Figure 162   Proof of Stokes' formula for a simplex

We will prove formula (3) for a simplex. Notice that a k-dimensional oriented cube can be mapped onto a k-dimensional simplex so that:

1. The interior of the cube goes diffeomorphically, with its orientation preserved, onto the interior of the simplex;
2. The interiors of some $(k-1)$-dimensional faces of the cube go diffeomorphically, with their orientations preserved, onto the interiors of the faces of the simplex; the images of the remaining $(k-1)$-dimensional faces of the cube lie in the $(k-2)$-dimensional faces of the simplex.

For example, for $k = 2$ such a map of the cube $0 \le x_1, x_2 \le 1$ onto the triangle is given by the formula $y_1 = x_1$, $y_2 = x_1 x_2$ (Figure 162). Then,

[58] A two-dimensional simplex is a triangle, a three-dimensional simplex is a tetrahedron, a k-dimensional simplex is the convex hull of $k + 1$ points in $\mathbb{R}^n$ which do not lie in any $k - 1$-dimensional plane.

EXAMPLE: $\{x \in \mathbb{R}^k : x_i \ge 0 \text{ and } \sum_{i=1}^{k} x_i \le 1\}$.

formula (3) for the simplex follows from formula (3) for the cube and the change of variables theorem (cf. Section 35C).

EXAMPLE 1. Consider the 1-form

$$\omega^1 = p_1\, dq_1 + \cdots + p_n\, dq_n = \mathbf{p}\, d\mathbf{q}$$

on $\mathbb{R}^{2n}$ with coordinates $p_1, \ldots, p_n, q_1, \ldots, q_n$. Then $d\omega^1 = dp_1 \wedge dq_1 + \cdots + dp_n \wedge dq_n = d\mathbf{p} \wedge d\mathbf{q}$, so

$$\iint_{c_2} d\mathbf{p} \wedge d\mathbf{q} = \int_{\partial c_2} \mathbf{p}\, d\mathbf{q}.$$

In particular, if c_2 is a closed surface ($\partial c_2 = 0$), then $\iint_{c_2} d\mathbf{p} \wedge d\mathbf{q} = 0$.

E *Example 2—Vector analysis*

In a three-dimensional oriented riemannian space M, every vector field $\mathbf{A}$ corresponds to a 1-form ω_A^1 and a 2-form ω_A^2. Therefore, exterior differentiation can be considered as an operation on vectors.

Exterior differentiation of 0-forms (functions), 1-forms, and 2-forms correspond to the operations of gradient, curl, and divergence defined by the relations

$$df = \omega_{\mathbf{grad}\, f}^1 \qquad d\omega_A^1 = \omega_{\mathbf{curl}\, A}^2 \qquad d\omega_A^2 = (\operatorname{div} \mathbf{A})\omega^3$$

(the form ω^3 is the volume element on M). Thus, it follows from (3) that

$$f(y) - f(x) = \int_l \mathbf{grad}\, f\, d\mathbf{l} \quad \text{if } \partial l = y - x$$

$$\int_l \mathbf{A}\, d\mathbf{l} = \iint_S \mathbf{curl}\, \mathbf{A} \cdot d\mathbf{n} \quad \text{if } \partial S = l$$

$$\iint_S \mathbf{A}\, d\mathbf{n} = \iiint_D (\operatorname{div} \mathbf{A})\omega^3 \quad \text{if } \partial D = S.$$

PROBLEM 6. Show that

$$\operatorname{div}[\mathbf{A}, \mathbf{B}] = (\mathbf{curl}\, \mathbf{A}, \mathbf{B}) - (\mathbf{curl}\, \mathbf{B}, \mathbf{A}),$$

$$\mathbf{curl}\, a\mathbf{A} = [\mathbf{grad}\, a, \mathbf{A}] + a\, \mathbf{curl}\, \mathbf{A},$$

$$\operatorname{div} a\mathbf{A} = (\mathbf{grad}\, a, \mathbf{A}) + a\, \operatorname{div} \mathbf{A}.$$

Hint. By the formula for differentiating the product of forms,

$$d(\omega_{[\mathbf{A},\mathbf{B}]}^2) = d(\omega_A^1 \wedge \omega_B^1) = d\omega_A^1 \wedge \omega_B^1 - \omega_A^1 \wedge d\omega_B^1.$$

PROBLEM 7. Show that $\mathbf{curl}\, \mathbf{grad} = \operatorname{div} \mathbf{curl} = 0$.
Hint. $dd = 0$.

F *Appendix 1: Vector operations in triply orthogonal systems*

Let x_1, x_2, x_3 be a triply orthogonal coordinate system on M, $ds^2 = E_1\,dx_1^2 + E_2\,dx_2^2 + E_3\,dx_3^2$ and $\mathbf{e}_i$ the coordinate unit vectors (cf. Section 34E).

PROBLEM 8. Given the components of a vector field $\mathbf{A} = A_1\mathbf{e}_1 + A_2\mathbf{e}_2 + A_3\mathbf{e}_3$, find the components of its curl.

Solution. According to Section 34E

$$\omega_{\mathbf{A}}^1 = A_1\sqrt{E_1}\,dx_1 + A_2\sqrt{E_2}\,dx_2 + A_3\sqrt{E_3}\,dx_3.$$

Therefore,

$$d\omega_{\mathbf{A}}^1 = \left(\frac{\partial A_3\sqrt{E_3}}{\partial x_2} - \frac{\partial A_2\sqrt{E_2}}{\partial x_3}\right)dx_2 \wedge dx_3 + \cdots = \omega_{\text{curl A}}^2.$$

According to Section 34E, we have

$$\text{curl } \mathbf{A} = \frac{1}{\sqrt{E_2 E_3}}\left(\frac{\partial A_3\sqrt{E_3}}{\partial x_2} - \frac{\partial A_2\sqrt{E_2}}{\partial x_3}\right)\mathbf{e}_1 + \cdots = \frac{1}{\sqrt{E_1 E_2 E_3}}\begin{vmatrix}\sqrt{E_1}\,\mathbf{e}_1 & \sqrt{E_2}\,\mathbf{e}_2 & \sqrt{E_3}\,\mathbf{e}_3 \\ \dfrac{\partial}{\partial x_1} & \dfrac{\partial}{\partial x_2} & \dfrac{\partial}{\partial x_3} \\ A_1\sqrt{E_1} & A_2\sqrt{E_2} & A_3\sqrt{E_3}\end{vmatrix}.$$

In particular, in cartesian, cylindrical, and spherical coordinates on $\mathbb{R}^3$,

$$\text{curl } \mathbf{A} = \left(\frac{\partial A_z}{\partial y} - \frac{\partial A_y}{\partial z}\right)\mathbf{e}_x + \left(\frac{\partial A_x}{\partial z} - \frac{\partial A_z}{\partial x}\right)\mathbf{e}_y + \left(\frac{\partial A_y}{\partial x} - \frac{\partial A_x}{\partial y}\right)\mathbf{e}_z$$

$$= \frac{1}{r}\left(\frac{\partial A_z}{\partial \varphi} - \frac{\partial r A_\varphi}{\partial z}\right)\mathbf{e}_r + \left(\frac{\partial A_r}{\partial z} - \frac{\partial A_z}{\partial r}\right)\mathbf{e}_\varphi + \frac{1}{r}\left(\frac{\partial r A_\varphi}{\partial r} - \frac{\partial A_r}{\partial \varphi}\right)\mathbf{e}_z$$

$$= \frac{1}{R\cos\theta}\left(\frac{\partial A_\theta}{\partial \varphi} - \frac{\partial A_\varphi \cos\theta}{\partial \theta}\right)\mathbf{e}_R + \frac{1}{R}\left(\frac{\partial A_R}{\partial \theta} - \frac{\partial R A_\theta}{\partial R}\right)\mathbf{e}_\varphi + \frac{1}{R}\left(\frac{\partial R A_\varphi}{\partial R} - \frac{1}{\cos\theta}\frac{\partial A_R}{\partial \varphi}\right)\mathbf{e}_\theta.$$

PROBLEM 9. Find the divergence of the field $\mathbf{A} = A_1\mathbf{e}_1 + A_2\mathbf{e}_2 + A_3\mathbf{e}_3$.

Solution. $\omega_{\mathbf{A}}^2 = A_1\sqrt{E_2 E_3}\,dx_2 \wedge dx_3 + \cdots$. Therefore,

$$d\omega_{\mathbf{A}}^2 = \frac{\partial}{\partial x_1}(A_1\sqrt{E_2 E_3})\,dx_1 \wedge dx_2 \wedge dx_3 + \cdots.$$

By the definition of divergence,

$$d\omega_{\mathbf{A}}^2 = \text{div } \mathbf{A}\sqrt{E_1 E_2 E_3}\,dx_1 \wedge dx_2 \wedge dx_3.$$

This means

$$\text{div } \mathbf{A} = \frac{1}{\sqrt{E_1 E_2 E_3}}\left(\frac{\partial}{\partial x_1}A_1\sqrt{E_2 E_3} + \frac{\partial}{\partial x_2}A_2\sqrt{E_3 E_1} + \frac{\partial}{\partial x_3}A_3\sqrt{E_1 E_2}\right).$$

In particular, in cartesian, cylindrical, and spherical coordinates on $\mathbb{R}^3$:

$$\operatorname{div} \mathbf{A} = \frac{\partial A_x}{\partial x} + \frac{\partial A_y}{\partial y} + \frac{\partial A_z}{\partial z} = \frac{1}{r}\left(\frac{\partial r A_r}{\partial r} + \frac{\partial A_\varphi}{\partial \varphi}\right) + \frac{\partial A_z}{\partial z}$$

$$= \frac{1}{R^2 \cos\theta}\left(\frac{\partial R^2 \cos\theta A_R}{\partial R} + \frac{\partial R A_\varphi}{\partial \varphi} + \frac{\partial R \cos\theta A_\theta}{\partial \theta}\right).$$

PROBLEM 10. The Laplace operator on M is the operator $\Lambda = \operatorname{div} \mathbf{grad}$. Find its expression in the coordinates x_i.

ANSWER.

$$\Delta f = \frac{1}{\sqrt{E_1 E_2 E_3}}\left[\frac{\partial}{\partial x_1}\left(\sqrt{\frac{E_2 E_3}{E_1}}\frac{\partial f}{\partial x_1}\right) + \cdots\right].$$

In particular, on $\mathbb{R}^3$

$$\Delta f = \frac{\partial^2 f}{\partial x^2} + \frac{\partial^2 f}{\partial y^2} + \frac{\partial^2 f}{\partial z^2} = \frac{\partial^2 f}{\partial r^2} + \frac{1}{r}\frac{\partial f}{\partial r} + \frac{1}{r^2}\frac{\partial^2 f}{\partial \varphi^2} + \frac{\partial^2 f}{\partial z^2}$$

$$= \frac{1}{R^2 \cos\theta}\left[\frac{\partial}{\partial R}\left(R^2 \cos\theta \frac{\partial f}{\partial R}\right) + \frac{\partial}{\partial \varphi}\left(\frac{1}{\cos\theta}\frac{\partial f}{\partial \varphi}\right) + \frac{\partial}{\partial \theta}\left(\cos\theta \frac{\partial f}{\partial \theta}\right)\right].$$

G Appendix 2: Closed forms and cycles

The flux of an incompressible fluid (without sources) across the boundary of a region D is equal to zero. We will formulate a higher-dimensional analogue to this obvious assertion. The higher-dimensional analogue of an incompressible fluid is called a *closed form*. The field $\mathbf{A}$ has no sources if $\operatorname{div} \mathbf{A} = 0$.

Definition. A differential form ω on a manifold M is *closed* if its exterior derivative is zero: $d\omega = 0$.

In particular, the 2-form ω_Λ^2 corresponding to a field $\mathbf{A}$ without sources is closed. Also, we have, by Stokes' formula (3):

Theorem. *The integral of a closed form ω^k over the boundary of any $(k + 1)$-dimensional chain c_{k+1} is equal to zero:*

$$\int_{\partial c_{k+1}} \omega^k = 0 \quad \text{if } d\omega^k = 0.$$

PROBLEM 11. Show that the differential of a form is always closed.

On the other hand, there are closed forms which are not differentials. For example, take for M the three-dimensional euclidean space $\mathbb{R}^3$ without O: $M = \mathbb{R}^3 - O$, with the 2-form being the flux of the field $\mathbf{A} = (1/R^2)\mathbf{e}_R$ (Figure 163). It is easy to convince oneself that $\operatorname{div} \mathbf{A} = 0$, so that our 2-form

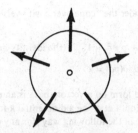

Figure 163   The field **A**

ω_A^2 is closed. At the same time, the flux over any sphere with center O is equal to 4π. We will show that the integral of the differential of a form over the sphere must be zero.

Definition. A *cycle* on a manifold M is a chain whose boundary is equal to zero.

The oriented surface of our sphere can be considered to be a cycle. It immediately follows from Stokes' formula (3) that

Theorem. *The integral of a differential over any cycle is equal to zero:*

$$\int_{c_{k+1}} d\omega^k = 0 \quad if \ \partial c_{k+1} = 0.$$

Thus, our 2-form ω_A^2 is not the differential of any 1-form.

The existence of closed forms on M which are not differentials is related to the topological properties of M. One can show that every closed k-form on a vector space is the differential of some $(k-1)$-form (Poincaré's lemma).

PROBLEM 12. Prove Poincaré's lemma for 1-forms.

Hint. Consider $\int_{x_0}^{x_1} \omega^1 = \varphi(x_1)$.

PROBLEM 13. Show that in a vector space the integral of a closed form over any cycle is zero.

Hint. Construct a $(k + 1)$-chain whose boundary is the given cycle (Figure 164).

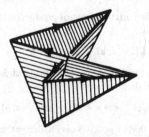

Figure 164   Cone over a cycle

197

Namely, for any chain c consider the "cone over c with vertex 0." If we denote the operation of constructing a cone by p, then

$$\partial \circ p + p \circ \partial = 1 \qquad \text{(the identity map)}.$$

Therefore, if the chain c is closed, $\partial(pc) = c$.

PROBLEM. Show that every closed form on a vector space is an exterior derivative.

Hint. Use the cone construction. Let ω^k be a differential k-form on $\mathbb{R}^n$. We define a $(k-1)$-form (the "co-cone over ω") $p\omega^k$ in the following way: for any chain c_{k-1}

$$\int_{c_{k-1}} p\omega^k = \int_{pc_{k-1}} \omega^k.$$

It is easy to see that the $(k-1)$-form $p\omega^k$ exists and is unique; its value on the vectors $\xi_1, \ldots, \xi_{k-1}$, tangent to $\mathbb{R}^n$ at x, is equal to

$$(p\omega)_x(\xi_1, \ldots, \xi_{k-1}) = \int_0^1 \omega_{tx}(x, t\xi_1, \ldots, t\xi_{k-1})dt.$$

It is easy to see that

$$d \circ p + p \circ d = 1 \qquad \text{(the identity map)}.$$

Therefore, if the form ω^k is closed, $d(p\omega^k) = \omega^k$.

PROBLEM. Let X be a vector field on M and ω a differential k-form. We define a differential $(k-1)$-form $i_X\omega$ (the *interior derivative* of ω by X) by the relation

$$(i_X\omega)(\xi_1, \ldots, \xi_{k-1}) = \omega(X, \xi_1, \ldots, \xi_{k-1}).$$

Prove the *homotopy formula*

$$i_X d + d i_X = L_X,$$

where L_X is the differentiation operator in the direction of the field X.

[The action of L_X on a form is defined, using the phase flow $\{g^t\}$ of the field X, by the relation

$$(L_X\omega)(\xi) = \left.\frac{d}{dt}\right|_{t=0} \omega(g^t_*\xi).$$

L_X is called the *Lie derivative* or *fisherman's derivative*: the flow carries all possible differential-geometric objects past the fisherman, and the fisherman sits there and differentiates them.]

Hint. We denote by H the "homotopy operator" associating to a k-chain $\gamma: \sigma \to M$ the $(k+1)$-chain $H\gamma: (I \times \sigma) \to M$ according to the formula $(H\gamma)(t, x) = g^t\gamma(x)$ (where $I = [0,1]$). Then

$$g^1\gamma - \gamma = \partial(H\gamma) + H(\partial\gamma).$$

PROBLEM. Prove the formula for differentiating a vector product on three-dimensional euclidean space (or on a riemannian manifold):

$$\mathbf{curl}[\mathbf{a}, \mathbf{b}] = \{\mathbf{a}, \mathbf{b}\} + \mathbf{a} \operatorname{div} \mathbf{b} - \mathbf{b} \operatorname{div} \mathbf{a}$$

(where $\{\mathbf{a}, \mathbf{b}\} = L_\mathbf{a}\mathbf{b}$ is the Poisson bracket of the vector fields, cf. Section 39).

Hint. If τ is the volume element, then

$$i_{\mathbf{curl}[\mathbf{a},\mathbf{b}]}\tau = di_{\mathbf{a}}i_{\mathbf{b}}\tau \qquad \operatorname{div} \mathbf{a} = di_{\mathbf{a}}\tau \quad \text{and} \quad \{\mathbf{a}, \mathbf{b}\} = L_\mathbf{a}\mathbf{b};$$

by using these relations and the fact that $d\tau = 0$, it is easy to derive the formula for $\mathbf{curl}[\mathbf{a}, \mathbf{b}]$ from the homotopy formula.

H Appendix 3: Cohomology and homology

The set of all k-forms on M is a vector space, the closed k-forms a subspace and the differentials of $(k - 1)$-forms a subspace of the subspace of closed forms. The quotient space

$$\frac{\{\text{closed forms}\}}{\{\text{differentials}\}} = H^k(M, \mathbb{R})$$

is called the k-th cohomology group of the manifold M. An element of this group is a class of closed forms differing from one another only by a differential.

PROBLEM 14. Show that for the circle S^1 we have $H^1(S^1, \mathbb{R}) = \mathbb{R}$.

The dimension of the space $H^k(M, \mathbb{R})$ is called the k-th Betti number of M.

PROBLEM 15. Find the first Betti number of the torus $T^2 = S^1 \times S^1$.

The flux of an incompressible fluid (without sources) over the surfaces of two concentric spheres is the same. In general, when integrating a closed form

Figure 165   Homologous cycles

over a k-dimensional cycle, we can replace the cycle with another one provided that their difference is the boundary of a $(k + 1)$-chain (Figure 165):

$$\int_a \omega^k = \int_b \omega^k,$$

if $a - b = \partial c_{k+1}$ and $d\omega^k = 0$.

Poincaré called two such cycles a and b homologous.

With a suitable definition[59] of the group of chains on a manifold M and its

[59] For this our group $\{c_k\}$ must be made smaller by identifying pieces which differ only by the choice of parametrization f or the choice of polyhedron D. In particular, we may assume that D is always one and the same simplex or cube. Furthermore, we must take every degenerate k-cell (D, f, Or) to be zero, i.e., $(D, f, \text{Or}) = 0$ if $f = f_2 \cdot f_1$, where $f_1 : D \to D'$ and D' has dimension smaller than k.

subgroups of cycles and boundaries (i.e., cycles homologous to zero), the quotient group

$$\frac{\{\text{cycles}\}}{\{\text{boundaries}\}} = H_k(M)$$

is called the *k-th homology group* of M.

An element of this group is a class of cycles homologous to one another.

The rank of this group is also equal to the *k-th Betti number* of M ("De Rham's Theorem").

Symplectic manifolds

8

A symplectic structure on a manifold is a closed nondegenerate differential 2-form. The phase space of a mechanical system has a natural symplectic structure.

On a symplectic manifold, as on a riemannian manifold, there is a natural isomorphism between vector fields and 1-forms. A vector field on a symplectic manifold corresponding to the differential of a function is called a hamiltonian vector field. A vector field on a manifold determines a phase flow, i.e., a one-parameter group of diffeomorphisms. The phase flow of a hamiltonian vector field on a symplectic manifold preserves the symplectic structure of phase space.

The vector fields on a manifold form a Lie algebra. The hamiltonian vector fields on a symplectic manifold also form a Lie algebra. The operation in this algebra is called the Poisson bracket.

37 Symplectic structures on manifolds

We define here symplectic manifolds, hamiltonian vector fields, and the standard symplectic structure on the cotangent bundle.

A Definition

Let M^{2n} be an even-dimensional differentiable manifold. A *symplectic structure* on M^{2n} is a closed nondegenerate differential 2-form ω^2 on M^{2n}:

$$d\omega^2 = 0 \quad \text{and} \quad \forall \xi \neq 0 \; \exists \eta : \omega^2(\xi, \eta) \neq 0 \qquad (\xi, \eta \in TM_x).$$

The pair (M^{2n}, ω^2) is called a *symplectic manifold*.

EXAMPLE. Consider the vector space $\mathbb{R}^{2n}$ with coordinates p_i, q_i and let $\omega^2 = \sum dp_i \wedge dq_i$.

PROBLEM. Verify that $(\mathbb{R}^{2n}, \omega^2)$ is a symplectic manifold. For $n = 1$ the pair $(\mathbb{R}^2, \omega^2)$ is the pair (the plane, area).

The following example explains the appearance of symplectic manifolds in dynamics. Along with the tangent bundle of a differentiable manifold, it is often useful to look at its dual—the cotangent bundle.

B *The cotangent bundle and its symplectic structure*

Let V be an n-dimensional differentiable manifold. A 1-form on the tangent space to V at a point $\mathbf{x}$ is called a *cotangent vector to V at* $\mathbf{x}$. The set of all cotangent vectors to V at $\mathbf{x}$ forms an n-dimensional vector space, dual to the tangent space $TV_{\mathbf{x}}$. We will denote this vector space of cotangent vectors by $T^*V_{\mathbf{x}}$ and call it the *cotangent space* to V at $\mathbf{x}$.

The union of the cotangent spaces to the manifold at all of its points is called the *cotangent bundle* of V and is denoted by T^*V. The set T^*V has a natural structure of a differentiable manifold of dimension $2n$. A point of T^*V is a 1-form on the tangent space to V at some point of V. If $\mathbf{q}$ is a choice of n local coordinates for points in V, then such a form is given by its n components $\mathbf{p}$. Together, the $2n$ numbers $\mathbf{p}, \mathbf{q}$ form a collection of local coordinates for points in T^*V.

There is a natural projection $f: T^*V \to V$ (sending every 1-form on $TV_{\mathbf{x}}$ to the point $\mathbf{x}$). The projection f is differentiable and surjective. The pre-image of a point $\mathbf{x} \in V$ under f is the cotangent space $T^*V_{\mathbf{x}}$.

Theorem. *The cotangent bundle T^*V has a natural symplectic structure. In the local coordinates described above, this symplectic structure is given by the formula*

$$\omega^2 = d\mathbf{p} \wedge d\mathbf{q} = dp_1 \wedge dq_1 + \cdots + dp_n \wedge dq_n.$$

PROOF. First, we define a distinguished 1-form on T^*V. Let $\xi \in T(T^*V)_p$ be a vector *tangent to the cotangent bundle at the point* $p \in T^*V_{\mathbf{x}}$ (Figure 166). The derivative $f_*: T(T^*V) \to TV$ of the natural projection $f: T^*V \to V$ takes ξ to a vector $f_*\xi$ tangent to V at $\mathbf{x}$. We define a 1-form ω^1 on T^*V by the relation $\omega^1(\xi) = p(f_*\xi)$. In the local coordinates described above, this form is $\omega^1 = \mathbf{p} \, d\mathbf{q}$. By the example in A, the closed 2-form $\omega^2 = d\omega^1$ is non-degenerate. $\qquad\square$

Remark. Consider a lagrangian mechanical system with configuration manifold V and lagrangian function L. It is easy to see that the lagrangian "generalized velocity" $\dot{\mathbf{q}}$ is a tangent vector to the configuration manifold V, and the "generalized momentum" $\mathbf{p} = \partial L/\partial \dot{\mathbf{q}}$ is a cotangent vector. Therefore, the "$\mathbf{p}, \mathbf{q}$" phase space of the lagrangian system is the cotangent bundle of the configuration manifold. The theorem above shows that the phase space of a mechanical problem has a natural symplectic manifold structure.

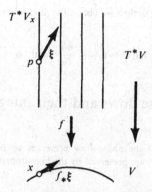

Figure 166 The 1-form **p** d**q** on the cotangent bundle

PROBLEM. Show that the Legendre transform does not depend on the coordinate system: it takes a function $L: TV \to \mathbb{R}$ on the tangent bundle to a function $H: T^*V \to \mathbb{R}$ on the cotangent bundle.

C Hamiltonian vector fields

A riemannian structure on a manifold establishes an isomorphism between the spaces of tangent vectors and 1-forms. A symplectic structure establishes a similar isomorphism.

Definition. To each vector ξ, tangent to a symplectic manifold (M^{2n}, ω^2) at the point **x**, we associate a 1-form ω^1_ξ on TM_x by the formula

$$\omega^1_\xi(\eta) = \omega^2(\eta, \xi) \qquad \forall \eta \in TM_x.$$

PROBLEM. Show that the correspondence $\xi \to \omega^1_\xi$ is an isomorphism between the 2n-dimensional vector spaces of vectors and of 1-forms.

EXAMPLE. In $\mathbb{R}^{2n} = \{(\mathbf{p}, \mathbf{q})\}$ we will identify vectors and 1-forms by using the euclidean structure $(\mathbf{x}, \mathbf{x}) = \mathbf{p}^2 + \mathbf{q}^2$. Then the correspondence $\xi \to \omega^1_\xi$ determines a transformation $\mathbb{R}^{2n} \to \mathbb{R}^{2n}$.

PROBLEM. Calculate the matrix of this transformation in the basis **p, q**.

ANSWER. $\begin{pmatrix} 0 & E \\ -E & 0 \end{pmatrix}$

We will denote by I the isomorphism $I: T^*M_x \to TM_x$ constructed above. Now let H be a function on a symplectic manifold M^{2n}. Then dH is a differential 1-form on M, and at every point there is a tangent vector to M associated to it. In this way we obtain a vector field $I\,dH$ on M.

Definition. The vector field $I\,dH$ is called a *hamiltonian vector field*; H is called the *hamiltonian function*.

EXAMPLE. If $M^{2n} = \mathbb{R}^{2n} = \{(\mathbf{p}, \mathbf{q})\}$, then we obtain the phase velocity vector field of Hamilton's canonical equations:

$$\dot{\mathbf{x}} = I\, dH(\mathbf{x}) \Leftrightarrow \dot{\mathbf{p}} = -\frac{\partial H}{\partial \mathbf{q}} \quad \text{and} \quad \dot{\mathbf{q}} = \frac{\partial H}{\partial \mathbf{p}}.$$

38 Hamiltonian phase flows and their integral invariants

Liouville's theorem asserts that the phase flow preserves volume. Poincaré found a whole series of differential forms which are preserved by the hamiltonian phase flow.

A Hamiltonian phase flows preserve the symplectic structure

Let (M^{2n}, ω^2) be a symplectic manifold and $H: M^{2n} \to \mathbb{R}$ a function. Assume that the vector field $I\, dH$ corresponding to H gives a 1-parameter group of diffeomorphisms $g^t: M^{2n} \to M^{2n}$:

$$\left.\frac{d}{dt}\right|_{t=0} g^t\mathbf{x} = I\, dH(\mathbf{x}).$$

The group g^t is called the *hamiltonian phase flow with hamiltonian function H*.

Theorem. *A hamiltonian phase flow preserves the symplectic structure*:

$$(g^t)^*\omega^2 = \omega^2.$$

In the case $n = 1$, $M^{2n} = \mathbb{R}^2$, this theorem says that the phase flow g^t preserves area (Liouville's theorem).

For the proof of this theorem, it is useful to introduce the following notation (Figure 167).

Let M be an arbitrary manifold, c a k-chain on M and $g^t: M \to M$ a one-parameter family of differentiable mappings. We will construct a $(k + 1)$-chain Jc on M, which we will call the *track of the chain c under the homotopy* g^t, $0 \le t \le \tau$.

Let (D, f, Or) be one of the cells in the chain c. To this cell will be associated a cell (D', f', Or') in the chain Jc, where $D' = I \times D$ is the direct product of the interval $0 \le t \le \tau$ and D; the mapping $f': D' \to M$ is obtained from $f: D \to M$ by the formula $f'(t, x) = g^t f(x)$; and the orientation Or' of the

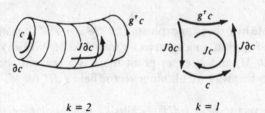

$$k = 2 \qquad\qquad k = 1$$

Figure 167 Track of a cycle under homotopy

space $\mathbb{R}^{k+1}$ containing D' is given by the frame $e_0, e_1, \ldots, e_k$, where e_0 is the unit vector of the t axis, and $e_1, \ldots, e_k$ is an oriented frame for D.

We could say that Jc is the chain swept out by c under the homotopy g^t, $0 \le t \le \tau$. The boundary of the chain Jc consists of "end-walls" made up of the initial and final positions of c, and "side surfaces" filled in by the boundary of c.

It is easy to verify that under the choice of orientation made above,

(1) $$\partial(Jc_k) = g^\tau c_k - c_k - J \, \partial c_k.$$

Lemma. *Let γ be a 1-chain in the symplectic manifold (M^{2n}, ω^2). Let g^t be a phase flow on M with hamiltonian function H. Then*

$$\frac{d}{d\tau} \int_{J\gamma} \omega^2 = \int_{g^\tau \gamma} dH.$$

PROOF. It is sufficient to consider a chain γ with one cell $f : [0, 1] \to M$. We introduce the notation

$$f'(s, t) = g^t f(s), \qquad \xi = \frac{\partial f'}{\partial s} \quad \text{and} \quad \eta = \frac{\partial f'}{\partial t} \in TM_{f'(s, t)}.$$

By the definition of the integral

$$\int_{J\gamma} \omega^2 = \int_0^1 \int_0^\tau \omega^2(\xi, \eta) dt \, ds.$$

But by the definition of the phase flow, η is a vector (at the point $f'(s, t)$) of the hamiltonian field with hamiltonian function H. By definition of a hamiltonian field, $\omega^2(\xi, \eta) = dH(\xi)$. Thus

$$\int_{J\gamma} \omega^2 = \int_0^\tau \left(\int_{g^t \gamma} dH \right) dt. \qquad \square$$

Corollary. *If the chain γ is closed ($\partial \gamma = 0$), then $\int_{J\gamma} \omega^2 = 0$.*

PROOF. $\int_\gamma dH = \int_{\partial \gamma} H = 0$. $\qquad \square$

PROOF OF THE THEOREM. We consider any 2-chain c. We have

$$0 \overset{1}{=} \int_{Jc} d\omega^2 \overset{2}{=} \int_{\partial Jc} \omega^2 \overset{3}{=} \left(\int_{g^\tau c} - \int_c - \int_{J \partial c} \right) \omega^2 \overset{4}{=} \int_{g^\tau c} \omega^2 - \int_c \omega^2$$

(1 since ω^2 is closed, 2 by Stokes' formula, 3 by formula (1), 4 by the corollary above with $\gamma = \partial c$). Thus the integrals of the form ω^2 on any chain c and on its image $g^\tau c$ are the same. $\qquad \square$

PROBLEM. Is every one-parameter group of diffeomorphisms of M^{2n} which preserves the symplectic structure a hamiltonian phase flow?

Hint. Cf. Section 40.

B *Integral invariants*

Let $g: M \rightarrow M$ be a differentiable map.

Definition. A differential k-form ω is called an *integral invariant* of the map g if the integrals of ω on any k-chain c and on its image under g are the same:

$$\int_{gc} \omega = \int_c \omega.$$

EXAMPLE. If $M = \mathbb{R}^2$ and $\omega^2 = dp \wedge dq$ is the area element, then ω^2 is an integral invariant of any map g with jacobian 1.

PROBLEM. Show that a form ω^k is an integral invariant of a map g if and only if $g^*\omega^k = \omega^k$.

PROBLEM. Show that if the forms ω^k and ω^l are integral invariants of the map g, then the form $\omega^k \wedge \omega^l$ is also an integral invariant of g.

The theorem in subsection A can be formulated as follows:

Theorem. *The form ω^2 giving the symplectic structure is an integral invariant of a hamiltonian phase flow.*

We now consider the exterior powers of ω^2,

$$(\omega^2)^2 = \omega^2 \wedge \omega^2 \qquad (\omega^2)^3 = \omega^2 \wedge \omega^2 \wedge \omega^2, \ldots.$$

Corollary. *Each of the forms $(\omega^2)^2, (\omega^2)^3, (\omega^2)^4, \ldots$ is an integral invariant of a hamiltonian phase flow.*

PROBLEM. Suppose that the dimension of the symplectic manifold (M^{2n}, ω^2) is $2n$. Show that $(\omega^2)^k = 0$ for $k > n$, and that $(\omega^2)^n$ is a nondegenerate $2n$-form on M^{2n}.

We define a volume element on M^{2n} using $(\omega^2)^n$. Then, a hamiltonian phase flow preserves volume, and we obtain Liouville's theorem from the corollary above.

EXAMPLE. Consider the symplectic coordinate space $M^{2n} = \mathbb{R}^{2n} = \{(\mathbf{p}, \mathbf{q})\}$, $\omega^2 = d\mathbf{p} \wedge d\mathbf{q} = \sum dp_i \wedge dq_i$. In this case the form $(\omega^2)^k$ is proportional to the form

$$\omega^{2k} = \sum_{i_1 < \cdots < i_k} dp_{i_1} \wedge \cdots \wedge dp_{i_k} \wedge dq_{i_1} \wedge \cdots \wedge dq_{i_k}.$$

The integral of ω^{2k} is equal to the sum of the oriented volumes of projections onto the coordinate planes $(p_{i_1}, \ldots, p_{i_k}, q_{i_1}, \ldots, q_{i_k})$.

A map $g: \mathbb{R}^{2n} \rightarrow \mathbb{R}^{2n}$ is called *canonical* if it has ω^2 as an integral invariant. A canonical map is generally called a *canonical transformation*. Each of the forms $\omega^4, \omega^6, \ldots, \omega^{2n}$ is an integral invariant of every canonical transformation. Therefore, *under a canonical transformation, the sum of the oriented areas of projections onto the coordinate planes* $(p_{i_1}, \ldots, p_{i_k}, q_{i_1}, \ldots, q_{i_k})$, $1 \le k \le n$, *is preserved.* In particular, *canonical transformations preserve volume.*

The hamiltonian phase flow given by the equations $\dot{\mathbf{p}} = -\partial H/\partial \mathbf{q}$, $\dot{\mathbf{q}} = \partial H/\partial \mathbf{p}$ consists of canonical transformations g^t.

The integral invariants considered above are also called *absolute integral invariants.*

Definition. A differential k-form ω is called a *relative integral invariant* of the map $g: M \to M$ if $\int_{gc} \omega = \int_c \omega$ for every *closed* k-chain c.

Theorem. *Let ω be a relative integral invariant of a map g. Then $d\omega$ is an absolute integral invariant of g.*

PROOF. Let c be a $k + 1$-chain. Then

$$\int_c d\omega \overset{1}{=} \int_{\partial c} \omega \overset{2}{=} \int_{g\partial c} \omega \overset{3}{=} \int_{\partial g c} \omega \overset{4}{=} \int_{gc} d\omega.$$

(1 and 4 are by Stokes' formula, 2 by the definition of relative invariant, and 3 by the definition of boundary). $\qquad\square$

EXAMPLE. A canonical map $g: \mathbb{R}^{2n} \to \mathbb{R}^{2n}$ has the 1-form

$$\omega^1 = \mathbf{p}\, d\mathbf{q} = \sum_{i=1}^n p_i\, dq_i \text{ as a relative integral invariant.}$$

In fact, every closed chain c on $\mathbb{R}^{2n}$ is the boundary of some chain σ, and we find

$$\int_{gc} \omega^1 \overset{1}{=} \int_{g\partial\sigma} \omega^1 \overset{2}{=} \int_{\partial g\sigma} \omega^1 \overset{3}{=} \int_{g\sigma} d\omega^1 \overset{4}{=} \int_\sigma d\omega^1 \overset{5}{=} \int_{\partial\sigma} \omega^1 \overset{6}{=} \int_c \omega^1;$$

(1 and 6 are by definition of σ, 2 by definition of ∂, 3 and 5 by Stokes' formula, and 4 since g is canonical and $d\omega^1 = d(p\, dq) = dq \wedge dq = \omega^2$).

PROBLEM. Let $d\omega^k$ be an absolute integral invariant of the map $g: M \to M$. Does it follow that ω^k is a relative integral invariant?

ANSWER. No, if there is a closed k-chain on M which is not a boundary.

C *The law of conservation of energy*

Theorem. *The function H is a first integral of the hamiltonian phase flow with hamiltonian function H.*

PROOF. The derivative of H in the direction of a vector $\boldsymbol{\eta}$ is equal to the value of dH on $\boldsymbol{\eta}$. By definition of the hamiltonian field $\boldsymbol{\eta} = I\, dH$ we find

$$dH(\boldsymbol{\eta}) = \omega^2(\boldsymbol{\eta}, I\, dH) = \omega^2(\boldsymbol{\eta}, \boldsymbol{\eta}) = 0. \qquad\square$$

PROBLEM. Show that the 1-form dH is an integral invariant of the phase flow with hamiltonian function H.

39 The Lie algebra of vector fields

Every pair of vector fields on a manifold determines a new vector field, called their Poisson bracket.[60] The Poisson bracket operation makes the vector space of infinitely differentiable vector fields on a manifold into a Lie algebra.

A Lie algebras

One example of a Lie algebra is a three-dimensional oriented euclidean vector space equipped with the operation of vector multiplication. The vector product is bilinear, skew-symmetric, and satisfies the *Jacobi identity*

$$[[A, B], C] + [[B, C], A] + [[C, A], B] = 0.$$

Definition. A *Lie algebra* is a vector space L, together with a bilinear skew-symmetric operation $L \times L \to L$ which satisfies the Jacobi identity.

The operation is usually denoted by square brackets and called the *commutator*.

PROBLEM. Show that the set of $n \times n$ matrices becomes a Lie algebra if we define the commutator by $[A, B] = AB - BA$.

B Vector fields and differential operators

Let M be a smooth manifold and $\mathbf{A}$ a smooth vector field on M: at every point $\mathbf{x} \in M$ we are given a tangent vector $\mathbf{A}(\mathbf{x}) \in TM_{\mathbf{x}}$. With every such vector field we associate the following two objects:

1. *The one-parameter group of diffeomorphisms* or *flow* $A^t \colon M \to M$ for which $\mathbf{A}$ is the velocity vector field (Figure 168):[61]

$$\left. \frac{d}{dt} \right|_{t=0} A^t \mathbf{x} = \mathbf{A}(\mathbf{x}).$$

2. The first-order differential operator $L_{\mathbf{A}}$. We refer here to the differentiation of functions in the direction of the field $\mathbf{A}$: for any function $\varphi \colon M \to \mathbb{R}$ the *derivative in the direction of* A is a new function $L_{\mathbf{A}}\varphi$, whose value at a point $\mathbf{x}$ is

$$(L_{\mathbf{A}}\varphi)(\mathbf{x}) = \left. \frac{d}{dt} \right|_{t=0} \varphi(A^t \mathbf{x}).$$

---

[60] Or Lie bracket [Trans. note].

[61] By theorems of existence, uniqueness, and differentiability in the theory of ordinary differential equations, the group A^t is defined if the manifold M is compact. In the general case the maps A^t are defined only in a neighborhood of $\mathbf{x}$ and only for small t; this is enough for the following constructions.

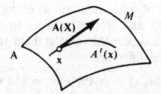

Figure 168　The group of diffeomorphisms given by a vector field

PROBLEM. Show that the operator L_A is linear:

$$L_A(\lambda_1\varphi_1 + \lambda_2\varphi_2) = \lambda_1 L_A\varphi_1 + \lambda_2 L_A\varphi_2 \qquad (\lambda_1, \lambda_2 \in \mathbb{R}).$$

Also, prove Leibniz's formula $L_A(\varphi_1\varphi_2) = \varphi_1 L_A\varphi_2 + \varphi_2 L_A\varphi_1$.

EXAMPLE. Let $(x_1, \ldots, x_n)$ be local coordinates on M. In this coordinate system the vector $A(x)$ is given by its components $(A_1(x), \ldots, A_n(x))$; the flow A^t is given by the system of differential equations

$$\dot{x}_1 = A_1(x), \ldots, \dot{x}_n = A_n(x)$$

and, therefore, the derivative of $\varphi = \varphi(x_1, \ldots, x_n)$ in the direction A is

$$L_A\varphi = A_1 \frac{\partial\varphi}{\partial x_1} + \cdots + A_n \frac{\partial\varphi}{\partial x_n}.$$

We could say that in the coordinates $(x_1, \ldots, x_n)$ the operator L_A has the form

$$L_A = A_1 \frac{\partial}{\partial x_1} + \cdots + A_n \frac{\partial}{\partial x_n}.$$

this is the general form of a first-order linear differential operator on coordinate space.

PROBLEM. Show that the correspondences between vector fields A, flows A^t, and differentiations L_A are one-to-one.

C The Poisson bracket of vector fields

Suppose that we are given two vector fields A and B on a manifold M. The corresponding flows A^t and B^s do not, in general, commute: $A^t B^s \neq B^s A^t$ (Figure 169).

PROBLEM. Find an example.
　　Solution. The fields $A = e_1$, $B = x_1 e_2$ on the (x_1, x_2) plane.

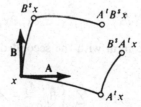

Figure 169　Non-commutative flows

To measure the degree of noncommutativity of the two flows A^t and B^s we consider the points $A^t B^s x$ and $B^s A^t x$. In order to estimate the difference between these points, we compare the value at them of some smooth function φ on the manifold M. The difference

$$\Delta(t; s; x) = \varphi(A^t B^s x) - \varphi(B^s A^t x)$$

is clearly a differentiable function which is zero for $s = 0$ and for $t = 0$. Therefore, the first term different from 0 in the Taylor series in s and t of Δ at 0 contains st, and the other terms of second order vanish. We will calculate this principal bilinear term of Δ at 0.

Lemma 1. *The mixed partial derivative $\partial^2 \Delta / \partial s \, \partial t$ at 0 is equal to the commutator of differentiation in the directions* $\mathbf{A}$ *and* $\mathbf{B}$:

$$\left. \frac{\partial^2}{\partial s \, \partial t} \right|_{s=t=0} \{ \varphi(A^t B^s x) - \varphi(B^s A^t x) \} = (L_{\mathbf{B}} L_{\mathbf{A}} \varphi - L_{\mathbf{A}} L_{\mathbf{B}} \varphi)(x).$$

PROOF. By the definition of $L_{\mathbf{A}}$,

$$\left. \frac{\partial}{\partial t} \right|_{t=0} \varphi(A^t B^s x) = (L_{\mathbf{A}} \varphi)(B^s x).$$

If we denote the function $L_{\mathbf{A}} \varphi$ by ψ, then by the definition of $L_{\mathbf{B}}$

$$\left. \frac{\partial}{\partial s} \right|_{s=0} \psi(B^s x) = (L_{\mathbf{B}} \psi) x.$$

Thus,

$$\left. \frac{\partial^2}{\partial s \partial t} \right|_{s=t=0} \varphi(A^t B^s x) = (L_{\mathbf{B}} L_{\mathbf{A}} \varphi) x. \qquad \square$$

We now consider the commutator of differentiation operators $L_{\mathbf{B}} L_{\mathbf{A}} - L_{\mathbf{A}} L_{\mathbf{B}}$. At first glance this is a *second*-order differential operator.

Lemma 2. *The operator $L_{\mathbf{B}} L_{\mathbf{A}} - L_{\mathbf{A}} L_{\mathbf{B}}$ is a first-order linear differential operator.*

PROOF. Let $(A_1, \ldots, A_n)$ and $(B_1, \ldots, B_n)$ be the components of the fields $\mathbf{A}$ and $\mathbf{B}$ in the local coordinate system $(x_1, \ldots, x_n)$ on M. Then

$$L_{\mathbf{B}} L_{\mathbf{A}} \varphi = \sum_{i=1}^{n} B_i \frac{\partial}{\partial x_i} \sum_{j=1}^{n} A_j \frac{\partial}{\partial x_j} \varphi = \sum_{i,j=1}^{n} B_i \frac{\partial A_j}{\partial x_i} \frac{\partial}{\partial x_j} \varphi + \sum_{i,j=1}^{n} B_i A_j \frac{\partial^2 \varphi}{\partial x_i \partial x_j}.$$

If we subtract $L_{\mathbf{A}} L_{\mathbf{B}} \varphi$, the term with the second derivatives of φ vanishes, and we obtain

$$(L_{\mathbf{B}} L_{\mathbf{A}} - L_{\mathbf{A}} L_{\mathbf{B}}) \varphi = \sum_{i,j=1}^{n} \left(B_i \frac{\partial A_j}{\partial x_i} - A_i \frac{\partial B_j}{\partial x_i} \right) \frac{\partial \varphi}{\partial x_j}. \qquad \square$$

Since every first-order linear differential operator is given by a vector field, our operator $L_B L_A - L_A L_B$ also corresponds to some vector field **C**.

Definition. The *Poisson bracket* or *commutator* of two vector fields **A** and **B** on a manifold M^{62} is the vector field **C** for which

$$L_C = L_B L_A - L_A L_B.$$

The Poisson bracket of two vector fields is denoted by

$$\mathbf{C} = [\mathbf{A}, \mathbf{B}].$$

PROBLEM. Suppose that the vector fields **A** and **B** are given by their components A_i, B_i in coordinates x_i. Find the components of the Poisson bracket.

Solution. In the proof of Lemma 2 we proved the formula

$$[\mathbf{A}, \mathbf{B}]_i = \sum_{i=1}^{n} B_i \frac{\partial A_j}{\partial x_i} - A_i \frac{\partial B_j}{\partial x_i}.$$

PROBLEM. Let A_1 be the linear vector field of velocities of a rigid body rotating with angular velocity ω_1 around 0, and A_2 the same thing with angular velocity ω_2. Find the Poisson bracket $[A_1, A_2]$.

D *The Jacobi identity*

Theorem. *The Poisson bracket makes the vector space of vector fields on a manifold M into a Lie algebra.*

PROOF. Linearity and skew-symmetry of the Poisson bracket are clear. We will prove the Jacobi identity. By definition of Poisson bracket, we have

$$L_{[[A,B],C]} = L_C L_{[A,B]} - L_{[A,B]} L_C$$
$$= L_C L_B L_A - L_C L_A L_B + L_A L_B L_C - L_B L_A L_C.$$

There will be 12 terms in all in the sum $L_{[[A,B],C]} + L_{[[B,C],A]} + L_{[[C,A],B]}$. Each term appears in the sum twice, with opposite signs. $\square$

E *A condition for the commutativity of flows*

Let **A** and **B** be vector fields on a manifold M.

Theorem. *The two flows A^t and B^s commute if and only if the Poisson bracket of the corresponding vector fields [A, B] is equal to zero.*

PROOF. If $A^t B^s \equiv B^s A^t$, then $[\mathbf{A}, \mathbf{B}] = 0$ by Lemma 1. If $[\mathbf{A}, \mathbf{B}] = 0$, then, by Lemma 1,

$$\varphi(A^t B^s x) - \varphi(B^s A^t x) = o(s^2 + t^2), \qquad s \to 0 \text{ and } t \to 0$$

---

[62] In many books the bracket is given the opposite sign. Our sign agrees with the sign of the commutator in the theory of Lie groups (cf. subsection F).

for any function φ at any point x. We will show that this implies $\varphi(A^t B^s x) = \varphi(B^s A^t x)$ for sufficiently small s and t. If we apply this to the local coordinates $(\varphi = x_1, \ldots, \varphi = x_n)$, we obtain $A^t B^s = B^s A^t$.

Consider the rectangle $0 \leq t \leq t_0, 0 \leq s \leq s_0$ (Figure 170) in the t, s-plane. To every path going from $(0, 0)$ to (t_0, s_0) and consisting of a finite number of intervals in the coordinate directions, we associate a product of transformations of the flows A^t and B^s. Namely, to each interval $t_1 \leq t \leq t_2$ we associate $A^{t_2 - t_1}$, and to each interval $s_1 \leq s \leq s_2$ we associate $B^{s_2 - s_1}$; the transformations are applied in the order in which the intervals occur in the path, beginning at $(0, 0)$. For example, the sides $(0 \leq t \leq t_0, s = 0)$ and $(t = t_0, 0 \leq s \leq s_0)$ corresponds to the product $B^{s_0} A^{t_0}$, and the sides $(t = 0, 0 \leq s \leq s_0)$ and $(s = s_0, 0 \leq t \leq t_0)$ to the product $A^{t_0} B^{s_0}$.

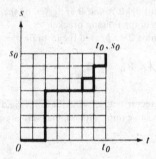

Figure 170   Proof of the commutativity of flows

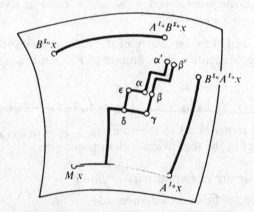

Figure 171   Curvilinear quadrilateral $\beta\gamma\delta\varepsilon\alpha$

In addition, we associate to each such path in the (t, s)-plane a path on the manifold M starting at the point x and composed of trajectories of the flows A^t and B^s (Figure 171). If a path in the (t, s)-plane corresponds to the product $A^{t_1} B^{s_1} \cdots A^{t_n} B^{s_n}$, then on the manifold M the corresponding path ends at the point $A^{t_1} B^{s_1} \cdots A^{t_n} B^{s_n} x$. Our goal will be to show that all these paths actually terminate at the one point $A^{t_0} B^{s_0} x = B^{s_0} A^{t_0} x$.

We partition the intervals $0 \leq t \leq t_0$ and $0 \leq s \leq s_0$ into N equal parts, so that the whole rectangle is divided into N^2 small rectangles. The passage from the sides $(0, 0) - (t_0, 0) - (t_0, s_0)$ to the sides $(0, 0) - (0, s_0) - (t_0, s_0)$ can be accomplished in N^2 steps, in each of which a pair of neighboring sides of a small rectangle is exchanged for the other pair (Figure 172). In general,

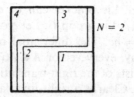

Figure 172   Going from one pair of sides to the other.

this small rectangle corresponds to a non-closed curvilinear quadrilateral $\beta\gamma\delta\varepsilon\alpha$ on the manifold M (Figure 171). Consider the distance[63] between its vertices α and β corresponding to the largest values of s and t. As we saw earlier, $\rho(\alpha, \beta) \leq C_1 N^{-3}$ (where the constant $C_1 > 0$ does not depend on N). Using the theorem of the differentiability of solutions of differential equations with respect to the initial data, it is not difficult to derive from this a bound on the distance between the ends α' and β' of the paths $x\delta\gamma\beta\beta'$ and $x\delta\varepsilon\alpha\alpha'$ on M: $\rho(\alpha', \beta') < C_2 N^{-3}$, where the constant $C_2 > 0$ again does not depend on N. But we broke up the whole journey from $B^{s_0}A^{t_0}x$ to $A^{t_0}B^{s_0}x$ into N^2 such pieces. Thus, $\rho(A^{t_0}B^{s_0}x, B^{s_0}A^{t_0}x) \leq N^2C_2N^{-3} \,\forall N$. Therefore, $A^{t_0}B^{s_0}x = B^{s_0}A^{t_0}x$. $\qquad\square$

F Appendix: Lie algebras and Lie groups

A *Lie group* is a group G which is a differentiable manifold, and for which the operations (product and inverse) are differentiable maps $G \times G \to G$ and $G \to G$.

The tangent space, TG_e, to a Lie group G at the identity has a natural Lie algebra structure; it is defined as follows:

For each tangent vector $A \in TG_e$ there is a one-parameter subgroup $A^t \subset G$ with velocity vector $A = (d/dt)|_{t=0}A^t$.

The degree of non-commutativity of two subgroups A^t and B^t is measured by the product $A^tB^sA^{-t}B^{-s}$. It turns out that there is one and only one subgroup C^r for which

$$\rho(A^tB^sA^{-t}B^{-s}, C^{st}) = o(s^2 + t^2) \quad \text{as } s \text{ and } t \to 0.$$

The corresponding vector $C = (d/dr)|_{r=0}C^r$ is called the *Lie bracket* $C = [A, B]$ of the vectors A and B. It can be verified that the operation of Lie bracket introduced in this way makes the space TG_e into a Lie algebra (i.e., the operation is bilinear, skew-symmetric, and satisfies the Jacobi identity). This algebra is called the *Lie algebra of the Lie group G*.

PROBLEM. Compute the bracket operation in the Lie algebra of the group SO(3) of rotations in three-dimensional euclidean space.

Lemma 1 shows that the Poisson bracket of vector fields can be defined as the Lie bracket for the "infinite-dimensional Lie group" of all diffeomorphisms[64] of the manifold M.

---

[63] In some riemannian metric on M.

[64] Our choice of sign in the definition of Poisson bracket was determined by this correspondence.

On the other hand, the Lie bracket can be defined using the Poisson bracket of vector fields on a Lie group G. Let $g \in G$. Right translation R_g is the map $R_g: G \to G$, $R_g h = hg$. The differential of R_g at the point e maps TG_e into TG_g. In this way, every vector $\mathbf{A} \in TG_e$ corresponds to a vector field on the group: it consists of the right translations $(R_g)_* \mathbf{A}$ and is called a *right-invariant vector field*. Clearly, a right-invariant vector field on a group is uniquely determined by its value at the identity.

PROBLEM. Show that the Poisson bracket of right-invariant vector fields on a Lie group G is a right-invariant vector field, and its value at the identity of the group is equal to the Lie bracket of the values of the original vector fields at the identity.

40 The Lie algebra of hamiltonian functions

The hamiltonian vector fields on a symplectic manifold form a subalgebra of the Lie algebra of all fields. The hamiltonian functions also form a Lie algebra: the operation in this algebra is called the Poisson bracket of functions. The first integrals of a hamiltonian phase flow form a subalgebra of the Lie algebra of hamiltonian functions.

A *The Poisson bracket of two functions*

Let (M^{2n}, ω^2) be a symplectic manifold. To a given function $H: M^{2n} \to \mathbb{R}$ on the symplectic manifold there corresponds a one-parameter group $g_H^t: M^{2n} \to M^{2n}$ of canonical transformations of M^{2n}—the phase flow of the hamiltonian function equal to H. Let $F: M^{2n} \to \mathbb{R}$ be another function on M^{2n}.

Definition. The *Poisson bracket* (F, H) of functions F and H given on a symplectic manifold (M^{2n}, ω^2) is the derivative of the function F in the direction of the phase flow with hamiltonian function H:

$$(F, H)(x) = \left. \frac{d}{dt} \right|_{t=0} F(g_H^t(x)).$$

Thus, the Poisson bracket of two functions on M is again a function on M.

Corollary 1. *A function F is a first integral of the phase flow with hamiltonian function H if and only if its Poisson bracket with H is identically zero*: $(F, H) \equiv 0$.

We can give the definition of Poisson bracket in a slightly different form if we use the isomorphism I between 1-forms and vector fields on a symplectic manifold (M^{2n}, ω^2). This isomorphism is defined by the relation (cf. Section 37)

$$\omega^2(\mathbf{\eta}, I\omega^1) = \omega^1(\mathbf{\eta}).$$

The velocity vector of the phase flow g_H^t is $I \, dH$. This implies

Corollary 2. *The Poisson bracket of the functions F and H is equal to the value of the 1-form dF on the velocity vector $I\,dH$ of the phase flow with hamiltonian function H:*

$$(F, H) = dF(I\,dH).$$

Using the preceding formula again, we obtain

Corollary 3. *The Poisson bracket of the functions F and H is equal to the "skew scalar product" of the velocity vectors of the phase flows with hamiltonian functions H and F:*

$$(F, H) = \omega^2(I\,dH, I\,dF).$$

It is now clear that

Corollary 4. *The Poisson bracket of the functions F and H is a skew-symmetric bilinear function of F and H:*

$$(F, H) = -(H, F)$$

and

$$(H, \lambda_1 F_1 + \lambda_2 F_2) = \lambda_1(H, F_1) + \lambda_2(H, F_2) \qquad (\lambda_i \in \mathbb{R}).$$

Although the arguments above are obvious, they lead to nontrivial deductions, including the following generalization of a theorem of E. Noether.

Theorem. *If a hamiltonian function H on a symplectic manifold (M^{2n}, ω^2) admits the one-parameter group of canonical transformations given by a hamiltonian F, then F is a first integral of the system with hamiltonian function H.*

PROOF. Since H is a first integral of the flow g_F^t, $(H, F) = 0$ (Corollary 1). Therefore, $(F, H) = 0$ (Corollary 4) and F is a first integral (Corollary 1). $\square$

PROBLEM 1. Compute the Poisson bracket of two functions F and H in the canonical coordinate space $\mathbb{R}^{2n} = \{(\mathbf{p}, \mathbf{q})\}$, $\omega^2(\boldsymbol{\xi}, \boldsymbol{\eta}) = (I\boldsymbol{\xi}, \boldsymbol{\eta})$.

Solution. By Corollary 3 we have

$$(F, H) = \sum_{i=1}^{n} \frac{\partial H}{\partial p_i}\frac{\partial F}{\partial q_i} - \frac{\partial H}{\partial q_i}\frac{\partial F}{\partial p_i}$$

(we use the fact that I is symplectic and has the form

$$I = \begin{pmatrix} 0 & -E \\ E & 0 \end{pmatrix}$$

in the basis $(\mathbf{p}, \mathbf{q})$).

PROBLEM 2. Compute the Poisson brackets of the basic functions p_i and q_j.

Solution. The gradients of the basic functions form a "symplectic basis": their skew-scalar products are

$$(p_i, p_j) = (p_i, q_j) = (q_i, q_j) = 0 \quad (\text{if } i \neq j) \qquad (q_i, p_i) = -(p_i, q_i) = 1.$$

PROBLEM 3. Show that the map $A: \mathbb{R}^{2n} \to \mathbb{R}^{2n}$ sending $(\mathbf{p}, \mathbf{q}) \to (\mathbf{P}(\mathbf{p}, \mathbf{q}), \mathbf{Q}(\mathbf{p}, \mathbf{q}))$ is canonical if and only if the Poisson brackets of any two functions in the variables $(\mathbf{p}, \mathbf{q})$ and $(\mathbf{P}, \mathbf{Q})$ coincide:

$$(F, H)_{\mathbf{p}, \mathbf{q}} = \frac{\partial H}{\partial \mathbf{p}} \frac{\partial F}{\partial \mathbf{q}} - \frac{\partial H}{\partial \mathbf{q}} \frac{\partial F}{\partial \mathbf{p}} = \frac{\partial H}{\partial \mathbf{P}} \frac{\partial F}{\partial \mathbf{Q}} - \frac{\partial H}{\partial \mathbf{Q}} \frac{\partial F}{\partial \mathbf{P}} = (F, H)_{\mathbf{P}, \mathbf{Q}}.$$

Solution. Let A be canonical. Then the symplectic structures $d\mathbf{p} \wedge d\mathbf{q}$ and $d\mathbf{P} \wedge d\mathbf{Q}$ coincide. But the definition of the Poisson bracket (F, H) was given invariantly in terms of the symplectic structure; it did not involve the coordinates. Therefore,

$$(F, H)_{\mathbf{p}, \mathbf{q}} = (F, H) = (F, H)_{\mathbf{P}, \mathbf{Q}}.$$

Conversely, suppose that the Poisson brackets $(P_i, Q_j)_{\mathbf{p}, \mathbf{q}}$ have the standard form of Problem 2. Then, clearly, $d\mathbf{P} \wedge d\mathbf{Q} = d\mathbf{p} \wedge d\mathbf{q}$, i.e., the map A is canonical.

PROBLEM 4. Show that the Poisson bracket of a product can be calculated by Leibniz's rule:

$$(F_1 F_2, H) = F_1(F_2, H) + F_2(F_1, H).$$

Hint. The Poisson bracket $(F_1 F_2, H)$ is the derivative of the product $F_1 F_2$ in the direction of the field $I \, dH$.

B *The Jacobi identity*

Theorem. *The Poisson bracket of three functions A, B, and C satisfies the Jacobi identity:*

$$((A, B), C) + ((B, C), A) + ((C, A), B) = 0.$$

Corollary (Poisson's theorem). *The Poisson bracket of two first integrals F_1, F_2 of a system with hamiltonian function H is again a first integral.*

PROOF OF THE COROLLARY. By the Jacobi identity,

$$((F_1, F_2), H) = (F_1, (F_2, H)) + (F_2, (H, F_1)) = 0 + 0,$$

as was to be shown. □

In this way, by knowing two first integrals we can find a third, fourth, etc. by a simple computation. Of course, not all the integrals we get will be essentially new, since there cannot be more than $2n$ independent functions on M^{2n}. Sometimes we may get functions of old integrals or constants, which may be zero. But sometimes we do obtain new integrals.

PROBLEM. Calculate the Poisson brackets of the components $p_1, p_2, p_3, M_1, M_2, M_3$ of the linear and angular momentum vectors of a mechanical system.

ANSWER. $(M_1, M_2) = M_3, (M_1, p_1) = 0, (M_1, p_2) = p_3, (M_1, p_3) = -p_2$. This implies

Theorem. *If two components, M_1 and M_2, of the angular momentum of some mechanical problem are conserved, then the third component is also conserved.*

PROOF OF THE JACOBI IDENTITY. Consider the sum

$$((A, B), C) + ((B, C), A) + ((C, A), B).$$

This sum is a "linear combination of second partial derivatives" of the functions A, B, and C. We will compute the terms in the second derivatives of A:

$$((A, B), C) + ((C, A), B) = (L_C L_B - L_B L_C)A,$$

where L_ξ is differentiation in the direction of ξ and $\mathbf{F}$ is the hamiltonian field with hamiltonian function F.

But, by Lemma 2, Section 39, the commutator of the differentiations $L_C L_B - L_B L_C$ is a *first*-order differential operator. This means that none of the second derivatives of A are contained in our sum. The same thing is true for the second derivatives of B and C. Therefore, the sum is zero. $\quad\square$

Corollary 5. *Let $\mathbf{B}$ and $\mathbf{C}$ be hamiltonian fields with hamiltonian functions B and C. Consider the Poisson bracket $[\mathbf{B}, \mathbf{C}]$ of the vector fields. This vector field is hamiltonian, and its hamiltonian function is equal to the Poisson bracket of the hamiltonian functions (B, C).*

PROOF. Set $(B, C) = D$. The Jacobi identity can be rewritten in the form

$$(A, D) = ((A, B), C) - ((A, C), B),$$

$$L_D = L_C L_B - L_B L_C \qquad L_D = L_{[\mathbf{B}, \mathbf{C}]},$$

as was to be shown. $\quad\square$

C The Lie algebras of hamiltonian fields, hamiltonian functions, and first integrals

A linear subspace of a Lie algebra is called a *subalgebra* if the commutator of any two elements of the subspace belongs to it. A subalgebra of a Lie algebra is itself a Lie algebra. The preceding corollary implies, in particular,

Corollary 6. *The hamiltonian vector fields on a symplectic manifold form a subalgebra of the Lie algebra of all vector fields.*

Poisson's theorem on first integrals can be re-formulated as

Corollary 7. *The first integrals of a hamiltonian phase flow form a subalgebra of the Lie algebra of all functions.*

The Lie algebra of hamiltonian functions can be mapped naturally onto the Lie algebra of hamiltonian vector fields. To do this, to every function H we associate the hamiltonian vector field $\mathbf{H}$ with hamiltonian function H.

Corollary 8. *The map of the Lie algebra of functions onto the Lie algebra of hamiltonian fields is an algebra homomorphism. Its kernel consists of the locally constant functions. If M^{2n} is connected, the kernel is one-dimensional and consists of constants.*

PROOF. Our map is linear. Corollary 5 says that our map carries the Poisson bracket of functions into the Poisson bracket of vector fields. The kernel consists of functions H for which $I\,dH \equiv 0$. Since I is an isomorphism, $dH \equiv 0$ and $H = \text{const.}$ ☐

Corollary 9. *The phase flows with hamiltonian functions H_1 and H_2 commute if and only if the Poisson bracket of the functions H_1 and H_2 is (locally) constant.*

PROOF. By the theorem in Section 39, E, it is necessary and sufficient that $[H_1, H_2] \equiv 0$, and by Corollary 8 this condition is equivalent to $d(H_1, H_2) \equiv 0$. ☐

We obtain yet another generalization of E. Noether's theorem: given a flow which commutes with the one under consideration, one can construct a first integral.

D Locally hamiltonian vector fields

Let (M^{2n}, ω^2) be a symplectic manifold and $g^t \colon M^{2n} \to M^{2n}$ a one-parameter group of diffeomorphisms preserving the symplectic structure. Will g^t be a hamiltonian flow?

EXAMPLE. Let M^{2n} be a two-dimensional torus T^2, a point of which is given by a pair of coordinates $(p, q)\bmod 1$. Let ω^2 be the usual area element $dp \wedge dq$. Consider the family of translations $g^t(p, q) = (p + t, q)$ (Figure 173). The maps g^t preserve the symplectic structure (i.e., area). Can we find a hamiltonian function corresponding to the vector field $(\dot{p} = 1, \dot{q} = 0)$? If $\dot{p} = -\partial H/\partial q$ and $\dot{q} = \partial H/\partial p$, we would have $\partial H/\partial p = 0$ and $\partial H/\partial q = -1$, i.e., $H = -q + C$. But q is only a *local* coordinate on T^2; there is no map $H \colon T^2 \to \mathbb{R}$ for which $\partial H/\partial p = 0$ and $\partial H/\partial q = 1$. Thus g^t is not a hamiltonian phase flow.

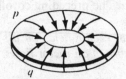

Figure 173   A locally hamiltonial field on the torus

Definition. A *locally hamiltonian vector field* on a symplectic manifold (M^{2n}, ω^2) is the vector field $I\omega^1$, where ω^1 is a closed 1-form on M^{2n}.

Locally, a closed 1-form is the differential of a function, $\omega^1 = dH$. However, in attempting to extend the function H to the whole manifold M^{2n} we may obtain a "many-valued hamiltonian function," since a closed 1-form on a non-simply-connected manifold may not be a differential (for example, the form dq on T^2). A phase flow given by a locally hamiltonian vector field is called a *locally hamiltonian* flow.

PROBLEM. Show that a one-parameter group of diffeomorphisms of a symplectic manifold preserves the symplectic structure if and only if it is a locally hamiltonian phase flow.

Hint. Cf. Section 38A.

PROBLEM. Show that in the symplectic space $\mathbb{R}^{2n}$, every one-parameter group of canonical diffeomorphisms (preserving $d\mathbf{p} \wedge d\mathbf{q}$) is a hamiltonian flow.

Hint. Every closed 1-form on $\mathbb{R}^{2n}$ is the differential of a function.

PROBLEM. Show that the locally hamiltonian vector fields form a sub-algebra of the Lie algebra of all vector fields. In addition, the Poisson bracket of two locally hamiltonian fields is actually a hamiltonian field, with a hamiltonian function uniquely[65] determined by the given fields ξ and η by the formula $H = \omega^2(\xi, \eta)$. Thus, the hamiltonian fields form an ideal in the Lie algebra of locally hamiltonian fields.

41 Symplectic geometry

A euclidean structure on a vector space is given by a symmetric bilinear form, and a symplectic structure by a skew-symmetric one. The geometry of a symplectic space is different from that of a euclidean space, although there are many similarities.

A *Symplectic vector spaces*

Let $\mathbb{R}^{2n}$ be an even-dimensional vector space.

Definition. A *symplectic linear structure* on $\mathbb{R}^{2n}$ is a nondegenerate[66] bilinear skew-symmetric 2-form given in $\mathbb{R}^{2n}$. This form is called the *skew-scalar product* and is denoted by $[\xi, \eta] = -[\eta, \xi]$. The space $\mathbb{R}^{2n}$, together with the symplectic structure $[\ ,\ ]$, is called a *symplectic vector space*.

EXAMPLE. Let $(p_1, \ldots, p_n, q_1, \ldots, q_n)$ be coordinate functions on $\mathbb{R}^{2n}$, and ω^2 the form

$$\omega^2 = p_1 \wedge q_1 + \cdots + p_n \wedge q_n.$$

Since this form is nondegenerate and skew-symmetric, it can be taken for a skew-scalar product: $[\xi, \eta] = \omega^2(\xi, \eta)$. In this way the coordinate space $\mathbb{R}^{2n} = \{(\mathbf{p}, \mathbf{q})\}$ receives a symplectic structure. This structure is called the *standard symplectic structure*. In the standard symplectic structure the skew-scalar product of two vectors ξ and η is equal to the sum of the oriented areas of the parallelogram (ξ, η) on the n coordinate planes (p_i, q_i).

Two vectors ξ and η in a symplectic space are called skew-orthogonal ($\xi \prec \eta$) if their skew-scalar product is equal to zero.

PROBLEM. Show that $\xi \prec \xi$: every vector is skew-orthogonal to itself.

The set of all vectors skew-orthogonal to a given vector η is called the skew-orthogonal complement to η.

[65] Not just up to a constant.

[66] A 2-form $[\ ,\ ]$ on $\mathbb{R}^{2n}$ is nondegenerate if $([\xi, \eta] = 0, \forall\eta) \Rightarrow (\xi = 0)$.

PROBLEM. Show that the skew-orthogonal complement to η is a $2n - 1$-dimensional hyperplane containing η.

Hint. If all vectors were skew-orthogonal to η, then the form [,] would be degenerate.

B *The symplectic basis*

A euclidean structure under a suitable choice of basis (it must be ortho-normal) is given by a scalar product in a particular standard form. In exactly the same way, a symplectic structure takes the standard form indicated above in a suitable basis.

PROBLEM. Find the skew-scalar product of the basis vectors $\mathbf{e}_{p_i}$ and $\mathbf{e}_{q_i}$ $(i = 1 \ldots, n)$ in the example presented above.

Solution. The relations

$$(1) \qquad [\mathbf{e}_{p_i}, \mathbf{e}_{p_j}] = [\mathbf{e}_{p_i}, \mathbf{e}_{q_j}] = [\mathbf{e}_{q_i}, \mathbf{e}_{q_j}] = 0 \qquad [\mathbf{e}_{p_i}, \mathbf{e}_{q_i}] = 1$$

follow from the definition of $p_1 \wedge q_1 + \cdots + p_n \wedge q_n$.

We now return to the general symplectic space.

Definition. A *symplectic basis* is a set of $2n$ vectors, $\mathbf{e}_{p_i}$, $\mathbf{e}_{q_i}$ $(i = 1, \ldots, n)$ whose scalar products have the form (1).

In other words, every basis vector is skew-orthogonal to all the basis vectors except one, associated to it; its product with the associated vector is equal to ± 1.

Theorem. *Every symplectic space has a symplectic basis. Furthermore, we can take any nonzero vector* $\mathbf{e}$ *for the first basis vector.*

PROOF. This theorem is entirely analogous to the corresponding theorem in euclidean geometry and is proved in almost the same way.

Since the vector $\mathbf{e}$ is not zero, there is a vector $\mathbf{f}$ not skew-orthogonal to it (the form [,] is nondegenerate). By choosing the length of this vector, we can insure that its skew-scalar product with $\mathbf{e}$ is equal to 1. In the case $n = 1$, the theorem is proved.

If $n > 1$, consider the skew-orthogonal complement D (Figure 174) to the pair of vectors $\mathbf{e}$, $\mathbf{f}$. D is the intersection of the skew-orthogonal comple-ments to $\mathbf{e}$ and $\mathbf{f}$. These two $2n - 1$-dimensional spaces do not coincide,

Figure 174 Skew-orthogonal complement

since **e** is not in the skew-orthogonal complement to **f**. Therefore, their intersection has even dimension $2n - 2$.

We will show that D is a symplectic subspace of $\mathbb{R}^{2n}$, i.e., that the skew-scalar product $[\ ,\ ]$ restricted to D is nondegenerate. If a vector $\xi \in D$ were skew-orthogonal to the whole subspace D, then since it would also be skew-orthogonal to **e** and to **f**, ξ would be skew-orthogonal to $\mathbb{R}^{2n}$, which contradicts the nondegeneracy of $[\ ,\ ]$ on $\mathbb{R}^{2n}$. Thus D^{2n-2} is symplectic.

Now if we adjoin the vectors **e** and **f** to a symplectic basis for D^{2n-2} we get a sympletic basis for $\mathbb{R}^{2n}$, and the theorem is proved by induction on n.

□

Corollary. *All symplectic spaces of the same dimension are isomorphic.*

If we take the vectors of a symplectic basis as coordinate unit vectors, we obtain a coordinate system p_i, q_i in which $[\ ,\ ]$ takes the standard form $p_1 \wedge q_1 + \cdots + p_n \wedge q_n$. Such a coordinate system is called *symplectic.*

C The symplectic group

To a euclidean structure we associated the orthogonal group of linear mappings which preserved the euclidean structure. In a symplectic space the symplectic group plays an analogous role.

Definition. A linear transformation $S: \mathbb{R}^{2n} \to \mathbb{R}^{2n}$ of the symplectic space $\mathbb{R}^{2n}$ to itself is called *symplectic* if it preserves the skew-scalar product:

$$[S\xi, S\eta] = [\xi, \eta], \qquad \forall \xi, \eta \in \mathbb{R}^{2n}.$$

The set of all symplectic transformations of $\mathbb{R}^{2n}$ is called the *symplectic group* and is denoted by $Sp(2n)$.

It is clear that the composition of two symplectic transformations is symplectic. To justify the term symplectic group, we must only show that a symplectic transformation is nonsingular; it is then clear that the inverse is also symplectic.

PROBLEM. Show that the group $Sp(2)$ is isomorphic to the group of real two-by-two matrices with determinant 1 and is homeomorphic to the interior of a solid three-dimensional torus.

Theorem. *A transformation $S: \mathbb{R}^{2n} \to \mathbb{R}^{2n}$ of the standard symplectic space* (**p, q**) *is symplectic if and only if it is linear and canonical, i.e., preserves the differential 2-form*

$$\omega^2 = dp_1 \wedge dq_1 + \cdots + dp_n \wedge dq_n.$$

PROOF. Under the natural identification of the tangent space to $\mathbb{R}^{2n}$ with $\mathbb{R}^{2n}$, the 2-form ω^2 goes to $[\ ,\ ]$. □

Corollary. *The determinant of any symplectic transformation is equal to* 1.

PROOF. We already know (Section 38B) that canonical maps preserve the exterior powers of the form ω^2. But its n-th exterior power is (up to a constant multiple) the volume element on $\mathbb{R}^{2n}$. This means that symplectic transformations S of the standard $\mathbb{R}^{2n} = \{(\mathbf{p}, \mathbf{q})\}$ preserve the volume element, so det $S = 1$. But since every symplectic linear structure can be written down in standard form in a symplectic coordinate system, the determinant of a symplectic transformation of any symplectic space is equal to 1. $\qquad\square$

Theorem. *A linear transformation* $S: \mathbb{R}^{2n} \to \mathbb{R}^{2n}$ *is symplectic if and only if it takes some* (*and therefore any*) *symplectic basis into a symplectic basis.*

PROOF. The skew-scalar product of any two linear combinations of basis vectors can be expressed in terms of skew-scalar products of basis vectors. If the transformation does not change the skew-scalar products of basis vectors, then it does not change the skew-scalar products of any vectors. $\qquad\square$

D *Planes in symplectic space*

In a euclidean space all planes are equivalent: each of them can be carried into any other one by a motion. We will now look at a symplectic vector space from this point of view.

PROBLEM. Show that a nonzero vector in a symplectic space can be carried into any other non-zero vector by a symplectic transformation.

PROBLEM. Show that not every two-dimensional plane of the symplectic space $\mathbb{R}^{2n}$ can be obtained from a given 2-plane by a symplectic transformation.

Hint. Consider the planes (p_1, p_2) and (p_1, q_1).

Definition. A k-dimensional plane (i.e., subspace) of a symplectic space is called *null*[67] if it is skew-orthogonal to itself, i.e., if the skew-scalar product of any two vectors of the plane is equal to zero.

EXAMPLE. The coordinate plane $(p_1, \ldots, p_k)$ in the symplectic coordinate system $\mathbf{p}, \mathbf{q}$ is null. (Prove it!)

PROBLEM. Show that any non-null two-dimensional plane can be carried into any other non-null two-plane by a symplectic transformation.

For calculations in symplectic geometry it may be useful to impose some euclidean structure on the symplectic space. We fix a symplectic coordinate system $\mathbf{p}, \mathbf{q}$ and introduce a euclidean structure using the coordinate scalar product

$$(\mathbf{x}, \mathbf{x}) = \sum p_i^2 + q_i^2, \quad \text{where } \mathbf{x} = \sum p_i \mathbf{e}_{p_i} + q_i \mathbf{e}_{q_i}.$$

---

[67] Null planes are also called *isotropic*, and for $k = n$, *lagrangian*.

The symplectic basis e_p, e_q is orthonormal in this euclidean structure. The skew-scalar product, like every bilinear form, can be expressed in terms of the scalar product by

$$(2) \qquad\qquad [\xi, \eta] = (I\xi, \eta)$$

where $I: \mathbb{R}^{2n} \to \mathbb{R}^{2n}$ is some operator. It follows from the skew-symmetry of the skew-scalar product that the operator I is skew-symmetric.

PROBLEM. Compute the matrix of the operator I in the symplectic basis e_p, e_q.

ANSWER.

$$\begin{pmatrix} 0 & -E \\ E & 0 \end{pmatrix}.$$

where E is the $n \times n$ identity matrix.

Thus, for $n = 1$ (in the p, q-plane), I is simply rotation by 90°, and in the general case I is rotation by 90° in each of the n planes p_i, q_i.

PROBLEM. Show that the operator I is symplectic and that $I^2 = -E_{2n}$.

Although the euclidean structures and the operator I are not invariantly associated to a symplectic space, they are often convenient.

The following theorem follows directly from (2).

Theorem. *A plane π of a symplectic space is null if and only if the plane $I\pi$ is orthogonal to π.*

Notice that the dimensions of the planes π and $I\pi$ are the same, since I is nonsingular. Hence

Corollary. *The dimension of a null plane in $\mathbb{R}^{2n}$ is less than or equal to n.*

This follows since the two k-dimensional planes π and $I\pi$ cannot be orthogonal if $k > n$.

We consider more carefully the n-dimensional null planes in the symplectic coordinate space $\mathbb{R}^{2n}$. An example of such a plane is the coordinate **p**-plane. There are in all $\binom{2n}{n}$ n-dimensional coordinate planes in $\mathbb{R}^{2n} = \{(\mathbf{p}, \mathbf{q})\}$.

PROBLEM. Show that there are 2^n null planes among the $\binom{2n}{n}$ n-dimensional coordinate planes: to each of the 2^n partitions of the set $(1, \ldots, n)$ into two parts $(i_1, \ldots, i_k)$, $(j_1, \ldots, j_{n-k})$ we associate the null coordinate plane $p_{i_1}, \ldots, p_{i_k}, q_{j_1}, \ldots, q_{j_{n-k}}$.

In order to study the generating functions of canonical transformations we need

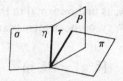

Figure 175   Construction of a coordinate plane σ transversal to a given plane π.

Theorem. *Every n-dimensional null plane π in the symplectic coordinate space $\mathbb{R}^{2n}$ is transverse[68] to at least one of the 2^n coordinate null planes.*

PROOF. Let P be the null plane $p_1, \ldots, p_n$ (Figure 175). Consider the intersection $\tau = \pi \cap P$. Suppose that the dimension of τ is equal to $k, 0 \leq k \leq n$. Like every k-dimensional subspace of the n-dimensional space, the plane τ is transverse to at least one $(n - k)$-dimensional coordinate plane in P, let us say the plane

$$\eta = (p_{i_1}, \ldots, p_{i_{n-k}}); \qquad \tau + \eta = P, \tau \cap \eta = 0.$$

We now consider the null n-dimensional coordinate plane

$$\sigma = (p_{i_1}, \ldots, p_{i_{n-k}}, q_{j_1}, \ldots, q_{j_k}), \qquad \eta = \sigma \cap P,$$

and show that our plane π is transverse to σ:

$$\pi \cap \sigma = 0.$$

We have

$$\left.\begin{array}{l} \tau \subset \pi, \pi \prec \pi \Rightarrow \tau \prec \pi \\ \eta \subset \sigma, \sigma \prec \sigma \Rightarrow \eta \prec \sigma \end{array}\right\} \Rightarrow (\tau + \eta) \prec (\pi \cap \sigma) \Rightarrow P \prec (\pi \cap \sigma).$$

But P is an n-dimensional null plane. Therefore, every vector skew-orthogonal to P belongs to P (cf. the corollary above). Thus $(\pi \cap \sigma) \subset P$. Finally,

$$\pi \cap \sigma = (\pi \cap P) \cap (\sigma \cap P) = \tau \cap \eta = 0,$$

as was to be shown. $\qquad\qquad\qquad\qquad\qquad\qquad\qquad\qquad\qquad\qquad\qquad\square$

PROBLEM. Let π_1 and π_2 be two k-dimensional planes in symplectic $\mathbb{R}^{2n}$. Is it always possible to carry π_1 to π_2 by a symplectic transformation? How many classes of planes are there which cannot be carried one into another?

ANSWER. $[k/2] + 1$, if $k \leq n$; $[(2n - k)/2] + 1$ if $k \geq n$.

E Symplectic structure and complex structure

Since $I^2 = -E$ we can introduce into our space $\mathbb{R}^{2n}$ not only a symplectic structure $[\ , \ ]$ and euclidean structure $(\ , \ )$, but also a complex structure, by defining multiplication by $i = \sqrt{-1}$ to be the action of I. The space $\mathbb{R}^{2n}$

[68] Two subspaces L_1 and L_2 of a vector space L are transverse if $L_1 + L_2 = L$. Two n-dimensional planes in $\mathbb{R}^{2n}$ are transverse if and only if they intersect only in 0.

is identified in this way with a complex space $\mathbb{C}^n$ (the coordinate space with coordinates $z_k = p_k + iq_k$). The linear transformations of $\mathbb{R}^{2n}$ which preserve the euclidean structure form the *orthogonal group* $O(2n)$; those preserving the complex structure form the *complex linear group* $GL(n, \mathbb{C})$.

PROBLEM. Show that transformations which are both orthogonal and symplectic are complex, that those which are both complex and orthogonal are symplectic, and that those which are both symplectic and complex are orthogonal; thus that the intersection of two of the three groups is equal to the intersection of all three:

$$O(2n) \cap Sp(2n) = Sp(2n) \cap GL(n, \mathbb{C}) = GL(n, \mathbb{C}) \cap O(2n).$$

This intersection is called the *unitary group* $U(n)$.

Unitary transformations preserve the hermitian scalar product $(\xi, \eta) + i[\xi, \eta]$; the scalar and skew-scalar products on $\mathbb{R}^{2n}$ are its real and imaginary parts.

42 Parametric resonance in systems with many degrees of freedom

During our investigation of oscillating systems with periodically varying parameters (cf. Section 25), we explained that parametric resonance depends on the behavior of the eigenvalues of a certain linear transformation ("the mapping at a period"). The dependence consists of the fact that an equilibrium position of a system with periodically varying parameters is stable if the eigenvalues of the mapping at a period have modulus less than 1, and unstable if at least one of the eigenvalues has modulus greater than 1.

The mapping at a period obtained from a system of Hamilton's equations with periodic coefficients is symplectic. The investigation in Section 25 of parametric resonance in a system with one degree of freedom relied on our analysis of the behavior of the eigenvalues of symplectic transformations of the plane. In this paragraph we will analyze, in an analogous way, the behavior of the eigenvalues of symplectic transformations in a phase space of any dimension. The results of this analysis (due to M. G. Krein) can be applied to the study of conditions for the appearance of parametric resonance in mechanical systems with many degrees of freedom.

A Symplectic matrices

Consider a linear transformation of a symplectic space, $S: \mathbb{R}^{2n} \to \mathbb{R}^{2n}$. Let $p_1, \ldots, p_n; q_1, \ldots, q_n$ be a symplectic coordinate system. In this coordinate system, the transformation is given by a matrix S.

Theorem. *A transformation is symplectic if and only if its matrix S in the symplectic coordinate system* $(\mathbf{p}, \mathbf{q})$ *satisfies the relation*

$$S'IS = I,$$

where

$$I = \begin{pmatrix} 0 & -E \\ E & 0 \end{pmatrix}$$

and S' is the transpose of S.

225

PROOF. The condition for being symplectic ($[S\xi, S\eta] = [\xi, \eta]$ for all ξ and η) can be written in terms of the scalar product by using the operator I, as follows:

$$(IS\xi, S\eta) = (I\xi, \eta), \qquad \forall \xi, \eta$$

or

$$(S'IS\xi, \eta) = (I\xi, \eta), \qquad \forall \xi, \eta,$$

as was to be shown. □

B *Symmetry of the spectrum of a symplectic transformation*

Theorem. *The characteristic polynomial of a symplectic transformation*

$$p(\lambda) = \det(S - \lambda E)$$

is reflexive,[69] *i.e.,* $p(\lambda) = \lambda^{2n} p(1/\lambda)$.

PROOF. We will use the facts that $\det S = \det I = 1$, $I^2 = -E$, and $\det A' = \det A$. By the theorem above, $S = -IS'^{-1}I$. Therefore,

$$p(\lambda) = \det(S - \lambda E) = \det(-IS'^{-1}I - \lambda E) = \det(-S'^{-1} + \lambda E)$$

$$= \det(-E + \lambda S)$$

$$= \lambda^{2n} \det\left(S - \frac{1}{\lambda} E\right) = \lambda^{2n} p\left(\frac{1}{\lambda}\right). \qquad □$$

Corollary. *If λ is an eigenvalue of a symplectic transformation, then $1/\lambda$ is also an eigenvalue.*

On the other hand, the characteristic polynomial is real; therefore, if λ is a complex eigenvalue, then $\bar{\lambda}$ is an eigenvalue different from λ. It follows that the roots λ of the characteristic polynomial lie symmetrically with respect to the real axis and to the unit circle (Figure 176). They come in 4-tuples,

$$\lambda, \bar{\lambda}, \frac{1}{\lambda}, \frac{1}{\bar{\lambda}} \qquad (|\lambda| \neq 1, \operatorname{Im} \lambda \neq 0),$$

and pairs lying on the real axis,

$$\lambda = \bar{\lambda} \qquad \frac{1}{\lambda} = \frac{1}{\bar{\lambda}},$$

---

[69] A reflexive polynomial is a polynomial $a_0 x^m + a_1 x^{m-1} + \cdots + a_m$ which has symmetric coefficients $a_0 = a_m$, $a_1 = a_{m-1}, \ldots$.

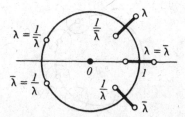

Figure 176   Distribution of the eigenvalues of a symplectic transformation

or on the unit circle,

$$\lambda = \frac{1}{\bar{\lambda}} \qquad \bar{\lambda} = \frac{1}{\lambda}.$$

It is not hard to verify that the multiplicities of all four points of a 4-tuple (or both points of a pair) are the same.

C Stability

Definition. A transformation S is called *stable* if

$$\forall \varepsilon > 0, \; \exists \delta > 0 : |\mathbf{x}| < \delta \Rightarrow |S^N \mathbf{x}| < \varepsilon, \qquad \forall N > 0.$$

PROBLEM. Show that if at least one of the eigenvalues of a symplectic transformation S does not lie on the unit circle, then S is unstable.

Hint. In view of the demonstrated symmetry, if one of the eigenvalues does not lie on the unit circle, then there exists an eigenvalue outside the unit circle $|\lambda| > 1$; in the corresponding invariant subspace, S is an "expansion with a rotation."

PROBLEM. Show that if all the eigenvalues of a linear transformation are distinct and lie on the unit circle, then the transformation is stable.

Hint. Change to a basis of eigenvectors.

Definition. A symplectic transformation S is called *strongly stable* if every symplectic transformation sufficiently close[70] to S is stable.

In Section 25 we established that $S: \mathbb{R}^2 \to \mathbb{R}^2$ is strongly stable if $\lambda_{1,2} = e^{\pm i\alpha}$ and $\lambda_1 \neq \lambda_2$.

Theorem. *If all 2n eigenvalues of a symplectic transformation S are distinct and lie on the unit circle, then S is strongly stable.*

PROOF. We enclose the $2n$ eigenvalues λ in $2n$ non-intersecting neighborhoods, symmetric with respect to the unit circle and the real axis (Figure 177). The $2n$ roots of the characteristic polynomial depend continuously on the elements of the matrix of S. Therefore, if the matrix S_1 is sufficiently close to S,

---

[70] S_1 is "sufficiently close" to S if the elements of the matrix of S_1 in a fixed basis differ from the elements of the matrix of S in the same basis by less than a sufficiently small number ε.

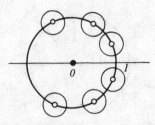

Figure 177 Behavior of simple eigenvalues under a small change of the symplectic transformation

exactly one eigenvalue λ_1 of the matrix of S_1 will lie in each of the $2n$ neighborhoods of the $2n$ points of λ. But if one of the points λ_1 did not lie on the unit circle, for example, if it lay outside the unit circle, then by the theorem in subsection B, there would be another point λ_2, $|\lambda_2| < 1$ in the same neighborhood, and the total number of roots would be greater than $2n$, which is not possible.

Thus all the roots of S_1 lie on the unit circle and are distinct, so S_1 is stable. □

We might say that an eigenvalue λ of a symplectic transformation can leave the unit circle only by colliding with another eigenvalue (Figure 178); at the same time, the complex-conjugate eigenvalues will collide, and from the two pairs of roots on the unit circle we obtain one 4-tuple (or pair of real λ).

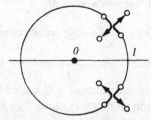

Figure 178 Behavior of multiple eigenvalues under a small change of the symplectic transformation

It follows from the results of Section 25 that the condition for parametric resonance to arise in a linear canonical system with a periodically changing hamilton function is precisely that the corresponding symplectic transformation of phase space should cease to be stable. It is clear from the theorem above that this can happen only after a collision of eigenvalues on the unit circle. In fact, as M. G. Krein noticed, not every such collision is dangerous.

It turns out that the eigenvalues λ with $|\lambda| = 1$ are divided into two classes: *positive* and *negative*. When two roots with the same sign collide, the roots "go through one another," and cannot leave the unit circle. On the other hand, when two roots with different signs collide, they generally leave the unit circle.

M.G. Krein's theory goes beyond the limits of this book; we will formulate the basic results here in the form of problems.

PROBLEM. Let λ and $\bar{\lambda}$ be simple (multiplicity 1) eigenvalues of a symplectic transformation S with $|\lambda| = 1$. Show that the two-dimensional invariant plane π_λ corresponding to λ, $\bar{\lambda}$, is non-null.

Hint. Let ξ_1 and ξ_2 be complex eigenvectors of S with eigenvalues λ_1 and λ_2. Then if $\lambda_1 \lambda_2 \neq 1$, the vectors ξ_1 and ξ_2 are skew-orthogonal: $[\xi_1, \xi_2] = 0$.

Let ξ be a real vector of the plane π_λ, where Im $\lambda > 0$ and $|\lambda| = 1$. The eigenvalue λ is called *positive* if $[S\xi, \xi] > 0$.

PROBLEM. Show that this definition is correct, i.e., it does not depend on the choice of $\xi \neq 0$ in the plane π_λ.

Hint. If the plane π_λ contained two non-collinear skew-orthogonal vectors, it would be null.

In the same way, an eigenvalue λ of multiplicity k with $|\lambda| = 1$ is of definite sign if the quadratic form $[S\xi, \xi]$ is (positive or negative) definite on the invariant $2k$-dimensional subspace corresponding to λ, $\bar{\lambda}$.

PROBLEM. Show that S is strongly stable if and only if all the eigenvalues λ lie on the unit circle and are of definite sign.

Hint. The quadratic form $[S\xi, \xi]$ is invariant with respect to S.

43 A symplectic atlas

In this paragraph we prove Darboux's theorem, according to which every symplectic manifold has local coordinates $\mathbf{p}$, $\mathbf{q}$ in which the symplectic structure can be written in the simplest way: $\omega^2 = d\mathbf{p} \wedge d\mathbf{q}$.

A *Symplectic coordinates*

Recall that the definition of manifold includes a compatibility condition for the charts of an atlas. This is a condition on the maps $\varphi_i^{-1}\varphi_j$ going from one chart to another. The maps $\varphi_i^{-1}\varphi_j$ are maps of a region of coordinate space.

Definition. An atlas of a manifold M^{2n} is called *symplectic* if the standard symplectic structure $\omega^2 = d\mathbf{p} \wedge d\mathbf{q}$ is introduced into the coordinate space $\mathbb{R}^{2n} = \{(\mathbf{p}, \mathbf{q})\}$, and the transfer from one chart to another is realized by a canonical (i.e., ω^2-preserving) transformation[71] $\varphi_i^{-1}\varphi_j$.

PROBLEM. Show that a symplectic atlas defines a symplectic structure on M^{2n}.

The converse is also true: every symplectic manifold has a symplectic atlas. This follows from the following theorem.

---

[71] Complex-analytic manifolds, for example, are defined analogously; there must be a complex-analytic structure on coordinate space, and the transfer from one chart to another must be complex analytic.

B Darboux's theorem

Theorem. *Let ω^2 be a closed nondegenerate differential 2-form in a neighborhood of a point $\mathbf{x}$ in the space $\mathbb{R}^{2n}$. Then in some neighborhood of $\mathbf{x}$ one can choose a coordinate system $(p_1, \ldots, p_n; q_1, \ldots, q_n)$ such that the form has the standard form:*

$$\omega^2 = \sum_{i=1}^{n} dp_i \wedge dq_i.$$

This theorem allows us to extend to all symplectic manifolds any assertion of a local character which is invariant with respect to canonical transformations and is proven for the standard phase space ($\mathbb{R}^{2n}$, $\omega^2 = d\mathbf{p} \wedge d\mathbf{q}$).

C *Construction of the coordinates p_1 and q_1*

For the first coordinate p_1 we take a non-constant linear function (we could have taken any differentiable function whose differential is not zero at the point $\mathbf{x}$). For simplicity we will assume that $p_1(\mathbf{x}) = 0$.

Let $\mathbf{P}_1 = I\,dp_1$ denote the hamiltonian field corresponding to the function p_1 (Figure 179). Note that $\mathbf{P}_1(\mathbf{x}) \neq 0$; therefore, we can draw a hyperplane N^{2n-1} through the point $\mathbf{x}$ which does not contain the vector $\mathbf{P}_1(\mathbf{x})$ (we could have taken any surface transverse to $\mathbf{P}_1(\mathbf{x})$ as N^{2n-1}).

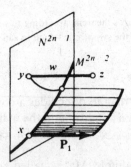

Figure 179   Construction of symplectic coordinates

Consider the hamiltonian flow P_1^t with hamiltonian function p_1. We consider the time t necessary to go from N to the point $\mathbf{z} = P_1^t(\mathbf{y})$ ($\mathbf{y} \in N$) under the action of P_1^t as a function of the point $\mathbf{z}$. By the usual theorems in the theory of ordinary differential equations, this function is defined and differentiable in a neighborhood of the point $\mathbf{x} \in \mathbb{R}^{2n}$. Denote it by q_1. Note that $q_1 = 0$ on N and that the derivative of q_1 in the direction of the field $\mathbf{P}_1$ is equal to 1. Thus the Poisson bracket of the functions q_1 and p_1 we constructed is equal to 1:

$$(q_1, p_1) \equiv 1.$$

D Construction of symplectic coordinates by induction on n

If $n = 1$, the construction is finished. Let $n > 1$. We will assume that Darboux's theorem is already proved for $\mathbb{R}^{2n-2}$. Consider the set M given by the equations $p_1 = q_1 = 0$. The differentials dp_1 and dq_1 are linearly independent at x since $\omega^2(I\,dp_1,\,I\,dq_1) = (q_1, p_1) \equiv 1$. Thus, by the implicit function theorem, the set M is a manifold of dimension $2n - 2$ in a neighborhood of x; we will denote it by M^{2n-2}.

Lemma. *The symplectic structure ω^2 on $\mathbb{R}^{2n}$ induces a symplectic structure on some neighborhood of the point x on M^{2n-2}.*

PROOF. For the proof we need only the nondegeneracy of ω^2 on TM_x. Consider the symplectic vector space $T\mathbb{R}_x^{2n}$. The vectors $\mathbf{P}_1(x)$ and $\mathbf{Q}_1(x)$ of the hamiltonian vector fields with hamiltonian functions p_1 and q_1 belong to $T\mathbb{R}_x^{2n}$. Let $\xi \in TM_x$. The derivatives of p_1 and q_1 in the direction ξ are equal to zero. This means that $dp_1(\xi) = \omega^2(\xi, \mathbf{P}_1) = 0$ and $dq_1(\xi) = \omega^2(\xi, \mathbf{Q}_1) = 0$. Thus TM_x is the skew-orthogonal complement to $\mathbf{P}_1(x)$, $\mathbf{Q}_1(x)$. By Section 41B, the form ω^2 on TM_x is nondegenerate. $\qquad\square$

By the induction hypothesis there are symplectic coordinates in a neighborhood of the point x on the symplectic manifold $(M^{2n-2}, \omega^2|_M)$. Denote them by $p_i, q_i\ (i = 2, \ldots, n)$. We extend the functions $p_2, \ldots, q_n$ to a neighborhood of x in $\mathbb{R}^{2n}$ in the following way. Every point z in a neighborhood of x in $\mathbb{R}^{2n}$ can be uniquely represented in the form $z = P_1^t Q_1^s w$, where $w \in M^{2n-2}$, and s and t are small numbers. We set the values of the coordinates $p_2, \ldots, q_n$ at z equal to their values at the point w (Figure 179). The $2n$ functions $p_1, \ldots, p_n, q_1, \ldots, q_n$ thus constructed form a local coordinate system in a neighborhood of x in $\mathbb{R}^{2n}$.

E Proof that the coordinates constructed are symplectic

Denote by P_i^t and $Q_i^t\ (i = 1, \ldots, n)$ the hamiltonian flows with hamiltonian functions p_i and q_i, and by $\mathbf{P}_i$ and $\mathbf{Q}_i$ the corresponding vector fields. We will compute the Poisson brackets of the functions $p_1, \ldots, q_n$. We already saw in C that $(q_1, p_1) \equiv 1$. Therefore, the flows P_1^t and Q_1^s commute: $P_1^t Q_1^s = Q_1^s P_1^t$.

Recalling the definitions of $p_2, \ldots, q_n$ we see that each of these functions is invariant with respect to the flows P_1^t and Q_1^t. Thus the Poisson brackets of p_1 and q_1 with all $2n - 2$ functions $p_i, q_i\ (i > 1)$ are equal to zero.

The map $P_1^t Q_1^s$ therefore commutes with all $2n - 2$ flows $P_i^t, Q_i^s\ (i > 1)$. Consequently, it leaves each of the $2n - 2$ vector fields $\mathbf{P}_i, \mathbf{Q}_i\ (i > 1)$ fixed. $P_1^t Q_1^s$ preserves the symplectic structure ω^2 since the flows P_1^t and Q_1^s are hamiltonian; therefore, the values of the form ω^2 on the vectors of any two of the $2n - 2$ fields $\mathbf{P}_i$, $\mathbf{Q}_i$ $(i > 1)$ are the same at the points $\mathbf{z} = P_1^t Q_1^s \mathbf{w} \in \mathbb{R}^{2n}$ and $\mathbf{w} \in M^{2n-2}$. But these values are equal to the values of the Poisson brackets of the corresponding hamiltonian functions. Thus, the values of the Poisson bracket of any two of the $2n - 2$ coordinates p_i, q_i $(i > 1)$ at the points $\mathbf{z}$ and $\mathbf{w}$ are the same if $\mathbf{z} = P_1^t Q_1^s \mathbf{w}$.

The functions p_1 and q_1 are first integrals of each of the $2n - 2$ flows P_i^t, Q_i^s $(i > 1)$. Therefore, each of the $2n - 2$ fields $\mathbf{P}_i$, $\mathbf{Q}_i$ is tangent to the level manifold $p_1 = q_1 = 0$. But this manifold is M^{2n-2}. Therefore, each of the $2n - 2$ fields $\mathbf{P}_i$, $\mathbf{Q}_i$ $(i > 1)$ is tangent to M^{2n-2}. Consequently, these fields are hamiltonian fields on the symplectic manifold $(M^{2n-2}, \omega^2|_M)$, and the corresponding hamiltonian functions are $p_i|_M$, $q_i|_M$ $(i > 1)$. Thus, in the whole space $(\mathbb{R}^{2n}, \omega^2)$, the Poisson bracket of any two of the $2n - 2$ co-ordinates p_i, q_i $(i > 1)$ considered on M^{2n-2} is the same as the Poisson bracket of these coordinates in the symplectic space $(M^{2n-2}, \omega^2|_M)$.

But, by our induction hypothesis, the coordinates on M^{2n-2} $(p_i|_M, q_i|_M;$ $i > 1)$ are symplectic. Therefore, in the whole space $\mathbb{R}^{2n}$, the Poisson brackets of the constructed coordinates have the standard values

$$(p_i, p_j) \equiv (p_i, q_j) \equiv (q_i, q_j) \equiv 0 \quad \text{and} \quad (q_i, p_i) \equiv 1.$$

The Poisson brackets of the coordinates $\mathbf{p}$, $\mathbf{q}$ on $\mathbb{R}^{2n}$ have the same form if $\omega^2 = \sum dp_i \wedge dq_i$. But a bilinear form ω^2 is determined by its values on pairs of basis vectors. Therefore, the Poisson brackets of the coordinate functions determine the shape of ω^2 uniquely. Thus

$$\omega^2 = dp_1 \wedge dq_1 + \cdots + dp_n \wedge dq_n,$$

and Darboux's theorem is proved. $\qquad\qquad\qquad\qquad\qquad\qquad\qquad\qquad\square$

Canonical formalism 9

The coordinate point of view will predominate in this chapter. The technique of generating functions for canonical transformations, developed by Hamilton and Jacobi, is the most powerful method available for integrating the differential equations of dynamics. In addition to this technique, the chapter contains an "odd-dimensional" approach to hamiltonian phase flows.

This chapter is independent of the previous one. It contains new proofs of several of the results in Chapter 8, as well as an explanation of the origin of the theory of symplectic manifolds.

44 The integral invariant of Poincaré–Cartan

In this section we look at the geometry of 1-forms in an odd-dimensional space.

A *A hydrodynamical lemma*

Let $\mathbf{v}$ be a vector field in three-dimensional oriented euclidean space $\mathbb{R}^3$, and $\mathbf{r} = \mathbf{curl}\ \mathbf{v}$ its curl. The integral curves of $\mathbf{r}$ are called *vortex lines*. If γ_1 is any closed curve in $\mathbb{R}^3$ (Figure 180), the vortex lines passing through the points of γ_1 form a tube called a *vortex tube*.

Let γ_2 be another curve encircling the same vortex tube, so that $\gamma_1 - \gamma_2 = \partial\sigma$, where σ is a 2-cycle representing a part of the vortex tube. Then:

Stokes' lemma. *The field $\mathbf{v}$ has equal circulation along the curves γ_1 and γ_2:*

$$\oint_{\gamma_1} \mathbf{v}\ d\mathbf{l} = \oint_{\gamma_2} \mathbf{v}\ d\mathbf{l}.$$

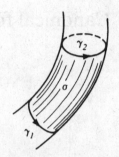

Figure 180   Vortex tube

PROOF. By Stokes' formula, $\int_{\gamma_1} \mathbf{v}\, d\mathbf{l} - \int_{\gamma_2} \mathbf{v}\, d\mathbf{l} = \iint_\sigma \mathbf{curl}\ \mathbf{v}\, d\mathbf{n} = 0$, since $\mathbf{curl}\ \mathbf{v}$ is tangent to the vortex tube.   □

B The multi-dimensional Stokes' lemma

It turns out that Stokes' lemma generalizes to the case of any odd-dimensional manifold M^{2n+1} (in place of $\mathbb{R}^3$). To formulate this generalization we replace our vector field by a differential form.

The circulation of a vector field $\mathbf{v}$ is the integral of the 1-form ω^1 $(\omega^1(\xi) = (\mathbf{v}, \xi))$. To the curl of $\mathbf{v}$ there corresponds the 2-form $\omega^2 = d\omega^1$ $(d\omega^1(\xi, \eta) = (\mathbf{r}, \xi, \eta))$. It is clear from these formulas that there is a direction

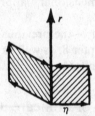

Figure 181   Axis invariantly connected with a 2-form in an odd-dimensional space

at every point (namely, the direction of $\mathbf{r}$, Figure 181), having the property that the circulation of $\mathbf{v}$ along the boundary of every "infinitesimal square" containing $\mathbf{r}$ is equal to zero:

$$d\omega^1(\mathbf{r}, \eta) = 0, \qquad \forall \eta.$$

In fact, $d\omega^1(\mathbf{r}, \eta) = (\mathbf{r}, \mathbf{r}, \eta) = 0$.

Remark. Passing from the 2-form $\omega^2 = d\omega^1$ to the vector field $\mathbf{r} = \mathbf{curl}\ \mathbf{v}$ is *not an invariant operation*: it depends on the euclidean structure of $\mathbb{R}^3$. Only the direction[72] of $\mathbf{r}$ is invariantly associated with ω^2 (and, therefore, with the 1-form ω^1). It is easy to verify that, if $\mathbf{r} \neq 0$, then the direction of $\mathbf{r}$ is uniquely determined by the condition that $\omega^2(\mathbf{r}, \eta) = 0$ for all η.

[72] I.e., the unoriented line in $T\mathbb{R}^3$ with direction vector $\mathbf{r}$.

The algebraic basis for the multi-dimensional Stokes' lemma is the existence of an axis for every rotation of an odd-dimensional space.

Lemma. Let ω^2 be an exterior algebraic 2-form on the odd-dimensional vector space $\mathbb{R}^{2n+1}$. Then there is a vector $\xi \neq 0$ such that

$$\omega^2(\xi, \eta) = 0, \qquad \forall \eta \in \mathbb{R}^{2n+1}.$$

PROOF. A skew-symmetric form ω^2 is given by a skew-symmetric matrix A

$$\omega^2(\xi, \eta) = (A\xi, \eta)$$

of odd order $2n + 1$. The determinant of such a matrix is equal to zero, since

$$A' = -A \quad \det A = \det A' = \det(-A) = (-1)^{2n+1} \det A = -\det A.$$

Thus the determinant of A is zero. This means A has an eigenvector $\xi \neq 0$ with eigenvalue 0, as was to be shown. $\qquad\square$

A vector ξ for which $\omega^2(\xi, \eta) = 0, \forall \eta$ is called a *null vector for the form ω^2*. The null vectors of ω^2 clearly form a linear subspace. The form ω^2 is called *nonsingular* if the dimension of this space is the minimal possible (i.e., 1 for an odd-dimensional space $\mathbb{R}^{2n+1}$ or 0 for an even-dimensional space).

PROBLEM. Consider the 2-form $\omega^2 = dp_1 \wedge dq_1 + \cdots + dp_n \wedge dq_n$ on an even-dimensional space $\mathbb{R}^{2n}$ with coordinates $p_1, \ldots, p_n; q_1, \ldots, q_n$. Show that ω^2 is nonsingular.

PROBLEM. On an odd-dimensional space $\mathbb{R}^{2n+1}$ with coordinates $p_1, \ldots, p_n; q_1, \ldots, q_n; t$, consider the 2-form $\omega^2 = \sum dp_i \wedge dq_i - \omega^1 \wedge dt$, where ω^1 is any 1-form on $\mathbb{R}^{2n+1}$. Show that ω^2 is nonsingular.

If ω^2 is a nonsingular form on an odd-dimensional space $\mathbb{R}^{2n+1}$, then the null vectors ξ of ω^2 all lie on a line. This line is invariantly associated to the form ω^2.

Now let M^{2n+1} be an odd-dimensional differentiable manifold and ω^1 a 1-form on M. By the lemma above, at every point $\mathbf{x} \in M$ there is a direction (i.e., a straight line $\{c\xi\}$ in the tangent space $TM_\mathbf{x}$) having the property that the integral of ω^1 along the boundary of an "infinitesimal square containing this direction" is equal to zero:

$$d\omega^1(\xi, \eta) = 0, \qquad \forall \eta \in TM_\mathbf{x}.$$

Suppose further that the 2-form $d\omega^1$ is nonsingular. Then the direction ξ is uniquely determined. We call it the "vortex direction" of the form ω^1.

The integral curves of the field of vortex directions are called the *vortex lines* (or *characteristic lines*) of the form ω^1.

Let γ_1 be a closed curve on M. The vortex lines going out from points of γ_1 form a "vortex tube." We have

The multi-dimensional Stokes' lemma. *The integrals of a 1-form ω^1 along any two curves encircling the same vortex tube are the same: $\oint_{\gamma_1} \omega^1 = \oint_{\gamma_2} \omega^1$, if $\gamma_1 - \gamma_2 = \partial\sigma$, where σ is a piece of the vortex tube.*

PROOF. By Stokes' formula

$$\oint_{\gamma_1} \omega^1 - \oint_{\gamma_2} \omega^1 = \int_{\partial\sigma} \omega^1 = \int_\sigma d\omega^1.$$

But the value of $d\omega^1$ on any pair of vectors tangent to the vortex tube is equal to zero. (These two vectors lie in a 2-plane containing the vortex direction, and $d\omega^1$ vanishes on this plane.) Thus, $\int_\sigma d\omega^1 = 0$. $\square$

C Hamilton's equations

All the basic propositions of hamiltonian mechanics follow directly from Stokes' lemma.

For M^{2n+1} we will take the "extended phase space $\mathbb{R}^{2n+1}$" with coordinates $p_1, \ldots, p_n; q_1, \ldots, q_n; t$. Suppose we are given a function $H = H(\mathbf{p}, \mathbf{q}, t)$. Then we can construct[73] the 1-form

$$\omega^1 = \mathbf{p}\, d\mathbf{q} - H dt \qquad (\mathbf{p}\, d\mathbf{q} = p_1\, dq_1 + \cdots + p_n\, dq_n).$$

We apply Stokes' lemma to ω^1 (Figure 182).

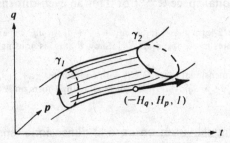

Figure 182 Hamiltonian field and vortex lines of the form $\mathbf{p}\, d\mathbf{q} - H\, dt$.

Theorem. *The vortex lines of the form $\omega^1 = \mathbf{p}\, d\mathbf{q} - H dt$ on the $2n + 1$-dimensional extended phase space $\mathbf{p}, \mathbf{q}, t$ have a one-to-one projection onto the t axis, i.e., they are given by functions $\mathbf{p} = \mathbf{p}(t), \mathbf{q} = \mathbf{q}(t)$. These functions satisfy the system of canonical differential equations with hamiltonian function H:*

$$(1) \qquad \frac{d\mathbf{p}}{dt} = -\frac{\partial H}{\partial \mathbf{q}}, \qquad \frac{d\mathbf{q}}{dt} = \frac{\partial H}{\partial \mathbf{p}}.$$

In other words, the vortex lines of the form $\mathbf{p}\, d\mathbf{q} - H\, dt$ are the trajectories of the phase flow in the extended phase space, i.e., the integral curves of the canonical equations (1).

---

[73] The form ω^1 seems here to appear out of thin air. In the following paragraph we will see how the idea of using this form arose from optics.

PROOF. The differential of the form $\mathbf{p} \, d\mathbf{q} - H \, dt$ is equal to

$$d\omega^1 = \sum_{i=1}^{n} \left(dp_i \wedge dq_i - \frac{\partial H}{\partial p_i} \, dp_i \wedge dt - \frac{\partial H}{\partial q_i} \, dq_i \wedge dt \right).$$

It is clear from this expression that the matrix of the 2-form $d\omega^1$ in the coordinates $\mathbf{p}, \mathbf{q}, t$ has the form

$$A = \begin{bmatrix} 0 & -E & H_{\mathbf{p}} \\ E & 0 & H_{\mathbf{q}} \\ -H_{\mathbf{p}} & -H_{\mathbf{q}} & 0 \end{bmatrix},$$

where

$$E = \underbrace{\begin{bmatrix} 1 & & \\ & \ddots & \\ & & 1 \end{bmatrix}}_{n}, H_{\mathbf{p}} = \frac{\partial H}{\partial \mathbf{p}}, H_{\mathbf{q}} = \frac{\partial H}{\partial \mathbf{q}}$$

(verify this!).

The rank of this matrix is $2n$ (the upper left $2n$-corner is non-degenerate); therefore, $d\omega^1$ is nonsingular. It can be verified directly that the vector $(-H_{\mathbf{q}}, H_{\mathbf{p}}, 1)$ is an eigenvector of A with eigenvalue 0 (do it!). This means that it gives the direction of the vortex lines of the form $\mathbf{p} \, d\mathbf{q} - H \, dt$. But the vector $(-H_{\mathbf{q}}, H_{\mathbf{p}}, 1)$ is also the velocity vector of the phase flow of (1). Thus the integral curves of (1) are the vortex lines of the form $\mathbf{p} \, d\mathbf{q} - H \, dt$, as was to be shown. $\square$

D A theorem on the integral invariant of Poincaré–Cartan

We now apply Stokes' lemma. We obtain the fundamental

Theorem. *Suppose that the two curves γ_1 and γ_2 encircle the same tube of phase trajectories of (1). Then the integrals of the form $\mathbf{p} \, d\mathbf{q} - H \, dt$ along them are the same:*

$$\oint_{\gamma_1} \mathbf{p} \, d\mathbf{q} - H \, dt = \oint_{\gamma_2} \mathbf{p} \, d\mathbf{q} - H \, dt.$$

The form $\mathbf{p} \, d\mathbf{q} - H \, dt$ is called the *integral invariant of Poincaré–Cartan*.[74]

PROOF. The phase trajectories are the vortex lines of the form $\mathbf{p} \, d\mathbf{q} - H \, dt$, and the integrals along closed curves contained in the same vortex tube are the same by Stokes' lemma. $\square$

[74] In the calculus of variations $\int \mathbf{p} \, d\mathbf{q} - H \, dt$ is called *Hilbert's invariant integral.*

237

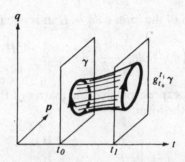

Figure 183   Poincaré's integral invariant

We will consider, in particular, curves consisting of simultaneous states, i.e., lying in the planes $t = $ const (Figure 183). Along such curves, $dt = 0$ and $\oint \mathbf{p}\, d\mathbf{q} - H\, dt = \oint \mathbf{p}\, d\mathbf{q}$. From the preceding theorem we obtain the important:

Corollary 1. *The phase flow preserves the integral of the form* $\mathbf{p}\, d\mathbf{q} = p_1\, dq_1 + \cdots + p_n\, dq_n$ *on closed curves.*

PROOF. Let $g_{t_0}^{t_1}: \mathbb{R}^{2n} \to \mathbb{R}^{2n}$ be the transformation of the phase space $(\mathbf{p}, \mathbf{q})$ realized by the phase flow from time t_0 to t_1 (i.e., $g_{t_0}^{t}(\mathbf{p}_0, \mathbf{q}_0)$ is the solution to the canonical equations (1) with initial conditions $\mathbf{p}(t_0) = \mathbf{p}_0, \mathbf{q}(t_0) = \mathbf{q}_0$). Let γ be any closed curve in the space $\mathbb{R}^{2n} \subset \mathbb{R}^{2n+1}$ $(t = t_0)$. Then $g_{t_0}^{t_1}\gamma$ is a closed curve in the space $\mathbb{R}^{2n}$ $(t = t_1)$, contained in the same tube of phase trajectories in $\mathbb{R}^{2n+1}$. Since $dt = 0$ on γ and on $g_{t_0}^{t_1}\gamma$ we find by the preceding theorem that $\int_{\gamma} \mathbf{p}\, d\mathbf{q} = \int_{g_{t_0}^{t_1}\gamma} \mathbf{p}\, d\mathbf{q}$, as was to be shown.   □

The form $\mathbf{p}\, d\mathbf{q}$ is called *Poincaré's relative integral invariant*. It has a simple geometric meaning. Let σ be a two-dimensional oriented chain and $\gamma = \partial\sigma$. Then, by Stokes' formula, we find

$$\oint_{\gamma} \mathbf{p}\, d\mathbf{q} = \iint_{\sigma} d\mathbf{p} \wedge d\mathbf{q}.$$

Thus we have proved the important:

Corollary 2. *The phase flow preserves the sum of the oriented areas of the projections of a surface onto the n coordinate planes* (p_i, q_i):

$$\iint_{\sigma} d\mathbf{p} \wedge d\mathbf{q} = \iint_{g_{t_0}^{t_1}\sigma} d\mathbf{p} \wedge d\mathbf{q}.$$

In other words, the 2-form $\omega^2 = d\mathbf{p} \wedge d\mathbf{q}$ *is an absolute integral invariant of the phase flow.*

EXAMPLE. For $n = 1$, ω^2 is area, and we obtain Liouville's theorem: the phase flow preserves area.

E Canonical transformations

Let g be a differentiable mapping of the phase space $\mathbb{R}^{2n} = \{(\mathbf{p}, \mathbf{q})\}$ to $\mathbb{R}^{2n}$.

Definition. The mapping g is called *canonical*, or a *canonical transformation*, if g preserves the 2-form $\omega^2 = \sum dp_i \wedge dq_i$.

It is clear from the argument above that this definition can be written in any of three equivalent forms:

1. $g^*\omega^2 = \omega^2$ (g preserves the 2-form $\sum dp_i \wedge dq_i$);
2. $\iint_\sigma \omega^2 = \iint_{g\sigma} \omega^2$, $\forall \sigma$ (g preserves the sum of the areas of the projections of any surface);
3. $\oint_\gamma \mathbf{p}\, d\mathbf{q} = \oint_{g\gamma} \mathbf{p}\, d\mathbf{q}$ (the form $\mathbf{p}\, d\mathbf{q}$ is a relative integral invariant of g).

PROBLEM. Show that definitions (1) and (2) are equivalent to (3) if the domain of the map in question is a simply connected region in the phase space $\mathbb{R}^{2n}$: in the general case $3 \Rightarrow 2 \Leftrightarrow 1$.

The corollaries above can now be formulated as:

Theorem. *The transformation of phase space induced by the phase flow is canonical.*[75]

Let $g\colon \mathbb{R}^{2n} \to \mathbb{R}^{2n}$ be a canonical transformation: g preserves the form ω^2. Then g also preserves the exterior square of ω^2:

$$g^*(\omega^2 \wedge \omega^2) = \omega^2 \wedge \omega^2 \quad \text{and} \quad g^*(\omega^2)^k = (\omega^2)^k.$$

The exterior powers of the form $\sum dp_i \wedge dq_i$ are proportional to the forms

$$\omega^4 = \sum_{i<j} dp_i \wedge dp_j \wedge dq_i \wedge dq_j,$$

$$\omega^{2k} = \sum_{i_1 < \cdots < i_k} dp_{i_1} \wedge \cdots \wedge dp_{i_k} \wedge dq_{i_1} \wedge \cdots \wedge dq_{i_k}.$$

Thus we have proved

Theorem. *Canonical transformations preserve the integral invariants* $\omega^4, \ldots, \omega^{2n}$.

Geometrically, the integral of the form ω^{2k} is the sum of the oriented volumes of the projections onto the coordinate planes $(p_{i_1}, \ldots, p_{i_k}, q_{i_1}, \ldots, q_{i_k})$. In particular, ω^{2n} is proportional to the volume element, and we obtain:

Corollary. *Canonical transformations preserve the volume element in phase space:*

the volume of gD is equal to the volume of D, for any region D.

---

[75] The proof of this theorem which is presented in the excellent book by Landau and Lifshitz (*Mechanics*, Pergamon, Oxford, 1960) is incorrect.

In particular, applying this to the phase flow we obtain

Corollary. *The phase flow* (1) *has as integral invariants the forms* $\omega^2, \omega^4, \ldots, \omega^{2n}$.

The last of these invariants is the phase volume, so we have again proved Liouville's theorem.

45 Applications of the integral invariant of Poincaré–Cartan

In this paragraph we prove that canonical transformations preserve the form of Hamilton's equations, that a first integral of Hamilton's equations allows us to reduce immediately the order of the system by two and that motion in a natural lagrangian system proceeds along geodesics of the configuration space provided with a certain riemannian metric.

A *Changes of variables in the canonical equations*

The invariant nature of the connection between the form $\mathbf{p}\, d\mathbf{q} - H\, dt$ and its curl lines gives rise to a way of writing the equations of motion in any system of $2n + 1$ coordinates in extended phase space $\{(\mathbf{p}, \mathbf{q}, t)\}$.

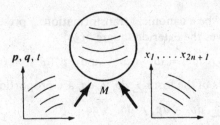

Figure 184   Change of variables in Hamilton's equations

Let $(x_1, \ldots, x_{2n+1})$ be coordinate functions in some chart of extended phase space (considered as a manifold M^{2n+1}, Figure 184). The coordinates $(\mathbf{p}, \mathbf{q}, t)$ can be considered as giving another chart on M. The form $\omega^1 = \mathbf{p}\, d\mathbf{q} - H\, dt$ can be considered as a differential 1-form on M. Invariantly associated (not depending on the chart) to this form is a family of lines on M — the vortex lines. In the chart $(\mathbf{p}, \mathbf{q}, t)$, these lines are represented as the trajectories of the phase flow

$$(1) \qquad \frac{d\mathbf{p}}{dt} = -\frac{\partial H}{\partial \mathbf{q}} \quad, \quad \frac{d\mathbf{q}}{dt} = \frac{\partial H}{\partial \mathbf{p}}$$

with hamiltonian function $H(\mathbf{p}, \mathbf{q}, t)$.

Suppose that in the coordinates $(x_1, \ldots, x_{2n+1})$ the form ω^1 is written as

$$\mathbf{p}\, d\mathbf{q} - H\, dt = X_1\, dx_1 + \cdots + X_{2n+1}\, dx_{2n+1}.$$

Theorem. *In the chart* (x_i), *the trajectories of* (1) *are represented by the vortex lines of the form* $\sum X_i \, dx_i$.

PROOF. The curl lines of the forms $\sum X_i \, dx_i$ and $\mathbf{p} \, d\mathbf{q} - H \, dt$ are the images in two different charts of the vortex lines of the same form on M. But the integral curves of (1) are the vortex lines of $\mathbf{p} \, d\mathbf{q} - H \, dt$. Thus, their images in the chart (x_i) are the vortex lines of the form $\sum X_i \, dx_i$. $\qquad\square$

Corollary. *Let* $(P_1, \ldots, P_n; Q_1, \ldots, Q_n; T)$ *be a coordinate system on the extended phase space* $(\mathbf{p}, \mathbf{q}, t)$ *and* $K(\mathbf{P}, \mathbf{Q} \, T)$ *and* $S(\mathbf{P}, \mathbf{Q}, T)$ *functions such that*

$$\mathbf{p} \, d\mathbf{q} - H \, dt = \mathbf{P} \, d\mathbf{Q} - K \, dT + dS$$

(the left- and right-hand sides are forms on extended phase space).

Then the trajectories of the phase flow (1) *are represented in the chart* $(\mathbf{P}, \mathbf{Q}, T)$ *by the integral curves of the canonical equations*

$$(2) \qquad \frac{d\mathbf{P}}{dT} = -\frac{\partial K}{\partial \mathbf{Q}} \qquad \frac{d\mathbf{Q}}{dT} = \frac{\partial K}{\partial \mathbf{P}}.$$

PROOF. By the theorem above, the trajectories of (1) are represented by the vortex lines of the form $\mathbf{P} \, d\mathbf{Q} - K \, dT + dS$. But dS has no influence on the vortex lines (since $ddS = 0$). Therefore, the images of the trajectories of (1) are the vortex lines of the form $\mathbf{P} \, d\mathbf{Q} - K \, dT$. According to Section 44C, the vortex lines of such a form are integral curves of the canonical equations (2). $\qquad\square$

In particular, let $g: \mathbb{R}^{2n} \to \mathbb{R}^{2n}$ be a canonical transformation of phase space taking a point with coordinates $(\mathbf{p}, \mathbf{q})$ to a point with coordinates $(\mathbf{P}, \mathbf{Q})$. The functions $\mathbf{P}(\mathbf{p}, \mathbf{q})$ and $\mathbf{Q}(\mathbf{p}, \mathbf{q})$ can be considered as new coordinates on phase space.

Theorem. *In the new coordinates* $(\mathbf{P}, \mathbf{Q})$ *the canonical equations* (1) *have the canonical form*[76]

$$(3) \qquad \frac{d\mathbf{P}}{dt} = -\frac{\partial K}{\partial \mathbf{Q}} \qquad \frac{d\mathbf{Q}}{dt} = \frac{\partial K}{\partial \mathbf{P}}$$

with the same hamiltonian function: $K(\mathbf{P}, \mathbf{Q}, t) = H(\mathbf{p}, \mathbf{q}, t)$.

---

[76] In some textbooks the property of preserving the canonical form of Hamilton's equations is taken as the definition of a canonical transformation. This definition is not equivalent to the generally accepted one mentioned above. For example, the transformation $P = 2p$, $Q = q$, which is not canonical by our definition, preserves the hamiltonian form of the equations of motion. This confusion appears even in the excellent textbook by Landau and Lifshitz (*Mechanics*, Oxford, Pergamon, 1960); in Section 45 of this book they show that every transformation which preserves the canonical equations is canonical in our sense.

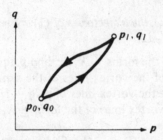

Figure 185   Closedness of the form $\mathbf{p}\,d\mathbf{q} - \mathbf{P}\,d\mathbf{Q}$

PROOF. Consider the 1-form $\mathbf{p}\,d\mathbf{q} - \mathbf{P}\,d\mathbf{Q}$ on $\mathbb{R}^{2n}$. For any closed curve γ we have (Figure 185)

$$\oint_\gamma \mathbf{p}\,d\mathbf{q} - \mathbf{P}\,d\mathbf{Q} = \oint_\gamma \mathbf{p}\,d\mathbf{q} - \oint_\gamma \mathbf{P}\,d\mathbf{Q} = 0$$

since g is canonical. Therefore, $\int_{\mathbf{p}_0,\mathbf{q}_0}^{\mathbf{p}_1,\mathbf{q}_1} \mathbf{p}\,d\mathbf{q} - \mathbf{P}\,d\mathbf{Q} = S$ does not depend on the path of integration but only on the endpoint $(\mathbf{p}_1, \mathbf{q}_1)$ (for a fixed initial point $(\mathbf{p}_0, \mathbf{q}_0)$). Thus $dS = \mathbf{p}\,d\mathbf{q} - \mathbf{P}\,d\mathbf{Q}$. Consequently, in the extended phase space, we have

$$\mathbf{p}\,d\mathbf{q} - H\,dt = \mathbf{P}\,d\mathbf{Q} - H\,dt + dS.$$

Thus, the theorem above is applicable, and (2) is transformed to (3).   □

PROBLEM. Let $g(t): \mathbb{R}^{2n} \to \mathbb{R}^{2n}$ be a canonical transformation of phase space depending on the parameter t, $g(t)(\mathbf{p}, \mathbf{q}) = (\mathbf{P}(\mathbf{p}, \mathbf{q}, t), \mathbf{Q}(\mathbf{p}, \mathbf{q}, t))$. Show that in the variables $\mathbf{P}, \mathbf{Q}, t$ the canonical equations (1) have the canonical form with new hamiltonian function

$$K(\mathbf{P}, \mathbf{Q}, t) = H(\mathbf{p}, \mathbf{q}, t) + \frac{\partial S}{\partial t},$$

where

$$S(\mathbf{p}_1, \mathbf{q}_1, t) = \int_{\mathbf{p}_0, \mathbf{q}_0}^{\mathbf{p}_1, \mathbf{q}_1} \mathbf{p}\,d\mathbf{q} - \mathbf{P}\,d\mathbf{Q} \quad (d\mathbf{Q} \text{ for fixed } t)$$

B Reduction of order using the energy integral

Suppose now that the hamiltonian function $H(\mathbf{p}, \mathbf{q})$ does not depend on time. Then the canonical equations (1) have a first integral: $H(\mathbf{p}(t), \mathbf{q}(t)) = \text{const}$. It turns out that by using this integral we can reduce the dimension $(2n + 1)$ of the extended phase space by two, thereby reducing the problem to integration of a system of canonical equations in a $(2n - 1)$-dimensional space.

We assume that (in some region) the equation $h = H(p_1, \ldots, p_n; q_1, \ldots, q_n)$ can be solved for p_1:

$$p_1 = K(\mathbf{P}, \mathbf{Q}, T; h),$$

where $\mathbf{P} = (p_2, \ldots, p_n)$; $\mathbf{Q} = (q_2, \ldots, q_n)$; $T = -q_1$. Then we find

$$\mathbf{p}\, d\mathbf{q} - H\, dt = \mathbf{P}\, d\mathbf{Q} - K\, dT - d(Ht) + t\, dH.$$

Now let γ be an integral curve of the canonical equations (1) lying on the $2n$-dimensional surface $H(\mathbf{p}, \mathbf{q}) = h$ in $\mathbb{R}^{2n+1}$. Then γ is a vortex line of the form $\mathbf{p}\, d\mathbf{q} - H\, dt$ (Figure 186). We project the extended phase space $\mathbb{R}^{2n+1} = \{(\mathbf{p}, \mathbf{q}, t)\}$ onto the phase space $\mathbb{R}^{2n} = \{(\mathbf{p}, \mathbf{q})\}$. The surface $H = h$ is projected onto a $(2n-1)$-dimensional manifold $M^{2n-1} : H(\mathbf{p}, \mathbf{q}) = h$ in $\mathbb{R}^{2n}$, and γ is projected to a curve $\bar{\gamma}$ lying on this submanifold. The variables $\mathbf{P}, \mathbf{Q}, T$ form local coordinates on M^{2n-1}.

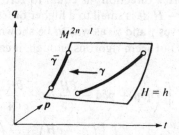

Figure 186   Lowering the order of a hamiltonian system

PROBLEM. Show that the curve $\bar{\gamma}$ is a vortex line of the form $\mathbf{p}\, d\mathbf{q} = \mathbf{P}\, d\mathbf{Q} - K\, dT$ on M^{2n-1}.
Hint. $d(Ht)$ does not affect the vortex lines, and dH is zero on M.

But the vortex lines of $\mathbf{P}\, d\mathbf{Q} - K\, dT$ satisfy Hamilton's equations (2). Thus we have proved

Theorem. *The phase trajectories of the equations* (1) *on the surface* M^{2n-1}, $H = h$, *satisfy the canonical equations*

$$\frac{dp_i}{dq_1} = \frac{\partial K}{\partial q_i} \qquad \frac{dq_i}{dq_1} = -\frac{\partial K}{\partial p_i}, \qquad (i = 2, \ldots, n),$$

where the function $K(p_2, \ldots, p_n; q_2, \ldots, q_n; T, h)$ *is defined by the equation* $H(K, p_2, \ldots, p_n; -T, q_2, \ldots, q_n) = h$.

C *The principle of least action in phase space*

In the extended phase space $\{(\mathbf{p}, \mathbf{q}, t)\}$, we consider an integral curve of the canonical equations (1) connecting the points $(\mathbf{p}_0, \mathbf{q}_0, t_0)$ and $(\mathbf{p}_1, \mathbf{q}_1, t_1)$.

Theorem. *The integral* $\int \mathbf{p}\, d\mathbf{q} - H\, dt$ *has* γ *as an extremal under variations of* γ *for which the ends of the curve remain in the* n-*dimensional subspaces* $(t = t_0, \mathbf{q} = \mathbf{q}_0)$ *and* $(t = t_1, \mathbf{q} = \mathbf{q}_1)$.

PROOF. The curve γ is a vortex line of the form $\mathbf{p}\, d\mathbf{q} - H\, dt$ (Figure 187). Therefore, the integral of $\mathbf{p}\, d\mathbf{q} - H\, dt$ over an "infinitely small parallelogram

243

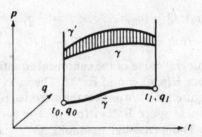

Figure 187 Principle of least action in phase space

passing through the vortex direction" is equal to zero. In other words, the increment $\int_{\gamma'} - \int_{\gamma} \mathbf{p}\, d\mathbf{q} - H\, dt$ is small to a higher order in comparison with the difference of the curves γ and γ', as was to be shown.

If this argument does not seem rigorous enough, it can be replaced by the computation

$$\delta \int_{\gamma} (\mathbf{p}\dot{\mathbf{q}} - H)dt = \int_{\gamma} \left(\dot{\mathbf{q}}\delta\mathbf{p} + \mathbf{p}\delta\dot{\mathbf{q}} - \frac{\partial H}{\partial \mathbf{p}}\,\delta\mathbf{p} - \frac{\partial H}{\partial \mathbf{q}}\,\delta\mathbf{q} \right)dt$$

$$= \mathbf{p}\,\delta\mathbf{q}\Big|_0^1 + \int_{\gamma} \left[\left(\dot{\mathbf{q}} - \frac{\partial H}{\partial \mathbf{p}} \right)\delta\mathbf{p} - \left(\dot{\mathbf{p}} + \frac{\partial H}{\partial \mathbf{q}} \right)\delta\mathbf{q} \right]dt.$$

We see that the integral curves of Hamilton's equations are the only extremals of the integral $\int \mathbf{p}\, d\mathbf{q} - H\, dt$ in the class of curves γ whose ends lie in the n-dimensional subspaces $(t = t_0, \mathbf{q} = \mathbf{q}_0)$ and $(t = t_1, \mathbf{q} = \mathbf{q}_1)$ of extended phase space. $\qquad\qquad\Box$

Remark. The principle of least action in Hamilton's form is a particular case of the principle considered above. Along extremals, we have

$$\int_{t_0, \mathbf{q}_0}^{t_1, \mathbf{q}_1} \mathbf{p}\, d\mathbf{q} - H\, dt = \int_{t_0}^{t_1} (\mathbf{p}\dot{\mathbf{q}} - H)\, dt = \int_{t_0}^{t_1} L\, dt$$

(since the lagrangian L and the hamiltonian H are Legendre transforms of one another). Now let $\bar{\gamma}$ (Figure 188) be the projection of the extremal γ onto the $\mathbf{q}, t$ plane. To any nearby curve $\bar{\gamma}'$ connecting the same points $(t_0, \mathbf{q}_0)$ and $(t_1, \mathbf{q}_1)$ in the $\mathbf{q}, t$ plane we associate a curve γ' in the

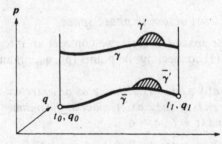

Figure 188 Comparison curves for the principles of least action in the configuration and phase spaces

phase space $(\mathbf{p}, \mathbf{q}, t)$ by setting $\mathbf{p} = \partial L/\partial \dot{\mathbf{q}}$. Then, along γ', too, $\int_{\gamma'} \mathbf{p} \, d\mathbf{q} - H \, dt = \int_{\gamma'} L \, dt$. But by the theorem above, $\delta \int_\gamma \mathbf{p} \, d\mathbf{q} - H \, dt = 0$ for any variation curve γ (with boundary conditions $(t = t_0, \mathbf{q} = \mathbf{q}_0)$ and $(t = t_1, \mathbf{q} = \mathbf{q}_1)$. In particular, this is true for variations of the special form taking γ to γ'. Thus γ is an extremal of $\int L \, dt$, as was to be shown.

In the theorem above we are allowed to compare γ with a significantly wider class of curves γ' than in Hamilton's principle: there are no restrictions placed on the relation of $\mathbf{p}$ with $\dot{\mathbf{q}}$. Surprisingly, one can show that the two principles are nevertheless equivalent: an extremal in the narrower class of variations ($\mathbf{p} = \partial L/\partial \dot{\mathbf{q}}$) is an extremal under all variations. The explanation is that, for fixed $\dot{\mathbf{q}}$, the value $\mathbf{p} = \partial L/\partial \dot{\mathbf{q}}$ is an extremal of $\mathbf{p}\dot{\mathbf{q}} - H$ (cf. the definition of the Legendre transform, Section 14).

D The principle of least action in the Maupertuis–Euler–Lagrange–Jacobi form

Suppose now that the hamiltonian function $H(\mathbf{p}, \mathbf{q})$ does not depend on time. Then $H(\mathbf{p}, \mathbf{q})$ is a first integral of Hamilton's equations (1). We project the surface $H(\mathbf{p}, \mathbf{q}) = h$ from the extended phase space $\{(\mathbf{p}, \mathbf{q}, t)\}$ to the space $\{(\mathbf{p}, \mathbf{q})\}$. We obtain a $(2n - 1)$-dimensional surface $H(\mathbf{p}, \mathbf{q}) = h$ in $\mathbb{R}^{2n}$, which we already studied in subsection B and which we denoted by M^{2n-1}.

The phase trajectories of the canonical equations (1) beginning on the surface M^{2n-1} lie entirely in M^{2n-1}. They are the vortex lines of the form $\mathbf{p} \, d\mathbf{q} = \mathbf{P} \, d\mathbf{Q} - K \, dT$ (in the notation of B) on M^{2n-1}. By the theorem in subsection C, the curves (1) on M^{2n-1} are extremals for the variational principle corresponding to this form. Therefore, we have proved

Theorem. *If the hamiltonian function $H = H(\mathbf{p}, \mathbf{q})$ does not depend on time, then the phase trajectories of the canonical equations (1) lying on the surface $M^{2n-1}: H(\mathbf{p}, \mathbf{q}) = h$ are extremals of the integral $\int \mathbf{p} \, d\mathbf{q}$ in the class of curves lying on M^{2n-1} and connecting the subspaces $\mathbf{q} = \mathbf{q}_0$ and $\mathbf{q} = \mathbf{q}_1$.*

We now consider the projection onto the $\mathbf{q}$-space of an extremal lying on the surface $M^{2n-1}: H(\mathbf{p}, \mathbf{q}) = h$. This curve connects the points $\mathbf{q}_0$ and $\mathbf{q}_1$. Let γ be another curve connecting the points $\mathbf{q}_0$ and $\mathbf{q}_1$ (Figure 189). The curve γ is the projection of some curve $\hat{\gamma}$ on M^{2n-1}. Specifically, we

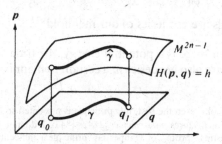

Figure 189 Maupertuis' principle

parametrize γ by τ, $a \leq \tau \leq b$, $\gamma(a) = \mathbf{q}_0$, $\gamma(b) = \mathbf{q}_1$. Then at every point $\mathbf{q}$ of γ there is a velocity vector $\dot{\mathbf{q}} = d\gamma(\tau)/d\tau$, and the corresponding momentum $\mathbf{p} = \partial L/\partial \dot{\mathbf{q}}$. If the parameter τ is chosen so that $H(\mathbf{p}, \mathbf{q}) = h$, then we obtain a curve $\hat{\gamma}: \mathbf{q} = \gamma(\tau)$, $\mathbf{p} = \partial L/\partial \dot{\mathbf{q}}$ on the surface M^{2n-1}. Applying the theorem above to the curve $\hat{\gamma}$ on M^{2n-1}, we obtain

Corollary. *Among all curves* $\mathbf{q} = \gamma(\tau)$ *connecting the two points* $\mathbf{q}_0$ *and* $\mathbf{q}_1$ *on the plane* $\mathbf{q}$ *and parametrized so that the hamiltonian function has a fixed value* $H(\partial L/\partial \dot{\mathbf{q}}, \mathbf{q}) = h$, *the trajectory of the equations of dynamics* (1) *is an extremal of the integral of "reduced action"*

$$\int_\gamma \mathbf{p} \, d\mathbf{q} = \int_\gamma \mathbf{p}\dot{\mathbf{q}} \, d\tau = \int_\gamma \frac{\partial L}{\partial \dot{\mathbf{q}}}(\tau)\dot{\mathbf{q}}(\tau) \, d\tau.$$

This is also the principle of least action of Maupertuis (Euler–Lagrange–Jacobi).[77] It is important to note that the interval $a \leq \tau \leq b$ parametrizing the curve γ is not fixed and can be different for different curves being compared. On the other hand, the energy (the hamiltonian function) must be the same. We note also that the principle determines the shape of a trajectory but not the time: in order to determine the time we must use the energy constant.

The principle above takes a particularly simple form in the case when the system represents inertial motion on a smooth manifold.

Theorem. *A point mass confined to a smooth riemannian manifold moves along geodesic lines (i.e., along extremals of the length $\int ds$).*

PROOF. In this case,

$$H = L = T = \frac{1}{2}\left(\frac{ds}{d\tau}\right)^2 \quad \text{and} \quad \frac{\partial L}{\partial \dot{\mathbf{q}}}\dot{\mathbf{q}} = 2T = \left(\frac{ds}{d\tau}\right)^2.$$

Therefore, in order to guarantee a fixed value of $H = h$, the parameter must be chosen proportional to the length $d\tau = ds/\sqrt{2h}$. The reduced action integral is then equal to

$$\int_\gamma \frac{\partial L}{\partial \dot{\mathbf{q}}}\dot{\mathbf{q}} \, d\tau = \int_\gamma \sqrt{2h} \, ds = \sqrt{2h}\int_\gamma ds;$$

therefore, extremals are geodesics of our manifold. $\qquad\square$

In the case when there is a potential energy, the trajectories of the equations of dynamics are also geodesics in a certain riemannian metric.

---

[77] "In almost all textbooks, even the best, this principle is presented so that it is impossible to understand." (C. Jacobi, *Lectures on Dynamics*, 1842–1843). I do not choose to break with tradition. A very interesting "proof" of Maupertuis' principle is in Section 44 of the mechanics textbook of Landau and Lifshitz (*Mechanics*, Oxford, Pergamon, 1960).

Let ds^2 be a riemannian metric on configuration space which gives the kinetic energy (so that $T = \frac{1}{2}(ds/d\tau)^2$). Let h be a constant.

Theorem. *In the region of configuration space where* $U(\mathbf{q}) < h$ *we define a riemannian metric by the formula*

$$d\rho = \sqrt{h - U(\mathbf{q})}\, ds.$$

Then the trajectories of the system with kinetic energy $T = \frac{1}{2}(ds/d\tau)^2$, *potential energy* $U(\mathbf{q})$, *and total energy* h *will be geodesic lines of the metric* $d\rho$.

PROOF. In this case $L = T - U$, $H = T + U$, and $(\partial L/\partial \dot{\mathbf{q}})\dot{\mathbf{q}} = 2T = (ds/d\tau)^2 = 2(h - U)$. Therefore, in order to guarantee a fixed value of $H = h$, the parameter τ must be chosen proportional to length: $d\tau = ds/\sqrt{2(h - U)}$. The reduced action integral will then be equal to

$$\int_\gamma \frac{\partial L}{\partial \dot{\mathbf{q}}}\dot{\mathbf{q}}\, d\tau = \int_\gamma \sqrt{2(h - U)}\, ds = \sqrt{2}\int_\gamma d\rho.$$

By Maupertuis' principle, the trajectories are geodesics in the metric $d\rho$, as was to be shown. $\qquad\square$

Remark 1. The metric $d\rho$ is obtained from ds by a "stretching" depending on the point $\mathbf{q}$ but not depending on the direction. Therefore, angles in the metric $d\rho$ are the same as angles in the metric ds. On the boundary of the region $U \leq h$ the metric $d\rho$ has a singularity: the closer we come to the boundary, the smaller the ρ-length becomes. In particular, the length of any curve lying in the boundary ($U = h$) is equal to zero.

Remark 2. If the initial and endpoints of a geodesic γ are sufficiently close, then the extremum of length is a *minimum*. This justifies the name "principle of *least* action." In general, an extremum of the action is not necessarily a minimum, as we see by considering geodesics on the unit sphere (Figure 190). Every arc of a great circle is a geodesic, but only those with length less than π are minimal: the arc $NS'M$ is shorter than the great circle arc NSM.

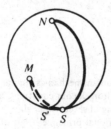

Figure 190   Non-minimal geodesic

Remark 3. If h is larger than the maximum value of U on the configuration space, then the metric $d\rho$ has no singularities; therefore, we can apply topological theorems about geodesics on riemannian manifolds to the study of mechanical systems. For example, we consider the torus T^2 with some riemannian metric. Among all closed curves on T^2 making m rotations

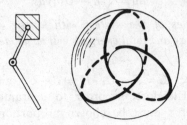

Figure 191   Periodic motion of a double pendulum

around the parallel and n around the meridian, there exists a curve of shortest length (Figure 191). This curve is a closed geodesic (for a proof see books on the calculus of variations or "Morse theory"). On the other hand, the torus T^2 is the configuration space of a planar double pendulum. Therefore,

Theorem. *For any integers m and n there is a periodic motion of the double pendulum under which one segment makes m rotations while the other segment makes n rotations.*

Furthermore, such periodic motions exist for any sufficiently large values of the constant h (h must be larger than the potential energy at the highest position).

As a last example we consider a rigid body fastened at a stationary point and located in an arbitrary potential field. The configuration space $(SO(3))$ is not simply connected: there exist non-contractible curves in it. The above arguments imply

Theorem. *In any potential force field, there exists at least one periodic motion of the body. Furthermore, there exist periodic motions for which the total energy h is arbitrarily large.*

46 Huygens' principle

The fundamental notions of hamiltonian mechanics (momenta, the hamiltonian function H, the form $\mathbf{p}\,d\mathbf{q} - H\,dt$ and the Hamilton–Jacobi equations, all of which we will be concerned with below) arose by the transforming of several very simple and natural notions of geometric optics, guided by a particular variational principle—that of Fermat, into general variational principles (and in particular into Hamilton's principle of stationary action, $\delta \int L\,dt = 0$).

A *Wave fronts*

We consider briefly[78] the fundamental notions of geometric optics. According to the extremal principle of Fermat, *light travels from a point* q_0 *to a point* q_1 *in the shortest possible time.* The speed of the light can depend both on the point q (an "inhomogeneous medium") and on the direction of the ray (in an "anisotropic medium," such as a crystal). The characteristics of a medium can be described by giving a surface (the "indicatrix") in the tangent space at each point q. To do this, we take in every direction the velocity vector of the propagation of light at the given point in the given direction (Figure 192).

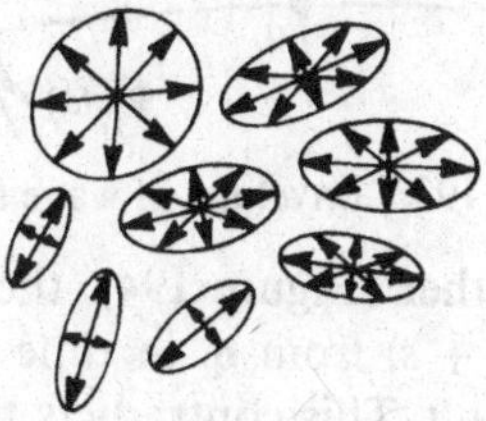

Figure 192   An anisotropic, inhomogeneous medium

Now let $t > 0$. We look at the set of all points q to which light from a given point q_0 can travel in time less than or equal to t. The boundary of this set, $\Phi_{q_0}(t)$, is called the *wave front* of the point q_0 after time t and consists of points to which light can travel in time t and not faster.

There is a remarkable relation, discovered by Huygens, between the wave fronts corresponding to different values of t. (Figure 193)

Huygens' theorem. *Let* $\Phi_{q_0}(t)$ *be the wave front of the point* q_0 *after time* t. *For every point* q *of this front, consider the wave front after time* s, $\Phi_q(s)$. *Then the wave front of the point* q_0 *after time* $s + t$, $\Phi_{q_0}(s + t)$, *will be the envelope of the fronts* $\Phi_q(s)$, $q \in \Phi_{q_0}(t)$.

Proof. Let $q_{t+s} \in \Phi_{q_0}(t + s)$. Then there exists a path from q_0 to q_{t+s} along which the time of travel of light equals $t + s$, and there is none shorter. We look at the point q_t on this path, to which light travels in time t. No shorter path from q_0 to q_t can exist; otherwise, the path $q_0 q_{t+s}$ would not be the shortest. Therefore, the point q_t lies on the front $\Phi_{q_0}(t)$. In exactly the same way light travels the path $q_t q_{t+s}$ in time s, and there is no shorter path from q_t to q_{t+s}. Therefore, the point q_{t+s} lies on the front of the point q_t at time s, $\Phi_{q_t}(s)$. We will show that the fronts $\Phi_{q_t}(s)$ and $\Phi_{q_0}(t + s)$ are tangent. In

---

[78] We will not pursue rigor here, and will assume that all determinants are different from zero, etc. The proofs of the subsequent theorems do not depend on the semi-heuristic arguments of this paragraph. It should be noted that the appropriate lagrangian for geometric optics is homogeneous of order 1 in the velocities. To apply the Legendre transform, and to make the analogy with mechanics in the following section, we should square this lagrangian, which does not affect the indicatrix surface where the value is 1. In fact, the real meaning of Huygens' principle is best expressed in contact geometry (see Appendix 4 or the author's *Singularities of Caustics and Wave Fronts*, Kluwer 1990).

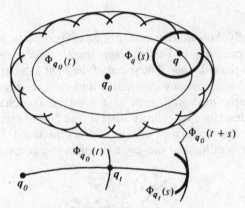

Figure 193   Envelope of wave fronts

fact, if they crossed each other (Figure 194), then it would be possible to reach some points of $\Phi_{q_0}(t + s)$ from q_t in time less than s, and therefore from q_0 in time less than $s + t$. This contradicts the definition of $\Phi_{q_0}(t + s)$; and so the fronts $\Phi_{q_t}(s)$ and $\Phi_{q_0}(t + s)$ are tangent at the point q_{t+s}, as was to be proved. $\qquad\square$

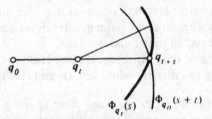

Figure 194   Proof of Huygens' theorem

The theorem which has been proved is called *Huygens' principle*. It is clear that the point q_0 could be replaced by a curve, surface, or, in general, by a closed set, the three-dimensional space $\{q\}$ by any smooth manifold, and propagation of light by the propagation of any disturbance transmitting itself "locally."

Huygens' principle reduces to two descriptions of the process of propagation. First, we can trace the *rays*, i.e., the shortest paths of the propagation of light. In this case the local character of the propagation is given by a velocity vector $\dot{\mathbf{q}}$. If the direction of the ray is known, then the magnitude of the velocity vector is given by the characteristics of the medium (the indicatrix).

On the other hand, we can trace the wave fronts. Assuming that we are given a riemannian metric on the space $\{q\}$, we can talk about the *velocity of motion of the wave front*. We look, for example, at the propagation of light in a medium filling ordinary euclidean space. Then one can characterize the motion of the wave front by a vector $\mathbf{p}$ perpendicular to the front, which will be constructed in the following manner.

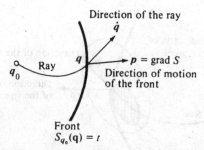

Figure 195   Direction of a ray and direction of motion of the wave front

For every point $\mathbf{q}_0$ we define the function $S_{\mathbf{q}_0}(\mathbf{q})$ as the *optical length of the path from* $\mathbf{q}_0$ *to* $\mathbf{q}$, i.e., the least time of the propagation of light from $\mathbf{q}_0$ to $\mathbf{q}$. The level set $\{\mathbf{q}: S_{\mathbf{q}_0}(\mathbf{q}) = t\}$ is nothing other than the wave front $\Phi_{\mathbf{q}_0}(t)$ (Figure 195). The gradient of the function S (in the sense of the metric mentioned above) is perpendicular to the wave front and characterizes the motion of the wave front. In this connection, the *bigger* the gradient, the *slower* the front moves. Therefore, Hamilton called the vector

$$\mathbf{p} = \frac{\partial S}{\partial \mathbf{q}}$$

the *vector of normal slowness of the front.*

The direction of the ray $\dot{\mathbf{q}}$ and the direction of motion of the front $\mathbf{p}$ do not coincide in an anisotropic medium. However, they are related to one another by a simple relationship, easily derived from Huygens' principle. Recall that the characteristics of the medium are at every point described by a surface of velocity vectors of light—the indicatrix.

Definition. The direction of the hyperplane tangent to the indicatrix at the point $\mathbf{v}$ is called *conjugate* to the direction $\mathbf{v}$ (Figure 196).

Theorem. *The direction of the wave front* $\Phi_{\mathbf{q}_0}(t)$ *at the point* $\mathbf{q}_t$ *is conjugate to the direction of the ray* $\dot{\mathbf{q}}$.

PROOF. We look (Figure 197) at points $\mathbf{q}_\tau$ of the ray $\mathbf{q}_0 \mathbf{q}_t$, $0 \leq \tau \leq t$. Take ε very small. Then the front $\Phi_{\mathbf{q}_{t-\varepsilon}}(\varepsilon)$ differs by quantities of order $O(\varepsilon^2)$ from the indicatrix at the point $\mathbf{q}_t$, contracted by ε. By Huygens' principle, this front $\Phi_{\mathbf{q}_{t-\varepsilon}}(\varepsilon)$ is tangent to the front $\Phi_{\mathbf{q}_0}(t)$ at the point $\mathbf{q}_t$. Passing to the limit as $\varepsilon \to 0$, we obtain the theorem. $\qquad\square$

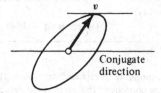

Figure 196   Conjugate hyperplane

251

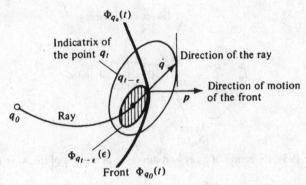

Figure 197   Conjugacy of the direction of a wave and of the front

If the auxiliary metric used to define the vector **p** is changed, the natural velocity of the motion of the front, i.e. both the magnitude and direction of the vector **p**, will be changed. However, the differential form $\mathbf{p}\,d\mathbf{q} = dS$ on the space $\{\mathbf{q}\} = \mathbb{R}^3$ is defined in a way which is independent of the auxiliary metric; its value depends only on the chosen fronts (or rays). On the hyperplane conjugate to the velocity vector of a ray, this form is equal to zero, and its value on the velocity vector is equal to 1.[79]

B *The optical-mechanical analogy*

We return now to mechanics. Here the trajectories of motion are also extremals of a variational principle, and one can construct mechanics as the geometric optics of a many-dimensional space, as Hamilton did; we will not develop this construction in full detail, but will only enumerate those optical concepts which led Hamilton to basic mechanical concepts.

Optics	Mechanics
Optical medium	Extended configuration space $\{(\mathbf{q}, t)\}$
Fermat's principle	Hamilton's principle $\delta \int L\,dt = 0$
Rays	Trajectories $\mathbf{q}(t)$
Indicatrices	Lagrangian L
Normal slowness vector **p** of the front	Momentum **p**
Expression of **p** in terms of the velocity of the ray, $\dot{\mathbf{q}}$	Legendre transformation
1-form $\mathbf{p}\,d\mathbf{q}$	1-form $\mathbf{p}\,d\mathbf{q} - H\,dt$

[79] In this way, the vectors **p** corresponding to various fronts passing through a given point are not arbitrary, but are subject to one condition: the permissible values of **p** fill a hypersurface in $\{\mathbf{p}\}$-space which is dual to the indicatrix of velocities.

The optical length of the path $S_{q_0}(\mathbf{q})$ and Huygens' principle have not yet been used. Their mechanical analogues are the *action function* and the *Hamilton–Jacobi equation*, to which we now turn.

C Action as a function of coordinates and time

Definition. The *action function* $S(\mathbf{q}, t)$ is the integral

$$S_{q_0, t_0}(\mathbf{q}, t) = \int_\gamma L \, dt$$

along the extremal γ connecting the points $(\mathbf{q}_0, t_0)$ and $(\mathbf{q}, t)$.

In order for this definition to be correct, we must take several precautions: we must require that the extremals going from the point $(\mathbf{q}_0, t_0)$ do not intersect elsewhere, but instead form a so-called "central field of extremals" (Figure 198). More precisely, we associate to every pair $(\dot{\mathbf{q}}_0, t)$ a point $(\mathbf{q}, t)$ which is the end of the extremal with initial condition $\mathbf{q}(0) = \mathbf{q}_0$, $\dot{\mathbf{q}}(0) = \dot{\mathbf{q}}_0$. We say that an extremal γ is *contained in a central field* if the mapping $(\dot{\mathbf{q}}_0, t) \to (\mathbf{q}, t)$ is nondegenerate (at the point corresponding to the extremal γ under consideration, and therefore in some neighborhood of it).

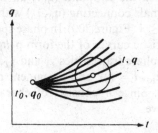

Figure 198   A central field of extremals

It can be shown that for $|t - t_0|$ small enough the extremal γ is contained in a central field.[80]

We now look at a sufficiently small neighborhood of the endpoint $(\mathbf{q}, t)$ of our extremal. Every point of this neighborhood is connected to $(\mathbf{q}_0, t_0)$ by a unique extremal of the central field under consideration. This extremal depends differentiably on the endpoint $(\mathbf{q}, t)$. Therefore, in the indicated neighborhood the action function is correctly defined

$$S_{q_0, t_0}(\mathbf{q}, t) = \int_\gamma L \, dt.$$

In geometric optics we were looking at the differential of the optical length of a path. It is natural here to look at the differential of the action function.

---

[80] PROBLEM. Show that this is not true for large $t - t_0$. *Hint.* $\ddot{q} = -q$ (Figure 199).

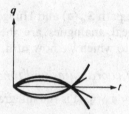

Figure 199 Extremal with a focal point which is not contained in any central field

Theorem. *The differential of the action function (for a fixed initial point) is equal to*

$$dS = \mathbf{p}\, d\mathbf{q} - H\, dt$$

where $\mathbf{p} = \partial L/\partial \dot{\mathbf{q}}$ *and* $H = \mathbf{p}\dot{\mathbf{q}} - L$ *are defined with the help of the terminal velocity* $\dot{\mathbf{q}}$ *of the trajectory* γ.

PROOF. We lift every extremal from $(\mathbf{q}, t)$-space to the extended phase space $\{(\mathbf{p}, \mathbf{q}, t)\}$, setting $\mathbf{p} = \partial L/\partial \dot{\mathbf{q}}$, i.e., replacing the extremal by a phase trajectory. We then get an $n + 1$-dimensional manifold in the extended phase space consisting of phase trajectories, i.e., characteristic curves of the form $\mathbf{p}\, d\mathbf{q} - H\, dt$. We now give the endpoint $(\mathbf{q}, t)$ an increment $(\Delta\mathbf{q}, \Delta t)$, and consider the set of extremals connecting $(\mathbf{q}_0, t_0)$ with points of the segment $\mathbf{q} + \theta\Delta\mathbf{q}, t + \theta\Delta t, 0 \leq \theta \leq 1$ (Figure 200). In phase space we get a quadrangle σ composed of characteristic curves of the form $\mathbf{p}\, d\mathbf{q} - H\, dt$, the boundary of which consists of two phase trajectories γ_1 and γ_2, a segment of a curve α lying in the space $(\mathbf{q} = \mathbf{q}_0, t = t_0)$, and a segment of a curve β projecting to the segment $(\Delta\mathbf{q}, \Delta t)$. Since σ consists of characteristic curves of the form $\mathbf{p}\, d\mathbf{q} - H\, dt$, we have

$$0 = \iint_\sigma d(\mathbf{p}\, d\mathbf{q} - H\, dt) = \int_{\partial\sigma} \mathbf{p}\, d\mathbf{q} - H\, dt$$

$$= \int_{\gamma_1} - \int_{\gamma_2} + \int_\beta - \int_\alpha \mathbf{p}\, d\mathbf{q} - H\, dt.$$

But, on the segment α, we have $d\mathbf{q} = 0, dt = 0$. On the phase trajectories γ_1 and γ_2, $\mathbf{p}\, d\mathbf{q} - H\, dt = L\, dt$ (Section 45C). So, the difference $\int_{\gamma_2} - \int_{\gamma_1} \mathbf{p}\, d\mathbf{q} - H\, dt$

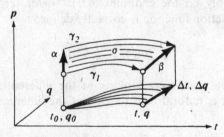

Figure 200 Calculation of the differential of the action function

is equal to the increase of the action function, and we find

$$\int_\beta \mathbf{p}\, d\mathbf{q} - H\, dt = S(\mathbf{q} + \Delta\mathbf{q}, t + \Delta t) - S(\mathbf{q}, t).$$

If now $\Delta\mathbf{q} \to 0$, $\Delta t \to 0$, then

$$\int_\beta \mathbf{p}\, d\mathbf{q} - H\, dt = \mathbf{p}\Delta\mathbf{q} - H\Delta t + o(\Delta t, \Delta\mathbf{q})$$

which proves the theorem. $\qquad\qquad\qquad\qquad\qquad\qquad\qquad\qquad$ $\square$

The form $\mathbf{p}\, d\mathbf{q} - H\, dt$ was formerly introduced to us artificially. We see now, by carrying out the optical-mechanical analogue, that it arises from examining the action function corresponding to the optical length of a path.

D The Hamilton–Jacobi equation

Recall that the "vector of normal slowness $\mathbf{p}$" cannot be altogether arbitrary: it is subject to one condition, $\mathbf{p}\dot{\mathbf{q}} = 1$, following from Huygens' principle. An analogous condition restricts the gradient of the action function S.

Theorem. *The action function satisfies the equation*

$$(1) \qquad\qquad \frac{\partial S}{\partial t} + H\!\left(\frac{\partial S}{\partial \mathbf{q}}, \mathbf{q}, t\right) = 0.$$

This nonlinear first-order partial differential equation is called the *Hamilton–Jacobi equation.*

PROOF. It is sufficient to notice that, by the previous theorem,

$$\frac{\partial S}{\partial t} = -H(\mathbf{p}, \mathbf{q}, t) \qquad \mathbf{p} = \frac{\partial S}{\partial \mathbf{q}}. \qquad\qquad \square$$

The relation just established between trajectories of mechanical systems ("rays") and partial differential equations ("wave fronts") can be used in two directions.

First, solutions of Equation (1) can be used for integrating the ordinary differential equations of dynamics. Jacobi's method of integrating Hamilton's canonical equations, presented in the next section, consists of just this.

Second, the relation of the ray and wave points of view allows one to reduce integration of the partial differential equations (1) to integration of a hamiltonian system of ordinary differential equations.

Let us go into this in a little more detail. For the Hamilton–Jacobi equation (1), the Cauchy problem is

$$(2) \qquad\qquad S(\mathbf{q}, t_0) = S_0(\mathbf{q}) \qquad \frac{\partial S}{\partial t} + H\!\left(\frac{\partial S}{\partial \mathbf{q}}, \mathbf{q}, t\right) = 0.$$

255

In order to construct a solution to this problem, we look at the hamiltonian system

$$\dot{\mathbf{p}} = -\frac{\partial H}{\partial \mathbf{q}} \qquad \dot{\mathbf{q}} = \frac{\partial H}{\partial \mathbf{p}}.$$

We consider the initial conditions (Figure 201):

$$\mathbf{q}(t_0) = \mathbf{q}_0 \qquad \mathbf{p}(t_0) = \frac{\partial S_0}{\partial \mathbf{q}}\bigg|_{\mathbf{q}_0}.$$

The solution corresponding to these equations is represented in $(\mathbf{q}, t)$-space by the curve $\mathbf{q} = \mathbf{q}(t)$, which is the extremal of the principle $\delta \int L \, dt = 0$ (where the lagrangian $L(\mathbf{q}, \dot{\mathbf{q}}, t)$ is the Legendre transformation with respect to $\mathbf{p}$ of the hamiltonian function $H(\mathbf{p}, \mathbf{q}, t)$). This extremal is called the *characteristic* of problem (2), emanating from the point $\mathbf{q}_0$.

If the value t_1 is sufficiently close to t_0, then the characteristics emanating from points close to $\mathbf{q}_0$ do not intersect for $t_0 \leq t \leq t_1$, $|\mathbf{q} - \mathbf{q}_0| < R$. Furthermore, the values of $\mathbf{q}_0$ and t can be taken as coordinates for points in the region $|\mathbf{q} - \mathbf{q}_{0*}| < R$, $t_0 \leq t \leq t_1$ (Figure 201).

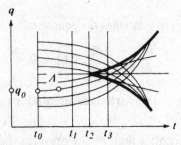

Figure 201   Characteristics for a solution of Cauchy's problem for the Hamilton-Jacobi equation

We now construct the "action function with initial condition S_0":

$$(3) \qquad S(A) = S_0(\mathbf{q}_0) + \int_{\mathbf{q}_0, t_0}^{A} L(\mathbf{q}, \dot{\mathbf{q}}, t) dt$$

(integrating along the characteristic leading to A).

Theorem. *The function* (3) *is a solution of problem* (2).

PROOF. The initial condition is clearly fulfilled. The fact that the Hamilton-Jacobi equation is satisfied is verified just as in the theorem on differentials of action functions (Figure 202).

By Stokes' lemma, $\int_{\gamma_1} - \int_{\gamma_2} + \int_{\beta} - \int_{\alpha} \mathbf{p} \, d\mathbf{q} - H \, dt = 0$. But on α, $H \, dt = 0$ and $\mathbf{p} = \partial S_0 / d\mathbf{q}$, so

$$\int_{\alpha} \mathbf{p} \, d\mathbf{q} - H \, dt = \int_{\alpha} \mathbf{p} \, d\mathbf{q} = \int_{\alpha} dS_0 = S_0(\mathbf{q}_0 + \Delta \mathbf{q}) - S_0(\mathbf{q}_0).$$

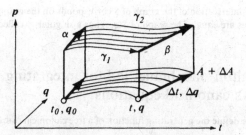

Figure 202   The action function as a solution of the Hamilton–Jacobi equation

Further, γ_1 and γ_2 are phase trajectories, so

$$\int_{\gamma_{1,2}} \mathbf{p}\, d\mathbf{q} - H\, dt = \int_{\gamma_{1,2}} L\, dt.$$

So

$$\int_\beta \mathbf{p}\, d\mathbf{q} - H\, dt = \left[S_0(\mathbf{q}_0 + \Delta\mathbf{q}) + \int_{\gamma_2} L\, dt \right] - \left[S_0(\mathbf{q}_0) + \int_{\gamma_1} L\, dt \right]$$

$$= S(A + \Delta A) - S(A).$$

For $\Delta t, \Delta \mathbf{q} \to 0$, we get $\partial S/\partial t = -H$, $\partial S/\partial \mathbf{q} = \mathbf{p}$, which proves the theorem. $\square$

PROBLEM. Show the uniqueness of the solution to problem (2).
   Hint. Differentiate S along the characteristics.

PROBLEM. Solve the Cauchy problem (2) for

$$H = \frac{p^2}{2} \qquad S_0 = \frac{q^2}{2}.$$

PROBLEM. Draw a graph of the multiple-valued "functions" $S(q)$ and $p(q)$ for $t = t_3$ (Figure 201).

ANSWER. Cf. Figure 203.

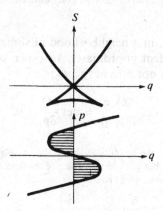

Figure 203   A typical singularity of a solution of the Hamilton–Jacobi equation

The point of self-intersection of the graph of S corresponds on the graph of p to the *Maxwell line*: the shaded areas are equal. The graph of $S(q, t)$ has a singularity called a *swallowtail* at the point $(0, t_2)$.

47 The Hamilton–Jacobi method for integrating Hamilton's canonical equations

In this paragraph we define the generating function of a free canonical transformation.

The idea of the Hamilton–Jacobi method consists of the following. Under canonical changes of coordinates, the canonical form of the equations of motion is preserved, as is the hamiltonian function (Section 45A). Therefore, if we succeed in finding a canonical transformation which reduces the hamiltonian function to a form such that the canonical equations can be integrated, then we can also integrate the original canonical equations. It turns out that the problem of constructing such a canonical transformation reduces to the determination of a sufficiently large number of solutions to the Hamilton–Jacobi partial differential equation. The generating function of the desired canonical transformation must satisfy this equation.

Before turning to the apparatus of generating functions, we remark that it is unfortunately noninvariant and it uses, in an essential way, the coordinate structure in phase space $\{(\mathbf{p}, \mathbf{q})\}$. It is necessary to use the apparatus of partial derivatives, in which even the notation is ambiguous.[81]

A *Generating functions*

Suppose that the $2n$ functions $\mathbf{P}(\mathbf{p}, \mathbf{q})$ and $\mathbf{Q}(\mathbf{p}, \mathbf{q})$ of the $2n$ variables $\mathbf{p}$ and $\mathbf{q}$ give a canonical transformation $g: \mathbb{R}^{2n} \to \mathbb{R}^{2n}$. Then the 1-form $\mathbf{p} \, d\mathbf{q} - \mathbf{P} \, d\mathbf{Q}$ is an exact differential (Section 45A):

$$(1) \qquad \mathbf{p} \, d\mathbf{q} - \mathbf{P} \, d\mathbf{Q} = dS(\mathbf{p}, \mathbf{q}).$$

PROBLEM. Show the converse: if this form is an exact differential, then the transformation is canonical.

We now assume that, in a neighborhood of some point $(\mathbf{p}_0, \mathbf{q}_0)$, we can take $(\mathbf{Q}, \mathbf{q})$ as independent coordinates. In other words, we assume that the following jacobian is not zero at $(\mathbf{p}_0, \mathbf{q}_0)$:

$$\det \frac{\partial(\mathbf{Q}, \mathbf{q})}{\partial(\mathbf{p}, \mathbf{q})} = \det \frac{\partial \mathbf{Q}}{\partial \mathbf{p}} \neq 0.$$

---

[81] It is important to note that the quantity $\partial u / \partial x$ on the x, y-plane depends not only on the function which is taken for x, but also on the choice of the function y: in new variables (x, z) the value of $\partial u / \partial x$ will be different. One should write

$$\left. \frac{\partial u}{\partial x} \right|_{y = \text{const}} \qquad \left. \frac{\partial u}{\partial x} \right|_{z = \text{const}}.$$

Such canonical transformations will be called *free*. In this case, the function S can be expressed locally in these coordinates:

$$S(\mathbf{p}, \mathbf{q}) = S_1(\mathbf{Q}, \mathbf{q}).$$

Definition. The function $S_1(\mathbf{Q}, \mathbf{q})$ is called a *generating function* of our canonical transformation g.

We emphasize that S_1 is not a function on the phase space $\mathbb{R}^{2n}$: it is a function on a region in the direct product $\mathbb{R}^n_\mathbf{q} \times \mathbb{R}^n_\mathbf{Q}$ of two n-dimensional coordinate spaces, whose points are denoted by $\mathbf{q}$ and $\mathbf{Q}$. It follows from (1) that the "partial derivatives" of S_1 are

(2) $$\frac{\partial S_1(\mathbf{Q}, \mathbf{q})}{\partial \mathbf{q}} = \mathbf{p} \quad \text{and} \quad \frac{\partial S_1(\mathbf{Q}, \mathbf{q})}{\partial \mathbf{Q}} = -\mathbf{P}.$$

Conversely, every function S_1 gives a canonical transformation g by formulas (2).

Theorem. *Let $S_1(\mathbf{Q}, \mathbf{q})$ be a function given on a neighborhood of some point $(\mathbf{Q}_0, \mathbf{q}_0)$ of the direct product of two n-dimensional euclidean spaces. If*

$$\det \left. \frac{\partial^2 S_1}{\partial \mathbf{Q}\, \partial \mathbf{q}} \right|_{\mathbf{Q}_0, \mathbf{q}_0} \neq 0,$$

then S_1 is a generating function of some free canonical transformation.

PROOF. Consider the equation for the $\mathbf{Q}$ coordinates:

$$\frac{\partial S_1(\mathbf{Q}, \mathbf{q})}{\partial \mathbf{q}} = \mathbf{p}.$$

By the implicit function theorem this equation can be solved to determine a function $\mathbf{Q}(\mathbf{p}, \mathbf{q})$ in a neighborhood of the point

$$\left(\mathbf{q}_0, \mathbf{p}_0 = \left(\frac{\partial S_1(\mathbf{Q}, \mathbf{q})}{\partial \mathbf{q}} \right) \bigg|_{\mathbf{Q}_0, \mathbf{q}_0} \right)$$

(with $\mathbf{Q}(\mathbf{p}_0, \mathbf{q}_0) = \mathbf{Q}_0$). In fact, the determinant we need here is

$$\det \left(\frac{\partial^2 S_1(\mathbf{Q}, \mathbf{q})}{\partial \mathbf{Q}\, \partial \mathbf{q}} \right) \bigg|_{\mathbf{Q}_0, \mathbf{q}_0},$$

and this is different from zero by hypothesis.

We now consider the function

$$\mathbf{P}_1(\mathbf{Q}, \mathbf{q}) = -\frac{\partial}{\partial \mathbf{Q}} S_1(\mathbf{Q}, \mathbf{q}),$$

and set

$$\mathbf{P}(\mathbf{p}, \mathbf{q}) = \mathbf{P}_1(\mathbf{Q}(\mathbf{p}, \mathbf{q}), \mathbf{q}).$$

Then the local map $g\colon \mathbb{R}^{2n} \to \mathbb{R}^{2n}$ sending the point $(\mathbf{p}, \mathbf{q})$ to the point $(\mathbf{P}(\mathbf{p}, \mathbf{q}), \mathbf{Q}(\mathbf{p}, \mathbf{q}))$ will be canonical with generating function S_1, since by construction

$$\mathbf{p}\, d\mathbf{q} - \mathbf{P}\, d\mathbf{Q} = \frac{\partial S_1(\mathbf{Q}, \mathbf{q})}{\partial \mathbf{q}}\, d\mathbf{q} + \frac{\partial S_1(\mathbf{Q}, \mathbf{q})}{\partial \mathbf{Q}}\, d\mathbf{Q}.$$

It is free, since $\det(\partial \mathbf{Q}/\partial \mathbf{p}) = \det(\partial^2 S_1(\mathbf{Q}, \mathbf{q})/\partial \mathbf{Q}\, \partial \mathbf{q})^{-1} \neq 0.$ $\square$

The transformation $g\colon \mathbb{R}^{2n} \to \mathbb{R}^{2n}$ is given in general by $2n$ functions of $2n$ variables. We see that a canonical transformation is given entirely by *one* function of $2n$ variables—its generating function. It is easy to see how useful generating functions are in all calculations related to canonical transformations. This becomes even more so as the number of variables, $2n$, becomes large.

B The Hamilton–Jacobi equation for generating functions

We notice that canonical equations in which the hamiltonian function depends only on the variable $\mathbf{Q}$ are easy to integrate. If $H = K(\mathbf{Q}, t)$, then the canonical equations have the form

$$(3) \qquad \dot{\mathbf{Q}} = 0 \qquad \dot{\mathbf{P}} = \frac{\partial K}{\partial \mathbf{Q}}$$

from which we have immediately

$$\mathbf{Q}(t) = \mathbf{Q}(0) \qquad \mathbf{P}(t) = \mathbf{P}(0) + \int^{t} \frac{\partial K}{\partial \mathbf{Q}}\bigg|_{\mathbf{Q}(0)} dt.$$

We will now look for a canonical transformation reducing the hamiltonian $H(\mathbf{p}, \mathbf{q})$ to the form $K(\mathbf{Q})$. To this end we will look for a generating function of such a transformation, $S(\mathbf{Q}, \mathbf{q})$. From (2) we obtain the condition

$$(4) \qquad H\left(\frac{\partial S(\mathbf{Q}, \mathbf{q})}{\partial \mathbf{q}}, \mathbf{q}, t\right) = K(\mathbf{Q}, t)$$

where *after* differentiation we must substitute $\mathbf{q}(\mathbf{P}, \mathbf{Q})$ for $\mathbf{q}$. We notice that for fixed $\mathbf{Q}$, Equation (4) has the form of the Hamilton–Jacobi equation.

Jacobi's theorem. *If a solution $S(\mathbf{Q}, \mathbf{q})$ is found to the Hamilton–Jacobi equation (4), depending on n parameters*[82] *Q_i and such that $\det(\partial^2 S/\partial \mathbf{Q}\partial \mathbf{q}) \neq 0$, then the canonical equations*

$$(5) \qquad \dot{\mathbf{p}} = -\frac{\partial H}{\partial \mathbf{q}} \quad \text{and} \quad \dot{\mathbf{q}} = \frac{\partial H}{\partial \mathbf{p}}$$

can be solved explicitly by quadratures. The functions $\mathbf{Q}(\mathbf{p}, \mathbf{q})$ determined by the equations $\partial S(\mathbf{Q}, \mathbf{q})/\partial \mathbf{q} = \mathbf{p}$ are first integrals of the equation (5).

---

[82] An n-parameter family of solutions of (4) is called a *complete integral* of the equation.

PROOF. Consider the canonical transformation with generating function $S(\mathbf{Q}, \mathbf{q})$. By (2) we have $\mathbf{p} = (\partial S/\partial \mathbf{q})(\mathbf{Q}, \mathbf{q})$, from which we can determine $\mathbf{Q}(\mathbf{p}, \mathbf{q})$. We calculate the function $H(\mathbf{p}, \mathbf{q})$ in the new coordinates $\mathbf{P}$, $\mathbf{Q}$. We have $H(\mathbf{p}, \mathbf{q}) = H((\partial S/\partial \mathbf{q})(\mathbf{Q}, \mathbf{q}), \mathbf{q})$. In order to find the hamiltonian function in the new coordinates we must substitute into this expression (after differentiation) for $\mathbf{q}$ its expression in terms of $\mathbf{P}$ and $\mathbf{Q}$. However, by (4), this expression does not depend on $\mathbf{P}$ at all, so we have simply

$$H(\mathbf{p}, \mathbf{q}) = K(\mathbf{Q}).$$

Thus, in the new variables, Equation (5) has the form (3), from which Jacobi's theorem follows directly. □

Jacobi's theorem reduces solving the system of ordinary differential equations (5) to finding a complete integral of the partial differential equation (4). It may appear surprising that this "reduction" from the simple to the complicated provides an effective method for solving concrete problems. Nevertheless, it turns out that this is the most powerful method known for exact integration, and many problems which were solved by Jacobi cannot be solved by other methods.

C Examples

We consider the problem of attraction by two fixed centers. Interest in this problem has grown recently in connection with the study of the motion of artificial earth satellites. It is fairly clear that two close centers of attraction on the z-axis approximate attraction by an ellipsoid slightly extended along the z-axis. Unfortunately, the earth is not prolate, but oblate. To overcome this difficulty, one must place the centers at imaginary points at distances $\pm i\varepsilon$ from the origin along the z-axis. Analytic formulas for the solution are true, of course, in the complex region. In this way we obtain an approximation to the earth's field of gravity, in which the equations of motion can be exactly integrated and which is closer to reality than the keplerian approximation in which the earth is a point.

For simplicity we will consider only the planar problem of attraction by two fixed points with equal masses. The success of Jacobi's method is based on the adoption of a suitable coordinate system, called elliptic coordinates. Suppose that the distance between the fixed points O_1 and O_2 is $2c$ (Figure

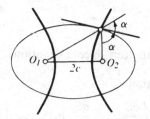

Figure 204   Elliptic coordinates

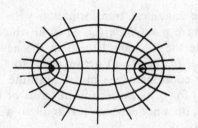

Figure 205   Confocal ellipses and hyperbolas

204), and that the distances of a moving mass from them are r_1 and r_2, respectively. The elliptic coordinates ξ, η are defined as the sum and difference of the distances to the points O_1 and O_2: $\xi = r_1 + r_2$, $\eta = r_1 - r_2$.

PROBLEM. Express the hamiltonian function in elliptic coordinates.

Solution. The lines ξ = const are ellipses with foci at O_1 and O_2; the lines η = const are hyperbolas with the same foci (Figure 205). They are mutually orthogonal; therefore,

$$ds^2 = a^2 \, d\xi^2 + b^2 \, d\eta^2.$$

We will find the coefficients a and b. For motion along an ellipse we have $dr_1 = ds \cos \alpha$ and $dr_2 = -ds \cos \alpha$, so $d\eta = 2 \cos \alpha \, ds$. For motion along a hyperbola we have $dr_1 = ds \sin \alpha$ and $dr_2 = ds \sin \alpha$, so $d\xi = 2 \sin \alpha \, ds$. Thus $a = (2 \sin \alpha)^{-1}$ and $b = (2 \cos \alpha)^{-1}$. Furthermore, from the triangle $O_1 M O_2$ we find $r_1^2 + r_2^2 + 2r_1 r_2 \cos 2\alpha = 4c^2$, which implies

$$\cos^2 \alpha - \sin^2 \alpha = \frac{4c^2 - r_1^2 - r_2^2}{2r_1 r_2},$$

$$\cos^2 \alpha + \sin^2 \alpha = \frac{2r_1 r_2}{2r_1 r_2},$$

$$\cos^2 \alpha = \frac{4c^2 - (r_1 - r_2)^2}{4r_1 r_2} \qquad \sin^2 \alpha = \frac{(r_1 + r_2)^2 - 4c^2}{4r_1 r_2}.$$

But if $ds^2 = \sum a_i^2 \, dq_i^2$, then

$$T = \sum a_i^2 \frac{\dot{q}_i^2}{2}, \quad p_i = a_i^2 \dot{q}_i, \quad H = \sum \frac{p_i^2}{2a_i^2} + U.$$

Thus,

$$H = p_\xi^2 \frac{(r_1 + r_2)^2 - 4c^2}{2r_1 r_2} + p_\eta^2 \frac{4c^2 - (r_1 - r_2)^2}{2r_1 r_2} - \frac{k}{r_1} - \frac{k}{r_2}.$$

But $r_1 + r_2 = \xi$, $r_1 - r_2 = \eta$, $4r_1 r_2 = \xi^2 - \eta^2$. Therefore, finally,

$$H = 2p_\xi^2 \frac{\xi^2 - 4c^2}{\xi^2 - \eta^2} + 2p_\eta^2 \frac{4c^2 - \eta^2}{\xi^2 - \eta^2} - \frac{4k\xi}{\xi^2 - \eta^2}.$$

We will now solve the Hamilton–Jacobi equation.

Definition. If, in the equation

$$\Phi_1\left(\frac{\partial S}{\partial q_1}, \ldots, \frac{\partial S}{\partial q_n}; q_1, \ldots, q_n\right) = 0,$$

the variable q_1 and derivative $\partial S/\partial q_1$ appear only in the form of a combination $\varphi(\partial S/\partial q_1, q_1)$, then we say that *the variable q_1 is separable.*

In this case it is useful to look for a solution of the equation of the form

$$S = S_1(q_1) + S'(q_2, \ldots, q_n).$$

By setting $\varphi(\partial S_1/\partial q_1, q_1) = c_1$ in this equation, we obtain an equation for S' with a smaller number of variables

$$\Phi_2\left(\frac{\partial S'}{\partial q_2}, \ldots, \frac{\partial S'}{\partial q_n}, q_2, \ldots, q_n; c_1\right) = 0.$$

Let $S' = S'(q_2, \ldots, q_n; c_1, \mathbf{c})$ be a family of solutions to this equation depending on the parameters c_i. The functions $S_1(q_1, c_1) + S'$ will satisfy the desired equation if S_1 satisfies the ordinary differential equation $\varphi(\partial S_1/\partial q_1, q_1) = c_1$. This equation is easy to solve; we express $\partial S_1/\partial q_1$ in terms of q_1 and c_1 to obtain $\partial S_1/\partial q_1 = \psi(q_1, c_1)$, from which $S_1 = \int^{q_1} \psi(q_1, c_1) dq_1$.

If one of the variables, say q_2, is separable in the new equation (with Φ_2) we can repeat this procedure and (in the most favorable case) we can find a solution of the original equation depending on n constants

$$S_1(q_1; c_1) + S_2(q_2; c_1, c_2) + \cdots + S_n(q_n; c_1, \ldots, c_n).$$

In this case we say that the variables are *completely separable.*

If the variables are completely separable, then a solution depending on n parameters of the Hamilton–Jacobi equation, $\Phi_1(\partial S/\partial \mathbf{q}, \mathbf{q}) = 0$, is found by quadratures. But then the corresponding system of canonical equations can also be integrated by quadratures (Jacobi's theorem).

We apply the above to the problem of two fixed centers. The Hamilton–Jacobi equation (4) has the form

$$\left(\frac{\partial S}{\partial \xi}\right)^2 (\xi^2 - 4c^2) + \left(\frac{\partial S}{\partial \eta}\right)^2 (4c^2 - \eta^2) = K(\xi^2 - \eta^2) + 4k\xi.$$

We can separate variables by, for instance, setting

$$\left(\frac{\partial S}{\partial \xi}\right)^2 (\xi^2 - 4c^2) - 4k\xi - K\xi^2 = c_1$$

and

$$\left(\frac{\partial S}{\partial \eta}\right)^2 (4c^2 - \eta^2) + K\eta^2 = -c_1.$$

Then we find the complete integral of Equation (4) in the form

$$S(\xi, \eta; c_1, c_2) = \int \sqrt{\frac{c_1 + c_2 \xi^2 + 4k\xi}{\xi^2 - 4c^2}}\, d\xi + \int \sqrt{\frac{-c_1 - c_2 \eta^2}{4c^2 - \eta^2}}\, d\eta.$$

Jacobi's theorem now gives an explicit expression, in terms of elliptic integrals, for motion in the problem of two fixed centers. A more detailed investigation of this motion can be found in Charlier's book "Die Mechanik des Himmels," Berlin, Leipzig, W. de Gruyter & Co., 1927.

Another application of the problem of the attraction of two fixed centers is *the study of motion with fixed pull in a field with one attracting center.*

This is a question of the motion of a point mass under the action of a newtonian attraction of a fixed center and one more force ("pull") of constant magnitude and direction. This problem can be looked at as the limiting case of the problem of attraction by two fixed centers. In the passage to the limit, one center goes off to infinity in the direction of the thrust force (during which its mass must grow proportionally to the square of the distance moved in order to guarantee constant pull).

This limiting case of the problem of the attraction of two fixed centers can be integrated explicitly (in elliptic functions). We can convince ourselves of this by passing to a limit or by directly separating variables in the problem of motion with constant pull in a field with one center. The coordinates in which the variables are separated in this problem are obtained as the limit of elliptic coordinates as one of the centers approaches infinity. They are called parabolic coordinates and are given by the formulas

$$u = r - x \qquad v = r + x$$

(the pull is directed along the x-axis).

A description of the trajectories of a motion with constant pull (many of which are very intricate) can be found in V. V. Beletskii's book "Sketches on the motion of celestial bodies," Nauka, 1972.

As one more example we consider the problem of geodesics on a triaxial ellipsoid.[83] Here Jacobi's elliptic coordinates λ_1, λ_2, and λ_3 are helpful, where the λ_i are the roots of the equation

$$\frac{x_1^2}{a_1 + \lambda} + \frac{x_2^2}{a_2 + \lambda} + \frac{x_3^2}{a_3 + \lambda} = 1, \qquad \lambda_1 > \lambda_2 > \lambda_3;$$

x_1, x_2, and x_3 are cartesian coordinates. We will not carry out the computations showing that the variables are separable (they can be found, for example, in Jacobi's "Lectures on dynamics"), but will mention only the result: we will describe the behavior of the geodesics.

The surfaces λ_1 = const, λ_2 = const, and λ_3 = const are surfaces of second degree, called *confocal quadrics.* The first of these is an ellipsoid, the second a hyperboloid of one sheet, and the third a hyperboloid of two sheets. The ellipsoid can degenerate into the interior of an ellipse, the one-sheeted hyperboloid either into the exterior of an ellipse or into the part of a plane

---

[83] The problem of geodesics on an ellipsoid and the closely related problem of ellipsoidal billiards have found application in a series of recent results in physics connected with laser devices.

between the branches of a hyperbola, and the two-sheeted hyperboloid either into the part of a plane outside the branches of a hyperbola or into a plane.

Suppose that the ellipsoid under consideration is one of the ellipsoids in the family with semi-axes $a > b > c$. Each of the three ellipses $x_1 = 0$, $x_2 = 0$, and $x_3 = 0$ is a closed geodesic. A geodesic starting from a point of the largest ellipse (with semiaxes a and b) in a direction close to the direction of the ellipse (Figure 206), is alternately tangent to the two closed lines of intersection of the ellipsoid with the one-sheeted hyperboloid of our family $\lambda = \text{const.}$[84] This geodesic is either closed or is dense in the area

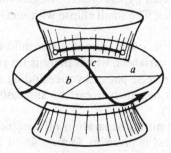

Figure 206   Geodesic on a triaxial ellipsoid

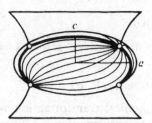

Figure 207   Geodesics emanating from an umbilical point

between the two lines of intersection. As the slope of the geodesic increases, the hyperboloids collapse down to the region "inside" the hyperbola which intersects our ellipsoid in its four "umbilical points." In the limiting case we obtain geodesics passing through the umbilical points (Figure 207).

It is interesting to note that all the geodesics starting at an umbilical point again converge at the opposite umbilical point, and all have the same length between the two umbilical points. Only one of these geodesics is closed, namely, the middle ellipse with semi-axes a and c. If we travel along any other geodesic passing through an umbilical point in any direction, we will approach this ellipse asymptotically.

Finally, geodesics which intersect the largest ellipse even more "steeply" (Figure 208) are alternately tangent to the two lines of intersection of our

[84] These lines of intersection of the confocal surfaces are also *lines of curvature* of the ellipsoid.

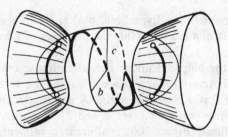

Figure 208   Geodesics of an ellipsoid which are tangent to a two-sheeted hyperboloid

ellipsoid with a two-sheeted hyperboloid.[85] In general, they are dense in the region between these lines. The small ellipse with semi-axes b and c is among these geodesics.

"The main difficulty in integrating a given differential equation lies in introducing convenient variables, which there is no rule for finding. Therefore, we must travel the reverse path and after finding some notable substitution, look for problems to which it can be successfully applied." (Jacobi, "Lectures on dynamics").

A list of problems admitting separation of variables in spherical, elliptic, and parabolic coordinates is given in Section 48 of Landau and Lifshitz's "Mechanics" (Oxford, Pergamon, 1960).

48 Generating functions

In this paragraph we construct the apparatus of generating functions for non-free canonical transformations.

A *The generating function $S_2(\mathbf{P}, \mathbf{q})$*

Let $f: \mathbb{R}^{2n} \to \mathbb{R}^{2n}$ be a canonical transformation with $g(\mathbf{p}, \mathbf{q}) = (\mathbf{P}, \mathbf{Q})$. By the definition of canonical transformation the differential form on $\mathbb{R}^{2n}$

$$\mathbf{p}\, d\mathbf{q} - \mathbf{P}\, d\mathbf{Q} = dS$$

is the total differential of some function $S(\mathbf{p}, \mathbf{q})$. A canonical transformation is free if we can take $\mathbf{q}, \mathbf{Q}$ as $2n$ independent coordinates. In this case the function S expressed in the coordinates $\mathbf{q}$ and $\mathbf{Q}$ is called a generating function $S_1(\mathbf{q}, \mathbf{Q})$. Knowing this function alone, we can find all $2n$ functions giving the transformation from the relations

(1) $$\mathbf{p} = \frac{\partial S_1(\mathbf{q}, \mathbf{Q})}{\partial \mathbf{q}} \quad \text{and} \quad \mathbf{P} = -\frac{\partial S_1(\mathbf{q}, \mathbf{Q})}{\partial \mathbf{Q}}.$$

It is far from the case that all canonical transformations are free. For example, in the case of the identity transformation $\mathbf{q}$ and $\mathbf{Q} = \mathbf{q}$ are dependent. Therefore, the identity transformation cannot be given by a generating

[85] These are also lines of curvature.

function $S_1(\mathbf{q}, \mathbf{Q})$. We can, however, obtain generating functions of another form by means of the Legendre transformation. Suppose, for instance, that we can take $\mathbf{P}, \mathbf{q}$ as independent local coordinates on $\mathbb{R}^{2n}$ (i.e., the determinant $\det(\partial(\mathbf{P}, \mathbf{q})/\partial(\mathbf{p}, \mathbf{q})) = \det(\partial\mathbf{P}/\partial\mathbf{p})$ is not zero). Then we have

$$\mathbf{p}\,d\mathbf{q} - \mathbf{P}\,d\mathbf{Q} = dS \quad \text{and} \quad \mathbf{p}\,d\mathbf{q} + \mathbf{Q}\,d\mathbf{P} = d(\mathbf{PQ} + S).$$

The quantity $\mathbf{PQ} + S$, expressed in terms of $(\mathbf{P}, \mathbf{q})$, is also called a generating function

$$S_2(\mathbf{P}, \mathbf{q}) = \mathbf{PQ} + S(\mathbf{p}, \mathbf{q}).$$

For this function, we find

(2) $$\mathbf{p} = \frac{\partial S_2(\mathbf{P}, \mathbf{q})}{\partial \mathbf{q}} \quad \text{and} \quad \mathbf{Q} = \frac{\partial S_2(\mathbf{P}, \mathbf{q})}{\partial \mathbf{P}}.$$

Conversely, if $S_2(\mathbf{P}, \mathbf{q})$ is any function for which the determinant

$$\det\left(\frac{\partial^2 S_2(\mathbf{P}, \mathbf{q})}{\partial \mathbf{q}\,\partial \mathbf{P}}\right)\Bigg|_{\mathbf{P}_0, \mathbf{q}_0}$$

is not zero, then in a neighborhood of the point

$$\left(\mathbf{p}_0 = \left(\frac{\partial S_2(\mathbf{P}, \mathbf{q})}{\partial \mathbf{q}}\right)\Bigg|_{\mathbf{P}_0, \mathbf{q}_0}, \mathbf{q}_0\right)$$

we can solve the first group of equations (2) for $\mathbf{P}$ and obtain a function $\mathbf{P}(\mathbf{p}, \mathbf{q})$ (where $\mathbf{P}(\mathbf{p}_0, \mathbf{q}_0) = \mathbf{P}_0$). After this, the second group of equations (2) determine $\mathbf{Q}(\mathbf{p}, \mathbf{q})$, and the map $(\mathbf{p}, \mathbf{q}) \to (\mathbf{P}, \mathbf{Q})$ is canonical (prove this!).

PROBLEM. Find a generating function S_2 for the identity map $\mathbf{P} = \mathbf{p}, \mathbf{Q} = \mathbf{q}$.

ANSWER. $\mathbf{Pq}$.

Remark. The generating function $S_2(\mathbf{P}, \mathbf{q})$ is convenient also because there are no minus signs in the formulas (2), and they are easy to remember if we remember that the generating function of the identity transformation is $\mathbf{Pq}$.

B 2^n generating functions

Unfortunately, the variables $\mathbf{P}, \mathbf{q}$ cannot always be chosen for local co-ordinates either; however, we can always choose *some* set of n new co-ordinates

$$\mathbf{P}_i = (P_{i_1}, \ldots, P_{i_k}) \qquad \mathbf{Q}_j = (Q_{j_1}, \ldots, Q_{j_{n-k}})$$

so that together with the old $\mathbf{q}$ we obtain $2n$ independent coordinates.

Here $(i_1, \ldots, i_k)(j_1, \ldots, j_{n-k})$ is any partition of the set $(1, \ldots, n)$ into two non-intersecting parts; so there are in all 2^n cases.

Theorem. *Let* $g: \mathbb{R}^{2n} \to \mathbb{R}^{2n}$ *be a canonical transformation given by the functions* $\mathbf{P}(\mathbf{p}, \mathbf{q})$ *and* $\mathbf{Q}(\mathbf{p}, \mathbf{q})$. *In a neighborhood of every point* $(\mathbf{p}_0, \mathbf{q}_0)$ *at least one of the* 2^n *sets of functions* $(\mathbf{P}_i, \mathbf{Q}_j, \mathbf{q})$ *can be taken as independent coordinates on* $\mathbb{R}^{2n}$:

$$\det \frac{\partial(\mathbf{P}_i, \mathbf{Q}_j, \mathbf{q})}{\partial(\mathbf{p}_i, \mathbf{p}_j, \mathbf{q})} = \det \frac{\partial(\mathbf{P}_i, \mathbf{Q}_j)}{\partial(\mathbf{p}_i, \mathbf{p}_j)} \neq 0.$$

In a neighborhood of such a point, the canonical transformation g *can be reconstructed from the function*

$$S_3(\mathbf{P}_i, \mathbf{Q}_j, \mathbf{q}) = (\mathbf{P}_i \mathbf{Q}_i) + \int \mathbf{p} \, d\mathbf{q} - \mathbf{P} \, d\mathbf{Q}$$

by the relations

(3) $$\mathbf{p} = \frac{\partial S_3}{\partial \mathbf{q}}, \quad \mathbf{Q}_i = \frac{\partial S_3}{\partial \mathbf{P}_i}, \quad \text{and} \quad \mathbf{P}_j = -\frac{\partial S_3}{\partial \mathbf{Q}_j}.$$

Conversely, if $S_3(\mathbf{P}_i, \mathbf{Q}_j, \mathbf{q})$ *is any function for which the determinant* $\det(\partial^2 S_3 / \partial \mathbf{P} \, \partial \mathbf{q})|_{\mathbf{P}_0, \mathbf{q}_0}$ $(\mathbf{P} = \mathbf{P}_i, \mathbf{Q}_j)$ *is not zero, then the relations* (3) *give a canonical transformation in a neighborhood of the point* $\mathbf{p}_0, \mathbf{q}_0$.

PROOF. The proof of this theorem is almost the same as the one carried out above in the particular case $k = n$. We need only verify that the determinant $\det[(\partial(\mathbf{P}_i, \mathbf{Q}_j)/\partial(\mathbf{p}_i, \mathbf{p}_j))]$ is not zero for one of the 2^n.sets $(\mathbf{P}_i, \mathbf{Q}_j, \mathbf{q})$.

We consider the differential of our transformation g at the point $(\mathbf{p}_0, \mathbf{q}_0)$. By identifying the tangent space to $\mathbb{R}^{2n}$ with $\mathbb{R}^{2n}$, we can consider dg as a symplectic transformation $S: \mathbb{R}^{2n} \to \mathbb{R}^{2n}$.

Consider the coordinate $\mathbf{p}$-plane P in $\mathbb{R}^{2n}$ (Figure 209). This is a null n-plane, and its image SP is also a null plane. We project the plane SP onto the coordinate plane $\sigma = \{(\mathbf{p}_i, \mathbf{q}_j)\}$ parallel to the remaining coordinate axes, i.e., in the direction of the n-dimensional null coordinate plane $\bar{\sigma} = \{(\mathbf{p}_j, \mathbf{q}_i)\}$. We denote the projection operator by $TS: P \to \sigma$.

The condition $\det(\partial(\mathbf{P}_i, \mathbf{Q}_j)/\partial(\mathbf{p}_i, \mathbf{p}_j)) \neq 0$ means that $T: SP \to \sigma$ is nonsingular. The operator S is nonsingular. Therefore, TS is nonsingular if and only if $T: SP \to \sigma$ is nonsingular. In other words, the null plane SP must be transverse to the null coordinate plane $\bar{\sigma}$. But we showed in

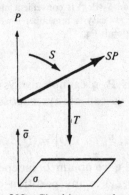

Figure 209   Checking non-degeneracy

Section 41 that at least one of the 2^n null coordinate planes is transverse to SP. This means that one of our 2^n determinants is nonzero, as was to be shown. □

PROBLEM. Show that this system of 2^n types of generating functions is minimal: given any one of the 2^n determinants, there exists a canonical transformation for which only this determinant is nonzero.[86]

C Infinitesimal canonical transformations

We now consider a canonical transformation which is close to the identity. Its generating function can be taken close to the generating function $\mathbf{Pq}$ of the identity. We look at a family of canonical transformations g_ε depending differentiably on the parameter ε, such that the generating functions have the form

$$(4) \qquad \mathbf{Pq} + \varepsilon S(\mathbf{P}, \mathbf{q}; \varepsilon) \qquad \mathbf{p} = \mathbf{P} + \varepsilon \frac{\partial S}{\partial \mathbf{q}} \qquad \mathbf{Q} = \mathbf{q} + \varepsilon \frac{\partial S}{\partial \mathbf{P}}.$$

An *infinitesimal canonical transformation* is an equivalence class of families g_ε, two families g_ε and h_ε being equivalent if their difference is small of higher than first order, $|g_\varepsilon - h_\varepsilon| = O(\varepsilon^2)$, $\varepsilon \to 0$.

Theorem. *An infinitesimal canonical transformation satisfies Hamilton's differential equations*

$$\frac{d\mathbf{P}}{d\varepsilon}\bigg|_{\varepsilon=0} = -\frac{\partial H}{\partial \mathbf{q}} \qquad \frac{d\mathbf{Q}}{d\varepsilon}\bigg|_{\varepsilon=0} = \frac{\partial H}{\partial \mathbf{p}}$$

with hamiltonian function $H(\mathbf{p}, \mathbf{q}) = S(\mathbf{p}, \mathbf{q}, 0)$.

PROOF. The result follows from formula (4): $\mathbf{P} \to \mathbf{p}$ as $\varepsilon \to 0$. □

Corollary. *A one-parameter group of transformations of phase space $\mathbb{R}^{2n}$ satisfies Hamilton's canonical equations if and only if the transformations are canonical.*

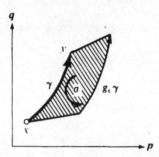

Figure 210   Geometric meaning of Hamilton's function

[86] The number of kinds of generating functions in different textbooks ranges from 4 to 4^n.

The hamiltonian function H is called the "generating function of the infinitesimal canonical transformation." We notice that unlike the generating function S, the function H is a function of points of phase space, invariantly associated to the transformation.

The function H has a simple geometric meaning. Let x and y be two points in $\mathbb{R}^{2n}$ (Figure 210), γ a curve connecting them, and $\partial\gamma = y - x$. Consider the images of the curve γ under the transformations g_τ, $0 \leq \tau \leq \varepsilon$; they form a band $\sigma(\varepsilon)$. Now consider the integral of the form $\omega^2 = \sum dp_i \wedge dq_i$ over the 2-chain σ, using the fact that $\partial\sigma = g_\varepsilon\gamma - \gamma + g_\tau x - g_\tau y$.

PROBLEM. Show that

$$\lim_{\varepsilon \to 0} \frac{1}{\varepsilon} \iint_{\sigma(\varepsilon)} \omega^2 = H(x) - H(y)$$

exists and does not depend on the representative of the class g_ε.

From this result we once more obtain the well-known

Corollary. *Under canonical transformations the canonical equations retain their form, with the same hamiltonian function.*

PROOF. We computed the variation of the hamiltonian function using only an infinitesimal canonical transformation and the symplectic structure of $\mathbb{R}^{2n}$—the form ω^2. $\square$

Introduction to perturbation theory 10

Perturbation theory consists of a very useful collection of methods for finding approximate solutions of "perturbed" problems which are close to completely solvable "unperturbed" problems. These methods can be easily justified if we are investigating motion over a small interval of time. Relatively little is known about how far we can trust the conclusions of perturbation theory in investigating motion over large or infinite intervals of time.

We will see that the motion in many "unperturbed" integrable problems. turns out to be conditionally periodic. In the study of unperturbed problems, and even more so in the study of the perturbed problems, special symplectic coordinates, called "action-angle" variables, are useful. In conclusion, we will prove a theorem justifying perturbation theory for single-frequency systems and will prove the adiabatic invariance of action variables in such systems.

49 Integrable systems

In order to integrate a system of $2n$ ordinary differential equations, we must know $2n$ first integrals. It turns out that if we are given a canonical system of differential equations, it is often sufficient to know only n first integrals – each of them allows us to reduce the order of the system not just by one, but by two.

A Liouville's theorem on integrable systems

Recall that a function F is a first integral of a system with hamiltonian function H if and only if the Poisson bracket

$$(H, F) \equiv 0$$

is identically equal to zero.

Definition. Two functions F_1 and F_2 on a symplectic manifold are *in involution* if their Poisson bracket is equal to zero.

Liouville proved that if, in a system with n degrees of freedom (i.e., with a $2n$-dimensional phase space), n independent first integrals in involution are known, then the system is integrable by quadratures.

Here is the exact formulation of this theorem: Suppose that we are given n functions in involution on a symplectic $2n$-dimensional manifold

$$F_1, \ldots, F_n \qquad (F_i, F_j) \equiv 0, \qquad i, j = 1, 2, \ldots, n.$$

Consider a level set of the functions F_i

$$M_f = \{x: F_i(x) = f_i, i = 1, \ldots, n\}.$$

Assume that the n functions F_i are independent on M_f (i.e., the n 1-forms dF_i are linearly independent at each point of M_f). Then

1. M_f is a smooth manifold, invariant under the phase flow with hamiltonian function $H = F_1$.
2. If the manifold M_f is compact and connected, then it is diffeomorphic to the n-dimensional torus

$$T^n = \{(\varphi_1, \ldots, \varphi_n) \bmod 2\pi\}.$$

3. The phase flow with hamiltonian function H determines a conditionally periodic motion on M_f, i.e., in angular coordinates $\boldsymbol{\varphi} = (\varphi_1, \ldots, \varphi_n)$ we have

$$\frac{d\boldsymbol{\varphi}}{dt} = \boldsymbol{\omega}, \qquad \boldsymbol{\omega} = \boldsymbol{\omega}(\mathbf{f}).$$

4. The canonical equations with hamiltonian function H can be integrated by quadratures.

Before proving this theorem, we note a few of its corollaries.

Corollary 1. *If, in a canonical system with two degrees of freedom, a first integral F is known which does not depend on the hamiltonian H, then the system is integrable by quadratures; a compact connected two-dimensional submanifold of the phase space $H = h$, $F = f$ is an invariant torus, and motion on it is conditionally periodic.*

PROOF. F and H are in involution since F is a first integral of a system with hamiltonian function H. ☐

As an example with three degrees of freedom, we consider a heavy symmetric Lagrange top fixed at a point on its axis. Three first integrals are immediately obvious: H, M_z, and M_3. It is easy to verify that the integrals

M_z and M_3 are in involution. Furthermore, the manifold $H = h$ in the phase space is compact. Therefore, we can immediately say, without any calculations, that for the majority of initial conditions[87] the motion of the top is conditionally periodic: the phase trajectories fill up the three-dimensional torus $H = c_1$, $M_z = c_2$, $M_3 = c_3$. The corresponding three frequencies are called frequencies of fundamental rotation, precession, and nutation.

Other examples arise from the following observation: *if a canonical system can be integrated by the method of Hamilton–Jacobi, then it has n first integrals in involution.* The method consists of a canonical transformation $(\mathbf{p}, \mathbf{q}) \to (\mathbf{P}, \mathbf{Q})$ such that the Q_i are first integrals. But the functions Q_i and Q_j are clearly in involution.

In particular, the observation above applies to the problem of attraction by two fixed centers. Other examples are easily found. In fact, the theorem of Liouville formulated above covers all the problems of dynamics which have been integrated to the present day.

B *Beginning of the proof of Liouville's theorem*

We turn now to the proof of the theorem. Consider the level set of the integrals:

$$M_f = \{x : F_i = f_i, i = 1, \ldots, n\}.$$

By hypothesis, the n 1-forms dF_i are linearly independent at each point of M_f; therefore, by the implicit function theorem, M_f is an n-dimensional submanifold of the $2n$-dimensional phase space.

Lemma 1. *On the n-dimensional manifold M_f there exist n tangent vector fields which commute with one another and which are linearly independent at every point.*

PROOF. The symplectic structure of phase space defines an operator I taking 1-forms to vector fields. This operator I carries the 1-form dF_i to the field $I\,dF_i$ of phase velocities of the system with hamiltonian function F_i. We will show that *the n fields $I\,dF_i$ are tangent to M_f, commute, and are independent.*

The independence of the $I\,dF_i$ at every point of M_f follows from the independence of the dF_i and the nonsingularity of the isomorphism I. The fields $I\,dF_i$ commute with one another, since the Poisson brackets of their hamiltonian functions (F_i, F_j) are identically 0. For the same reason, the derivative of the function F_i in the direction of the field $I\,dF_j$ is equal to zero for any $i, j = 1, \ldots, n$. Thus the fields $I\,dF_i$ are tangent to M_f, and Lemma 1 is proved. □

[87] The singular level sets, where the integrals are not functionally independent, constitute the exception.

We notice that we have proved even more than Lemma 1:

1'. The manifold M_f is invariant with respect to each of the n commuting phase flows g_i^t with hamiltonian functions F_i: $g_i^t g_j^s = g_j^s g_i^t$.

1''. The manifold M_f is null (i.e., the 2-form ω^2 is zero on $TM_f|_x$).

This is true since the n vectors $I\, dF_i|_x$ are skew-orthogonal to one another $((F_i, F_j) \equiv 0)$ and form a basis of the tangent plane to the manifold M_f at the point x.

C Manifolds on which the action of the group $\mathbb{R}^n$ is transitive

We will now use the following topological proposition (the proof is completed in Section D).

Lemma 2. *Let M^n be a compact connected differentiable n-dimensional manifold, on which we are given n pairwise commutative and linearly independent at each point vector fields. Then M^n is diffeomorphic to an n-dimensional torus.*

PROOF. We denote by g_i^t, $i = 1, \ldots, n$, the one-parameter groups of diffeomorphisms of M corresponding to the n given vector fields. Since the fields commute, the groups g_i^t and g_j^s commute. Therefore, we can define an action g of the commutative group $\mathbb{R}^n = \{\mathbf{t}\}$ on the manifold M by setting

$$g^{\mathbf{t}}: M \to M \qquad g^{\mathbf{t}} = g_1^{t_1} \cdots g_n^{t_n}, \qquad (\mathbf{t} = (t_1, \ldots, t_n) \in \mathbb{R}^n).$$

Clearly, $g^{\mathbf{t}+\mathbf{s}} = g^{\mathbf{t}} g^{\mathbf{s}}$, $\mathbf{t}, \mathbf{s} \in \mathbb{R}^n$. Now fix a point $x_0 \in M$. Then we have a map

$$g: \mathbb{R}^n \to M \qquad g(\mathbf{t}) = g^{\mathbf{t}} x_0.$$

(The point x_0 moves along the trajectory of the first flow for time t_1, along the second flow for time t_2, etc.)

PROBLEM 1. Show that the map g (Figure 211) of a sufficiently small neighborhood V of the point $0 \in \mathbb{R}^n$ gives a chart in a neighborhood of x_0: every point $x_0 \in M$ has a neighborhood U $(x_0 \in U \subset M)$ such that g maps V diffeomorphically onto U.

Hint. Apply the implicit function theorem and use the linear independence of the fields at x_0.

PROBLEM 2. Show that $g: \mathbb{R}^n \to M$ is onto.

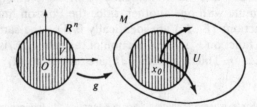

Figure 211   Problem 1

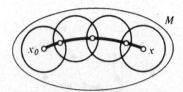

Figure 212   Problem 2

Hint. Connect a point $x \in M$ with x_0 by a curve (Figure 212), cover the curve by a finite number of the neighborhoods U of the preceding problem and define **t** as the sum of shifts t_i corresponding to pieces of the curve.

We note that the map $g: \mathbb{R}^n \to M^n$ cannot be one-to-one since M^n is compact and $\mathbb{R}^n$ is not. We will examine the set of pre-images of $x_0 \in M^n$.

Definition. The *stationary group* of the point x_0 is the set Γ of points $\mathbf{t} \in \mathbb{R}^n$ for which $g^t x_0 = x_0$.

PROBLEM 3. Show that Γ is a subgroup of the group $\mathbb{R}^n$, independent of the point x_0.

Solution. If $g^s x_0 = x_0$ and $g^t x_0 = x_0$, then $g^{s+t} x_0 = g^s g^t x_0 = g^s x_0 = x_0$ and $g^{-t} x_0 = g^{-t} g^t x_0 = x_0$. Therefore, Γ is a subgroup of $\mathbb{R}^n$. If $x = g^r x_0$ and $\mathbf{t} \in \Gamma$, then $g^t x = g^{t+r} x_0 = g^r g^t x_0 = g^r x_0 = x$.

In this way the stationary group Γ is a well-defined subgroup of $\mathbb{R}^n$ independent of the point x_0. In particular, the point $\mathbf{t} = 0$ clearly belongs to Γ.

PROBLEM 4. Show that, in a sufficiently small neighborhood V of the point $0 \in \mathbb{R}^n$, there is no point of the stationary group other than $\mathbf{t} = 0$.

Hint. The map $g: V \to U$ is a diffeomorphism.

PROBLEM 5. Show that, in the neighborhood $\mathbf{t} + V$ of any point $\mathbf{t} \in \Gamma \subset \mathbb{R}^n$, there is no point of the stationary group Γ other than $\mathbf{t}$. (Figure 213)

Thus the points of the stationary group Γ lie in $\mathbb{R}^n$ *discretely*. Such subgroups are called *discrete subgroups*.

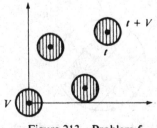

Figure 213   Problem 5

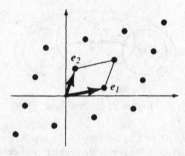

Figure 214   A discrete subgroup of the plane

EXAMPLE. Let $e_1, \ldots, e_k$ be k linearly independent vectors in $\mathbb{R}^n$, $0 \leq k \leq n$. The set of all their integral linear combinations (Figure 214)

$$m_1 e_1 + \cdots + m_k e_k, \qquad m_i \in \mathbb{Z} = (\ldots, -2, -1, 0, 1, \ldots)$$

forms a discrete subgroup of $\mathbb{R}^n$. For example, the set of all integral points in the plane is a discrete subgroup of the plane.

D Discrete subgroups in $\mathbb{R}^n$

We will now use the algebraic fact that the example above includes all discrete subgroups of $\mathbb{R}^n$. More precisely, we will prove

Lemma 3. *Let Γ be a discrete subgroup of $\mathbb{R}^n$. Then there exist k ($0 \leq k \leq n$) linearly independent vectors $e_1; \ldots, e_k \in \Gamma$ such that Γ is exactly the set of all their integral linear combinations.*

PROOF. We will consider $\mathbb{R}^n$ with some euclidean structure. We always have $0 \in \Gamma$. If $\Gamma = \{0\}$ the lemma is proved. If not, there is a point $e_0 \in \Gamma$, $e_0 \neq 0$ (Figure 215). Consider the line $\mathbb{R}e_0$. We will show that among the elements of Γ on this line, there is a point e_1 which is closest to 0. In fact, in the disk of radius $|e_0|$ with center 0, there are only a finite number of points of Γ (as we saw above, every point x of Γ has a neighborhood V of standard size which does not contain any other point of Γ). Among the finite number of points of Γ inside this disc and lying on the line $\mathbb{R}e_0$, the point closest to 0 will be the closest point to 0 on the whole line. The integral multiples of this point e_1 ($m e_1$, $m \in \mathbb{Z}$) constitute the intersection of the line $\mathbb{R}e_0$ with Γ.

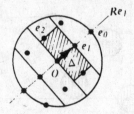

Figure 215   Proof of the lemma on discrete subgroups

In fact, the points me_1 divide the line into pieces of length $|e_1|$. If there were a point $e \in \Gamma$ inside one of these pieces $(me_1, (m + 1)e_1)$, then the point $e - me_1 \in \Gamma$ would be closer to 0 than e_1.

If there are no points of Γ off the line $\mathbb{R}e_1$, the lemma is proved. Suppose there is a point $e \in \Gamma, e \notin \mathbb{R}e_1$. We will show that there is a point $e_2 \in \Gamma$ closest to the line $\mathbb{R}e_1$ (but not lying on the line). We project e orthogonally onto $\mathbb{R}e_1$. The projection lies in exactly one interval $\Delta = \{\lambda e_1\}$, $m \leq \lambda < m + 1$. Consider the right circular cylinder C with axis Δ and radius equal to the distance from Δ to e. In this cylinder lie a finite (nonempty) number of points of the group Γ. Let e_2 be the closest one to the axis $\mathbb{R}e_1$ not lying on the axis.

PROBLEM 6. Show that the distance from this axis to any point e of Γ not lying on $\mathbb{R}e_1$ is greater than or equal to the distance of e_2 from $\mathbb{R}e_1$.

 Hint. By a shift of me_1 we can move the projection of e onto the axis interval Δ.

The integral linear combinations of e_1 and e_2 form a lattice in the plane $\mathbb{R}e_1 + \mathbb{R}e_2$.

PROBLEM 7. Show that there are no points of Γ on the plane $\mathbb{R}e_1 + \mathbb{R}e_2$ other than integral linear combinations of e_1 and e_2.

 Hint. Partition the plane into parallelograms (Figure 216) $\Delta = \{\lambda_1 e_1 + \lambda_2 e_2\}$, $m_i \leq \lambda_i < m_i + 1$. If there were an $e \in \Delta$ with $e \neq m_1 e_1 + m_2 e_2$, then the point $e - m_1 e_1 - m_2 e_2$ would be closer to $\mathbb{R}e_1$ than e_2.

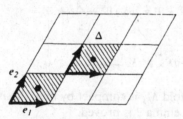

Figure 216   Problem 7

If there are no points of Γ outside the plane $\mathbb{R}e_1 + \mathbb{R}e_2$, the lemma is proved. Suppose that there is a point $e \in \Gamma$ outside this plane. Then there exists a point $e_3 \in \Gamma$ closest to $\mathbb{R}e_1 + \mathbb{R}e_2$; the points $m_1 e_1 + m_2 e_2 + m_3 e_3$ exhaust Γ in the three-dimensional space $\mathbb{R}e_1 + \mathbb{R}e_2 + \mathbb{R}e_3$. If Γ is not exhausted by these, we take the closest point to this three-dimensional space, etc.

PROBLEM 8. Show that this closest point always exists.

 Hint. Take the closest of the finite number of points in a "cylinder" C.

Note that the vectors $e_1, e_2, e_3, \ldots$ are linearly independent. Since they all lie in $\mathbb{R}^n$, there are $k \leq n$ of them.

PROBLEM 9. Show that Γ is exhausted by the integral linear combinations of $e_1, \ldots, e_k$.

Hint. Partition the plane $\Re e_1 + \cdots + \Re e_k$ into parallelepipeds Δ and show that there cannot be a point of Γ in any Δ. If there is an $e \in \Gamma$ outside the plane $\Re e_1 + \cdots + \Re e_k$, the construction is not finished.

Thus Lemma 3 is proved. ☐

It is now easy to prove Lemma 2: M_f is diffeomorphic to a torus T^n. Consider the direct product of k circles and $n - k$ straight lines:

$$T^k \times \mathbb{R}^{n-k} = \{(\varphi_1, \ldots, \varphi_k; y_1, \ldots, y_{n-k})\}, \qquad \boldsymbol{\varphi} \bmod 2\pi,$$

together with the natural map $p: \mathbb{R}^{2n} \to T^k \times \mathbb{R}^{n-k}$,

$$p(\boldsymbol{\varphi}, \mathbf{y}) = (\boldsymbol{\varphi} \bmod 2\pi, \mathbf{y}).$$

The points $\mathbf{f}_1, \ldots, \mathbf{f}_k \in \mathbb{R}^n$ ($\mathbf{f}_i$ has coordinates $\varphi_i = 2\pi$, $\varphi_j = 0$, $\mathbf{y} = 0$) are mapped to 0 under this map.

Let $\mathbf{e}_1, \ldots, \mathbf{e}_k \in \Gamma \subset \mathbb{R}^n$ be the generators of the group Γ (cf. Lemma 3). We map the vector space $\mathbb{R}^n = \{(\boldsymbol{\varphi}, \mathbf{y})\}$ onto the space $\mathbb{R}^n = \{\mathbf{t}\}$ so that the vectors $\mathbf{f}_i$ go to $\mathbf{e}_i$. Let $A: \mathbb{R}^n \to \mathbb{R}^n$ be such an isomorphism.

We now note that $\mathbb{R}^n = \{(\boldsymbol{\varphi}, \mathbf{y})\}$ gives charts for $T^k \times \mathbb{R}^{n-k}$, and $\mathbb{R}^n = \{\mathbf{t}\}$ gives charts for our manifold M_f.

PROBLEM 10. Show that the map of charts $A: \mathbb{R}^n \to \mathbb{R}^n$ gives a diffeomorphism $\tilde{A}: T^k \times \mathbb{R}^{n-k} \to M_f$.

$$
\begin{array}{ccc}
\mathbb{R}^n = \{(\boldsymbol{\varphi}, \mathbf{y})\} & \xrightarrow{\ A\ } & \mathbb{R}^n = \{\mathbf{t}\} \\
\downarrow{\scriptstyle p} & & \downarrow{\scriptstyle g} \\
T^k \times \mathbb{R}^{n-k} & \xrightarrow{\ \tilde{A}\ } & M_f
\end{array}
$$

But, since the manifold M_f is compact by hypothesis, $k = n$ and M_f is an n-dimensional torus. Lemma 2 is proved. ☐

In view of Lemma 1, the first two statements of the theorem are proved. At the same time, we have constructed angular coordinates $\varphi_1, \ldots, \varphi_n \bmod 2\pi$ on M_f.

PROBLEM 11. Show that under the action of the phase flow with hamiltonian H the angular coordinates $\boldsymbol{\varphi}$ vary uniformly with time

$$\dot{\varphi}_i = \omega_i \qquad \omega_i = \omega_i(\mathbf{f}) \qquad \boldsymbol{\varphi}(t) = \boldsymbol{\varphi}(0) + \boldsymbol{\omega} t.$$

In other words, motion on the invariant torus M_f is conditionally periodic.

Hint. $\boldsymbol{\varphi} = A^{-1}\mathbf{t}$.

Of all the assertions of the theorem, only the last remains to be proved: that the system can be integrated by quadratures.

50 Action-angle variables

We show here that, under the hypotheses of Liouville's theorem, we can find symplectic co-ordinates $(\mathbf{I}, \boldsymbol{\varphi})$ such that the first integrals $\mathbf{F}$ depend only on $\mathbf{I}$, and $\boldsymbol{\varphi}$ are angular coordinates on the torus M_f.

A *Description of action-angle variables*

In Section 49 we studied one particular compact connected level manifold of the integrals: $M_f = \{x: \mathbf{F}(x) = \mathbf{f}\}$; it turned out that M_f was an n-dimensional torus, invariant with respect to the phase flow. We chose angular coordinates φ_i on M so that the phase flow with hamiltonian function $H = F_1$ takes an especially simple form:

$$\frac{d\boldsymbol{\varphi}}{dt} = \boldsymbol{\omega}(\mathbf{f}) \qquad \boldsymbol{\varphi}(t) = \boldsymbol{\varphi}(0) + \boldsymbol{\omega}t.$$

We will now look at a neighborhood of the n-dimensional manifold M_f in $2n$-dimensional phase space.

PROBLEM. Show that the manifold M_f has a neighborhood diffeomorphic to the direct product of the n-dimensional torus T^n and the disc D^n in n-dimensional euclidean space.

Hint. Take the functions F_i and the angles φ_i constructed above as coordinates. In view of the linear independence of the dF_i, the functions F_i and φ_i $(i = 1, \ldots, n)$ give a diffeomorphism of a neighborhood of M_f onto the direct product $T^n \times D^n$.

In the coordinates $(\mathbf{F}, \boldsymbol{\varphi})$ the phase flow with hamiltonian function $H = F_1$ can be written in the form of the simple system of $2n$ ordinary differential equations

$$(1) \qquad \frac{d\mathbf{F}}{dt} = 0 \qquad \frac{d\boldsymbol{\varphi}}{dt} = \boldsymbol{\omega}(\mathbf{F}),$$

which is easily integrated: $\mathbf{F}(t) = \mathbf{F}(0)$, $\boldsymbol{\varphi}(t) = \boldsymbol{\varphi}(0) + \boldsymbol{\omega}(\mathbf{F}(0))t$.

Thus, in order to integrate explicitly the original canonical system of differential equations, it is sufficient to find the variables $\boldsymbol{\varphi}$ in explicit form. It turns out that this can be done using only quadratures. A construction of the variables $\boldsymbol{\varphi}$ is given below.

We note that the variables $(\mathbf{F}, \boldsymbol{\varphi})$ are not, in general, symplectic co-ordinates. It turns out that there are functions of $\mathbf{F}$, which we will denote by $\mathbf{I} = \mathbf{I}(\mathbf{F})$, $\mathbf{I} = (I_1, \ldots, I_n)$, such that the variables $(\mathbf{I}, \boldsymbol{\varphi})$ are symplectic coordinates: the original symplectic structure ω^2 is expressed in them by the usual formula

$$\omega^2 = \sum dI_i \wedge d\varphi_i.$$

The variables **I** are called action variables;[88] together with the angle variables **φ** they form the *action-angle system of canonical coordinates* in a neighborhood of $M_{\mathbf{f}}$.

The quantities I_i are first integrals of the system with hamiltonian function $H = F_1$, since they are functions of the first integrals F_j. In turn, the variables F_i can be expressed in terms of **I** and, in particular, $H = F_1 = H(\mathbf{I})$. In action-angle variables the differential equations of our flow (1) have the form

(2)
$$\frac{d\mathbf{I}}{dt} = 0 \qquad \frac{d\boldsymbol{\varphi}}{dt} = \boldsymbol{\omega}(\mathbf{I}).$$

PROBLEM. Can the functions $\boldsymbol{\omega}(\mathbf{I})$ in (2) be arbitrary?

Solution. In the variables $(\mathbf{I}, \boldsymbol{\varphi})$, the equations of the flow (2) have the canonical form with hamiltonian function $H(\mathbf{I})$. Therefore, $\boldsymbol{\omega}(\mathbf{I}) = \partial H/\partial \mathbf{I}$; thus if the number of degrees of freedom is $n \geq 2$, the functions $\boldsymbol{\omega}(\mathbf{I})$ are not arbitrary, but satisfy the symmetry condition $\partial \omega_i/\partial I_j = \partial \omega_j/\partial I_i$.

Action-angle variables are especially important for perturbation theory; in Section 52 we will demonstrate their application to the theory of adiabatic invariants.

B *Construction of action-angle variables in the case of one degree of freedom*

A system with one degree of freedom in the phase plane (p, q) is given by the hamiltonian function $H(p, q)$.

EXAMPLE 1. The harmonic oscillator $H = \frac{1}{2}p^2 + \frac{1}{2}q^2$; or, more generally, $H = \frac{1}{2}a^2p^2 + \frac{1}{2}b^2q^2$.

EXAMPLE 2. The mathematical pendulum $H = \frac{1}{2}p^2 - \cos q$. In both cases we have a compact closed curve $M_h(H = h)$, and the conditions of the theorem of Section 49 for $n = 1$ are satisfied.

In order to construct the action-angle variables, we will look for a canonical transformation $(p, q) \to (I, \varphi)$ satisfying the two conditions:

(3)
$$1. \ I = I(h),$$
$$2. \ \oint_{M_h} d\varphi = 2\pi.$$

PROBLEM. Find the action-angle variables in the case of the simple harmonic oscillator $H = \frac{1}{2}p^2 + \frac{1}{2}q^2$.

Solution. If r, φ are polar coordinates, then $dp \wedge dq = r\, dr \wedge d\varphi = d(r^2/2) \wedge d\varphi$. Therefore, $I = H = (p^2 + q^2)/2$.

---

[88] It is not hard to see that **I** has the dimensions of action.

In order to construct the canonical transformation $p, q \to I, \varphi$ in the general case, we will look for its generating function $S(I, q)$:

(4) $\qquad p = \dfrac{\partial S(I, q)}{\partial q} \qquad \varphi = \dfrac{\partial S(I, q)}{\partial I} \qquad H\left(\dfrac{\partial S(I, q)}{\partial q}, q\right) = h(I).$

We first assume that the function $h(I)$ is known and invertible, so that every curve M_h is determined by the value of I ($M_h = M_{h(I)}$). Then for a fixed value of I we have from (4)

$$dS|_{I = \text{const}} = p \, dq.$$

This relation determines a well-defined differential 1-form dS on the curve $M_{h(I)}$.

Integrating this 1-form on the curve $M_{h(I)}$ we obtain (in a neighborhood of a point q_0) a function

$$S(I, q) = \int_{q_0}^{q} p \, dq.$$

This function will be the generating function of the transformation (4) in a neighborhood of the point (I, q_0). The first of the conditions (3) is satisfied automatically: $I = I(h)$. To verify the second condition, we consider the behavior of $S(I, q)$ "in the large." After a circuit of the closed curve $M_{h(I)}$ the integral of $p \, dq$ increases by

$$\Delta S(I) = \oint_{M_{h(I)}} p \, dq,$$

equal to the area Π enclosed by the curve $M_{h(I)}$. Therefore, the function S is a "multiple-valued function" on $M_{h(I)}$: it is determined up to addition of integral multiples of Π. This term has no effect on the derivative $\partial S(I, q)/\partial q$; but it leads to the multi-valuedness of $\varphi = \partial S/\partial I$. This derivative turns out to be defined only up to multiples of $d \, \Delta S(I)/dI$. More precisely, the formulas (4) define a 1-form $d\varphi$ on the curve $M_{h(I)}$, and the integral of this form on $M_{h(I)}$ is equal to $d \, \Delta S(I)/dI$.

In order to fulfill the second condition, $\oint_{M_h} d\varphi = 2\pi$, we need that

$$\frac{d}{dI} \Delta S(I) = 2\pi \qquad I = \frac{\Delta S}{2\pi} = \frac{\Pi}{2\pi},$$

where $\Pi = \oint_{M_h} p \, dq$ is the area bounded by the phase curve $H = h$.

Definition. The *action variable* in the one-dimensional problem with hamiltonian function $H(p, q)$ is the quantity $I(h) = (1/2\pi)\Pi(h)$.

Finally, we arrive at the following conclusion. Let $d\Pi/dh \neq 0$. Then the inverse $I(h)$ of the function $h(I)$ is defined.

Theorem. *Set* $S(I, q) = \int_{q_0}^{q} p \, dq|_{H = h(I)}$. *Then formulas* (4) *give a canonical transformation* $p, q \to I, \varphi$ *satisfying conditions* (3).

Thus, the action-angle variables in the one-dimensional case are constructed.

PROBLEM. Find S and I for a harmonic oscillator.

ANSWER. If $H = \frac{1}{2}a^2 p^2 + \frac{1}{2}b^2 q^2$ (Figure 217), then M_h is the ellipse bounding the area $\Pi(h) = \pi(\sqrt{2h}/a)(\sqrt{2h}/b) = 2\pi h/ab = 2\pi h/\omega$. Thus for a harmonic oscillator the action variable is the ratio of energy to frequency. The angle variable φ is, of course, the phase of oscillation.

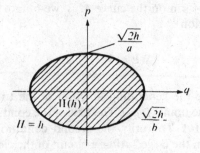

Figure 217   Action variable for a hamonic oscillator

PROBLEM. Show that the period T of motion along the closed curve $H = h$ on the phase plane p, q is equal to the derivative with respect to h of the area bounded by this curve:

$$T = \frac{d\Pi(h)}{dh}.$$

Solution. In action-angle variables the equations of motion (2) give

$$\dot{\varphi} = \frac{\partial H}{\partial I} = \left(\frac{dI}{dh}\right)^{-1} = 2\pi\left(\frac{d\Pi}{dh}\right)^{-1} \qquad T = \frac{2\pi}{\dot{\varphi}} = \frac{d\Pi}{dh}.$$

C *Construction of action-angle variables in* $\mathbb{R}^{2n}$

We turn now to systems with n degrees of freedom given in $\mathbb{R}^{2n} = \{(\mathbf{p}, \mathbf{q})\}$ by a hamiltonian function $H(\mathbf{p}, \mathbf{q})$ and having n first integrals in involution $F_1 = H, F_2, \ldots, F_n$. We will not repeat the reasoning which brought us to the choice of $2\pi I = \oint p \, dq$ in the one-dimensional case, but will immediately define n action variables $\mathbf{I}$.

Let $\gamma_1, \ldots, \gamma_n$ be a basis for the one-dimensional cycles on the torus $M_{\mathbf{f}}$ (the increase of the coordinate φ_i on the cycle γ_j is equal to 2π if $i = j$ and 0 if $i \neq j$). We set

(5) $$I_i(\mathbf{f}) = \frac{1}{2\pi} \oint_{\gamma_i} \mathbf{p} \, d\mathbf{q}.$$

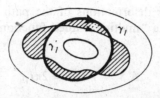

Figure 218   Independence of the curve of integration for the action variable

PROBLEM. Show that this integral does not depend on the choice of the curve γ_i representing the cycle (Figure 218).

Hint. In Section 49 we showed that the 2-form $\omega^2 = \sum dp_i \wedge dq_i$ on the manifold M_f is equal to zero. By Stokes' formula,

$$\oint_\gamma - \oint_{\gamma'} \mathbf{p}\, d\mathbf{q} = \iint_\sigma d\mathbf{p} \wedge d\mathbf{q} = 0,$$

where $\partial\sigma = \gamma - \gamma'$.

Definition. The n quantities $I_i(\mathbf{f})$ given by formula (5) are called the *action variables.*

We assume now that, for the given values f_i of the n integrals F_i, the n quantities I_i are independent: $\det(\partial\mathbf{I}/\partial\mathbf{f})|_{\mathbf{f}} \neq 0$. Then in a neighborhood of the torus M_f we can take the variables $\mathbf{I}, \boldsymbol{\varphi}$ as coordinates.

Theorem. *The transformation* $\mathbf{p}, \mathbf{q} \rightarrow \mathbf{I}, \boldsymbol{\varphi}$ *is canonical, i.e.,*

$$\sum dp_i \wedge dq_i = \sum dI_i \wedge d\varphi_i.$$

We outline the proof of this theorem. Consider the differential 1-form $\mathbf{p}\, d\mathbf{q}$ on M_f. Since the manifold M_f is null (Section 49) this 1-form on M_f is closed: its exterior derivative $\omega^2 = d\mathbf{p} \wedge d\mathbf{q}$ is identically equal to zero on M_f. Therefore (Figure 219),

$$S(x) = \int_{x_0}^{x} \mathbf{p}\, d\mathbf{q}|_{M_f}$$

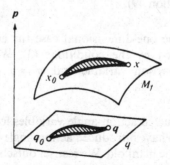

Figure 219   Independence of the path for the integral of $\mathbf{p}\, d\mathbf{q}$ on M_f

does not change under deformations of the path of integration (Stokes' formula). Thus $S(x)$ is a "multiple-valued function" on $M_{\mathfrak{f}}$, with periods equal to

$$\Delta_i S = \int_{\gamma_i} dS = 2\pi I_i.$$

Now let x_0 be a point on $M_{\mathfrak{f}}$, in a neighborhood of which the n variables $\mathbf{q}$ are coordinates on $M_{\mathfrak{f}}$, such that the submanifold $M_{\mathfrak{f}} \subset \mathbb{R}^{2n}$ is given by n equations of the form $\mathbf{p} = \mathbf{p}(\mathbf{I}, \mathbf{q}), \mathbf{q}(x_0) = \mathbf{q}_0$. In a simply connected neighborhood of the point $\mathbf{q}_0$ a single-valued function is defined,

$$S(\mathbf{I}, \mathbf{q}) = \int_{q_0}^{q} \mathbf{p}(\mathbf{I}, \mathbf{q}) d\mathbf{q},$$

and we can use it as the generating function of a canonical transformation $\mathbf{p}, \mathbf{q} \to \mathbf{I}, \boldsymbol{\varphi}$:

$$\mathbf{p} = \frac{\partial S}{\partial \mathbf{q}} \qquad \boldsymbol{\varphi} = \frac{\partial S}{\partial \mathbf{I}}.$$

It is not difficult to verify that these formulas actually give a canonical transformation, not only in a neighborhood of the point under consideration, but also "in the large" in a neighborhood of $M_{\mathfrak{f}}$. The coordinates $\boldsymbol{\varphi}$ will be multiple-valued with periods

$$\Delta_i \varphi_j = \Delta_i \frac{\partial S}{\partial I_j} = \frac{\partial}{\partial I_j} \Delta_i S = \frac{\partial}{\partial I_j} 2\pi I_i = 2\pi \delta_{ij},$$

as was to be shown. $\qquad\qquad\qquad\qquad\qquad\qquad\qquad\qquad\qquad\square$

We now note that all our constructions involve only "algebraic" operations (inverting functions) and "quadrature"—calculation of the integrals of known functions. In this way the problem of integrating a canonical system with $2n$ equations, of which n first integrals in involution are known, is solved by quadratures, which proves the last assertion of Liouville's theorem (Section 49). $\qquad\qquad\qquad\qquad\qquad\square$

Remark 1. Even in the one-dimensional case the action-angle variables are not uniquely defined by the conditions (3). We could have taken $I' = I + \text{const}$ for the action variable and $\varphi' = \varphi + c(I)$ for the angle variable.

Remark 2. We constructed action-angle variables for systems with phase space $\mathbb{R}^{2n}$. We could also have introduced action-angle variables for a system on an arbitrary symplectic manifold. We restrict outselves here to one simple example (Figure 220).

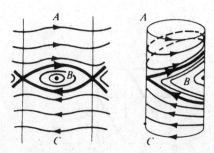

Figure 220   Action-angle variables on a symplectic manifold

We could have taken the phase space of a pendulum $(H = \frac{1}{2}p^2 - \cos q)$ to be, instead of the plane $\{(p, q)\}$, the surface of the cylinder $\mathbb{R}^1 \times S^1$ obtained by identifying angles q differing by an integral multiple of 2π.

The critical level lines $H = \pm 1$ divide the cylinder into three parts, A, B, and C, each of which is diffeomorphic to the direct product $\mathbb{R}^1 \times S^1$. We can introduce action-angle variables into each part. In the bounded part (B) the closed trajectories represent the oscillation of the pendulum; in the unbounded parts they represent rotation.

Remark 3. In the general case, as in the example analyzed above, the equations $F_i = f_i$ cease to be independent for some values of f_i, and M_f ceases to be a manifold. Such critical values of $\mathbf{f}$ correspond to separatrices dividing the phase space of the integrable problem into parts corresponding to the parts A, B, and C above. In some of these parts the manifolds M_f can be unbounded (parts A and C in the plane $\{(p, q)\}$); others are stratified into n-dimensional invariant tori M_f; in a neighborhood of such a torus we can introduce action-angle variables.

51 Averaging

In this paragraph we show that time averages and space averages are equal for systems undergoing conditionally periodic motion.

A Conditionally periodic motion

In the earlier sections of this book, we have frequently encountered conditionally periodic motion: Lissajous figures, precession, nutation, rotation of a top, etc.

Definition. Let T^n be the n-dimensional torus and $\boldsymbol{\varphi} = (\varphi_1, \ldots, \varphi_n)$ mod 2π angular coordinates. Then by a *conditionally periodic motion* we mean a one-parameter group of diffeomorphisms $T^n \to T^n$ given by the differential equations (Figure 221):

$$\dot{\boldsymbol{\varphi}} = \boldsymbol{\omega}, \qquad \boldsymbol{\omega} = (\omega_1, \ldots, \omega_n) = \text{const.}$$

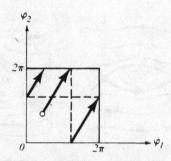

Figure 221   Conditionally periodic motion

These differential equations are easily integrated:

$$\boldsymbol{\varphi}(t) = \boldsymbol{\varphi}(0) + \boldsymbol{\omega}t.$$

Thus the trajectories in the chart $\{\boldsymbol{\varphi}\}$ are straight lines. A trajectory on the torus is called a *winding* of the torus.

EXAMPLE. Let $n = 2$. If $\omega_1/\omega_2 = k_1/k_2$, the trajectories are closed; if ω_1/ω_2 is irrational, then trajectories on the torus are dense (cf. Section 16).

The quantities $\omega_1, \ldots, \omega_n$ are called the *frequencies* of the conditionally periodic motion. The frequencies are called *independent* if they are linearly independent over the field of rational numbers: if $\mathbf{k} \in \mathbb{Z}^n$ [89] and $(\mathbf{k}, \boldsymbol{\omega}) = 0$, then $\mathbf{k} = 0$.

B Space average and time average

Let $f(\boldsymbol{\varphi})$ be an integrable function on the torus T^n.

Definition. The *space* average of a function f on the torus T^n is the number

$$\bar{f} = (2\pi)^{-n} \int_0^{2\pi} \cdots \int_0^{2\pi} f(\boldsymbol{\varphi}) d\varphi_1 \cdots d\varphi_n.$$

Consider the value of the function $f(\boldsymbol{\varphi})$ on the trajectory $\boldsymbol{\varphi}(t) = \boldsymbol{\varphi}_0 + \boldsymbol{\omega}t$. This is a function of time, $f(\boldsymbol{\varphi}_0 + \boldsymbol{\omega}t)$. We consider its average.

Definition. The *time* average of the function f on the torus T^n is the function

$$f^*(\boldsymbol{\varphi}_0) = \lim_{T \to \infty} \frac{1}{T} \int_0^T f(\boldsymbol{\varphi}_0 + \boldsymbol{\omega}t) dt$$

(defined where the limit exists).

Theorem on the averages. *The time average exists everywhere, and coincides with the space average if f is continuous (or merely Riemann integrable) and the frequencies ω_i are independent.*

[89] $\mathbf{k} = (k_1, \ldots, k_n)$ with integral k_i.

PROBLEM. Show that if the frequencies are dependent, then the time average can differ from the space average.

Corollary 1. *If the frequencies are independent, then every trajectory $\{\varphi(t)\}$ is dense on the torus T^n.*

PROOF. Assume the contrary. Then in some neighborhood D of some point of the torus, there is no point of the trajectory $\varphi(t)$. It is easy to construct a continuous function f equal to zero outside D and with space average equal to 1. The time average $f^*(\varphi_0)$ on the trajectory $\varphi(t)$ is equal to $0 \neq 1$. This contradicts the assertion of the theorem. □

Corollary 2. *If the frequencies are independent, then every trajectory is uniformly distributed on the torus T^n.*

This means that the time the trajectory spends in a neighborhood D is proportional to the measure of D.

More precisely, let D be a (Jordan) measurable region of T^n. We denote by $\tau_D(T)$ the amount of time that the interval $0 \leq t \leq T$ of the trajectory $\varphi(t)$ is inside of D. Then

$$\lim_{T \to \infty} \frac{\tau_D(T)}{T} = \frac{\text{mes } D}{(2\pi)^n}.$$

PROOF. We apply the theorem to the characteristic function f of the set D (f is Riemann integrable since D is Jordan measurable). Then $\int_0^T f(\varphi(t))dt = \tau_D(T)$, and $\bar{f} = (2\pi)^{-n}$ mes D, and the corollary follows immediately from the theorem. □

Corollary. *In the sequence*

$$1, 2, 4, 8, 1, 3, 6, 1, 2, 5, 1, 2, \ldots$$

of first digits of the numbers 2^n, the number 7 appears $(\log 8 - \log 7)/(\log 9 - \log 8)$ times as often as 8.

The theorem on averages may be found implicitly in the work of Laplace, Lagrange, and Gauss on celestial mechanics; it is one of the first "ergodic theorems." A rigorous proof was given only in 1909 by P. Bohl, W. Sierpiński, and H. Weyl in connection with a problem of Lagrange on the mean motion of the earth's perihelion. Below we reproduce H. Weyl's proof.

C *Proof of the theorem on averages*

Lemma 1. *The theorem is true for exponentials $f = e^{i(\mathbf{k}, \varphi)}$, $\mathbf{k} \in \mathbb{Z}^n$.*

PROOF. If $\mathbf{k} = 0$, then $\bar{f} = f = f^* = 1$ and the theorem is obvious. If $\mathbf{k} \neq 0$, then $\bar{f} = 0$. On the other hand,

$$\int_0^T e^{i(\mathbf{k}, \varphi_0 + \omega t)} \, dt = e^{i(\mathbf{k}, \varphi_0)} \frac{e^{i(\mathbf{k}, \omega)T} - 1}{i(\mathbf{k}, \omega)}.$$

287

Therefore, the time average is

$$\lim_{T \to \infty} \frac{e^{i(\mathbf{k}, \, \boldsymbol{\varphi}_0)}}{i(\mathbf{k}, \, \boldsymbol{\omega})} \frac{e^{i(\mathbf{k}, \, \boldsymbol{\omega})T} - 1}{T} = 0. \qquad \square$$

Lemma 2. *The theorem is true for trigonometric polynomials*

$$f = \sum_{|\mathbf{k}| < N} f_{\mathbf{k}} e^{i(\mathbf{k}, \, \boldsymbol{\varphi})}.$$

PROOF. Both the time and space averages depend linearly on f, and therefore agree by Lemma 1. $\qquad \square$

Lemma 3. *Let f be a real continuous (or at least Riemann integrable) function. Then, for any $\varepsilon > 0$, there exist two trigonometric polynomials P_1 and P_2 such that $P_1 < f < P_2$ and $(1/(2\pi)^n)\int_{T^n}(P_2 - P_1)d\varphi \le \varepsilon$.*

PROOF. Suppose first that f is continuous. By the Weierstrass theorem, we can approximate f by a trigonometric polynomial P with $|f - P| < \tfrac{1}{2}\varepsilon$. The polynomials $P_1 = P - \tfrac{1}{2}\varepsilon$ and $P_2 = P + \tfrac{1}{2}\varepsilon$ are the ones we are looking for.

If f is not continuous but Riemann integrable, then there are two continuous functions f_1 and f_2 such that $f_1 < f < f_2$ and $(2\pi)^{-n} \int (f_2 - f_1)d\varphi < \tfrac{1}{3}\varepsilon$ (Figure 222 corresponds to the characteristic function of an interval). By approximating f_1 and f_2 by polynomials $P_1 < f_1 < f_2 < P_2$, $(2\pi)^{-n} \int (P_2 - f_2)d\varphi < \tfrac{1}{3}\varepsilon$, $(2\pi)^{-n} \int (f_1 - P_1)d\varphi < \tfrac{1}{3}\varepsilon$, we obtain what we need. Lemma 3 is proved. $\qquad \square$

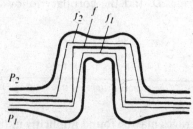

Figure 222 Approximation of the function f by trigonometric polynomials P_1 and P_2

It is now easy to finish the proof of the theorem. Let $\varepsilon > 0$. Then, by Lemma 3, there are trigonometric polynomials $P_1 < f < P_2$ with $(2\pi)^{-n} \int (P_2 - P_1)d\varphi < \varepsilon$.

For any T, we then have

$$\frac{1}{T} \int_0^T P_1(\varphi(t))dt < \frac{1}{T} \int_0^T f(\varphi(t))dt < \frac{1}{T} \int_0^T P_2(\varphi(t))dt.$$

By Lemma 2, for $T > T_0(\varepsilon)$,

$$\left| \bar{P}_i - \frac{1}{T} \int_0^T P_i(\varphi(t))dt \right| < \varepsilon \qquad (i = 1, 2).$$

Furthermore, $\bar{P}_1 < \bar{f} < \bar{P}_2$ and $\bar{P}_2 - \bar{P}_1 < \varepsilon$. Therefore, $\bar{P}_2 - \bar{f} < \varepsilon$ and $\bar{f} - \bar{P}_1 < \varepsilon$; therefore, for $T > T_0(\varepsilon)$,

$$\left| \frac{1}{T} \int_0^T f(\boldsymbol{\varphi}(t)) dt - \bar{f} \right| < 2\varepsilon,$$

as was to be proved. $\qquad\qquad\qquad\qquad\qquad\qquad\qquad\qquad\qquad\qquad\qquad\square$

PROBLEM. A two-dimensional oscillator with kinetic energy $T = \frac{1}{2}\dot{x}^2 + \frac{1}{2}\dot{y}^2$ and potential energy $U = \frac{1}{2}x^2 + y^2$ performs an oscillation with amplitudes $a_x = 1$ and $a_y = 1$. Find the time average of the kinetic energy.

PROBLEM.[90] Let ω_k be independent, $a_k > 0$. Calculate

$$\lim_{t \to \infty} \frac{1}{t} \arg \sum_{k=1}^{3} a_k e^{i\omega_k t}.$$

ANSWER. $(\omega_1 \alpha_1 + \omega_2 \alpha_2 + \omega_3 \alpha_3)/\pi$, where α_1, α_2, and α_3 are the angles of the triangle with sides a_k (Figure 223).

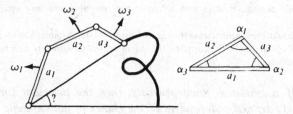

Figure 223 Problem on mean motion of perihelia

D Degeneracies

So far we have considered the case when the frequencies $\boldsymbol{\omega}$ are independent. An integral vector $\mathbf{k} \in \mathbb{Z}^n$ is called a *relation among the frequencies* if $(\mathbf{k}, \boldsymbol{\omega}) = 0$.

PROBLEM. Show that the set of all relations between a given set of frequencies $\boldsymbol{\omega}$ is a subgroup Γ of the lattice $\mathbb{Z}^n$.

We saw in Section 49 that such a subgroup consists entirely of linear combinations of r independent vectors $\mathbf{k}_i$, $1 \leq r \leq n$. We say that there are r *(independent) relations* among the frequencies.[91]

---

[90] Lagrange showed that the investigation of the average motion of the perihelion of a planet reduces to a similar problem. The solution of this problem can be found in the work of H. Weyl. The eccentricity of the earth's orbit varies as the modulus of an analogous sum. Ice ages appear to be related to these changes in eccentricity.

[91] Show that the number r does not depend on the choice of independent vectors $\mathbf{k}_i$.

289

PROBLEM. Show that the closure of a trajectory $\{\varphi(t) = \varphi_0 + \omega t\}$ (on T^n) is a torus of dimension $n - r$ if there are r independent relations among the frequencies ω; in this case the motion on T^{n-r} is conditionally periodic with $n - r$ independent frequencies.

We turn now to the integrable hamiltonian system given in action-angle variables I, φ by the equations

$$\dot{I} = 0 \qquad \dot{\varphi} = \omega(I), \quad \text{where } \omega(I) = \frac{\partial H}{\partial I}.$$

Every n-dimensional torus $I = \textbf{const}$ in the $2n$-dimensional phase space is invariant, and motion on it is conditionally periodic.

Definition. A system is called *nondegenerate* if the determinant

$$\det \frac{\partial \omega}{\partial I} = \det \frac{\partial^2 H}{\partial I^2}$$

is not zero.

PROBLEM. Show that, if a system is nondegenerate, then in any neighborhood of any point there is a conditionally periodic motion with n frequencies, and also with any smaller number of frequencies.

Hint. We can take the frequencies ω themselves instead of the variables I as local coordinates. In the space of collections of frequencies, the set of points ω with any number of relations $r(0 \leq r < n)$ is dense.

Corollary. *If a system is nondegenerate, then the invariant tori $I = \textbf{const}$ are uniquely defined, independent of the choice of action-angle coordinates I, φ, the construction of which always involves some arbitrariness.*[92]

PROOF. The tori $I = \textbf{const}$ can be defined as the closures of the phase trajectories corresponding to the independent ω. $\qquad \square$

We note incidentally that, for the majority of values I, the frequencies ω will be independent.

PROBLEM. Show that the set of I for which the frequencies $\omega(I)$ in a nondegenerate system are dependent has Lebesgue measure equal to zero.

Hint. Show first that

$$\text{mes } \{\omega : \exists k \neq 0, (\omega, k) = 0\} = 0.$$

On the other hand, in degenerate systems we can construct systems of action-angle variables such that the tori $I = \textbf{const}$ will be different in different systems. This is the case because the closures of trajectories in a degenerate system are tori of dimension $k < n$, and they can be contained in different ways in n-dimensional tori.

[92] For example, we can always write the substitution $I' = I$, $\varphi' = \varphi + S_1(I)$, or I_1, I_2: $\varphi_1, \varphi_2 \rightarrow I_1 + I_2, I_2$: $\varphi_1, \varphi_2 - \varphi_1$.

EXAMPLE 1. The planar harmonic oscillator $\ddot{\mathbf{x}} = -\mathbf{x}$; $n = 2$, $k = 1$. Separation of variables in cartesian and polar coordinates leads to different action-angle variables and different tori.

EXAMPLE 2. Keplerian planar motion ($U = -1/r$), $n = 2$, $k = 1$. Here, too, separation of variables in polar and in elliptic coordinates leads to different $\mathbf{I}$.

52 Averaging of perturbations

Here we show the adiabatic invariance of the action variable in a system with one degree of freedom.

A Systems close to integrable ones

We have considered a great many integrable systems (one-dimensional problems, the two-body problem, small oscillations, the Euler and Lagrange cases of the motion of a rigid body with a fixed point, etc.). We studied the characteristics of phase trajectories in these systems: they turned out to be "windings of tori," densely filling up the invariant tori in phase space; every trajectory is uniformly distributed on this torus.

One should not conclude from this that integrability is the typical situation. Actually, the properties of trajectories in many-dimensional systems can be highly diverse and not at all similar to the properties of conditionally periodic motions. In particular, the closure of a trajectory of a system with n degrees of freedom can fill up complicated sets of dimension greater than n in $2n$-dimensional phase space; a trajectory could even be dense and uniformly distributed on a whole $(2n - 1)$-dimensional manifold given by the equation $H = h$.[93] One may call such systems "nonintegrable" since they do not admit single-valued first integrals independent of H. The study of such systems is still far from complete; it constitutes a problem in "ergodic theory."

One approach to nonintegrable systems is to study systems which are close to integrable ones. For example, the problem of the motion of planets around the sun is close to the integrable problem of the motion of non-interacting points around a stationary center; other examples are the problem of the motion of a slightly asymmetric heavy top and the problem of nonlinear oscillations close to an equilibrium position (the nearby integrable problem is linear). The following method is especially fruitful in the investigation of these and similar problems.

B The averaging principle

Let $\mathbf{I}$, $\boldsymbol{\varphi}$ be action-angle variables in an integrable ("unperturbed") system with hamiltonian function $H_0(\mathbf{I})$:

$$\dot{\mathbf{I}} = 0 \qquad \dot{\boldsymbol{\varphi}} = \boldsymbol{\omega}(\mathbf{I}) \qquad \boldsymbol{\omega}(\mathbf{I}) = \frac{\partial H_0}{\partial \mathbf{I}}.$$

[93] For example, inertial motion on a manifold of negative curvature has this property.

As the nearby "perturbed" system we take the system

(1) $$\dot{\boldsymbol{\varphi}} = \boldsymbol{\omega}(\mathbf{I}) + \varepsilon\mathbf{f}(\mathbf{I}, \boldsymbol{\varphi}) \qquad \dot{\mathbf{I}} = \varepsilon\mathbf{g}(\mathbf{I}, \boldsymbol{\varphi}),$$

where $\varepsilon \ll 1$.

We will ignore for a while that the system is hamiltonian and consider an arbitrary system of differential equations in the form (1) given on the direct product $T^k \times G$ of the k-dimensional torus $T^k = \{\boldsymbol{\varphi} = (\varphi_1, \ldots, \varphi_k) \bmod 2\pi\}$ and a region G in l-dimensional space $G \subset \mathbb{R}^l = \{\mathbf{I} = (I_1, \ldots, I_l)\}$. For $\varepsilon = 0$ the motion in (1) is conditionally periodic with at most k frequencies and with k-dimensional invariant tori.

The *averaging principle for system* (1) consists of its replacement by another system, called the averaged system:

(2) $$\dot{\mathbf{J}} = \varepsilon\bar{\mathbf{g}}(\mathbf{J}) \qquad \bar{\mathbf{g}}(\mathbf{J}) = (2\pi)^{-k} \int_0^{2\pi} \cdots \int_0^{2\pi} \mathbf{g}(\mathbf{J}, \boldsymbol{\varphi})d\varphi_1, \ldots, d\varphi_k$$

in the l-dimensional region $G \subset \mathbb{R}^l = \{\mathbf{J} = (J_1, \ldots, J_l)\}$.

We claim that system (2) is a "good approximation" to system (1).

We note that this principle is neither a theorem, an axiom, nor a definition, but rather a physical proposition, i.e., a vaguely formulated and, strictly speaking, untrue assertion. Such assertions are often fruitful sources of mathematical theorems.

This averaging principle may be found explicitly in the work of Gauss (in studying the perturbations of planets on one another, Gauss proposed to distribute the mass of each planet around its orbit proportionally to time and to replace the attraction of each planet by the attraction of the ring so obtained). Nevertheless, a satisfactory description of the connection between the solutions of systems (1) and (2) in the general case has not yet been found.

In replacing system (1) by system (2) we discard the term $\varepsilon\tilde{\mathbf{g}}(\mathbf{I}, \boldsymbol{\varphi}) = \varepsilon\mathbf{g}(\mathbf{I}, \boldsymbol{\varphi}) - \varepsilon\bar{\mathbf{g}}(\mathbf{I})$ on the right-hand side. This term has order ε as does the remaining term $\varepsilon\bar{\mathbf{g}}$. In order to understand the different roles of the terms $\bar{\mathbf{g}}$ and $\tilde{\mathbf{g}}$ in $\mathbf{g}$, we consider the simplest example.

PROBLEM. Consider the case $k = l = 1$,

$$\dot{\varphi} = \omega \neq 0 \qquad \dot{I} = \varepsilon g(\varphi).$$

Show that for $0 < t < 1/\varepsilon$,

$$|I(t) - J(t)| < c\varepsilon, \quad \text{where } J(t) = I(0) + \varepsilon\bar{g}t.$$

Solution

$$I(t) - I(0) = \int_0^t \varepsilon g(\varphi_0 + \omega t)dt = \int_0^t \varepsilon\bar{g}\, dt + \frac{\varepsilon}{\omega} \int_0^{\omega t} \tilde{g}(\varphi)d\varphi = \varepsilon\bar{g}t + \frac{\varepsilon}{\omega} h(\omega t)$$

where $h(\varphi) = \int_0^{\varphi} \tilde{g}(\varphi)d\varphi$ is a periodic, and therefore bounded, function.

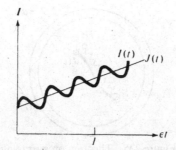

Figure 224   Evolution and oscillation

Thus the variation in I with time consists of two parts: an oscillation of order ε depending on $\tilde{g}$ and a systematic "evolution" with velocity $\varepsilon\bar{g}$ (Figure 224).

The averaging principle is based on the assertion that in the general case the motion of system (1) can be divided into the "evolution" (2) and small oscillations. In its general form, this assertion is invalid and the principle itself is untrue. Nevertheless, we will apply the principle to the hamiltonian system (1):

$$\dot{\boldsymbol{\varphi}} = -\frac{\partial}{\partial \mathbf{I}}(H_0(\mathbf{I}) + \varepsilon H_1(\mathbf{I}, \boldsymbol{\varphi})) \qquad \dot{\mathbf{I}} = \frac{\partial}{\partial \boldsymbol{\varphi}}(H_0(\mathbf{I}) + \varepsilon H_1(\mathbf{I}, \boldsymbol{\varphi})).$$

For the right-hand side of the averaged system (2) we then obtain

$$\bar{\mathbf{g}} = (2\pi)^{-n}\int_0^{2\pi}\frac{\partial}{\partial \boldsymbol{\varphi}}H_1(\mathbf{I}, \boldsymbol{\varphi})d\boldsymbol{\varphi} = 0.$$

In other words, there is no evolution in a nondegenerate hamiltonian system.

One variant of this entirely nonrigorous deduction leads to the so-called Laplace theorem: The semi-major axes of the keplerian ellipses of the planets have no secular perturbations.

The discussion above suffices to convince us of the importance of the averaging principle; we now formulate a theorem justifying this principle in one very particular case—that of single-frequency oscillations ($k = 1$). This theorem shows that the averaging principle correctly describes evolution over a large interval of time ($0 < t < 1/\varepsilon$).

C Averaging in a single-frequency system

Consider the system of $l + 1$ differential equations

(1)
$$\left.\begin{array}{l}\dot{\varphi} = \omega(\mathbf{I}) + \varepsilon f(\mathbf{I}, \varphi) \\ \dot{\mathbf{I}} = \varepsilon \mathbf{g}(\mathbf{I}, \varphi)\end{array}\right\} \qquad \begin{array}{l}\varphi \bmod 2\pi \in S^1, \\ \mathbf{I} \in G \subset \mathbb{R}^l,\end{array}$$

where $f(\mathbf{I}, \varphi + 2\pi) \equiv f(\mathbf{I}, \varphi)$ and $\mathbf{g}(\mathbf{I}, \varphi + 2\pi) \equiv \mathbf{g}(\mathbf{I}, \varphi)$, together with the "averaged" system of l equations

(2)
$$\dot{\mathbf{J}} = \varepsilon\bar{\mathbf{g}}(\mathbf{J}), \quad \text{where } \bar{\mathbf{g}}(\mathbf{J}) = \frac{1}{2\pi}\int_0^{2\pi}\mathbf{g}(\mathbf{J}, \varphi)d\varphi.$$

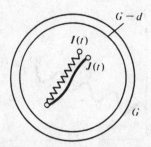

Figure 225   Theorem on averaging

We denote by $\mathbf{I}(t)$, $\varphi(t)$ the solution of system (1) with initial conditions $\mathbf{I}(0)$, $\varphi(0)$, and by $\mathbf{J}(t)$ the solution of system (2) with the same initial conditions $\mathbf{J}(0) = \mathbf{I}(0)$ (Figure 225).

Theorem. *Suppose that*:

1. *the functions ω, f, and $\mathbf{g}$ are defined for $\mathbf{I}$ in a bounded region G, and in this region they are bounded, together with their derivatives up to second order*:
$$\|\omega, f, \mathbf{g}\|_{C^2(G \times S^1)} < c_1;$$

2. *in the region G, we have*
$$\omega(\mathbf{I}) > c > 0;$$

3. *for $0 \le t \le 1/\varepsilon$, a neighborhood of radius d of the point $\mathbf{J}(t)$ belongs to G:*
$$\mathbf{J}(t) \in G - d.$$

Then for sufficiently small ε ($0 < \varepsilon < \varepsilon_0$)
$$|\mathbf{I}(t) - \mathbf{J}(t)| < c_9\varepsilon, \quad \text{for all } t, 0 \le t \le \frac{1}{\varepsilon},$$

where the constant $c_9 > 0$ depends on c_1, c, and d, but not on ε.

Some applications of this theorem will be given below ("adiabatic invariants"). We remark that the basic idea of the proof of this theorem (a change of variables diminishing the perturbation) is more important than the theorem itself; this is one of the basic ideas in the theory of ordinary differential equations; it is encountered in elementary courses as the "method of variation of constants."

D Proof of the theorem on averaging

In place of the variables $\mathbf{I}$ we will introduce new variables $\mathbf{P}$

(3) $$\mathbf{P} = \mathbf{I} + \varepsilon\mathbf{k}(\mathbf{I}, \varphi),$$

where the function $\mathbf{k}$, 2π-periodic in φ, will be chosen so that the vector $\mathbf{P}$ will satisfy a simpler differential equation.

294

By (1) and (3), the rate of change of $P(t)$ is

(4) $\quad \dot{\mathbf{P}} = \dot{\mathbf{I}} + \varepsilon \dfrac{\partial \mathbf{k}}{\partial \mathbf{I}} \dot{\mathbf{I}} + \varepsilon \dfrac{\partial \mathbf{k}}{\partial \varphi} \dot{\varphi} = \varepsilon \left[\mathbf{g}(\mathbf{I}, \varphi) + \dfrac{\partial \mathbf{k}}{\partial \varphi} \omega(\mathbf{I}) \right] + \varepsilon^2 \dfrac{\partial \mathbf{k}}{\partial \mathbf{I}} \mathbf{g} + \varepsilon^2 \dfrac{\partial \mathbf{k}}{\partial \varphi} f.$

We assume that the substitution (3) can be inverted, so that

(5) $\qquad\qquad\qquad \mathbf{I} = \mathbf{P} + \varepsilon \mathbf{h}(\mathbf{P}, \varphi, \varepsilon)$

(where the functions $\mathbf{h}$ are 2π-periodic in φ).

Then (4) and (5) imply that $P(t)$ satisfies the system of equations

(6) $\qquad\qquad \dot{\mathbf{P}} = \varepsilon \left[\mathbf{g}(\mathbf{P}, \varphi) + \dfrac{\partial \mathbf{k}}{\partial \varphi} \omega(\mathbf{P}) \right] + \mathbf{R},$

where the "remainder term" $\mathbf{R}$ is small of second order with respect to ε:

(7) $\qquad\qquad |\mathbf{R}| < c_2 \varepsilon^2, \qquad c_2(c_1, c_3, c_4) > 0,$

if only

(8) $\quad \|\omega\|_{C^2} < c_1 \quad \|f\|_{C^2} < c_1 \quad \|\mathbf{g}\|_{C^2} < c_1 \quad \|\mathbf{k}\|_{C^2} < c_3 \quad \|\mathbf{h}\|_{C^2} < c_4.$

We will now try to choose the change of variables (3) so that the term involving ε in (6) becomes zero. For $\mathbf{k}$ we get the equation

$$\dfrac{\partial \mathbf{k}}{\partial \varphi} = -\dfrac{1}{\omega} \mathbf{g}.$$

In general, such an equation is not solvable in the class of functions $\mathbf{k}$ periodic in φ. In fact, the average value (with respect to φ) of the left-hand side is always equal to 0, and the average value of the right-hand side can be different from 0. Therefore, we cannot choose $\mathbf{k}$ in such a way as to kill the entire term involving ε in (6). However, we can kill the entire "periodic" part of $\mathbf{g}$,

$$\tilde{\mathbf{g}}(\mathbf{P}, \varphi) = \mathbf{g}(\mathbf{P}, \varphi) - \bar{\mathbf{g}}(\mathbf{P}),$$

by setting

(9) $\qquad\qquad \mathbf{k}(\mathbf{P}, \varphi) = -\displaystyle\int_0^\varphi \dfrac{\tilde{\mathbf{g}}(\mathbf{P}, \varphi)}{\omega(\mathbf{P})} d\varphi.$

So we define the function $\mathbf{k}$ by formula (9). Then, by hypotheses 1. and 2. of the theorem, the function $\mathbf{k}$ satisfies the estimate $\|\mathbf{k}\|_{C^2} < c_3$, where $c_3(c_1, c) > 0$. In order to establish the inequality (8), we must estimate $\mathbf{h}$. For this we must first show that the substitution (3) is invertible.

Fix a positive number α.

Lemma. *If ε is sufficiently small, then the restriction of the mapping (3)*[94]

$$\mathbf{I} \to \mathbf{I} + \varepsilon \mathbf{k}, \quad \text{where } |\mathbf{k}|_{C^2(G)} < c_3,$$

---

[94] For any fixed value of the parameter φ.

to the region $G - \alpha$ (consisting of points whose α-neighborhood is contained in G) is a diffeomorphism. The inverse diffeomorphism (5) in the region $G - 2\alpha$ satisfies the estimate $\|\mathbf{h}\|_{C^2} < c_4$ with some constant $c_4(\alpha, c_3) > 0$.

PROOF. The necessary estimate follows directly from the implicit function theorem. The only difficulty is in verifying that the map $\mathbf{I} \to \mathbf{I} + \varepsilon\mathbf{k}$ is one-to-one in the region $G - \alpha$. We note that the function $\mathbf{k}$ satisfies a Lipschitz condition (with some constant $L(\alpha, c_3)$) in $G - \alpha$. Consider two points $\mathbf{I}_1$, $\mathbf{I}_2$ in $G - \alpha$. For sufficiently small ε (namely, for $L\varepsilon < 1$) the distance between $\varepsilon\mathbf{k}(\mathbf{I}_1)$ and $\varepsilon\mathbf{k}(\mathbf{I}_2)$ will be smaller than $|\mathbf{I}_1 - \mathbf{I}_2|$. Therefore, $\mathbf{I}_1 + \varepsilon\mathbf{k}(\mathbf{I}_1) \neq \mathbf{I}_2 + \varepsilon\mathbf{k}(\mathbf{I}_2)$. Thus the map (3) is one-to-one on $G - \alpha$, and the lemma is proved. $\square$

It follows from the lemma that for ε small enough all the estimates (8) are satisfied. Thus the estimate (7) is also true.

We now compare the system of differential equations for $\mathbf{J}$

(2) $$\dot{\mathbf{J}} = \varepsilon\bar{\mathbf{g}}(\mathbf{J})$$

and for $\mathbf{P}$; the latter, in view of (9), takes the form

(6') $$\dot{\mathbf{P}} = \varepsilon\bar{\mathbf{g}}(\mathbf{P}) + \mathbf{R}.$$

Since the difference between the right sides is of order $\lesssim \varepsilon^2$ (cf. (7)), for time $t \lesssim 1/\varepsilon$ the difference $|\mathbf{P} - \mathbf{J}|$ between the solutions is of order ε (Figure 226). On the other hand, $|\mathbf{I} - \mathbf{P}| = \varepsilon|\mathbf{k}| \lesssim \varepsilon$. Thus, for $t \lesssim 1/\varepsilon$, the difference $|\mathbf{I} - \mathbf{J}|$ is of order $\lesssim \varepsilon$, as was to be proved. $\square$

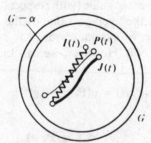

Figure 226   Proof of the theorem on averaging

To find an accurate estimate, we introduce the quantity

(10) $$z(t) = P(t) - J(t).$$

Then (6') and (9) imply

$$\dot{z} = \varepsilon(\bar{g}(P) - \bar{g}(J)) + R = \varepsilon\frac{\partial g}{\partial P} z + R',$$

where $|R'| < c_2\varepsilon^2 + c_5\varepsilon|z|$ if the segment (P, J) lies in $G - \alpha$. Under this assumption we find

(11) $$|\dot{z}| < c_6\varepsilon|z| + c_2\varepsilon^2 \quad (\text{where } c_6 = c_5 + c_1)$$

$$|z(0)| < c_3\varepsilon.$$

Lemma. *If* $|\dot{z}| \leq a|z| + b$ *and* $|z(0)| < d$ *for* $a, b, d, t > 0$, *then* $|z(t)| \leq (d + bt)e^{at}$.

PROOF. $|z(t)|$ is no greater than the solution $y(t)$ of the equation $\dot{y} = ay + b$, $y(0) = d$. Solving this equation, we find $y = Ce^{at}$, $\dot{C}e^{at} = b$, $\dot{C} = e^{-at}b$, $C(0) = d$, $C \leq d + bt$. $\qquad \square$

Now from (11) and the assumption that the segment $(\mathbf{P}, \mathbf{J})$ lies in $G - \alpha$ (Figure 226), we have

$$|z(t)| < (c_3 \varepsilon + c_2 \varepsilon^2 t)e^{c_6 \varepsilon t}.$$

From this it follows that, for $0 \leq t \leq 1/\varepsilon$,

$$|z(t)| < c_7 \varepsilon \qquad c_7 = (c_3 + c_2)e^{c_6}.$$

We see that, if $\alpha = d/3$ and ε is small enough, the entire segment $(\mathbf{P}(t), \mathbf{J}(t))(t \leq 1/\varepsilon)$ lies inside $G - \alpha$ and, therefore,

$$|\mathbf{P}(t) - \mathbf{J}(t)| < c_8 \varepsilon \quad \text{for all } 0 \leq t \leq \frac{1}{\varepsilon}.$$

On the other hand, $|\mathbf{P}(t) - \mathbf{I}(t)| < |\varepsilon \mathbf{k}| < c_3 \varepsilon$. Thus, for all t with $0 \leq t \leq 1/\varepsilon$,

$$|\mathbf{I}(t) - \mathbf{J}(t)| < c_9 \varepsilon \qquad c_9 = c_8 + c_3 > 0$$

and the theorem is proved. $\qquad \square$

E *Adiabatic invariants*

Consider a hamiltonian system with one degree of freedom, with hamiltonian function $H(p, q; \lambda)$ depending on a parameter λ. As an example, we can take a pendulum:

$$H = \frac{p^2}{2l^2} + lg\frac{q^2}{2};$$

as the parameter λ we can take the length l or the acceleration of gravity g. Suppose that the parameter changes slowly with time. It turns out that in the limit as the rate of change of the parameter approaches 0, there is a remarkable asymptotic phenomenon: two quantities, generally independent, become functions of one another.

Assume, for example, that the length of the pendulum changes slowly (in comparison with its characteristic oscillations). Then the amplitude of its oscillation becomes a function of the length of the pendulum. If we very slowly increase by a factor of two the length of the pendulum and then very slowly decrease it to the original value, then at the end of this process the amplitude of the oscillation will be the same as it was at the start.

Furthermore, it turns out that the ratio of the energy H of the pendulum to the frequency ω changes very little under a slow change of the parameter, although the energy and frequency themselves may change a lot. Quantities such as this ratio, which change little under slow changes of parameter, are called by physicists *adiabatic invariants*.

It is easy to see that the adiabatic invariance of the ratio of the energy of a pendulum to its frequency is an assertion of a physical character, i.e., it is untrue without further assumptions. In fact, if we vary the length of a pendulum arbitrarily slowly, but chose the phase of oscillation under which

297

Figure 227   Adiabatic change in the length of a pendulum

the length increases and decreases, we can set the pendulum swinging (parametric resonance). In view of this, physicists have suggested formulating the definition of adiabatic invariance as follows: the person changing the parameters of the system must not see what state the system is in (Figure 227). Giving this definition a rigorous mathematical meaning is a very delicate and as yet unsolved problem. Fortunately, we can get along with a surrogate. The assumption of ignorance of the internal state of the system on the part of the person controlling the parameter may be replaced by the requirement that the change of parameter must be smooth, i.e., twice continuously differentiable.

More precisely, let $H(p, q; \lambda)$ be a fixed, twice continuously differentiable function of λ. Set $\lambda = \varepsilon t$ and consider the resulting system with slowly varying parameter $\lambda = \varepsilon t$:

$$(*) \qquad \dot{p} = -\frac{\partial H}{\partial q} \qquad \dot{q} = \frac{\partial H}{\partial p}, \qquad H = H(p, q; \varepsilon t).$$

Definition. The quantity $I(p, q; \lambda)$ is an *adiabatic invariant* of the system (*) if for every $\kappa > 0$ there is an $\varepsilon_0 > 0$ such that if $0 < \varepsilon < \varepsilon_0$ and $0 < t < 1/\varepsilon$, then

$$|I(p(t), q(t); \varepsilon t) - I(p(0), q(0); 0)| < \kappa.$$

Clearly, every first integral is also an adiabatic invariant. It turns out that every one-dimensional system (*) has an adiabatic invariant. Namely, the adiabatic invariant is the action variable in the corresponding problem with constant coefficients.

Assume that the phase trajectories of the system with hamiltonian $H(p, q; \lambda)$ are closed. We define a function $I(p, q; \lambda)$ in the following way. For fixed λ there is a phase portrait corresponding to the hamiltonian function $H(p, q; \lambda)$ (Figure 228). Consider the closed phase trajectory passing through a point (p, q). It bounds some region in the phase plane. We denote the area of this region by $2\pi I(p, q; \lambda)$. $I = \text{const}$ on every phase trajectory (for given λ). Clearly, I is nothing but the action variable (cf. Section 50).

Theorem. *If the frequency $\omega(I, \lambda)$ of the system (*) is nowhere zero, then $I(p, q; \lambda)$ is an adiabatic invariant.*

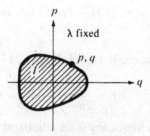

Figure 228   Adiabatic invariant of a one-dimensional system

F *Proof of the adiabatic invariance of action*

For fixed λ we can introduce action-angle variables I, φ into the system (*) by a canonical transformation depending on λ: $p, q \to I, \varphi$; $\dot{\varphi} = \omega(I, \lambda)$, $\dot{I} = 0$; $\omega(I, \lambda) = \partial H_0/\partial I$, $H_0 = H_0(I, \lambda)$.

We denote by $S(I, q; \lambda)$ the (multiple-valued) generating function of this transformation:

$$p = \frac{\partial S}{\partial q} \qquad \varphi = \frac{\partial S}{\partial I}.$$

Now let $\lambda = \varepsilon t$. Since the change from variables p, q to variables I, φ is now performed by a *time dependent* canonical transformation, the equations of motion in the new variables I, φ have the hamiltonian form, but with hamiltonian function (cf. Section 45A)

$$K = H_0 + \frac{\partial S}{\partial t} = H_0 + \varepsilon \frac{\partial S}{\partial \lambda}.$$

PROBLEM. Show that $\partial S(I, q; \lambda)/\partial \lambda$ is a single-valued function on the phase plane.
   Hint. S is determined up to the addition of multiples of $2\pi I$.

In this way we obtain the equations of motion in the form

$$\dot{\varphi} = \omega(I, \lambda) + \varepsilon f(I, \varphi; \lambda) \qquad f = \frac{\partial^2 S}{\partial I \, \partial \lambda},$$

$$\dot{I} = \varepsilon g(I, \varphi; \lambda) \qquad g = -\frac{\partial^2 S}{\partial \varphi \, \partial \lambda},$$

$$\dot{\lambda} = \varepsilon$$

Since $\omega \neq 0$, the averaging theorem (Section 52C) is applicable. The averaged system has the form

$$\dot{J} = \varepsilon \bar{g} \qquad \dot{\Lambda} = \varepsilon.$$

But $g = (\partial/\partial\varphi)(\partial S/\partial\lambda)$, and $\partial S/\partial\lambda$ is a single-valued function on the circle $I = \text{const}$. Therefore, $\bar{g} = (2\pi)^{-1} \int g \, d\varphi = 0$, and in the averaged system J does not change at all: $J(t) = J(0)$.

By the averaging theorem, $|I(t) - I(0)| < c\varepsilon$ for all t with $0 \le t \le 1/\varepsilon$, as was to be proved. $\square$

EXAMPLE. For a harmonic oscillator (cf. Figure 217),

$$H = \frac{a^2}{2}p^2 + \frac{b^2}{2}q^2 \qquad I = \frac{1}{2\pi}\pi\frac{\sqrt{2h}}{a}\frac{\sqrt{2h}}{b} = \frac{h}{\omega}, \qquad \omega = ab,$$

i.e., the ratio of energy to frequency is an adiabatic invariant.

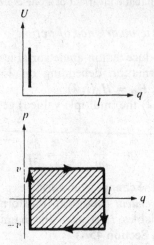

Figure 229   Adiabatic invariant of an absolutely elastic ball between slowly changing walls

PROBLEM. The length of a pendulum is slowly doubled ($l = l_0(1 + \varepsilon t)$, $0 \le t \le 1/\varepsilon$). How does the amplitude q_{max} of the oscillations vary?

Solution. $I = \frac{1}{2}l^{3/2}g^{1/2}q_{max}^2$; therefore,

$$q_{max}(t) = q_{max}(0)\left(\frac{l(0)}{l(t)}\right)^{3/4}.$$

As a second example, consider the motion of a perfectly elastic rigid ball of mass 1 between perfectly elastic walls whose separation l slowly varies (Figure 229). We may consider that a point is moving in an "infinitely deep rectangular potential well," and that the phase trajectories are rectangles of area $2vl$, where v is the velocity of the ball. In this case the product vl of the velocity of the ball and the distance between the walls turns out to be an adiabatic invariant.[95] Thus if we make the walls twice as close together, the velocity of the ball doubles, and if we separate the walls, the velocity decreases.

[95] This does not formally follow from the theorem, since the theorem concerns smooth systems without shocks. The proof of the adiabatic invariance of vl in this system is an instructive elementary problem.

Appendix 1: Riemannian curvature

From a sheet of paper, one can form a cone or a cylinder, but it is impossible to obtain a piece of a sphere without folding, stretching, or cutting. The reason lies in the difference between the "intrinsic geometries" of these surfaces: no part of the sphere can be isometrically mapped onto the plane.

The invariant which distinguishes riemannian metrics is called riemannian curvature. The riemannian curvature of a plane is zero, and the curvature of a sphere of radius R is equal to R^{-2}. If one riemannian manifold can be isometrically mapped to another, then the riemannian curvature at corresponding points is the same. For example, since a cone or cylinder is locally isometric to the plane, the riemannian curvature of the cone or cylinder at any point is equal to zero. Therefore, no region of a cone or cylinder can be mapped isometrically to a sphere.

The riemannian curvature of a manifold has a very important influence on the behavior of geodesics on it, i.e., on motion in the corresponding dynamical system. If the riemannian curvature of a manifold is positive (as on a sphere or ellipsoid), then nearby geodesics oscillate about one another in most cases, and if the curvature is negative (as on the surface of a hyperboloid of one sheet), geodesics rapidly diverge from one another.

In this appendix we define riemannian curvature and briefly discuss the properties of geodesics on manifolds of negative curvature. A further treatment of riemannian curvature can be found in the book, "Morse Theory" by John Milnor, Princeton University Press, 1963, and a treatment of geodesics on manifolds of negative curvature in D. V. Anosov's book, "Geodesic flows on closed riemannian manifolds with negative curvature," Proceedings of the Steklov Institute of Mathematics, No. 90 (1967), Am. Math. Soc., 1969.

A Parallel translation on surfaces

The definition of riemannian curvature is based on the construction of parallel translation of vectors along curves on a riemannian manifold.

We begin with the case when the given riemannian manifold is two-dimensional, i.e., a surface, and the given curve is a geodesic on this surface. [See do Carmo, Manfredo Perdigao, "Differential Geometry of Curves and Surfaces," Prentice-Hall, 1976. (Translator's note)]

Parallel translation of a vector tangent to the surface along a geodesic on this surface is defined as follows: the point of origin of the vector moves along the geodesic, and the vector itself moves continuously so that its angle with the geodesic and its length remain constant. By translating to the endpoint of the geodesic all vectors tangent to the surface at the initial point, we obtain a map from the tangent plane at the initial point to the tangent plane at the endpoint. This map is linear and isometric.

We now define *parallel translation of a vector on a surface along a broken line* consisting of several geodesic arcs (Figure 230). In order to translate a vector along a broken line, we translate it from the first vertex to the second

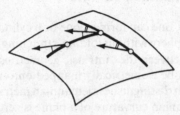

Figure 230   Parallel translation along a broken geodesic

along the first geodesic arc, then translate this vector along the second arc to the next vertex, etc.

PROBLEM. Given a vector tangent to the sphere at one vertex of a spherical triangle with three right angles, translate this vector around the triangle and back to the same vertex.

ANSWER. As a result of this translation the tangent plane to the sphere at the initial vertex will be turned by a right angle.

Finally, *parallel translation of a vector along any smooth curve on a surface* is defined by a limiting procedure, in which the curve is approximated by broken lines consisting of geodesic arcs.

PROBLEM. Translate a vector directed towards the North Pole and located at Leningrad (latitude $\lambda = 60°$) around the 60th parallel and back to Leningrad, moving to the east.

ANSWER. The vector turns through the angle $2\pi (1 - \sin \lambda)$, i.e., approximately 50° to the west. Thus the size of the angle of rotation is proportional to the area bounded by our parallel, and the direction of rotation coincides with the direction the origin of the vector is going around the North Pole.

Hint. It is sufficient to translate the vector along the same circle on the cone formed by the tangent lines to the meridian, going through all the points of the parallel (Figure 231). This cone then can be unrolled onto the plane, after which parallel translation on its surface becomes ordinary parallel translation on the plane.

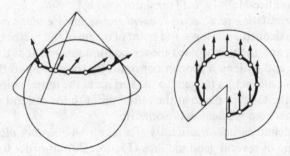

Figure 231   Parallel translation on the sphere

EXAMPLE. We consider the upper half-plane $y > 0$ of the plane of complex numbers $z = x + iy$ with the metric

$$ds^2 = \frac{dx^2 + dy^2}{y^2}.$$

It is easy to compute that the geodesics of this two-dimensional riemannian manifold are circles and straight lines perpendicular to the x-axis. Linear fractional transformations with real coefficients

$$z \rightarrow \frac{az + b}{cz + d}$$

are isometric transformations of our manifold, which is called the *Lobachevsky plane*.

PROBLEM. Translate a vector directed along the imaginary axis at the point $z = i$ to the point $z = t + i$ along the horizontal line $(dy = 0)$ (Figure 232).

ANSWER. Under translation by t the vector turns t radians in the direction from the y-axis towards the x-axis.

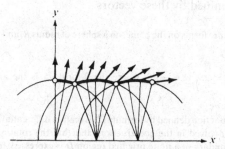

Figure 232   Parallel translation on the Lobachevsky plane

B *The curvature form*

We will now define the riemannian curvature at each point of a two-dimensional riemannian manifold (i.e., a surface). For this purpose, we choose an orientation of our surface in a neighborhood of the point under consideration and consider parallel translation of vectors along the boundary of a *small* region D on our surface. It is easy to calculate that the result of such a translation is rotation by a *small* angle. We denote this angle by $\varphi(D)$ (the sign of the angle is fixed by the choice of orientation of the surface).

If we divide the region D into two parts D_1 and D_2, the result of parallel translation along the boundary of D can be obtained by first going around one part, and then the other. Thus,

$$\varphi(D) = \varphi(D_1) + \varphi(D_2),$$

i.e., the angle φ is an additive function of regions. When we change the direction of travel along the boundary, the angle φ changes sign. It is natural therefore to represent $\varphi(D)$ as the integral over D of a suitable 2-form. Such

303

a 2-form in fact exists; it is called the *curvature form*, and we denote it by Ω. Thus we define the curvature form Ω by the relation

$$(1) \qquad\qquad \varphi(D) = \int_D \Omega.$$

The value of Ω on a pair of tangent vectors ξ, η in TM_x can be defined in the following way. We identify a neighborhood of the point 0 in the tangent space to M at x with a neighborhood of the point x on M (using, for example, some local coordinate system). We can then construct on M the parallelogram Π_ε spanned by the vectors $\varepsilon\xi$, $\varepsilon\eta$, at least for sufficiently small ε.

Now the value of the curvature form on our vectors is defined by the formula

$$(2) \qquad\qquad \Omega(\xi, \eta) = \lim_{\varepsilon \to 0} \frac{\varphi(\Pi_\varepsilon)}{\varepsilon^2}.$$

In other words, the value of the curvature form on a pair of tangent vectors is equal to the angle of rotation under translation along the infinitely small parallelogram determined by these vectors.

PROBLEM. Find the curvature forms on the plane, on a sphere of radius R, and on the Lobachevsky plane.

ANSWER. $\Omega = 0$, $\Omega = R^{-2} dS$, $\Omega = -dS$, where the 2-form dS is the area element on our oriented surface.

PROBLEM. Show that the function defined by formula (2) is really a differential 2-form, independent of the arbitrary choice involved in the construction, and that the rotation of a vector under translation along the boundary of a finite oriented region D is expressed, in terms of this form, by formula (1).

PROBLEM. Show that the integral of the curvature form over any convex surface in three-dimensional euclidean space is equal to 4π.

C *The riemannian curvature of a surface*

We note that every differential 2-form on a two-dimensional oriented riemannian manifold M can be written in the form ρdS, where dS is the oriented area element and ρ is a scalar function uniquely determined by the choice of metric and orientation.

In particular, the curvature form can be written in the form

$$\Omega = KdS,$$

where $K: M \to \mathbb{R}$ is a smooth function on M and dS is the area element.

The value of the function K at a point x is called the *riemannian curvature of the surface at* x.

PROBLEM. Calculate the riemannian curvature of the euclidean space, the sphere of radius R, and the Lobachevsky plane.

ANSWER. $K = 0$, $K = R^{-2}$, $K = -1$.

PROBLEM. Show that the riemannian curvature does not depend on the orientation of the manifold, but only on its metric.

Hint. The 2-forms Ω and dS both change sign under a change of orientation.

PROBLEM. Show that, for surfaces in ordinary three-dimensional euclidean space, the riemannian curvature at every point is equal to the product of the inverses of the principal radii of curvature (with minus sign if the centers of curvature lie on opposite sides of the surface).

We note that the sign of a manifold's curvature at a point does not depend on the orientation of the manifold; this sign may be defined without using the orientation at all.

Namely, on manifolds of *positive* curvature, a vector parallel translated around the boundary of a small region turns around its origin in the *same* direction as the point on the boundary goes around the region; on manifolds of *negative* curvature the direction of rotation is *opposite*.

We note further that the value of the curvature at a point is determined by the metric in a neighborhood of this point, and therefore is preserved under bending: the curvature is the same at corresponding points of isometric surfaces. Hence, riemannian curvature is also called intrinsic curvature.

The formulas for computing curvature in terms of components of the metric in some coordinate system involve the second derivatives of the metric and are rather complicated: cf. the problems in Section G below.

D Higher-dimensional parallel translation

The construction of parallel translation on riemannian manifolds of dimension greater than two is somewhat more complicated than the two-dimensional construction presented above. The reason is that in these dimensions the direction of the vector being translated is no longer determined by the condition that the angle with a geodesic be invariant. In fact, the vector could rotate around the direction of the geodesic while preserving its angle with the geodesic.

The refinement which we must introduce into the construction of parallel translation along a geodesic is the choice of a two-dimensional plane passing through the tangent to the geodesic, which must contain the translated vector. This choice is made in the following (unfortunately complicated) way.

At the initial point of a geodesic the needed plane is the plane spanned by the vector to be translated and the direction vector of the geodesic. We look at all geodesics proceeding from the initial point, in directions lying in this plane. The set of all such geodesics (close to the initial point) forms a smooth surface which contains the geodesic along which we intend to translate the vector (Figure 233).

Consider a new point on the geodesic at a small distance Δ from the initial point. The tangent plane at the new point to the surface described above contains the direction of the geodesic at this new point. We take this new

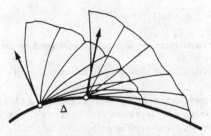

Figure 233   Parallel translation in space

point as the initial point and use its tangent plane to construct a new surface (formed by the bundle of geodesics emanating from the new point). This surface contains the original geodesic. We move along the original geodesic again by Δ and repeat the construction from the beginning.

After a finite number of steps we can reach any point of the original geodesic. As a result of our work we have, at every point of the geodesic, a tangent plane containing the direction of the geodesic. This plane depends on the length Δ of the steps in our construction. As $\Delta \to 0$ the family of tangent planes obtained converges (as can be calculated) to a definite limit. As a result we have a field of two-dimensional tangent planes along our geodesic containing the direction of the geodesic and determined in an intrinsic manner by the metric on the manifold.

Now parallel translation of our vector along a geodesic is defined as in the two-dimensional case: under translation the vector must remain in the planes described above; its length and its angle with the direction of the geodesic must be preserved. Parallel translation along any curve is defined using approximations by geodesic polygons, as in the two-dimensional case.

PROBLEM. Show that parallel translation of vectors from one point of a riemannian manifold to another along a fixed path is a linear isometric operator from the tangent space at the first point to the tangent space at the second point.

PROBLEM. Parallel translate any vector along the line

$$x_1 = t \qquad x_2 = 0 \qquad y = 1 \qquad (0 \le t \le \tau)$$

in a Lobachevsky space with metric

$$ds^2 = \frac{dx_1^2 + dx_2^2 + dy^2}{y^2}.$$

ANSWER. Vectors in the directions of the x_1 and y axes are rotated by angle τ in the plane spanned by them (rotation is in the direction from the y-axis towards the x_1-axis); vectors in the x_2-direction are carried parallel to themselves in the sense of the euclidean metric.

E *The curvature tensor*

We now consider, as in the two-dimensional case, parallel translation along small closed paths beginning and ending at a point of a riemannian manifold. Parallel translation along such a path returns vectors to the original tangent space. The map of the tangent space to itself thus obtained is a small rotation (an orthogonal transformation close to the identity).

In the two-dimensional case we characterized this rotation by one number—the angle of rotation φ. In higher dimensions a skew-symmetric operator plays the role of φ. Namely, any orthogonal operator A which is close to the identity can be written in a natural way in the form

$$A = e^{\Phi} = E + \Phi + \frac{\Phi^2}{2!} + \cdots,$$

where Φ is a small skew-symmetric operator.

PROBLEM. Compute Φ if A is a rotation of the plane through a small angle φ.

ANSWER.

$$A = \begin{pmatrix} \cos\varphi & \sin\varphi \\ -\sin\varphi & \cos\varphi \end{pmatrix} \qquad \Phi = \begin{pmatrix} 0 & \varphi \\ -\varphi & 0 \end{pmatrix}.$$

Unlike in the two-dimensional case, the function Φ is not generally additive (since the orthogonal group of n-space for $n > 2$ is not commutative). Nevertheless, we can construct a curvature form using Φ, describing the "infinitely small rotation caused by parallel translation around an infinitely small parallelogram" in the same way as in the two-dimensional case, i.e., using formula (2).

Thus, let ξ and η in TM_x be vectors tangent to the riemannian manifold M at the point x. Construct a small curvilinear parallelogram Π_ε on M (the sides of the parallelogram Π_ε are obtained from the vectors $\varepsilon\xi$ and $\varepsilon\eta$ by a coordinate identification of a neighborhood of zero in TM_x with a neighborhood of x in M). We will look at parallel translation along the sides of the parallelogram Π_ε (we begin the circuit at ξ).

The result of translation will be an orthogonal transformation of TM_x, close to the identity. It differs from the identity transformation by a quantity of order ε^2 and has the form

$$A_\varepsilon(\xi, \eta) = E + \varepsilon^2\Omega + o(\varepsilon^2),$$

where Ω is a skew-symmetric operator depending on ξ and η. Therefore, we can define a function Ω of pairs of vectors ξ, η in the tangent space at x with values in the space of skew-symmetric operators on TM_x by the formula

$$\Omega(\xi, \eta) = \lim_{\varepsilon \to 0} \frac{A_\varepsilon(\xi, \eta) - E}{\varepsilon^2}.$$

PROBLEM. Show that the function Ω is a differential 2-form (with values in the skew-symmetric operators on TM_x) and does not depend on the choice of coordinates we used to identify TM_x and M.

The form Ω is called the *curvature tensor* of the riemannian manifold. We could say that the curvature tensor describes the infinitesimal rotation in the tangent space obtained by parallel translation around an infinitely small parallelogram.

F Curvature in a two-dimensional direction

Consider a two-dimensional subspace L in the tangent space to a riemannian manifold at some point. We take geodesics emanating from this point in all the directions in L. These geodesics form a smooth surface close to our point. The surface constructed lies in the riemannian manifold and has an induced riemannian metric.

By the *curvature of a riemannian manifold M in the direction of a 2-plane L in the tangent space to M at a point x*, we mean the riemannian curvature at x of the surface described above.

PROBLEM. Find the curvatures of a three-dimensional sphere of radius R and of Lobachevsky space in all possible two-dimensional directions.

ANSWER. R^{-2}, -1.

In general, the curvatures of a riemannian manifold in different two-dimensional directions are different. Their dependence on the direction is described by formula (3) below.

Theorem. *The curvature of a riemannian manifold in the two-dimensional direction determined by a pair of orthogonal vectors ξ, η of length 1 can be expressed in terms of the curvature tensor Ω by the formula*

$$(3) \qquad\qquad K = \langle \Omega(\xi, \eta)\xi, \eta \rangle,$$

where the brackets denote the scalar product giving the riemannian metric.

The proof is obtained by comparing the definitions of the curvature tensor and of curvature in a two-dimensional direction. We will not go into it in a rigorous way. It is possible to take formula (3) for the definition of the curvature K.

G Covariant differentiation

Connected with parallel translation along curves in a riemannian manifold is a particular differential calculus—so-called covariant differentiation, or the riemannian connection. We define this differentiation in the following way.

Let ξ be a vector tangent to a riemannian manifold M at a point x, and v a vector field given on M in a neighborhood of x. The covariant derivative of the field v in the direction ξ is defined by using any curve passing through x with velocity ξ. After moving along this curve for a small interval of time t, we find ourselves at a new point $x(t)$. We take the vector field v at this point $x(t)$ and parallel translate it backwards along the curve to the original point x. We obtain a vector depending on t in the tangent space to M at x. For $t = 0$ this vector is $v(x)$, and for other t it changes according to the non-parallelness of the vector field v along our curve in the direction ξ.

Appendix I: Riemannian curvature

Consider the derivative of the resulting vector with respect to t, evaluated at $t = 0$. This derivative is a vector in the tangent space TM_x. It is called the *covariant derivative of the field v along ξ* and is denoted by $\nabla_\xi v$. It is easy to verify that the vector $\nabla_\xi v$ does not depend on the choice of curve specified in the definition, but only on ξ and v.

PROBLEM 1. Prove the following properties of covariant differentiation:

1. $\nabla_\xi v$ is a bilinear function of ξ and v.
2. $\nabla_\xi fv = (L_\xi f)v + f(x)\nabla_\xi v$, where f is a smooth function and $L_\xi f$ is the derivative of f in the direction of the vector ξ in TM_x.
3. $L_\xi \langle v, w \rangle = \langle \nabla_\xi v, w(x) \rangle + \langle v(x), \nabla_\xi w \rangle$.
4. $\nabla_{v(x)} w - \nabla_{w(x)} v = [w, v](x)$ (where $L_{[w, v]} = L_v L_w - L_w \dot{L}_v$).

PROBLEM 2. Show that the curvature tensor can be expressed in terms of covariant differentiation in the following way:

$$\Omega(\xi_0, \eta_0)\zeta_0 = -\nabla_\xi \nabla_\eta \zeta + \nabla_\eta \nabla_\xi \zeta + \nabla_{[\eta, \xi]}\zeta,$$

where ξ, η, ζ are any vector fields whose values at the point under consideration are ξ_0, η_0, and ζ_0.

PROBLEM 3. Show that the curvature tensor satisfies the following identities:

$$\Omega(\xi, \eta)\zeta + \Omega(\eta, \zeta)\xi + \Omega(\zeta, \xi)\eta = 0$$

$$\langle \Omega(\xi, \eta)\alpha, \beta \rangle = \langle \Omega(\alpha, \beta)\xi, \eta \rangle.$$

PROBLEM 4. Suppose that the riemannian metric is given in local coordinates $x_1, \ldots, x_n$ by the symmetric matrix g_{ij}:

$$ds^2 = \sum g_{ij} dx_i dx_j.$$

Denote by $e_1, \ldots, e_n$ the coordinate vector fields (so that differentiation in the direction e_i is $\partial_i = \partial/\partial x_i$). Then covariant derivatives can be calculated using the formulas in Problem 1 and the following formulas:

$$\nabla_{e_i} e_j = \sum_k \Gamma_{ij}^k e_k \qquad \Gamma_{ij}^k = \sum_l \tfrac{1}{2}(\partial_i g_{jl} + \partial_j g_{il} - \partial_l g_{ij})g^{lk},$$

where (g^{lk}) is the inverse matrix to (g_{kl}).

By using the expression for the curvature tensor in terms of the connection in Problem 2, we also obtain an explicit formula for the curvature. The numbers $R_{ijkl} = \langle \Omega(e_i, e_j)e_k, e_l \rangle$ are called the components of the curvature tensor.

H The Jacobi equation

The riemannian curvature of a manifold is closely connected with the behavior of its geodesics. In particular, let us consider a geodesic passing through some point in some direction, and alter slightly the initial conditions, i.e., the initial point and initial direction. The new initial conditions determine a new geodesic. At first this geodesic differs very little from the original geodesic. To investigate the divergence it is useful to linearize the differential equation of geodesics close to the original geodesic. The second-order linear

309

differential equation thus obtained ("the variational equation" for the equation of geodesics) is called the *Jacobi equation*; it is convenient to write it in terms of covariant derivatives and curvature tensors.

We denote by $x(t)$ a point moving along a geodesic in the manifold M with velocity (of constant magnitude) $v(t) \in TM_{x(t)}$. If the initial condition depends smoothly on a parameter α, then the geodesic also depends smoothly on the parameter. Consider the motion corresponding to a value of α. We denote the position of a point at time t on the corresponding geodesic by $x(t, \alpha) \in M$. We will assume that the initial geodesic corresponds to the zero value of the parameter, so that $x(t, 0) = x(t)$.

The *vector field of geodesic variation* is the derivative of the function $x(t, \alpha)$ with respect to α, evaluated at $\alpha = 0$; the value of this field at the point $x(t)$ is equal to

$$\left. \frac{d}{d\alpha} \right|_{\alpha = 0} x(t, \alpha) = \xi(t) \in TM_{x(t)}.$$

To write the variational equation, we define the *covariant derivative with respect to t* of a vector field $\zeta(t)$ given on the geodesic $x(t)$. To define this, we take the vector $\zeta(t + h)$, parallel translate it from the point $x(t + h)$ to $x(t)$ along the geodesic, differentiate the vector obtained in the tangent space $TM_{x(t)}$ with respect to h and evaluate at $h = 0$. The result is a vector in $TM_{x(t)}$, which is called the covariant derivative of the field $\zeta(t)$ with respect to t, and denoted by $D\zeta/Dt$.

Theorem *The vector field of geodesic variation satisfies the second-order linear differential equation*

(4)
$$\frac{D^2\xi}{Dt^2} = -\Omega(v, \xi)v,$$

where Ω is the curvature tensor, and $v = v(t)$ is the velocity vector of motion along the original geodesic.

Conversely, every solution of the differential equation (4) is a field of variation of the original geodesic.

Equation (4) is called the Jacobi equation.

PROBLEM. Prove the theorem above.

PROBLEM. Let M be a surface, $y(t)$ the magnitude of the component of the vector $\xi(t)$ in the direction normal to a given geodesic, and let the length of the vector $v(t)$ be equal to 1. Show that y satisfies the differential equation

(5)
$$\ddot{y} = -Ky,$$

where $K = K(t)$ is the riemannian curvature at the point $x(t)$

PROBLEM. Using Equation (5), compare the behavior of geodesics close to a given one on the sphere ($K = +R^{-2}$) and on the Lobachevsky plane ($K = -1$).

310

I Investigation of the Jacobi equation

In investigating the variational equations, it is useful to disregard the trivial variations, i.e., changes of the time origin and of the magnitude of the initial velocity of motion. To this end we decompose the variation vector ξ into components parallel and perpendicular to the velocity vector v. Then (since $\Omega(v, v) = 0$ and since the operator $\Omega(v, \xi)$ is skew-symmetric) for the normal component we again get the Jacobi equation, and for the parallel component we get the equation

$$\frac{D^2\xi}{Dt^2} = 0.$$

We now note that the Jacobi equation for the normal component can be written in the form of "Newton's equation"

$$\frac{D^2\xi}{Dt^2} = -\operatorname{grad} U,$$

where the quadratic form U of the vector ξ is expressed in terms of the curvature tensor and is proportional to the curvature K in the direction of the (ξ, v) plane:

$$U(\xi) = \tfrac{1}{2}\langle \Omega(v, \xi)v, \xi \rangle = \tfrac{1}{2}K\langle \xi, \xi \rangle \langle v, v \rangle.$$

Thus the behavior of the normal component of the variation vector of a geodesic with velocity 1 can be described by the equation of a (non-autonomous) linear oscillator whose potential energy is equal to the product of the curvature in the direction of the plane of velocity vectors and variations with the square of the length of the normal component of the variation.

In particular we consider the case when the curvature is *negative* in all two-dimensional directions containing the velocity vector of the geodesic (Figure 234). Then the divergence of nearby geodesics from the given one in

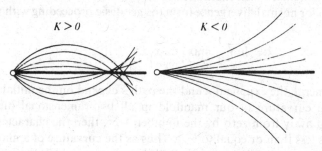

Figure 234   Nearby geodesics on manifolds of positive and negative curvature

the normal direction can be described by the equation of an oscillator with negative definite (and time-dependent) potential energy. Therefore, the normal component of divergence for nearby geodesics behaves like the divergence of a ball, located near the top of a hill, from the top. The equilibrium position of the ball at the top is unstable. This means that *geodesics near the given geodesic will diverge exponentially from it.*

If the potential energy of the newtonian equation we obtained did not depend on time, our conclusion would be rigorous. Let us assume further that the curvature in the different directions containing v is in the interval

$$-a^2 \leq K \leq -b^2, \quad \text{where } 0 < b < a.$$

Then solutions to the Jacobi equation for normal divergence will be linear combinations of exponential curves with exponent $\pm \lambda_i$, where the positive numbers λ_i are between a and b. Therefore, every solution to the Jacobi equation grows at least as fast as $e^{b|t|}$ as either $t \to +\infty$ or $t \to -\infty$; most solutions grow even faster, with rate $e^{a|t|}$.

The instability of an equilibrium position under negative definite potential energy is intuitively obvious also in the non-autonomous case. It can be proven by comparison with a corresponding autonomous system. As a result of such a comparison we may convince ourselves that under motion along a geodesic, all solutions of the Jacobi equation for normal divergence on a manifold of negative curvature grow at least as fast as an exponential function of the distance traveled, whose exponent is equal to the square root of the absolute value of the curvature in the two-dimensional direction for which this absolute value is minimal. In fact, most solutions grow even faster, but we cannot now assert that the exponent of growth for most solutions is determined by the direction in which the absolute value of the negative curvature is largest.

In summary, we can say that the behavior of geodesics on a manifold of negative curvature is characterized by exponential instability. For numerical estimates of this instability, it is useful to define the *characteristic path length* s as the average path length on which small errors in the initial conditions are increased e times.

More precisely, the characteristic path length s can be defined as the inverse of the exponent λ which characterizes the growth of the solution to the Jacobi equation for normal divergence from the geodesic proceeding with velocity 1:

$$\lambda = \varlimsup_{T \to \infty} \frac{1}{T} \max_{|t| < T} \max_{|\xi(0)| = 1} \ln|\xi(t)| \qquad s = \frac{1}{\lambda}.$$

In general, the exponent λ and the path s depend on the initial geodesic.

If the curvature of our manifold in all two-dimensional directions is bounded away from zero by the number $-b^2$, then the characteristic path length is less than or equal to b^{-1}. Thus as the curvature of a manifold gets more negative, the characteristic path length s, on which the instability of

312

geodesics is reduced to e-fold growth of error, gets smaller. In view of the exponential character of the growth of error, the course of a geodesic on a manifold of negative curvature is practically impossible to predict.

Assume, for example, that the curvature is negative and bounded away from zero by $-4m^{-2}$. The characteristic path length is less than or equal to half a meter, i.e., on a geodesic arc five meters long the error grows by approximately $e^{10} \sim 10^4$. Therefore, an error of a tenth of a millimeter in the initial conditions shows up in the form of a one-meter difference at the end of the geodesic.

J Geodesic flows on compact manifolds of negative curvature

Let M be a compact riemannian manifold whose curvature at every point in every two-dimensional direction is negative. (Such manifolds exist.) Consider the inertial motion of a point of mass 1 on M, without any external forces. The lagrangian function of this system is equal to the kinetic energy, which is equal to the total energy and is a first integral of the equations of motion.

If M has dimension n, then each energy level manifold has dimension $2n - 1$. This manifold is a submanifold of the tangent bundle of M. For example, we can fix the value of the energy at $\frac{1}{2}$ (which corresponds to initial velocity 1). Then the velocity vector of the point has length constantly equal to 1, and our level manifold turns out to be the fiber bundle

$$T_1 M \subset TM$$

consisting of the unit spheres in the tangent spaces to M at every point.

Thus, a point of the manifold $T_1 M$ is represented as a vector of length 1 at a point of M. By the Maupertuis–Jacobi principle, we can describe the motion of a point mass with fixed initial conditions in the following way: the point moves with velocity 1 along the geodesic determined by the indicated vector.

By the law of conservation of energy the manifold $T_1 M$ is an invariant manifold in the phase space of our system. Therefore, our phase flow determines a one-parameter group of diffeomorphisms on the $(2n - 1)$-dimensional manifold $T_1 M$. This group is called the *geodesic flow* on M. The geodesic flow can be described as follows: the transformation at time t carries the unit vector $\xi \in T_1 M$ located at the point x, to the unit velocity vector of the geodesic coming from x in the direction ξ, located at the point at distance t from x. We note that there is a naturally defined volume element on $T_1 M$ and that the geodesic flow preserves it (Liouville's theorem).

Up to now we have not used the negative curvature of the manifold M. But if we investigate the trajectories of the geodesic flow, it turns out that the negative curvature of M has a strong impact on the behavior of these trajectories (this is related to the exponential instability of geodesics on M).

Here are some properties of geodesic flows on manifolds of negative curvature (for further details, see the book of D. V. Anosov cited earlier).

1. Almost all phase trajectories are dense in the energy level manifold (the exceptional non-dense trajectories form a set of measure zero).
2. Uniform distribution: the amount of time which almost every trajectory spends in any region of the phase space $T_1 M$ is proportional to the volume of the region.
3. The phase flow g^t has the mixing property: if A and B are two regions, then

$$\lim_{t \to \infty} \text{mes}[(g^t A) \cap B] = \text{mes } A \text{ mes } B$$

(where mes denotes the volume, normalized by the condition that the whole space have measure 1).

From these properties of trajectories in phase space follow analogous statements about geodesics on the manifold itself. Physicists call these properties "stochastic": asymptotically for large t the trajectories behave as if the point were random. For example, the mixing property means that the probability of turning up in B at a time t long after exiting from A is proportional to the volume of B.

Thus, *the exponential instability of geodesics on manifolds of negative curvature leads to the stochasticity of the corresponding geodesic flow.*

K *Other applications of exponential instability*

The exponential instability property of geodesics on manifolds of negative curvature has been studied by many authors, beginning with Hadamard (and, in the case of constant curvature, also by Lobachevsky), but especially by E. Hopf. An unexpected discovery of the 1960s in this area was the surprising stability of exponentially unstable systems with respect to perturbations of the systems themselves.

Consider, for example, the vector field giving the geodesic flow on a compact surface of negative curvature. As we showed above, the phase curves of this flow are arranged in a complicated way: almost every one of them is dense in the three-dimensional energy level manifold. The flow has infinitely many closed trajectories, and the set of points on closed trajectories is also dense in the three-dimensional energy level manifold.

We now consider a nearby vector field. It turns out that, in spite of the complexity of the picture of phase curves, the entire picture with dense phase curves and infinitely many closed trajectories hardly changes at all if we pass to the nearby field. In fact, there is a homeomorphism close to the identity transformation which takes the phase curves of the unperturbed flow to the phase curves of the perturbed flow.

Thus our complicated phase flow has the same property of "structural stability" as a limit cycle, or a stable focus in the plane. We note that neither a center in the plane nor a winding of the torus has this property of structural stability: the topological type of the phase portrait in these cases changes for arbitrarily small changes in the vector field.

The existence of structurally stable systems with complicated motions, each of which is in itself exponentially unstable, is one of the basic discoveries of recent years in the theory of ordinary differential equations (the conjecture that geodesic flows on manifolds of negative curvature are structurally stable was made by S. Smale in 1961, and the proof was given by D. V. Anosov and published in 1967; the basic results on stochasticity of these flows were obtained by Ya. G. Sinai and D. V. Anosov, also in the 1960s).

Before these works most mathematicians believed that in systems of differential equations in "general form" only the simplest stable limiting behaviors were possible: equilibrium positions and cycles. If a system was more complicated (for example, if it was conservative), then it was assumed that after a small change in its equations (for example, after imposing small non-conservative perturbations) complicated motions are "dispersed" into simple ones. We now know that this is not so, and that in the function space of vector fields there are whole regions consisting of fields with more complicated behavior of phase curves.

The conclusions which follow from this are relevant to a wide range of phenomena, in which "stochastic" behavior of deterministic objects is observed.

Namely, suppose that in the phase space of some (non-conservative) system there is an attracting invariant manifold (or set) in which the phase curves have the property of exponential instability. We now know that systems with such a property are not exceptional: under small changes of the system this property must persist. What is seen by an experimenter observing motions of such a system?

The approach of phase curves to an attracting set will be interpreted as the establishment of some sort of limiting conditions. The further motion of a phase point near the attracting set will involve chaotic, unpredictable changes of "phase" of the limiting behavior, perceptible as "stochasticity" or "turbulence."

Unfortunately, no convincing analysis from this point of view has yet been developed for physical examples of a turbulent character. A primary example is the hydrodynamic instability of a viscous fluid, described by the so-called Navier–Stokes equations. The phase space of this problem is infinite-dimensional (it is the space of vector fields with divergence 0 in the domain of fluid flow), but the infinite-dimensionality of the problem is apparently not a serious obstacle, since the viscosity extinguishes the high harmonics (small vortices) faster and faster as the harmonics are higher and higher. As a result, the phase curves from the infinite-dimensional space seem to approach some finite-dimensional manifold (or set), to which the limit regime also belongs.

For large viscosity, we have a stable attracting equilibrium position in the phase space ("stable stationary flow"). As the viscosity decreases it loses stability; for example, a stable limit cycle can appear in phase space ("periodic flow") or a stable equilibrium position of a new type ("secondary stationary flow").[96] As the viscosity decreases further, more and more harmonics come into play, and the limit regime can become ever higher in dimension.

For small viscosity, the approach to a limit regime with exponentially unstable trajectories seems very likely. Unfortunately, the corresponding calculations have not yet been carried out due to the limited capacity of existing computers. However, the following general conclusion can be drawn without any calculations: turbulent phenomena may appear even if solutions exist and are unique; exponential instability, which is encountered even in deterministic systems with a finite number of degrees of freedom, is sufficient.

As one more example of an application of exponential instability we mention the proof announced by Ya. G. Sinai of the "ergodic hypothesis" of Boltzmann for systems of rigid balls. The hypothesis is that the phase flow corresponding to the motion of identical absolutely elastic balls in a box with elastic walls is ergodic on connected energy level sets. (Ergodicity means that almost every phase curve spends an amount of time in every measurable piece of the level set proportional to the measure of that piece.)

Boltzmann's hypothesis allows us to replace time averages by space averages, and was for a long time considered to be necessary to justify statistical mechanics. In reality, Boltzmann's hypothesis (in which it is a question of a limit as time approaches infinity) is not necessary for passing to the statistical limit (the number of pieces approaches infinity). However, Boltzmann's hypothesis inspired the entire analysis of the stochastic properties of dynamical systems (so-called ergodic theory), and its proof serves as a measure of the maturity of this theory.

The exponential instability of trajectories in Boltzmann's problem arises as a result of collisions of the balls with one another, and can be explained in the following way. For simplicity, we will consider a system of only two particles in the plane, and will represent a square box with reflection off the walls by the planar torus $\{(x, y) \bmod 1\}$. Then we can consider one of the particles as stationary (using the conservation of momentum); the other particle can be considered as a point.

In this way we arrive at the model problem of motion of a point on a toral billiard table with a circular wall in the middle from which the point is reflected according to the law "the angle of incidence is equal to the angle of reflection" (Figure 235).

To investigate this system we look at an analogous billiard table bounded on the outside by a planar convex curve (e.g., the motion of a point inside an ellipse). Motion on such a billiard table can be considered as the limiting case of the geodesic flow on the surface of an ellipsoid. Passage to the limit

[96] A more detailed account of loss of stability is given in "Lectures on bifurcations and versal families," *Russian Math. Surveys* **27**, no. 5 (1972), 55–123.

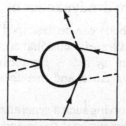

Figure 235   Torus-shaped billiard table with scattering by a circular wall

consists of decreasing the smallest axis of the ellipsoid to zero. As a result, geodesics on the ellipsoid become billiard trajectories on the ellipse. We discover from this that the ellipse can reasonably be thought of as two-sided and that, under every reflection, the geodesic goes from one side of the ellipse to the other.

We now return to our toral billiard table. Motion on it can be looked at as the limiting case of the geodesic flow on a smooth surface. This surface is obtained from looking at the torus with a hole as a two-sided surface, giving it some thickness and slightly smoothing the sharp edge. As a result we have a surface with the topology of a pretzel (a sphere with two handles).

After blowing up the ellipse into the ellipsoid we obtain a surface of positive curvature; after blowing up the torus with a hole we get a surface of negative curvature (in both cases the curvature is concentrated close to the edge, but the blowing up can be done so that the sign of the curvature does not change). Thus motion in our toral billiard table can be looked at as the limiting case of motion along geodesics on a surface of negative curvature.

Now, to prove Boltzmann's conjecture (in the simple case under consideration) it is sufficient to verify that the analysis of stochastic properties of geodesic flows on surfaces of negative curvature holds in the indicated limiting case.

A more detailed presentation of the proof turns out to be very complicated; it has been published only for the case of systems of two particles (Ya. G. Sinai, Dynamical systems with elastic reflections, Russian Mathematical Surveys, 25, no. 2 (1970), 137–189).

Appendix 2: Geodesics of left-invariant metrics on Lie groups and the hydrodynamics of ideal fluids

Eulerian motion of a rigid body can be described as motion along geodesics in the group of rotations of three-dimensional euclidean space provided with a left-invariant riemannian metric. A significant part of Euler's theory depends only upon this invariance, and therefore can be extended to other groups.

Among the examples involving such a generalized Euler theory are motion of a rigid body in a high-dimensional space and, especially interesting, the hydrodynamics of an ideal (incompressible and inviscid) fluid. In the latter case, the relevant group is the group of volume-preserving diffeomorphisms of the domain of fluid flow. In this example, the principle of least action implies that the motion of the fluid is described by the geodesics in the metric given by the kinetic energy. (If we wish, we can take this principle to be the mathematical definition of an ideal fluid.) It is easy to verify that this metric is (right) invariant.

Of course, extending results obtained for finite-dimensional Lie groups to the infinite-dimensional case should be done with care. For example, in three-dimensional hydrodynamics an existence and uniqueness theorem for solutions of the equations of motion has not yet been proved. Nevertheless, it is interesting to see what conclusions can be drawn by formally carrying over properties of geodesics on finite-dimensional Lie groups to the infinite-dimensional case. These conclusions take the character of *a priori* statements (identities, inequalities, etc.) which should be satisfied by all reasonable solutions. In some cases, the formal conclusions can then be rigorously justified directly, without infinite-dimensional analysis.

For example, the Euler equations of motion for a rigid body have as their analogue in hydrodynamics the Euler equations of motion of an ideal fluid. Euler's theorem on the stability of rotations around the large and small axes of the inertia ellipsoid corresponds in hydrodynamics to a slight generalization of Rayleigh's theorem on the stability of flows without inflection points of the velocity profile.

It is also easy to extract from Euler's formulas an explicit expression for the riemannian curvature of a group with a one-sided invariant metric. Applying this to hydrodynamics we find the curvature of the group of diffeomorphisms preserving the volume element. It is interesting to note that in sufficiently nice two-dimensional directions, the curvature turns out to be finite and, in many cases, negative. Negative curvature implies exponential instability of geodesics (cf. Appendix 1). In the case under consideration, the geodesics are motions of an ideal fluid; therefore the calculation of the curvature of the group of diffeomorphisms gives us some information on the instability of ideal fluid flow. In fact, the curvature determines the characteristic path length on which differences between initial conditions grow by e. Negative curvature leads to practical indeterminacy of the flow: on a path only a few times longer than the characteristic path length, a deviation in initial conditions grows 100 times larger.

In this appendix, we will briefly set out the results of calculations related to geodesics on groups with one-sided (right- or left-) invariant metrics. Proofs and further details can be found in the following places:

V. Arnold, Sur la géométrie différentielle des groupes de Lie de dimension infinie et ses applications à l'hydrodynamique des fluides parfaits. Annales de l'Institut Fourier, XVI, no. 1 (1966), 319–361.

V. I. Arnold, An a priori estimate in the theory of hydrodynamic stability, Izv. Vyssh. Uchebn. Zaved. Matematika 1966, no. 5 (54), 3–5. (Russian)

V. I. Arnold, The Hamiltonian nature of the Euler equations in the dynamics of a rigid body and of an ideal fluid, Uspekhi Matematicheskikh Nauk, 24 (1969), no. 3 (147) 225–226. (Russian)

L. A. Dikii, A remark on Hamiltonian systems connected with the rotation group, Functional Analysis and Its Applications, 6:4 (1972) 326–327.

D. G. Ebin, J. Marsden, Groups of diffeomorphisms and the motion of an incompressible fluid, Annals of Math. 92, no. 1 (1970), 102–163.

O. A. Ladyzhenskaya, On the local solvability of non-stationary problems for incompressible ideal and viscous fluids and vanishing viscosity, Boundary problems in mathematical physics, v. 5 (Zapiski nauchnikh seminarov LOMI, v. 21), "Nauka," 1971, 65–78. (Russian)

A. S. Mishchenko, Integrals of geodesic flows on Lie groups, Functional Analysis and Its Applications, 4, no. 3 (1970), 232–235.

A. M. Obukhov, On integral invariants in systems of hydrodynamic type, Doklady Acad. Nauk. 184, no. 2 (1969). (Russian)

L. D. Faddeev, Towards a stability theory of stationary planar-parallel flows of an ideal fluid, Boundary problems in mathematical physics, v. 5 (Zapiski nauchnikh seminarov LOMI, v. 21), "Nauka," 1971, 164–172. (Russian)

A Notation: The adjoint and co-adjoint representations

Let G be a real Lie group and $\mathfrak{g}$ its Lie algebra, i.e., the tangent space to the group at the identity provided with the commutator bracket operation $[\ ,\ ]$.

A Lie group acts on itself by left and right translation: every element g of the group G defines diffeomorphisms of the group onto itself:

$$L_g: G \to G \qquad L_g h = gh \qquad R_g = G \to G \qquad R_g h = hg.$$

The induced maps of the tangent spaces will be denoted by

$$L_{g*}: TG_h \to TG_{gh} \quad \text{and} \quad R_{g*}: TG_h \to TG_{hg}$$

for every h in G.

The diffeomorphism $R_{g^{-1}} L_g$ is an inner automorphism of the group. It leaves the group identity element fixed. Its derivative at the identity is a linear map from the algebra (i.e., the tangent space to the group at the identity) to itself. This map is denoted by

$$Ad_g: \mathfrak{g} \to \mathfrak{g} \qquad Ad_g = (R_{g^{-1}} L_g)_{*e}$$

and is called the *adjoint* representation of the group. It is easy to verify that Ad_g is an algebra homomorphism, i.e., that

$$Ad_g[\xi, \eta] = [Ad_g \xi, Ad_g \eta], \qquad \xi, \eta \in \mathfrak{g}.$$

It is also clear that $Ad_{gh} = Ad_g Ad_h$.

We can consider Ad as a map of the group into the space of linear operators on the algebra:

$$Ad(g) = Ad_g.$$

The map Ad is differentiable. Its derivative at the identity of the group is a linear map from the algebra $\mathfrak{g}$ to the space of linear operations on $\mathfrak{g}$. This map is denoted by ad, and its image on an element ξ in the algebra by ad_ξ. Thus ad_ξ is an endomorphism of the algebra space, and we have

$$ad = Ad_{*e} : \mathfrak{g} \to \text{End } \mathfrak{g} \qquad ad_\xi = \frac{d}{dt}\bigg|_{t=0} Ad_{e^{t\xi}},$$

where $e^{t\xi}$ is the one-parameter group with tangent vector ξ. From the formula written above it is easy to deduce an expression for ad in terms of the algebra alone:

$$ad_\xi \eta = [\xi, \eta].$$

We now consider the dual vector space $\mathfrak{g}^*$ to the Lie algebra $\mathfrak{g}$. This is the space of real linear functionals on the Lie algebra. In other words, $\mathfrak{g}^*$ is the cotangent space to the group at the identity, $\mathfrak{g}^* = T^*G_e$. The value of an element ξ of the cotangent space to the group at some point g on an element η of the tangent space at the same point will be denoted by round brackets:

$$(\xi, \eta) \in \mathbb{R}, \qquad \xi \in T^*G_g, \eta \in TG_g.$$

Left and right translation induce operators on the cotangent space dual to L_{g*} and R_{g*}. We denote them by

$$L_g^* : T^*G_{gh} \to T^*G_h \quad \text{and} \quad R_g^* : T^*G_{hg} \to T^*G_h$$

for every h in G. These operators are defined by the identities

$$(L_g^* \xi, \eta) \equiv (\xi, L_{g*} \eta) \quad \text{and} \quad (R_g^* \xi, \eta) \equiv (\xi, R_{g*} \eta).$$

The transpose operators Ad_g^*, where g runs through the Lie group G, form a representation of this group, i.e., they satisfy the relations

$$Ad_{gh}^* = Ad_h^* Ad_g^*.$$

This representation is called the *co-adjoint representation* of the group and plays an important role in all questions related to (left) invariant metrics on the group.

Consider the derivative of the operator Ad_g^* with respect to g at the identity. This derivative is a linear map from the algebra to the space of linear operators on the dual space to the algebra. This linear map is denoted by ad^*, and its image on an element ξ in the algebra is denoted by ad_ξ^*. Thus ad^* is a linear operator on the dual space to the algebra,

$$ad_\xi^*: \mathfrak{g}^* \to \mathfrak{g}^*.$$

It is easy to see that ad_ξ^* is the adjoint of ad_ξ:

$$(ad_\xi^* \eta, \zeta) \equiv (\eta, ad_\xi \zeta) \quad \text{for all } \eta \in \mathfrak{g}^*, \zeta \in \mathfrak{g}.$$

It is sometimes convenient to denote the action of ad^* by braces:

$$ad_\xi^* \eta = \{\xi, \eta\}, \quad \text{where } \xi \in \mathfrak{g}, \eta \in \mathfrak{g}^*.$$

Thus braces mean the bilinear function from $\mathfrak{g} \times \mathfrak{g}^*$ to $\mathfrak{g}^*$, related to commutation in the algebra by the identity

$$(\{\xi, \eta\}, \zeta) = (\eta, [\xi, \zeta]).$$

We consider now the orbits of the co-adjoint representation of the group in the dual space of the algebra. At each point of an orbit we have a natural symplectic structure (called the Kirillov form since A. A. Kirillov first used it to investigate representations of nilpotent Lie groups). Thus, the orbits of the co-adjoint representation are always even-dimensional. We also note that we obtain a series of examples of symplectic manifolds by looking at different Lie groups and all possible orbits.

The symplectic structure on the orbits of the co-adjoint representation is defined by the following construction. Let x be a point in the dual space to the algebra and ξ a vector tangent at this point to its orbit. Since $\mathfrak{g}^*$ is a vector space, we can consider the vector ξ, which really belongs to the tangent space to $\mathfrak{g}^*$ at x, as lying in $\mathfrak{g}^*$.

The vector ξ can be represented (in many ways) as the velocity vector of the motion of the point x under the co-adjoint action of the one-parameter group e^{at} with velocity vector $a \in \mathfrak{g}$. In other words, every vector tangent to the orbit of x in the co-adjoint representation of the group can be expressed in terms of a suitable vector a in the algebra by the formula

$$\xi = \{a, x\}, \quad a \in \mathfrak{g}, x \in \mathfrak{g}^*.$$

Now we are ready to define the value of the symplectic 2-form Ω on a pair of vectors ξ_1, ξ_2 tangent to the orbit of x. Namely, we express ξ_1 and ξ_2 in terms of algebra elements a_1 and a_2 by the formula above, and then obtain the scalar

$$\Omega(\xi_1, \xi_2) = (x, [a_1, a_2]), \quad x \in \mathfrak{g}^*, a_i \in \mathfrak{g}.$$

It is easy to verify that (1) the bilinear form Ω is well defined, i.e., its value does not depend on the choice of a_i; (2) Ω is skew-symmetric and therefore gives a differential 2-form Ω on the orbit; and (3) Ω is nondegenerate and closed (the proofs can be found, for instance, in Appendix 5). Thus the form Ω is a symplectic structure on an orbit of the co-adjoint representation.

B Left-invariant metrics

A riemannian metric on a Lie group G is called *left-invariant* if it is preserved by all left translations L_g, i.e., if the derivative of left translation carries every vector to a vector of the same length.

It is sufficient to give a left-invariant metric at one point of the group, for instance the identity; then the metric can be carried to the remaining points by left translations. Thus there are as many left-invariant riemannian metrics on a group as there are euclidean structures on the algebra.

A euclidean structure on the algebra is defined by a symmetric positive-definite operator from the algebra to its dual space. Thus, let $A: \mathfrak{g} \to \mathfrak{g}^*$ be a symmetric positive linear operator:

$$(A\xi, \eta) = (A\eta, \xi), \quad \text{for all } \xi, \eta \text{ in } \mathfrak{g}.$$

(It is not very important that A be positive, but in mechanical applications the quadratic form $(A\xi, \xi)$ is positive-definite.)

We define a symmetric operator $A_g: TG_g \to T^*G_g$ by left translation:

$$A_g \xi = L_{g^{-1}}^* A L_{g^{-1}*} \xi.$$

We thus obtain the following commutative diagram of linear operators:

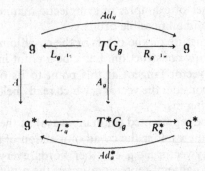

We will denote by angled brackets the scalar product determined by the operator A_g:

$$\langle \xi, \eta \rangle_g = (A_g \xi, \eta) = (A_g \eta, \xi) = \langle \eta, \xi \rangle_g.$$

This scalar product gives a riemannian metric on the group G, invariant under left translations. The scalar product in the algebra will be denoted simply by $\langle \ , \ \rangle$. We define an operation $B: \mathfrak{g} \times \mathfrak{g} \to \mathfrak{g}$ by the identity

$$\langle [a, b], c \rangle \equiv \langle B(c, a), b \rangle, \quad \text{for all } b \text{ in } \mathfrak{g}.$$

Clearly, this operation B is bilinear, and for fixed first argument is skew-symmetric in the second:

$$\langle B(c, a), b \rangle + \langle B(c, b), a \rangle = 0.$$

322

C Example

Let $G = SO(3)$ be the group of rotations of three-dimensional euclidean space, i.e. the configuration space of a rigid body fixed at a point. A motion of the body is then described by a curve $g = g(t)$ in the group. The Lie algebra of G is the three-dimensional space of angular velocities of all possible rotations. The commutator in this algebra is the usual vector product.

A rotation velocity $\dot{g}$ of the body is a tangent vector to the group at the point g. To get the angular velocity, we must carry this vector to the tangent space of the group at the identity, i.e. to the algebra. But this can be done in two ways: by left and right translation. As a result, we obtain two different vectors in the algebra:

$$\omega_c = L_{g^{-1}*}\dot{g} \in \mathfrak{g} \quad \text{and} \quad \omega_s = R_{g^{-1}*}\dot{g} \in \mathfrak{g}.$$

These two vectors are none other than the "angular velocity in the body" and the "angular velocity in space."

An element g of the group G corresponds to a position of the body obtained by the motion g from some initial state (corresponding to the identity element of the group and chosen arbitrarily). Let ω be an element of the algebra.

Let $e^{\omega t}$ be a one-parameter group of rotations with angular velocity ω; ω is the tangent vector to this one-parameter group at the identity. Now we look at the displacement

$$e^{\omega \tau}g, \quad \text{where } g = g(t) \in G, \ \omega \in \mathfrak{g}, \text{ and } \tau \ll 1,$$

obtained from the displacement g by a rotation with angular velocity ω after a small time τ. If the vector $\dot{g}$ coincides with the vector

$$\frac{d}{d\tau}\bigg|_{\tau=0} e^{\omega \tau}g,$$

then ω is called the *angular velocity relative to space* and is denoted by ω_s. Thus ω_s is obtained from $\dot{g}$ by *right* translation. In an analogous way we can show that the angular velocity in the body is the left translate of the vector $\dot{g}$ in the algebra.

The dual space $\mathfrak{g}^*$ to the algebra in our example is the space of angular momenta.

The kinetic energy of a body is determined by the vector of angular velocity in the body and does not depend on the position of the body in space. Therefore, kinetic energy gives a *left-invariant* riemannian metric on the group. The symmetric positive-definite operator $A_g: TG_g \to T^*G_g$ given by this metric is called the *moment of inertia operator* (or tensor). It is related to the kinetic energy by the formula $T = \frac{1}{2}\langle \dot{g}, \dot{g} \rangle_g = \frac{1}{2}\langle \omega_c, \omega_c \rangle = \frac{1}{2}(A\omega_c, \omega_c) = \frac{1}{2}(A_g\dot{g}, \dot{g})$, where $A:\mathfrak{g} \to \mathfrak{g}^*$ is the value of A_g for $g = e$. The image of the vector $\dot{g}$ under the action of the moment of inertia operator A_g is called the *angular momentum* and is denoted by $M = A_g\dot{g}$. The vector M lies in the cotangent space to the group at the point g, and it can be carried to the cotangent space to the group at the identity by both left and right translations.

We obtain two vectors

$$M_c = L_g^* M \in \mathfrak{g}^*$$

and

$$M_s = R_g^* M \in \mathfrak{g}^*$$

These vectors in the dual space to the algebra are none other than the angular momentum relative to the body (M_c) and the angular momentum relative to space (M_s). This follows easily from the expression for kinetic energy in terms of momentum and angular velocity:

$$T = \tfrac{1}{2}(M_c, \omega_c) = \tfrac{1}{2}(M, \dot{g}).$$

By the principle of least action, the motion of a rigid body under inertia (with no external forces) is a geodesic in the group of rotations with the left-invariant metric described above.

We will now look at a geodesic of an arbitrary left-invariant riemannian metric on an arbitrary Lie group as a motion of a "generalized rigid body" with configuration space G. Such a "rigid body with group G" is determined by its kinetic energy, i.e., a positive-definite quadratic form on the Lie algebra. More precisely, we will consider geodesics of a left-invariant metric on a group G given by a quadratic form $\langle \omega, \omega \rangle$ on the algebra as motions of a rigid body with group G and kinetic energy $\langle \omega, \omega \rangle/2$.

To every motion $t \to g(t)$ of our generalized rigid body we can associate four curves:

$$t \to \omega_c(t) \in \mathfrak{g} \qquad t \to \omega_s(t) \in \mathfrak{g}$$
$$t \to M_c(t) \in \mathfrak{g}^* \qquad t \to M_s(t) \in \mathfrak{g}^*,$$

called motions of the vectors of angular velocity and momentum in the body and in space. The differential equations which these curves satisfy were found by Euler for an ordinary rigid body. However, they are true in the most general case of an arbitrary group G, and we will call them the Euler equations for a generalized rigid body.

Remark. In the ordinary theory of a rigid body six different three-dimensional spaces $\mathbb{R}^3$, $\mathbb{R}^{3*}$, $\mathfrak{g}$, $\mathfrak{g}^*$, TG_g, and T^*G_g are identified. The fact that the dimensions of the space $\mathbb{R}^3$ in which the body moves and of the Lie algebra $\mathfrak{g}$ of its group of motions are the same is an accident related to the dimension 3; in the n-dimensional case, $\mathfrak{g}$ has dimension $n(n-1)/2$.

The identification of the Lie algebra $\mathfrak{g}$ with its dual space $\mathfrak{g}^*$ has a more profound basis. The fact is that on the group of rotations there exists (and is unique up to multiplication) a two-sided invariant riemannian metric. This metric gives once and for all a preferred isomorphism of the vector spaces $\mathfrak{g}$ and $\mathfrak{g}^*$ (and also of TG_g and T^*G_g). It allows us therefore to consider the vectors of angular velocity and momentum as lying in the same euclidean space. With this identification, the operation $\{\ ,\ \}$ is simply the commutator of the algebra, taken with a minus sign.

A two-sided invariant metric exists on any compact Lie group. Therefore, to study motions of rigid bodies with compact groups we may identify the spaces of angular velocities and momenta. However, we cannot make this identification for applications to non-compact (or infinite-dimensional) groups of diffeomorphisms.

D *Euler's equation*

The results of Euler (obtained by him in the particular case $G = SO(3)$) can be formulated as the following theorems on the motion of the vectors of angular velocity and momentum of a generalized rigid body with group G.

Theorem 1. *The vector of angular momentum relative to space is preserved under motion:*

$$\frac{dM_s}{dt} = 0.$$

Theorem 2. *The vector of angular momentum relative to the body satisfies Euler's equation*

$$\frac{dM_c}{dt} = \{\omega_c, M_c\}.$$

These theorems are proved for a generalized rigid body in the same way as for an ordinary rigid body.

Remark 1. The vector of angular velocity in the body, ω_c, can be expressed linearly in terms of the vector of angular momentum in the body, M_c, by using the inverse of the inertia operator: $\omega_c = A^{-1}M_c$. Therefore, Euler's equation can be considered as an equation for the vector of angular momentum in the body alone; its right-hand side is quadratic in M_c.

We can also express this result in the following way. Consider the phase flow of our rigid body. (Its phase space T^*G has dimension twice the dimension n of the group G or the space of angular momenta $\mathfrak{g}^*$.) Then this phase flow in a $2n$-dimensional manifold factors over the flow given by Euler's equation in the n-dimensional vector space $\mathfrak{g}^*$.

A factorization of a phase flow g^t on a manifold X over a phase flow f^t on a manifold Y is a smooth mapping π of X onto Y under which motions g^t are mapped to motions f^t, so that the following diagram commutes (i.e., $\pi g^t = f^t \pi$):

$$
\begin{array}{ccc}
X & \xrightarrow{\;g^t\;} & X \\
\downarrow{\scriptstyle \pi} & & \downarrow{\scriptstyle \pi} \\
Y & \xrightarrow{\;f^t\;} & Y
\end{array}
$$

In our case, $X = T^*G$ is the phase space of the body, $Y = \mathfrak{g}^*$ is the space of angular momenta. The projection $\pi \colon T^*G \to \mathfrak{g}^*$ is defined by left translation ($\pi M = L_g^* M$ for $M \in T^*G_g$), g^t is the phase flow of the body under consideration on the $2n$-dimensional space T^*G, and f^t is the phase flow of the Euler equation in the n-dimensional space of angular momenta $\mathfrak{g}^*$.

In other words, a motion of the vector of angular momentum relative to the body depends only on the initial position of the vector of angular momentum relative to the body and does not depend on the position of the body in the space.

Remark 2. The law of conservation of the vector of angular momentum relative to space can be expressed by saying that every component of this vector in some coordinate system on the space $\mathfrak{g}^*$ is conserved. We thus obtain a set of first integrals of the equations of motion of the rigid body. In particular, to every element of the Lie algebra $\mathfrak{g}$ there corresponds a linear function on the space $\mathfrak{g}^*$ and, therefore, a first integral. The Poisson brackets of first integrals given by functions on $\mathfrak{g}^*$ are themselves functions on $\mathfrak{g}^*$, as can be seen easily. We thus obtain an (infinite-dimensional) extension of the Lie algebra $\mathfrak{g}$, consisting of all functions on $\mathfrak{g}^*$. $\mathfrak{g}$ itself is included in this extension as the Lie algebra of linear functions on $\mathfrak{g}^*$. Of course, of all these first integrals of the phase flow in a $2n$-dimensional space only n are functionally independent. As the n independent integrals we can take, for example, n linear functions on $\mathfrak{g}^*$ which form a basis in $\mathfrak{g}$.

Because of possible infinite-dimensional applications, we would like to avoid coordinates and formulate statements about first integrals intrinsically. This can be done by reformulating Theorem 1 in the following way.

Theorem 3. *The orbits of the co-adjoint representation of a group in the dual space to the algebra are invariant manifolds for the flow in this space given by Euler's equation.*

PROOF. $M_c(t)$ is obtained from $M_s(t)$ by the action of the co-adjoint representation, and $M_s(t)$ remains fixed. $\qquad\qquad\qquad\qquad\qquad\square$

EXAMPLE. In the case of an ordinary rigid body, the orbits of the co-adjoint representation of the group in the space of momenta are the spheres $M_1^2 + M_2^2 + M_3^2 = \text{const}$. In this case Theorem 3 is reduced to the law of conservation of the length of the angular momentum. It consists of the fact that, if the initial point M_c lies on some orbit (i.e., in the given case on the sphere $M^2 = \text{const}$), then all the points of its trajectory under the action of Euler's equation lie on the same orbit.

We now return to the general case of an arbitrary group G and recall that each orbit of the co-adjoint representation has a symplectic structure (cf. subsection A). Furthermore, the kinetic energy of the body can be expressed in terms of the angular momentum relative to the body. As a result we obtain a quadratic form on the space of angular momenta

$$T = \tfrac{1}{2}(M_c, A^{-1}M_c).$$

Let us fix some one orbit V of the co-adjoint representation. We consider the kinetic energy as a function on this orbit:

$$H: V \to \mathbb{R}, \qquad H(M_c) = \tfrac{1}{2}(M_c, A^{-1}M_c).$$

Theorem 4. *On every orbit V of the co-adjoint representation, Euler's equation is hamiltonian with hamiltonian function H.*

PROOF. Every vector ξ tangent to V at a point M has the form $\xi = \{f, M\}$. where $f \in \mathfrak{g}$. In particular, the vector field on the right side of Euler's equation can be written in the form $X = \{dT, M\}$ (here the differential of the function T at a point M of the vector space $\mathfrak{g}^*$ is considered as a vector of the dual space to $\mathfrak{g}^*$, i.e., as an element of the Lie algebra $\mathfrak{g}$). It follows from the definitions of the symplectic structure Ω and the operation $\{\ ,\ \}$ (cf. subsection A) that for every vector ξ tangent to V at M,

$$\Omega(\xi, X) = (M, [f, dT]) = (dT, \{f, M\}) = (dH, \xi). \qquad \square$$

Euler's equation can be carried over from the dual space of the algebra to the algebra itself by inversion of the moment of inertia operator. As a result we obtain the following formulation of Euler's equation in terms of the operation B (section B).

Theorem 5. *The motion of the vector of angular velocity in the body is determined by the initial position of this vector and does not depend on the initial position of the body. The vector of angular velocity in the body satisfies an equation with quadratic right-hand side:*

$$\dot{\omega}_c = B(\omega_c, \omega_c).$$

We will call this equation Euler's equation for angular velocity. We notice that, under the action of the operator $A^{-1}: \mathfrak{g}^* \to \mathfrak{g}$, the orbits of the co-adjoint representation are carried to invariant manifolds of Euler's equation for angular velocity; these manifolds have symplectic structure, etc. However, unlike orbits in $\mathfrak{g}^*$, these invariant manifolds are not determined by the Lie group G itself, but depend also on the choice of rigid body (i.e., moment of inertia operator).

From the law of conservation of energy we have

Theorem 6. *Euler's equations (for momentum and angular velocity) have a quadratic first integral, whose value is equal to the kinetic energy*

$$T = \tfrac{1}{2}(M_c, A^{-1}M_c) = \tfrac{1}{2}(A\omega_c, \omega_c).$$

327

E *Stationary rotations and their stability*

A *stationary rotation* of a rigid body is a rotation for which the angular velocity in the body is constant (and thus also the angular velocity in space; it is easy to see that one implies the other). We know from the theory of an ordinary rigid body in $\mathbb{R}^3$ that stationary rotations are rotations around the major axes of the moment of inertia ellipsoid. Below, we formulate a generalization of this theorem to the case of a rigid body with any Lie group. We note that stationary rotations are geodesics of left-invariant metrics which are one-parameter subgroups. We note also that the directions of the major axes of the inertia ellipsoid can be determined by looking at the stationary points of the kinetic energy on the sphere of vectors of momentum of fixed length.

Theorem 7. *The angular momentum (respectively, angular velocity) of a stationary rotation with respect to the body is a critical point of the energy on the orbit of the co-adjoint representation (respectively on the image of the orbit under the action of the operator A^{-1}). Conversely, every critical point of the energy on an orbit determines a stationary rotation.*

The proof is a straightforward computation or application of Theorem 4.

We note that the partition of the space of momenta into orbits of the co-adjoint representation cannot be so easily constructed in the case of an arbitrary group as it was in the simple case of an ordinary rigid body; in that case it was the partition of three-dimensional space into spheres with center 0 and the point 0 itself. In the general case, the orbits can have different dimensions, and the partition into orbits at some points may not be a fibering; such a singularity already appeared in the three-dimensional case at the point 0.

We call a point M of the space of angular momenta a *regular point* if the partition of a neighborhood of M into orbits is diffeomorphic to a partition of euclidean space into parallel planes (in particular, all orbits near the point M have the same dimension). For example, for the group of rotations of three-dimensional space all points of the space of angular momenta are regular except the origin.

Theorem 8. *Suppose that a regular point M of the space of angular momenta is a critical point of the energy on an orbit of the co-adjoint representation, and that the second differential of the energy d^2H at this point is a (positive- or negative-)definite form. Then M is a (Liapunov) stable equilibrium position of Euler's equations.*

PROOF. It follows from the regularity of the orbits near this point that on every neighboring orbit there exists near M a point which is a conditional maximum or minimum of energy. $\square$

Theorem 9. *The second differential of the kinetic energy, restricted to the image of an orbit of the co-adjoint representation in the algebra, is given at a critical point $\omega \in \mathfrak{g}$ by the formula*

$$2d^2 H|_{\omega(\xi)} = \langle B(\omega, f), B(\omega, f) \rangle + \langle [f, \omega], B(\omega, f) \rangle,$$

where ξ is a tangent vector to this image, expressed in terms of f by the formula

$$\xi = B(\omega, f), \qquad f \in \mathfrak{g}.$$

F Riemannian curvature of a group with left-invariant metric

Let G be a Lie group provided with the left-invariant metric given by a scalar product $\langle \, , \, \rangle$ in the algebra. We note that the riemannian curvature of the group G at any point is determined by the curvature at the identity (since left translation maps the group to itself isometrically). Therefore, it is sufficient to calculate the curvature for two-dimensional planes lying in the Lie algebra.

Theorem 10. *The curvature of a group in the direction determined by an orthonormal pair of vectors ξ, η in the algebra is given by the formula*

$$K_{\xi, \eta} = \langle \delta, \delta \rangle + 2\langle \alpha, \beta \rangle - 3\langle \alpha, \alpha \rangle - 4\langle B_\xi, B_\eta \rangle,$$

where $2\delta = B(\xi, \eta) + B(\eta, \xi)$, $2\beta = B(\xi, \eta) - B(\eta, \xi)$, $2\alpha = [\xi, \eta]$, $2B_\xi = B(\xi, \xi)$, $2B_\eta = B(\eta, \eta)$, and where B is the operation defined in section B.

The proof is a tedious but straightforward calculation. It is based on the easily verified formula for covariant derivative

$$(\nabla_\xi \eta)_e = \tfrac{1}{2}([\xi, \eta] - B(\xi, \eta) - B(\eta, \xi)),$$

where ξ and η on the left are left-invariant vector fields and on the right are their values at the identity.

Remark 1. In the case of a two-sided invariant metric, the formula for curvature has the particularly simple form

$$K_{\xi, \eta} = \tfrac{1}{4}\langle [\xi, \eta], [\xi, \eta] \rangle.$$

Remark 2. The formula for the curvature of a group with a right-invariant riemannian metric coincides with the formula for the left-invariant case. In fact, a right-invariant metric on a group is a left-invariant metric on the group with the reverse multiplication law ($g_1 * g_2 = g_2 g_1$). Passage to the reverse group changes the signs of both the commutator and the operation B in the algebra. But, in every term of the formula for curvature, there is a product of two operations changing the sign. Therefore, the formula for curvature is the same in the right-invariant case.

In Euler's equation the right-hand side changes sign under passage to the right-invariant case.

G Application to groups of diffeomorphisms

Let D be a bounded region in a riemannian manifold. Consider the group of diffeomorphisms of D which preserve the volume element. We will denote this group by $SDiff\,D$.

The Lie algebra corresponding to the group $SDiff\,D$ consists of all vector fields with divergence 0 on D, tangent to the boundary (if it is not empty). We define the scalar product of two elements of this Lie algebra (i.e., two vector fields) as

$$\langle v_1, v_2 \rangle = \int_D (v_1 \cdot v_2) dx,$$

where $(\cdot)$ is the scalar product giving the riemannian metric on D, and dx is the riemannian volume element.

We now consider the flow of a uniform ideal (incompressible, non-viscous) fluid on the region D. Such a flow is described by a curve $t \to g_t$ in the group $SDiff\,D$. Namely, the diffeomorphism g_t is the map which carries every particle of the fluid from the place it was at time 0 to the place it is at time t. It turns out that the kinetic energy of the moving fluid is a right-invariant riemannian metric on the group of diffeomorphisms $SDiff\,D$.

Indeed, suppose that after time t the flow of the fluid gives a diffeomorphism g_t, and that the velocity at this moment of time is given by the vector field v. Then the diffeomorphism realized by the flow after time $t + \tau$ (where τ is small) will be $e^{v\tau}g_t$ up to a quantity small in comparison with τ (here $e^{v\tau}$ is the one-parameter group with vector v, i.e., the phase flow of the differential equation given by the field v). Therefore, the field of velocities v is obtained from the vector $\dot{g}$ tangent to the group at the point g by *right* translation. This also implies the right-invariance of the kinetic energy, which is by definition equal to

$$T = \tfrac{1}{2}\langle v, v \rangle$$

(we assume the density of the fluid to be 1).

The principle of least action (which in mathematical terms is the definition of an ideal fluid) asserts that flows of an ideal fluid are geodesics in the right-invariant metric just described on the group of diffeomorphisms.

Strictly speaking, an infinite-dimensional group of diffeomorphisms is not a manifold. Therefore the exact formulation of the definition above requires additional work: we must choose suitable functional spaces, prove a theorem on existence and uniqueness of solutions, etc. Up to now this has been done only in the case when the dimension of the region of the flow D is equal to 2. However, we will proceed as if these difficulties connected with infinite dimensions did not exist. Thus the following arguments are heuristic in character. It turns out that many of the results can be proved rigorously, independently of the theory of infinite-dimensional manifolds.

We will now indicate the form that the general formulas introduced above take in the case $G = SDiff\,D$, where D is a connected region with finite volume in a three-dimensional riemannian manifold. To do this we must first describe explicitly the bilinear operation $B: \mathfrak{g} \times \mathfrak{g} \to \mathfrak{g}$ defined in section B by the formula

$$\langle [a, b], c \rangle \equiv \langle B(c, a), b \rangle.$$

It is easy to verify that in the three-dimensional case the vector field $B(c, a)$ can be expressed in terms of the vector fields a and c of our Lie algebra by the formula

$$B(c, a) = (\text{curl } c) \wedge a + \text{grad } \alpha,$$

where $\wedge$ denotes the vector product, and α the single-valued function on D which is uniquely (up to a constant summand) determined by the condition $B \in \mathfrak{g}$ (i.e., the conditions div $B = 0$ and B is tangent to the boundary of D).

We note that the operation B does not depend on the choice of orientation, since the vector product and curl both change sign with a change of orientation.

Stationary flows. Euler's equation for "angular velocity" in the case $G = SDiff D$ has the form $\dot{v} = -B(v, v)$, since the metric is right-invariant. Therefore, in the case of the group of diffeomorphisms of three-dimensional space, it takes the form of "the equations of motion in Bernoulli's form"

$$\frac{\partial v}{\partial t} = v \wedge \text{curl } v + \text{grad } \alpha, \qquad \text{div } v = 0.$$

Euler's equation for momentum is written in the form of the "vorticity equation"

$$\frac{\partial \text{ curl } v}{\partial t} = [v, \text{curl } v].$$

In particular, the *vorticity of a stationary flow commutes with the field of velocities.*

This remark leads quickly to a topological classification of stationary flows of an ideal fluid in three-dimensional space.

Theorem 11. *Assume that the region D is bounded by a compact analytic surface, and that the field of velocities is analytic and not everywhere collinear with its curl. Then the region of the flow can be partitioned by an analytic submanifold into a finite number of cells, in each of which the flow is constructed in a standard way. Namely, the cells are of two types: those fibered into tori invariant under the flow and those fibered into surfaces invariant under the flow, diffeomorphic to the annulus $\mathbb{R} \times S^1$. On each of these tori the flow lines are either all closed or all dense, and on each annulus all the flow lines are closed.*

To prove this theorem we look at the "Bernoulli surfaces," i.e., the level surfaces of the function α. It follows from the condition for a flow to be stationary ($v \wedge \text{curl } v = -\text{grad } \alpha$) that both the flow lines and the vortex lines lie on the Bernoulli surface. Since the fields of velocity and vorticity commute, the group $\mathbb{R}^2$ acts on the closed Bernoulli surface, and it must be a torus (cf. the proof of Liouville's theorem in Section 49). An analogous calculation for the boundary conditions on the boundary of D shows that the non-closed Bernoulli surfaces consist of annuli with closed flow lines.

Remark. The analyticity of the field of velocities is not very essential, but it is important that the fields of velocity and vorticity not be collinear. Computer experiments conducted by M. Hénon show more complicated behavior than described in the theorem for the flow lines of a stationary flow on the three-dimensional torus; this field is given by the formulas

$$v_x = A \sin z + C \cos y \qquad v_y = B \sin x + A \cos z,$$

$$v_z = C \sin y + B \cos x.$$

The formulas are selected so that the vectors v and curl v are collinear. The results of Hénon's calculations suggest that some flow lines densely fill up a three-dimensional region.

I Isovortical fields

Two-dimensional hydrodynamics differs sharply from three-dimensional hydrodynamics. The essence of this difference is contained in the difference in the geometries of the orbits of the co-adjoint representation in the two- and three-dimensional cases. In the two-dimensional case the orbits are in some sense closed and behave, for example, like a family of level sets of a function (more precisely of several functions: actually even an infinite number of functions). In the three-dimensional case the orbits are more complicated; in particular, they are unbounded (and perhaps dense). The orbits of the co-adjoint representation of the group of diffeomorphisms of a three-dimensional riemannian manifold can be described in the following way. Let v_1 and v_2 be two vector fields of velocities of an incompressible fluid in the region D. We say that the fields v_1 and v_2 are *isovortical* if there is volume-preserving diffeomorphism $g: D \to D$ which carries every closed contour γ in D to a new contour such that the circulation of the first field along the original contour is equal to the circulation of the second field along the new contour:

$$\oint_\gamma v_1 = \oint_{g\gamma} v_2.$$

It is easy to verify that the image of an orbit of the co-adjoint representation in the algebra (under the action of the inverse of the inertia operator, A^{-1}) is none other than the set of fields *isovortical* to the given field.

In particular, Theorem 3 now takes the form of the following *law of conservation of circulation*:

Theorem 12. *The circulation of a field of velocities of an ideal fluid over a closed fluid contour does not change when the contour is carried by the flow to a new position.*

We note that if two fields of velocities of a three-dimensional ideal fluid on D are isovorticial, then the corresponding diffeomorphism carries the curl of the first field into the curl of the second:

$$g_* \text{ curl } v_1 = \text{curl } v_2.$$

Furthermore, the isovorticity of two fields can be defined as the equivalence of the fields of vorticity, *if the region of the flow is simply connected*. Therefore, the problem of the oribits of the co-adjoint representation in the three-dimensional case includes the problem of classifying *vector fields* with divergence zero up to volume-preserving diffeomorphisms. This last problem in three dimensions is hopelessly difficult.

We now consider the two-dimensional case. First, we translate the basic formulas into notation convenient for considering the two-dimensional case. We assume that the region D of the flow is two-dimensional and oriented. The metric and orientation give a symplectic structure on D; the vector field of velocities has divergence zero and is therefore hamiltonian. Therefore, this field is given by a hamiltonian function (many-valued, in general, if the region D is not simply connected). The hamiltonian function of a field of velocities is called the *stream function* in hydrodynamics, and is denoted by ψ. Thus

$$v = I \text{ grad } \psi,$$

where I is the operator of clockwise rotation by 90°.

The stream function of the commutator of two fields turns out to be the jacobian (or the Poisson bracket of hamiltonian formalism) of the stream functions of the original fields

$$\psi_{[v_1, v_2]} = J(\psi_1, \psi_2).$$

The vector field $B(c, a)$ is given, in the two-dimensional case, by the formula

$$B = -(\Delta\psi_c)\text{grad } \psi_a + \text{grad } \alpha,$$

where ψ_a and ψ_c are the stream functions of the fields a and c, and $\Delta = \text{div grad}$ is the laplacian.

In the particular case of the euclidean plane with cartesian coordinates x and y, the formulas for stream function, commutator and laplacian take the particularly simple form

$$v_x = \frac{\partial \psi}{\partial y} \qquad v_y = -\frac{\partial \psi}{\partial x}$$

$$\psi_{[v_1, v_2]} = \frac{\partial \psi_{v_1}}{\partial x}\frac{\partial \psi_{v_2}}{\partial y} - \frac{\partial \psi_{v_1}}{\partial y}\frac{\partial \psi_{v_2}}{\partial x}$$

$$\Delta = \frac{\partial^2}{\partial x^2} + \frac{\partial^2}{\partial y^2}.$$

The vorticity (or curl) of a two-dimensional field of velocities is the scalar function r such that the integral around any oriented region σ in D of the product of r with the oriented area element is equal to the circulation of the field of velocities around the boundary of σ:

$$\int_\sigma r \, dS = \oint_{\partial\sigma} v.$$

It is easy to compute an expression for the vorticity in terms of the stream function:

$$r = -\Delta\psi.$$

In the two-dimensional simply connected case, isovorticity of fields v_1 and v_2 means simply that the functions r_1 and r_2 (the vorticities of these fields) are carried to one another under a suitable volume-preserving diffeomorphism.

Under such conditions the two functions r_1 and r_2 have the same distribution function, i.e.,

$$\text{mes}\{x \in D : r_1(x) \leq c\} = \text{mes}\{x \in D : r_2(x) \leq c\},$$

for any number c. Therefore, if two fields are in the image of the same orbit of the co-adjoint representation, then a whole series of functionals are equal; for example, the integrals of all powers of the vorticity

$$\int_D r_1^k \, dS = \int_D r_2^k \, dS.$$

In particular, Euler's equations of motion of a two-dimensional ideal fluid

$$\frac{\partial v}{\partial t} + v\nabla v = -\text{grad} \; p \qquad \text{div } v = 0,$$

have an infinite collection of first integrals. For example, the integral of any power of the vorticity of the field of velocities

$$I_k = \iint_D \left(\frac{\partial v_2}{\partial x} - \frac{\partial v_1}{\partial y} \right)^k dx \wedge dy$$

is such a first integral.

The existence of these first integrals (i.e., the relatively simple structure of orbits of the co-adjoint representation) allows us to prove theorems on existence and uniqueness, etc. in the two-dimensional hydrodynamics of an ideal (and also of a viscous) fluid; the complicated geometry of orbits of the co-adjoint representation in the three-dimensional case (or, perhaps, insufficient information about these orbits) makes the foundations of three-dimensional hydrodynamics a very hard problem.

J *Stability of planar stationary flows*

Here we formulate general theorems about stationary rotations (Theorems 7, 8, and 9 above) for the case of a group of diffeomorphisms. We obtain in this way the following assertions:

1. A stationary flow of an ideal fluid is distinguished from all flows iso-vorticial to it by the fact that it is a conditional extremum (or critical point) of the kinetic energy.
2. If (i) the indicated critical point is actually an extremum, i.e., a local conditional maximum or minimum, (ii) it satisfies certain (generally satisfied) regularity conditions, and (iii) the extremum is non-degenerate (the second differential is positive- or negative-definite), then the stationary flow is stable (i.e., is a Liapunov stable equilibrium position of Euler's equation).
3. The formula for the second differential of the kinetic energy, on the tangent space to the manifold of fields which are isovorticial to a given one, has the following form in the two-dimensional case. Let D be a region in the euclidean plane with cartesian coordinates x and y. Consider a stationary flow with stream function $\psi = \psi(x, y)$. Then $2\,d^2H = \iint_D (\delta v)^2 + (\Delta\psi/\nabla\Delta\psi)(\delta r)^2\,dx\,dy$, where δv is the variation of the field of velocities (i.e., a vector of the tangent space indicated above), and $\delta r = \operatorname{curl} \delta v$.

We note that for a stationary flow, the gradient vectors of the stream function and its laplacian are collinear. Therefore the ratio $\nabla\psi/\nabla\Delta\psi$ makes sense. Furthermore, in a neighborhood of every point where the gradient of the vorticity is not zero, the stream function is a function of the vorticity function.

The assertions introduced above lead to the conclusion that the positive- or negative-definiteness of the quadratic form d^2H is a sufficient condition for stability of the stationary flow under consideration. This conclusion does not formally follow from Theorems 7, 8, and 9 since the application of any of our formulas in the infinite-dimensional case requires justification. Fortunately, we can justify the final conclusion about stability without justifying the intermediate constructions. Thus we can rigorously prove the following a priori bounds (expressing the stability of a stationary flow in terms of small perturbations of the initial velocity field).

Theorem 13. *Suppose that the stream function of a stationary flow, $\psi = \psi(x, y)$, in a region D is a function of the vorticity function (i.e., of the function $\Delta\psi$) not only locally, but globally. Suppose that the derivative of the stream function with respect to the vorticity satisfies the inequality*

$$c \leq \frac{\nabla\psi}{\nabla\Delta\psi} \leq C, \quad \text{where } 0 < c \leq C < \infty.$$

Appendix 2: Geodesics of left-invariant metrics on Lie groups

Let $\psi + \varphi(x, y, t)$ be the stream function of another flow, not necessarily stationary. Assume that, at the initial moment, the circulation of the velocity field of the perturbed flow (with flow function $\psi + \varphi$) around every boundary component of the region D is equal to the circulation of the original flow (with stream function ψ). Then the perturbation $\varphi = \varphi(x, y, t)$ at every moment of time is bounded in terms of the initial perturbation $\varphi_0 = \varphi(x, y, 0)$ by the formula

$$\iint_D (\nabla\varphi)^2 + c(\Delta\varphi)^2 \; dx \, dy \leq \iint_D (\nabla\varphi_0)^2 + C(\Delta\varphi_0)^2 \; dx \, dy.$$

If the stationary flow satisfies the inequality

$$c \leq -\frac{\nabla\psi}{\nabla\Delta\psi} \leq C, \qquad 0 < c < C < \infty,$$

then the perturbation φ is bounded in terms of φ_0 by the formula

$$\iint_D c(\Delta\varphi)^2 - (\nabla\varphi)^2 \; dx \, dy \leq \iint_D C(\Delta\varphi_0)^2 - (\nabla\varphi_0)^2 \; dx \, dy.$$

This theorem implies the stability of a stationary flow in the case of a positive-definite quadratic form

$$\iint_D (\nabla\varphi)^2 + \frac{\nabla\psi}{\nabla\Delta\psi}(\Delta\varphi)^2 \; dx \, dy$$

with respect to $\nabla\varphi$ (where φ is a constant function on every component of the boundary of D whose gradient flow is zero over every boundary component), and also in the case of a negative definite form

$$\iint_D (\nabla\varphi)^2 + \left(\max \frac{\nabla\psi}{\Delta\nabla\psi}\right)(\Delta\varphi)^2 \; dx \, dy.$$

EXAMPLE 1. Consider a planar parallel flow in the strip $Y_1 \leq y \leq Y_2$ in the (x, y)-plane with velocity profile $v(y)$ (i.e., with velocity field $(v(y), 0)$). Such a flow is stationary for any velocity profile. To make the region of the flow compact, we impose the condition that the velocity fields of all flows under consideration be periodic with period X in the x-coordinate.

The conditions of Theorem 13 are fulfilled if the velocity profile has no points of inflection (i.e., if $d^2v/dy^2 \neq 0$). We come to the conclusion that planar parallel flows of an ideal fluid with no inflection points in the velocity profile are stable.

The analogous proposition in the linearized problem is called *Rayleigh's theorem*.

336

We emphasize that in Theorem 13 it is not a question of stability "in a linear approximation," but of actual strict Liapunov stability (i.e., with respect to finite perturbations in the nonlinear problem). The difference between these two forms of stability is substantial in this case, since our problem has a hamiltonian character (cf. Theorem 4); for hamiltonian systems asymptotic stability is impossible, so stability in a linear approximation is always neutral and insufficient for a conclusion about the stability of an equilibrium position of the nonlinear problem.

EXAMPLE 2. Consider the planar-parallel flow on the torus

$$\{(x, y), x \bmod X, y \bmod 2\pi\}$$

with velocity field $v = (\sin y, 0)$, parallel to the x-axis. This field is determined by the stream function $\psi = -\cos y$ and has vorticity $r = -\cos y$. The velocity profile has two inflection points, but the stream function can be expressed as a function of the vorticity. The ratio $\nabla\psi/\nabla\Delta\psi$ is equal to minus one. By applying Theorem 13 we can convince ourselves of the stability of our stationary flow in the case when

$$\int_0^{2\pi} \int_0^X (\Delta\varphi)^2 \, dx \, dy \geq \int_0^{2\pi} \int_0^X (\nabla\varphi)^2 \, dx \, dy$$

for all functions φ of period X in x and 2π in y. It is easy to calculate that the last inequality is satisfied for $X \leq 2\pi$ and violated for $X > 2\pi$.

Thus Theorem 13 implies the stability of a sinusoidal stationary flow on a short torus, when the period in the direction of the basic flow (X) is less than the width of the flow (2π). On the other hand, we can directly verify that on a long torus (for $X > 2\pi$) our sinusoidal flow is unstable.[97] Thus, in this example, the sufficient condition for stability from Theorem 13 turns out to be necessary.

We should note that in general an indefinite quadratic form d^2H does not imply instability of the corresponding flow. In general, an equilibrium position of a hamiltonian system can be stable even though the hamiltonian function at this position is neither a maximum nor a minimum. The quadratic hamiltonian $H = p_1^2 + q_1^2 - p_2^2 - q_2^2$ is the simplest example of this kind.

K Riemannian curvature of a group of diffeomorphisms

The expression for the curvature of a Lie group provided with a one-sided-invariant metric, introduced in subsection E, makes sense also for the group $SDiff\,D$ of diffeomorphisms of a riemannian domain D. This group is the configuration space for an ideal fluid filling the domain D. The kinetic energy defines a right-invariant metric on $SDiff\,D$. The number which we obtain by formally applying the formula for the curvature of a Lie group to

[97] Cf., for example, the article of L. D. Meshalkin and Ya. G. Sinai, "Investigation of the stability of a stationary solution of a system of equations for the plane movement of an incompressible viscous liquid." *J. Applied Math. Mech.* **25** (1962), 1700–1705.

this infinite-dimensional group is naturally called the curvature of the group *SDiff D*.

Calculation of the curvature of a group of diffeomorphisms has been carried out completely only in the case of a flow on the two-dimensional torus with euclidean metric. Such a torus is obtained from the euclidean plane $\mathbb{R}^2$ by identifying points whose difference lies in some lattice (a discrete subgroup of the plane). An example of such a lattice is the set of points with integral coordinates. In general, to obtain an arbitrary lattice Γ we may replace the square lying at the basis of this special lattice by any parallelogram.

Now consider the Lie algebra of vector fields with divergence zero on the torus with a single-valued stream function. The corresponding group $S_0 Diff\ T^2$ consists of volume-preserving diffeomorphisms which leave the center of mass of the torus fixed. It is embedded in the group $SDiff\ T^2$ of all volume-preserving diffeomorphisms as a totally geodesic submanifold (i.e., a submanifold such that each of its geodesics is a geodesic in the ambient manifold).

The proof consists of the fact that if, at the initial moment, a velocity field of an ideal fluid has a single-valued stream function, then at all other moments of time the stream function will also be single-valued; this follows from the law of conservation of momentum.

We will now investigate the curvature of the group $S_0 Diff\ T^2$ in all possible two-dimensional directions passing through the identity of the group (the curvature of the group $SDiff\ T^2$ in every such direction is the same, since the submanifold $S_0 Diff\ T^2$ is totally geodesic).

Choose an orientation on $\mathbb{R}^2$. Then elements of the Lie algebra of the group $S_0 Diff\ T^2$ can be thought of as real functions on the torus having average value zero (a field with divergence zero is obtained from such a function by considering it to be a stream function). Therefore, a two-dimensional direction in the tangent space to the group $S_0 Diff\ T^2$ is determined by a pair of functions on the torus with average value zero.

We will give such a function by the set of its Fourier coefficients. It is convenient to carry out all calculations with Fourier series in the complex domain. We let e_k (where k, called a wave vector, is a point of the euclidean plane) denote the function whose value at a point x of our plane is equal to $e^{i(k, x)}$. Such a function determines a function on the torus if it is Γ-periodic, i.e., if adding a vector from the lattice Γ to x does not change the value of the function.

In other words, the scalar product (k, x) must be a multiple of 2π for all $x \in \Gamma$. All such vectors k belong to a lattice Γ^* on $\mathbb{R}^2$. The functions e_k, where $k \in \Gamma^*$, form a complete system in the space of complex functions on the torus.

We now complexify our Lie algebra, scalar product $\langle\ ,\ \rangle$, commutator $[\ ,\ ]$ and operation B in the algebra, as well as the riemannian connection and curvature tensor Ω, so that all these functions become (multi-) linear in the complex vector space of the complexified Lie algebra. The functions e_k (where $k \in \Gamma^*$, $k \neq 0$) form a basis of this vector space.

Theorem 14. *The explicit formulas for the scalar product, commutator, opera-tion B, connection, and curvature of a right-invariant metric on the group $S_0 \, Diff \, T^2$ have the following form:*

$$\langle e_k, e_l \rangle = 0 \quad for \; k + l \neq 0,$$

$$\langle e_k, e_{-k} \rangle = k^2 S;$$

$$[e_k, e_l] = (k \wedge l) e_{k+l};$$

$$B(e_k, e_l) = b_{k,l} e_{k+l}, \quad where \; b_{k,l} = (k \wedge l) \frac{k^2}{(k+l)^2};$$

$$\nabla_{e_k} e_l = d_{l,k+l} e_{k+l}, \quad where \; d_{u,v} = \frac{(v \wedge u)(u \cdot v)}{v^2};$$

$R_{k,l,m,n} = 0 \; if \; k + l + m + n \neq 0; \; if \; k + l + m + n = 0, \; then \; R_{k,l,m,n} = (a_{ln} a_{km} - a_{lm} a_{kn}) S, \; where \; a_{uv} = (u \wedge v)^2 / |u + v|.$

In these formulas, S is the area of the torus, and $u \wedge v$ the area of the parallelogram spanned by u and v (with respect to the chosen orientation of $\mathbb{R}^2$). The parentheses denote the euclidean scalar product in the plane, and angled brackets denote the scalar product in the Lie algebra.

The proof of this theorem is in the first article listed in the introduction to this appendix.

The formulas above allow us to calculate the curvature in any two-dimensional direction. These calculations show that in most directions the curvature is negative, but in a few it is positive. Consider, for instance, some fluid flow, i.e. a geodesic of our group. By Jacobi's equations, the stability of this geodesic is determined by the curvatures in the directions of all possible two-dimensional planes passing through the velocity vector of the geodesic at each of its points.

Assume now that the flow under consideration is stationary. Then the geodesic is a one-parameter subgroup of our group. From this it follows that the curvatures in the directions of all planes passing through velocity vectors of the geodesic at all of its points are equal to the curvatures in the corresponding planes going through the velocity vector of this geodesic at the initial moment of time (Proof: right translate to the identity element of the group). Thus the stability of a stationary flow depends only on the curvatures in the directions of those two-dimensional planes in the Lie algebra which contain the vector of the Lie algebra which is the velocity field of the stationary flow.

Consider, for example, the simplest parallel sinusoidal stationary flow. Such a flow is given by the stream function

$$\xi = \frac{e_k + e_{-k}}{2}.$$

Consider any other real vector of the algebra, $\eta = \sum x_l e_l$ (so $x_{-l} = \bar{x}_l$). We deduce easily from Theorem 14 that

Theorem 15. *The curvature of the group $S_0 Diff\, T^2$ in any two-dimensional plane containing the direction ξ is non-positive. Namely,*

$$\langle \Omega(\xi, \eta) \rangle = -\frac{S}{4} \sum_l a_{k,l}^2 |x_l + x_{l+2k}|^2.$$

From this formula it follows, in particular, that

1. The curvature is equal to zero only for those two-dimensional planes which consist of parallel flows in the same direction as ξ, so that $[\xi, \eta] = 0$;
2. The curvature in the plane defined by the flow functions $\xi = \cos kx$, $\eta = \cos lx$ is

$$K = \frac{k^2 + l^2}{4S} \sin^2 \alpha \sin^2 \beta,$$

   where S is the area of the torus, α is the angle between k and l, and β is the angle between $k + l$ and $k - l$;
3. In particular, the curvature of the group of diffeomorphisms of the torus $\{(x, y) \bmod 2\pi\}$ in directions determined by the velocity fields $(\sin y, 0)$ $(0, \sin x)$ is equal to

$$K = \frac{-1}{8\pi^2}.$$

L *Discussion*

It is natural to expect that the curvature of a group of diffeomorphisms is related to the stability of geodesics in this group (i.e. to the stability of flows of an ideal fluid) in the same way as the curvature of a finite-dimensional Lie group is related to the stability of geodesics on it. Namely, negative curvature causes exponential instability of geodesics. The characteristic path length (the average path length in which errors in the initial conditions grow e times) has order of magnitude $1/\sqrt{-K}$. Thus, knowing the curvatures of a group of diffeomorphisms allows us to estimate the time for which we can predict the development of the flow of an ideal fluid by means of an approximate initial velocity field before the error grows to a large order.

It should be emphasized that instability of a flow of an ideal fluid is here understood differently than in section K; it is a question of exponential instability of the *motion of the fluid*, not of its velocity field. It is possible for a stationary flow to be a Liapunov stable solution of Euler's equation while the corresponding motion of the fluid is exponentially unstable. The reason is that a small change in the velocity field of a fluid can induce an exponentially growing change in the motion of the fluid. In such a case (stability of the solution of Euler's equation and negative curvature of the group) we can predict the velocity field, but we cannot predict the motion of the fluid mass without a great loss of accuracy.

The formulas mentioned above for curvature can be used even for rough estimates of the time over which a long-term dynamical prediction of the weather is impossible, if we agree to a few simplifying assumptions. These simplifying assumptions consist of the following:

1. The earth has the shape of a torus obtained by factoring the plane by a square lattice.
2. The atmosphere is a two-dimensional homogeneous incompressible inviscid fluid.
3. The motion of the atmosphere is approximately a "tradewind current," parallel to the equator of the torus and having sinusoidal velocity profile.

To calculate the characteristic path length we must then estimate the curvature of the group $S_0 Diff T^2$ in directions containing the "tradewind current" ξ from Theorem 15. To do this we will look at T^2 as $\{(x, y) \bmod 2\pi\}$, $k = (0, 1)$. In other words, we look at 2π-periodic flows on the (x, y)-plane close to a stationary flow, parallel to the x-axis and with sinusoidal velocity profile

$$v = (\sin y, 0).$$

It is easy to see from the formula in Theorem 15 that the curvature of the group $S_0 Diff T^2$ in the planes containing our tradewind current v varies within the limits

$$-\frac{2}{S} < K < 0, \quad \text{where } S = 4\pi^2 \text{ is the area of the torus.}$$

Here the lower limit is obtained by a rather crude estimate. However, a direction with curvature $K = -1/2S$ certainty exists, and there are many other directions with curvature of approximately the same size. In order to make a rough estimate of the characteristic path length, we make the rough guess $K_0 = -1/2S$ as value of the "mean curvature."

If we agree to start from this value K_0 of the curvature, we obtain the characteristic path length

$$s = (\sqrt{-K_0})^{-1} = \sqrt{2S}.$$

The velocity of motion with respect to the group which corresponds to our tradewind current is equal to $\sqrt{S/2}$ (since the average square value of the sine is $\frac{1}{2}$). *Therefore, the time it takes for our flow to travel the characteristic path length is equal to 2. The fastest particles of the fluid go a distance of 2 after this time, i.e., $1/\pi$ of the entire orbit around the torus.*

Thus, if we take our value of the mean curvature, then the error grows by $e^\pi \approx 20$ after the time of one orbit of the fastest particle. Taking the value 100 km/hr as the maximal velocity of the tradewind current, we get 400 hours for the time of orbit, i.e., less than three weeks.

Thus, if at the initial moment the state of the weather was known with small error ε, then the order of magnitude of the error of prediction after n months would be

$$10^{kn}\varepsilon, \quad \text{where } k \approx \frac{30 \cdot 24}{400} \pi \log_{10} e \approx 2.5.$$

For example, to predict the weather two months in advance we must have initial data with five more digits of accuracy than the prediction accuracy. Practically, this means that calculating the weather for such a period is impossible.

It is clear that the estimates mentioned here are not very sharp, and the model we took is very simplified. The choice of the value of "mean curvature" also requires justification.

Appendix 3: Symplectic structures on algebraic manifolds

The symplectic manifolds of classical mechanics are most often phase spaces of lagrangian mechanical systems, i.e., cotangent bundles of configuration spaces.

An entirely different series of symplectic manifolds arises in algebraic geometry.

For example, any smooth complex algebraic manifold (given by a system of polynomial equations in complex projective space) has a natural symplectic structure.

The construction of a symplectic structure on an algebraic manifold is based on the fact that complex projective space itself has a particular symplectic structure, namely the imaginary part of its hermitian structure.

A *The hermitian structure of complex projective space*

Recall that n-dimensional complex projective space $\mathbb{C}P^n$ is the manifold of all complex lines passing through the point 0 in an $(n + 1)$-dimensional complex vector space $\mathbb{C}^{n+1}$. To construct a symplectic structure on $\mathbb{C}P^n$ we use the hermitian structure in the corresponding vector space $\mathbb{C}^{n+1}$.

Recall that a *hermitian scalar product* (or hermitian structure) on a complex vector space is a complex linear function on pairs of vectors, which (1) is linear in the first and anti-linear in the second variable, (2) changes its value to the complex conjugate when the arguments are interchanged, and (3) becomes a positive-definite real quadratic form if we take the arguments equal:

$$\langle \lambda\xi, \eta \rangle = \lambda \langle \xi, \eta \rangle \qquad \langle \eta, \xi \rangle = \overline{\langle \xi, \eta \rangle} \qquad \langle \xi, \xi \rangle > 0$$

for $\xi \neq 0$.

An example of a hermitian scalar product is

$$(1) \qquad \langle \xi, \eta \rangle = \sum \xi_k \bar{\eta}_k.$$

where ξ_k and η_k are the coordinates of the vectors ξ and η in some basis.

A basis for which a hermitian scalar product has the form (1) always exists, and is called a hermitian-orthonormal basis.

The real and imaginary parts of a hermitian scalar product are real bilinear forms. The first is symmetric, and the second skew-symmetric, and both are nondegenerate:

$$\langle \xi, \eta \rangle = (\xi, \eta) + i[\xi, \eta] \qquad (\xi, \eta) = (\eta, \xi) \qquad [\xi, \eta] = -[\eta, \xi].$$

The quadratic form (ξ, ξ) is positive-definite.

Thus a hermitian structure $\langle \, , \, \rangle$ on a complex vector space gives it a euclidean structure $(\, . \,)$ and a symplectic structure $[\, . \,]$. These two structures are related to the complex structure by the relation

$$[\xi, \eta] = (\xi, i\eta).$$

We will now define a riemannian metric on complex projective space. To do this, consider the unit sphere

$$S^{2n+1} = \{z \in \mathbb{C}^{n+1} : \langle z, z \rangle = 1\}$$

in the corresponding vector space $\mathbb{C}^{n+1}$. This sphere inherits the riemannian metric from $\mathbb{C}^{n+1}$. Every complex line intersects our sphere in a great circle.

Definition. The *distance* between two points of complex projective space is the distance between the two corresponding circles on the unit sphere.

We note that these two circles are parallel in the sense that the distance from any point of one of the circles to the other is the same (Proof: multiplication of z by $e^{i\varphi}$ preserves the metric on the sphere). This circumstance allows us at once to write down an explicit formula (2) for the riemannian metric on the complex projective space given by the construction defined above.

In fact, let p denote the mapping

$$p: \mathbb{C}^{n+1}\backslash 0 \to \mathbb{C}P^n,$$

taking a point $z \neq 0$ of the vector space $\mathbb{C}^{n+1}$ to the complex line passing through 0 and z.

Every vector ζ tangent to $\mathbb{C}P^n$ at the point pz can be represented (in many ways) as the image of a vector at the point z; under this map

$$\zeta = p_* \xi, \qquad \xi \in T\mathbb{C}_z^{n+1}.$$

Theorem. *The square of the length of a vector ζ in the riemannian metric defined above is given by the formula*

(2)
$$ds^2(\zeta) = \frac{\langle \xi, \xi \rangle \langle z, z \rangle - \langle \xi, z \rangle \langle z, \xi \rangle}{\langle z, z \rangle^2}.$$

PROOF. Assume first that the point z lies on the unit sphere S^{2n+1}.

Decompose the vector ξ into two components: one in the complex line determined by the vector z and the other in the hermitian-orthogonal direction. Note that hermitian-orthogonal to the vector z means euclidean-orthogonal to the vectors z and iz. The vector z is a euclidean normal vector to the sphere S^{2n+1} at z. The vector iz is a vector tangent to the circle in which the sphere intersects the complex line passing through z. Thus the component η of the vector ξ which is hermitian-orthogonal to z is tangent to the sphere S^{2n+1} and euclidean-orthogonal to the circle in which the sphere intersects the line pz.

By the definition of the metric on $\mathbb{C}P^n$, the riemannian square of the length of the vector ζ is equal to the euclidean square length of the component η of ξ which is hermitian-orthogonal to z.

We calculate the component η of ξ, hermitian-orthogonal to z. We write our decomposition as

$$\xi = cz + \eta, \quad \text{where } \langle \eta, z \rangle = 0.$$

By hermitian multiplication with z, we find

$$\langle \xi, z \rangle = c \langle z, z \rangle,$$

so

$$\eta = \frac{\langle z, z \rangle \xi - \langle \xi, z \rangle z}{\langle z, z \rangle}.$$

Calculating the hermitian square of the vector η, we find $\langle \eta, \eta \rangle = \langle \eta, \xi \rangle$ and

$$\langle \eta, \eta \rangle = \frac{\langle z, z \rangle \langle \xi, \xi \rangle - \langle \xi, z \rangle \langle z, \xi \rangle}{\langle z, z \rangle}.$$

Thus, formula (2) is proved for points z of the unit sphere. The general case follows from looking at the homothetic transformation $z \to z/|z|$. □

Note that our construction allows us to define not only a euclidean structure (2), but also a hermitian structure on the tangent space to $\mathbb{C}P^n$. Consider the hermitian-orthogonal complement H to the direction of the vector z in the space $T\mathbb{C}_z^{n+1}$, where $z \in S^{2n+1}$. The map $p_*: H \to T(\mathbb{C}P^n)_{pz}$ maps H isomorphically (as we showed above) onto the tangent space to $\mathbb{C}P^n$ and carries over the hermitian structure from H.

It is clear that the scalar square defined by this hermitian structure is given by formula (2). Therefore, the formula for the hermitian scalar product in the tangent space to $\mathbb{C}P^n$ can be written down without further calculations:

$$(3) \qquad \langle \zeta_1, \zeta_2 \rangle = \frac{\langle \xi_1, \xi_2 \rangle \langle z, z \rangle - \langle \xi_1, z \rangle \langle z, \xi_2 \rangle}{\langle z. z \rangle^2}$$

for any vectors ξ_1, ξ_2 in $T\mathbb{C}_z^{n+1}$ satisfying the relation $p_*\xi_k = \zeta_k \in T(\mathbb{C}P^n)_{pz}$. We note that in formula (3) the point z does not necessarily lie on the unit sphere.

The euclidean and hermitian structures (2) and (3) constructed on the tangent spaces to $\mathbb{C}P^n$ are not invariant under all projective transformations of the manifold $\mathbb{C}P^n$, but are invariant under those which are given by unitary (preserving the hermitian structure) linear transformations of the vector space $\mathbb{C}^{n+1}$.

B The symplectic structure of complex projective space

We consider the imaginary part of the hermitian form (3), taken with co-efficient $-1/\pi$ (the reason for taking this coefficient is explained in Problem 1, Section C):

$$(4) \qquad \Omega(\zeta_1, \zeta_2) = -\frac{1}{\pi} \operatorname{Im}\langle \zeta_1, \zeta_2 \rangle.$$

Like the imaginary part of any hermitian form, the real bilinear form Ω on the tangent space to complex projective space is skew-symmetric and non-degenerate.

Theorem. *The differential 2-form Ω gives a symplectic structure on complex projective space.*

PROOF. We need only verify that the form Ω is closed.

Consider the exterior derivative $d\Omega$ of the form Ω. This differential 3-form on $\mathbb{C}P^n$ is invariant with respect to mappings induced by unitary transformations of the space $\mathbb{C}^{n+1}$. It follows from this that it is equal to zero.

To see this, we look at a hermitian-orthonormal basis $e_1, \ldots, e_n$ of the tangent space to $\mathbb{C}P^n$ at some point z. Then the vectors $e_1, \ldots, e_n, ie_1, \ldots, ie_n$ form a euclidean-orthonormal $\mathbb{R}$-basis. We will show that the value of the form $d\Omega$ on any triple of these $\mathbb{R}$-basis vectors is equal to zero. (We assume that $n > 1$; for $n = 1$ there is nothing to prove.)

Note that in any triple of $\mathbb{R}$-basis vectors at least one is hermitian-orthogonal to the two others. Denote this vector by e. It is easy to construct a unitary transformation of the space $\mathbb{C}^{n+1}$

345

inducing a motion on $\mathbb{C}P^n$ which fixes the point z and the hermitian-orthogonal complement to e, and changes the direction of e.

The value of the form $d\Omega$ on our three vectors, e, f, and g is equal to its value on the triple $-e$, f, and g by the invariance of the form Ω, and is hence equal to zero. $\qquad\square$

Remark. Another method of constructing the same symplectic structure on complex projective space consists of the following. Consider small oscillations of a mathematical pendulum with an $(n + 1)$-dimensional configuration space. We make use of the integral of energy to decrease by 1 the degree of freedom of the system. The phase space obtained after this operation is $\mathbb{C}P^n$, and the symplectic structure on it agrees with the form Ω described above up to a factor.

One other method of constructing a symplectic structure on $\mathbb{C}P^n$ uses the fact that this space may be represented as one of the orbits of the co-adjoint representation of a Lie group, and on every such orbit there is always a standard symplectic structure (cf. Appendix 2, Section A). For the Lie group we can take the group of unitary (preserving the hermitian metric) operators in an $(n + 1)$-dimensional complex space. The orbits of the co-adjoint representation in this case are the same as of the adjoint representation. In the adjoint representation the operator of reflection through a hyperplane (which changes the sign of the first coordinate and leaves the others fixed) has $\mathbb{C}P^n$ as its orbit, since the reflection operator is uniquely determined by the complex line orthogonal to the hyperplane.

C *Symplectic structure on algebraic manifolds*

We will now obtain a symplectic structure on any complex submanifold M of complex projective space. Let $j: M \to \mathbb{C}P^n$ be an embedding of the complex manifold M into complex projective space. The riemannian, hermitian, and symplectic structures on projective space induce corresponding structures on M. For example, the symplectic structure on M is given by the formula

$$\Omega_M = j^*\Omega.$$

Theorem. *The differential form Ω_M gives a symplectic structure on the manifold M.*

PROOF. The nondegeneracy of the 2-form Ω_M follows from the fact that M is a *complex* submanifold. In fact, the quadratic form

$$(\xi, \xi) = \Omega_M(\xi, i\xi)$$

is positive-definite (it is induced by the riemannian metric on $\mathbb{C}P^n$). Therefore, the bilinear form $(\xi, \eta) = \Omega_M(\xi, i\eta)$ is nondegenerate. This means that the form Ω_M is also nondegenerate. The form Ω_M is closed since the form Ω is closed. $\qquad\square$

Remark. In the same way as for complex projective space, we define a hermitian structure on the tangent spaces of its complex submanifolds; the symplectic structure is the imaginary part.

A complex manifold with a hermitian metric whose imaginary part is a closed form (i.e. a symplectic structure) is called a *Kähler manifold* and its hermitian metric a *Kähler metric*. Many important results have been obtained in the geometry of Kähler manifolds; in particular, they have remarkable topological properties (cf., for example, A. Weil, "Variétés Kählériennes," Hermann, 1958).

Not all symplectic manifolds admit a Kähler structure.

PROBLEM 1. Calculate the symplectic structure Ω in the affine chart $w = z_1 : z_0$ of the projective line $\mathbb{C}P^1$.

ANSWER. $\Omega = (1/\pi)(dx \wedge dy)/(1 + x^2 + y^2)^2$, where $w = x + iy$. The coefficient in the definition of the form Ω is chosen to obtain the usual orientation of the complex line ($dx \wedge dy$) and so that the integral of the form Ω along the whole projective line is equal to 1.

PROBLEM 2. Show that the symplectic structure Ω in the affine chart $w_k = z_k z_0^{-1} (k = 1, \ldots, n)$ of the projective space $\mathbb{C}P^n = \{(z_0 : z_1 : \ldots : z_n)\}$ is given by the formula

$$\Omega = \frac{i}{2\pi} \frac{\sum_{0 \leq k < l \leq n} (w_k \, dw_l - w_l \, dw_k)(\bar{w}_k \, d\bar{w}_l - \bar{w}_l \, d\bar{w}_k)}{(\sum_{k=0}^{n} (w_k \bar{w}_k))^2}.$$

By convention, $w_0 = 1$.

Remark. Differential forms on a complex space with complex values (such as dw_k and $d\bar{w}_k$) are defined as complex linear functions of tangent vectors; if $w_k = x_k + iy_k$, then

$$dw_k = dx_k + i \, dy_k \qquad d\bar{w}_k = dx_k - i \, dy_k.$$

The space of such forms in $\mathbb{C}^n$ has complex dimension $2n$; the $2n$ forms dw_k, $d\bar{w}_k$ $(k = 1, \ldots, n)$, for example, form a $\mathbb{C}$-basis, or the $2n$ forms dx_k, dy_k.

Exterior multiplication is defined in the usual way and obeys the usual rules. For example,

$$dw \wedge d\bar{w} = (dx + idy) \wedge (dx - i \, dy) = -2i \, dx \wedge dy.$$

Let f be a real-smooth function on $\mathbb{C}^n$ (with complex values, in general). An example of such a function is $|w|^2 = \sum w_k \bar{w}_k$. The differential of the function f is a complex 1-form. Therefore, it can be decomposed in the basis dw_k, $d\bar{w}_k$. The coefficients of this decomposition are called the partial derivatives "with respect to w_k" and "with respect to $\bar{w}_k$":

$$df = \frac{\partial f}{\partial w} \, dw + \frac{\partial f}{\partial \bar{w}} \, d\bar{w}.$$

In calculating exterior derivatives it is also convenient to separate into differentiation d' with respect to the variable w and d'' with respect to the variable $\bar{w}$, so that $d = d' + d''$.

For example, for a function f

$$d'f = \frac{\partial f}{\partial w} \, dw \qquad d''f = \frac{\partial f}{\partial \bar{w}} \, d\bar{w}.$$

For the differential 1-form

$$\omega = \sum a_k \, dw_k + b_k \, d\bar{w}_k,$$

Appendix 3: Symplectic structures on algebraic manifolds

the operators d' and d'' are defined analogously:

$$d'\omega = \sum d'a_k \wedge dw_k + d'b_k \wedge d\bar{w}_k$$
$$d''\omega = \sum d''a_k \wedge dw_k + d''b_k \wedge d\bar{w}_k.$$

PROBLEM 3. Show that the symplectic structure Ω on the affine chart $(w_k = z_k z_0^{-1})$ of the projective space $\mathbb{C}P^n$ is given by the formula

$$\Omega = \frac{i}{2\pi} d'd'' \ln \sum_{k=0}^{n} |w_k|^2.$$

Appendix 4: Contact structures

An odd-dimensional manifold cannot admit a symplectic structure. The analogue of a symplectic structure for odd-dimensional manifolds is a little less symmetric, but also a very interesting structure—the contact structure.

The source of symplectic structures in mechanics are phase spaces (i.e., cotangent bundles to configuration manifolds), on which there is always a canonical symplectic structure. The source of contact structures are manifolds of contact elements of configuration spaces.

A contact element to an n-dimensional smooth manifold at some point is an $(n-1)$-dimensional plane tangent to the manifold at that point (i.e., an $(n-1)$-dimensional subspace of the n-dimensional tangent space at that point).

The set of all contact elements of an n-dimensional manifold has a natural smooth manifold structure of dimension $2n - 1$. It turns out that there is an interesting additional "contact structure" on this odd-dimensional manifold (we describe this below).

The manifold of contact elements of a riemannian n-dimensional manifold is closely related to the $(2n - 1)$-dimensional manifold of unit tangent vectors of this riemannian n-dimensional manifold, or to the $(2n - 1)$-dimensional energy level manifold of a point mass moving on the riemannian manifold under inertia. The contact structures on these $(2n - 1)$-dimensional manifolds are closely related to the symplectic structure on the $2n$-dimensional phase space of the point (i.e., the cotangent bundle of the original n-dimensional riemannian manifold).

A Definition of contact structure

Definition. A *contact structure* on a manifold is a smooth field of tangent hyperplanes[98] satisfying a nondegeneracy condition which will be formulated later.

To formulate this condition we examine what a field of hyperplanes looks like in general in a neighborhood of a point in an N-dimensional manifold.

EXAMPLE. Let $N = 2$. Then the manifold is a surface and a field of hyperplanes is a field of straight lines. Such a field in a neighborhood of a point is always constructed very simply, namely, as a field of tangents to a family of parallel lines in a plane. More precisely, one of the basic results of the local theory of ordinary differential equations is that it is possible to change any smooth field of tangent lines on a manifold into a field of tangents to a family of straight lines in euclidean space by using a diffeomorphism in a sufficiently small neighborhood of any point of the manifold.

If $N > 2$, then a hyperplane is not a line, and the question becomes significantly more complicated. For example, most fields of two-dimensional

---

[98] A hyperplane in a vector space is a subspace of dimension 1 less than the dimension of the space (i.e., the zero level set of a linear function which is not identically zero). A tangent hyperplane is a hyperplane in a tangent space.

tangent planes in ordinary three-dimensional space cannot be diffeo-morphically mapped onto a field of parallel planes. The reason is that there exist fields of tangent planes for which it is impossible to find "integral sur-faces," i.e., surfaces which have the prescribed tangent plane at each point.

The nondegeneracy condition for a field of hyperplanes which enters into the definition of contact structure consists of the stipulation that the field of hyperplanes must be maximally far from a field of tangents to a family of hyperplanes. In order to measure this distance, as well as to convince our-selves of the existence of fields without integral hypersurfaces, we must make a few constructions and calculations.[99]

B *Frobenius' integrability condition*

We will consider some point on an N-dimensional manifold and try to construct a surface passing through this point and tangent to a given field of $(N - 1)$-dimensional planes at each point (an integral surface).

To this end we introduce a coordinate system onto a neighborhood of this point so that at the point itself one coordinate surface is tangent to a plane of the field. We will call this plane the horizontal plane, and will call the coordinate axis not lying in it the vertical axis.

Construction of an integral surface. An integral surface, if one exists, is the graph of a function of $N - 1$ variables near the origin. To construct it, we can take some smooth path on the horizontal plane. Then the vertical lines over this path form a two-dimensional surface (cylinder); our field of planes intersects its tangent planes in a field of tangent lines. The integral surface we are looking for, if it exists, intersects this cylinder in an integral curve of the field of lines, starting at the origin. Such an integral curve always exists independent of whether an integral surface exists. Thus we can construct an integral surface over the horizontal plane by moving along smooth curves in the latter.

In order to obtain a smooth integral surface from all the integral curves we need the result of our construction to be independent of the path, deter-mined only by its endpoint. In particular, for a circuit of a closed path in a neighborhood of the origin in the horizontal plane, the integral curve on the cylinder must close up.

It is easy to construct examples of fields of planes for which such closure does not take place and, therefore, for which an integral surface does not exist. Such fields of planes are called *nonintegrable*.

Example of a nonintegrable field of planes. In order to give a field of planes and measure numerically the deviation from closure, we introduce the follow-ing notation. We note first of all that a field of hyperplanes can be given locally by a differential 1-form; a plane in the tangent space gives a 1-form up to

---

[99] From now on, we will omit the prefix "hyper-". If we wish, we may assume that we are in three-dimensional space and a hypersurface is an ordinary surface. The higher-dimensional case is analogous to the three-dimensional case.

multiplication by a nonzero constant. We will choose this constant so that the value of the form on the vertical basic vector is equal to 1.

This condition can be satisfied in some neighborhood of the origin since the plane of the field at zero does not contain the vertical direction. This condition determines the form uniquely (given the field of planes).

A field of planes in ordinary three-space which does not have an integral surface can be given, for example, by the 1-form

$$\omega = x\, dy + dz,$$

where x and y are the horizontal coordinates and z is the vertical. The proof of the fact that this field of planes is nonintegrable will be given below.

Construction of a 2-form measuring nonintegrability. With the help of the form giving the field, we can measure the degree of nonintegrability. This is done using the following construction (Figure 236).

Figure 236 Integral curves constructed for a non-integrable field of planes

Consider a pair of vectors emanating from the origin and lying in the horizontal plane of our coordinate system. Construct a parallelogram on them. We obtain two paths from the origin to the opposite vertex. Over each of these two paths we can construct an integral curve (with two sections) as described above. As a result, in general, there arise two different points over the vertex of the parallelogram opposite to the origin. The difference in the heights of these points is a function of our pair of vectors. This function is skew-symmetric and equal to zero if one of the vectors is equal to zero. Thus the linear part of the Taylor series of this function is zero at zero, and the quadratic part of its Taylor series is a bilinear skew-symmetric form on the horizontal plane.

If the field is integrable, then this 2-form is equal to zero. Therefore, this 2-form can be considered as a measure of the nonintegrability of the field.

The 2-form is well defined. We constructed the 2-form above with the help of coordinates. However, the value of our 2-form on a pair of tangent vectors does not depend on the coordinate system, but only on the 1-form used to give the field.

To convince ourselves of this, it is enough to prove the following.

Theorem. *The 2-form defined above agrees with the exterior derivative of the 1-form ω, $d\omega|_{\omega=0}$, on the null space of ω.*

PROOF. We will show that the difference in the heights of the two points obtained as a result of our two motions along the sides of the parallelogram is the same as the integral of the 1-form ω over the four sides of the parallelogram, up to a quantity small of third order with respect to the sides of the parallelogram.

To this end we note that the height of the rise of an integral curve along any path of length ε emanating from the origin has order ε^2, since at the origin the plane of the field is horizontal. Therefore, the integrals of the 2-form $d\omega$ over all four vertical areas over the sides of the parallelogram bounded by the integral curves and the horizontal plane, have order ε^3 if the sides are of order ε.

The integrals of the form ω along integral curves are exactly equal to zero. Therefore, by Stokes' formula, the increase in height along the integral curve lying over any of the sides of the parallelogram is equal to the integral of the 1-form ω along this side up to a quantity of third-order smallness.

Now the theorem follows directly from the definition of exterior differentiation. □

Some arbitrariness remains in the choice of the 1-form ω which we used to construct our 2-form. Namely, the form ω is defined by the field of planes only up to multiplication by a function f which is never zero. In other words, we could have started with the form $f\omega$. Then we would have obtained the 2-form

$$df\omega = f\,d\omega + df \wedge \omega,$$

which, on our plane, differs from the 2-form $d\omega$ by multiplication by the nonzero number $f(0)$.

Thus the 2-form constructed on the plane of the field is defined invariantly up to multiplication by a nonzero constant.

Condition for integrability of a field of planes

Theorem. *If a field of hyperplanes is integrable, then the 2-form constructed above on a plane of the field is equal to zero. Conversely, if the 2-form constructed on every plane of the field is equal to zero, then the field is integrable.*

PROOF. The first assertion of the theorem is clear by the construction of the 2-form. The proof of the second assertion can be carried out by exactly the same reasoning we used to prove the commutativity of phase flows for which the Poisson bracket of the velocity fields was equal to zero. We can simply refer to this commutativity, applying it to the integral curves arising over the lines of the coordinate directions in the horizontal plane. □

Theorem. *The integrability condition for a field of planes,*

$$d\omega = 0 \quad for \quad \omega = 0$$

is equivalent to the following condition of Frobenius:

$$\omega \wedge d\omega = 0.$$

PROOF. We consider the value of the 3-form above on any three distinct coordinate vectors. Only one of these vectors can be the vertical. Therefore, of all the terms entering into the definition of the value of the exterior product of the three vectors, only one is nonzero: the product of the value of the form ω on the vertical vector with the value of the form $d\omega$ on the pair of horizontal vectors. If the field given by the form is integrable, then the second factor is zero, so our 3-form is zero on arbitrary triples of vectors.

Conversely, if the 3-form is equal to zero for any vectors, then it is equal to zero for any triple of coordinate vectors, of which one is vertical and the other two horizontal. The value of the 3-form on such a triple is equal to the product of the value of ω on the vertical vector with the value of $d\omega$ on the pair of horizontal vectors. The first factor is not zero, so the second must be zero, and thus the form $d\omega$ is zero on a plane of the field. □

C Nondegenerate fields of hyperplanes

Definition. A field of hyperplanes is said to be *nondegenerate at a point* if the rank of the 2-form $d\omega|_{\omega=0}$ in the plane of the field passing through this point is equal to the dimension of the plane.

This means that for any nonzero vector in our plane, we can find another vector in the plane such that the value of the 2-form on this pair of vectors is not zero.

Definition. A field of planes is called *nondegenerate on a manifold* if it is non-degenerate at every point of the manifold.

Note that on an even-dimensional manifold there cannot be a nondegenerate field of hyperplanes; on such a manifold a hyperplane is odd-dimensional, and the rank of every skew-symmetric bilinear form on an odd-dimensional space is less than the dimension of the space (cf. Section 44).

Nondegenerate fields of hyperplanes do exist on odd-dimensional manifolds.

EXAMPLE. Consider a euclidean space of dimension $2m + 1$ with coordinates x, y, and z (where x and y are vectors in an m-dimensional space and z is a number). The 1-form

$$\omega = x\,dy + dz$$

defines a field of hyperplanes. The plane of the field passing through the origin has equation $dz = 0$. We take x and y as coordinates in this hyperplane. Therefore, in this plane of the field our 2-form can be written in the form

$$d\omega|_{\omega=0} = dx \wedge dy = dx_1 \wedge dy_1 + \cdots + dx_m \wedge dy_m.$$

The rank of this form is $2m$, so our field is nondegenerate at the origin, and thus also in a neighborhood of the origin (in fact, this field of planes is nondegenerate at all points of the space).

Now, finally, we can give the definition of a contact structure on a manifold: a contact structure on a manifold is a nondegenerate field of tangent hyperplanes.

D *The manifold of contact elements*

The term "contact structure" stems from the fact that there is always such a structure on a manifold of contact elements of a smooth n-manifold.

Definition. A hyperplane (dimension $n - 1$) tangent to a manifold at some point is called a *contact element*, and this point the *point of contact*.

The set of all contact elements of an n-dimensional manifold has the structure of a smooth manifold of dimension $2n - 1$.

In fact, the set of contact elements with a fixed point of contact is the set of all $(n - 1)$-dimensional subspaces of an n-dimensional vector space, i.e., a projective space of dimension $n - 1$. To give a contact element we must therefore give the n coordinates of the point of contact together with the $n - 1$ coordinates defining a point of an $(n - 1)$-dimensional projective space — $2n - 1$ coordinates in all.

The manifold of all contact elements of an n-dimensional manifold is a fiber bundle whose base is our manifold and whose fiber is $(n - 1)$-dimensional projective space.

Theorem. *The bundle of contact elements is the projectivization of the cotangent bundle: it can be obtained from the cotangent bundle by changing every cotangent n-dimensional vector space into an $(n - 1)$-dimensional projective space (a point of which is a line passing through the origin in the cotangent space).*

PROOF. A contact element is given by a 1-form on the tangent space, for which this element is a zero level set. This form is not zero, and it is determined up to multiplication by a nonzero number. But a form on the tangent space is a vector of the cotangent space. Therefore, a nonzero form on the tangent space, determined up to a multiplication by a nonzero number, is a nonzero vector of the cotangent space, determined up to a multiplication by a nonzero number, i.e., a point of the projectivized cotangent space. □

The contact structure on the manifold of contact elements. In the tangent space to the manifold of contact elements there is a distinguished hyperplane. It is called the *contact hyperplane* and is defined in the following way.

We fix a point of the $(2n - 1)$-dimensional manifold of contact elements on an n-dimensional manifold. We can think of this point as an $(n - 1)$-dimensional plane tangent to the original n-dimensional manifold.

Definition. A tangent vector to the manifold of contact elements at a fixed point belongs to the *contact hyperplane* if its projection onto the n-dimensional manifold lies in the $(n - 1)$-dimensional plane which is the given point of the manifold of contact elements.

In other words, a displacement of a contact element is tangent to the contact hyperplane if the velocity of the point of contact belongs to this contact element, no matter how the element turns.

EXAMPLE. We take some submanifold of our n-dimensional manifold and consider all $(n - 1)$-dimensional planes tangent to it (i.e., contact elements). The set of all such contact elements forms a smooth submanifold of the $(2n - 1)$-dimensional manifold of all contact elements. The dimension of this submanifold is equal to $n - 1$, no matter what the dimension of the original submanifold (which could be $(n - 1)$-dimensional, or have smaller dimension, down to a curve or even a point).

This $(n - 1)$-dimensional submanifold of the $(2n - 1)$-dimensional manifold of all contact elements is tangent at each of its points to the field of contact hyperplanes (by the definition of contact hyperplane). Thus the field of $(2n - 2)$-dimensional contact hyperplanes has an $(n - 1)$-dimensional integral manifold.

PROBLEM. Does this field of planes have integral manifolds of higher dimensions?

ANSWER. No.

PROBLEM. Is it possible to give the field of contact hyperplanes by a differential 1-form on the manifold of all contact elements?

ANSWER. No, even if the underlying n-dimensional manifold is a euclidean space (for example, the ordinary two-plane).

We will show below that the field of contact hyperplanes on the $(2n - 1)$-dimensional manifold of all contact elements of an n-dimensional manifold is nondegenerate. The proof uses the symplectic structure of the cotangent bundle. The manifold of contact elements is related by a simple construction to the space of the cotangent bundle (the projectivization of which is the manifold of contact elements). Moreover, the nondegeneracy of the field of contact planes of the projectivized bundle is closely related to the non-degeneracy of the 2-form giving the symplectic structure of the cotangent bundle.

The construction we are concerned with will be carried out below in a somewhat more general situation. Namely, for any odd-dimensional manifold with a contact structure we can construct its "symplectification"—a symplectic manifold whose dimension is one larger. The inter-relation between these two manifolds—the odd-dimensional contact manifold and the even-dimensional symplectic manifold—is the same as between the manifold of contact elements with its contact structure and the cotangent bundle with its symplectic structure.

E *Symplectification of a contact manifold*

Consider an arbitrary contact manifold, i.e., a manifold of odd dimension N with a nondegenerate field of tangent hyperplanes (of even dimension $N - 1$). We will call these planes *contact planes*. Every contact plane is tangent to the contact manifold at one point. We will call this point the *point of contact*.

Definition. A *contact form* is a linear form on the tangent space at the point of contact of the manifold such that its zero set is the contact plane.

It should be emphasized that the contact form is not a differential form but an algebraic linear form on one tangent space.

Definition. The *symplectification* of a contact manifold is the set of all contact forms on the contact manifold, provided with the structure of a symplectic manifold as defined below.

We note first of all that the set of all contact forms on a contact manifold has a natural structure of a smooth manifold of even dimension $N + 1$. Namely, we can consider the set of all contact forms as the space of a bundle over the original contact manifold. Projection onto the base is the mapping associating the contact form to the point of contact.

The fiber of this bundle is the set of contact forms with a common point of contact. All such forms are obtained from one another by multiplication by a nonzero number (so that they determine the same contact plane). Thus the fiber of our bundle is one-dimensional: it is the line minus a point.

We also note that the group of nonzero real numbers acts on the manifold of all contact forms by the operation of multiplication, i.e., the product of a contact form and a nonzero number is again a contact form. In this way the group acts on our bundle, leaving every fiber fixed (upon multiplication of a form by a number the point of contact is not changed).

Remark. So far we have not used the nondegeneracy of the field of planes. Nondegeneracy is needed only to insure that the manifold obtained by symplectification is symplectic.

EXAMPLE. Consider the manifold (of dimension $2n - 1$) of all contact elements of an n-dimensional smooth manifold. On the manifold of elements there is a field of hyperplanes (which we defined above and called the contact hyperplanes). Therefore, we can symplectify the manifold of contact elements.

As a result of symplectification we obtain a $2n$-dimensional manifold. This manifold is the space of the cotangent bundle of the original n-dimensional manifold without zero vectors. The action by the multiplicative group of real numbers on the fiber reduces to multiplication of vectors of the cotangent space by a number.

On the cotangent bundle there is a distinguished 1-form "$p \, dq$." There is an analogous 1-form on any manifold obtained by symplectification from a contact manifold.

The canonical 1-form on the symplectified space

Definition. The *canonical 1-form* in the symplectified space of a contact manifold is the differential 1-form α whose value on any vector ξ tangent to the symplectified space at some point p (Figure 237) is equal to the value on the projection of the vector ξ onto the tangent plane to the contact manifold of the 1-form on this tangent plane which *is* the point p:

$$\alpha(\xi) = p(\pi_* \xi),$$

where π is the projection of the symplectified space onto the contact manifold.

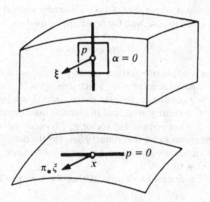

Figure 237   Symplectification of a contact manifold

Theorem. *The exterior derivative of the canonical 1-form on the symplectified space of a contact manifold is a nondegenerate 2-form.*

Corollary. *The symplectified space of a contact manifold has a symplectic structure which is canonically (i.e., uniquely, without arbitrariness) determined by the contact structure of the underlying odd-dimensional manifold.*

PROOF OF THEOREM. Since the assertions of the theorem are local, it is sufficient to prove it in a small neighborhood of a point of the manifold. In a small neighborhood of a point on a contact manifold, a field of contact planes can be given by a differential form ω on the contact manifold. We fix such a 1-form ω.

By the same token we can represent the symplectified space of the contact manifold over our neighborhood as the direct product of the neighborhood and the line minus a point. Namely, we associate to the pair (x, λ) where x is a point of the contact manifold and λ is a nonzero number   the contact form given by the differential 1-form $\lambda\omega$ on the tangent space at the point x. Thus in the part of the symplectified space we are considering, we have defined a function λ

357

whose values are nonzero numbers. It should be emphasized that λ is only a local coordinate on the symplectified manifold and that this coordinate is not defined canonically; it depends on the choice of differential 1-form ω. The canonical 1-form α can be written in our notation as

$$\alpha = \lambda\pi^*\omega$$

and does not depend on the choice of ω. The exterior derivative of the 1-form α thus has the form

$$d\alpha = d\lambda \wedge \pi^*\omega + \lambda\pi^*d\omega.$$

We will show that the 2-form $d\alpha$ is nondegenerate, i.e., that for any vector ξ tangent to the symplectification, we can find a vector η such that $d\alpha(\xi, \eta) \neq 0$. We select from vectors tangent to the symplectification, those of the following type. We call a vector ξ *vertical* if it is tangent to the fiber, i.e., if $\pi_*\xi = 0$. We call the vector ξ *horizontal* if it is tangent to a level surface of the function λ, i.e., if $d\lambda(\xi) = 0$. We call the vector ξ a *contact vector* if its projection onto the contact manifold lies in the contact plane, i.e., if $\omega(\pi_*\xi) = 0$ (in other words, if $\alpha(\xi) = 0$).

We calculate the value of the form $d\alpha$ on a pair of vectors (ξ, η):

$$d\alpha(\xi, \eta) = (d\lambda \wedge \pi^*\omega)(\xi, \eta) + (\lambda\pi^*d\omega)(\xi, \eta).$$

Assume that ξ is not a contact vector. For η, take a nonzero vertical vector, so that $\pi_*\eta = 0$. Then the second term is equal to zero, and the first term is equal to

$$-d\lambda(\eta)\omega(\pi_*\xi)$$

which is not zero since η is a nonzero vertical vector and ξ is not a contact vector. Thus if ξ is not a contact vector, we have found an η for which $d\alpha(\xi, \eta) \neq 0$.

Now assume that ξ is a contact vector and not vertical. Then for η we take any contact vector. Now the first term is entirely zero, and the second (and therefore the sum) is reduced to $\lambda\, d\omega(\pi_*\xi, \pi_*\eta)$. Since ξ is not vertical, the vector $\pi_*\xi$ lying in the contact plane is not zero. But the 2-form $d\omega$ is nondegenerate on the contact plane (by the definition of contact structure). Thus there is a contact vector η such that $d\omega(\pi_*\xi, \pi_*\eta) \neq 0$. Since $\lambda \neq 0$, we have found a vector η for which $d\alpha(\xi, \eta) \neq 0$.

Finally, if the vector ξ is nonzero and vertical, then for η we can take any vector which is not a contact vector. $\qquad\square$

Remark. The constructions of the 1-form α and the 2-form $d\alpha$ are valid for an arbitrary manifold with a field of hyperplanes, and do not depend on the condition of nondegeneracy. However, the 2-form $d\alpha$ will define a symplectic structure only in the case when the field of planes is nondegenerate.

PROOF. Assume that the field is degenerate, i.e., that there exists a nonzero vector ξ' in a plane of the field such that $d\omega(\xi', \eta') = 0$ for all vectors η' in this plane. For such a ξ', the quantity $d\omega(\xi', \eta')$ as a function of η' is a linear form, identically equal to zero on the plane of the field. Therefore there is a number μ not dependent on η' such that

$$d\omega(\xi', \eta') = \mu\omega(\eta')$$

for all vectors η' of the tangent space.

We now take for ξ a tangent vector to the symplectified manifold for which $\pi_*\xi = \xi'$. Such a vector ξ is determined up to addition of a vertical summand, and we will show that for a suitable choice of this summand we will have

$$d\alpha(\xi, \eta) = 0 \quad \text{for all } \eta.$$

The first term of the formula for $d\alpha$ is equal to $d\lambda(\xi)\omega(\pi_* \eta)$ (since $\omega(\pi_* \xi) = 0$). The second term is equal to $\lambda\, d\omega(\pi_* \xi, \pi_* \eta) = \lambda\mu\omega(\pi_* \eta)$. We choose the vertical component of the vector ξ so that $d\lambda(\xi) = -\lambda\mu$. Then ξ will be skew-orthogonal to all vectors η.

Thus if $d\alpha$ is a symplectic structure, then the underlying field of hyperplanes is a contact structure. $\qquad\square$

Corollary. *The field of contact hyperplanes defines a contact structure on the manifold of all contact elements of any smooth manifold.*

PROOF. The symplectification of the $(2n - 1)$-dimensional manifold of all contact elements on an n-dimensional smooth manifold, constructed with help of the field of $(2n - 2)$-dimensional contact planes, is by construction the space of the cotangent bundle of the underlying n-dimensional manifold without the zero cotangent vectors. The canonical 1-form α on the symplectification is, by its definition, the same 1-form on the cotangent bundle that we called "$p\,dq$" and which is fundamental in hamilton mechanics (cf. Section 37). Its derivative $d\alpha$ is therefore the form "$dp \wedge dq$" defining the usual symplectic structure of a phase space. Therefore the form $d\alpha$ is nondegenerate, and, by the preceding remark, the field of contact hyperplanes is nondegenerate. $\qquad\square$

F Contact diffeomorphisms and vector fields

Definition. A diffeomorphism of a contact manifold to itself is called a *contact diffeomorphism* if it preserves the contact structure, i.e., carries every plane of a given structure of a field of hyperplanes to a plane of the same field.

EXAMPLE. Consider the $(2n - 1)$-dimensional manifold of contact elements of an n-dimensional smooth manifold with its usual contact structure. To each contact element we can ascribe a "positive side" by choosing one of the halves into which this element divides the tangent space to the n-dimensional manifold.

We will call a contact element with a chosen side a (*transversally*) *oriented contact element*.

The oriented contact elements on our n-dimensional manifold form a $(2n - 1)$- dimensional smooth manifold with a natural contact structure (it is a double covering of the manifold of ordinary nonoriented contact elements).

Now assume that we are given a riemannian metric on the underlying n-dimensional manifold. Then there is a "geodesic flow"[100] on the manifold of oriented contact elements. The transformation after time t by this flow is defined as follows. We go out from the point of contact of a contact element along the geodesic orthogonal to it and directed to the side orienting the element. In the course of time t we will move the point of contact along the

[100] Strictly speaking, we need to require that the riemannian manifold be complete, i.e., geodesics can be continued without limit.

geodesic, keeping the element orthogonal to the geodesic. After time t we obtain a new oriented element. We have defined the geodesic flow of oriented contact elements.

Theorem. *The geodesic flow of oriented contact elements consists of contact diffeomorphisms.*

The proof of this theorem will not be presented since it is just a reformulation in new terms of Huygens' principle (cf. Section 46).

Definition. A vector field on a contact manifold is called a *contact vector field* if it is the velocity field of a one-parameter (local) group of contact diffeomorphisms.

Theorem. *The Poisson bracket of contact vector fields is a contact vector field. The contact vector fields form a subalgebra in the Lie algebra of all smooth vector fields on a contact manifold.*

The proof follows directly from the definitions.

G Symplectification of contact diffeomorphisms and fields

For every contact diffeomorphism of a contact manifold there is a canonically constructed symplectic diffeomorphism of its symplectification. This symplectic diffeomorphism commutes with the action of the multiplicative group of real numbers on the symplectified manifold and is defined by the following construction.

Recall that a point of the symplectified manifold is a contact form on the underlying contact manifold.

Definition. *The image of a contact form p* with point of contact x under the action of a contact diffeomorphism f of the contact manifold to itself is the form

$$f_! p = (f^*_{f(x)})^{-1} p.$$

In simple terms, we carry the form p from the tangent space at the point x to the tangent space at $f(x)$ using the diffeomorphism f (whose derivative at x determines an isomorphism between these two tangent spaces). The form $f_! p$ is a contact form since the diffeomorphism f is a contact diffeomorphism.

Theorem. *The mapping $f_!$ defined above of the symplectification of a contact manifold to itself is a symplectic diffeomorphism which commutes with the action of the multiplicative group of real numbers and preserves the canonical 1-form on the symplectification.*

PROOF. The assertion of the theorem follows from the fact that the canonical 1-form, the symplectic 2-form, and the action of the group of real numbers are all determined by the contact structure itself (for their construction we did not use coordinates or any other noninvariant tools), and the diffeomorphism f preserves the contact structure. It follows from this that f_1 preserves all that which was invariantly constructed using the contact structure, in particular the 1-form α, its derivative $d\alpha$, and the action of the group.

Theorem. *Every symplectic diffeomorphism of the symplectification of a contact manifold which commutes with the action of the multiplicative group (1) projects onto the underlying contact manifold as a contact diffeomorphism and (2) preserves the canonical 1-form α.*

PROOF. Every diffeomorphism which commutes with the action of the multiplicative group projects onto some diffeomorphism of the contact manifold. To show that this is a contact diffeomorphism it is sufficient to prove the second assertion of the theorem (since only those vectors for which $\alpha(\xi) = 0$ project onto the contact plane).

To prove the second assertion we express the integral of the form along any path γ in terms of the symplectic structure $d\alpha$:

$$\int_\gamma \alpha = \lim_{\varepsilon \to 0} \iint_{\sigma(\varepsilon)} d\alpha,$$

where the 2-chain $\sigma(\varepsilon)$ is obtained from γ by multiplication by all numbers in the interval $[\varepsilon, 1]$. The boundary of σ contains, besides γ, two vertical intervals and the path $\varepsilon\gamma$. The integrals of α over the vertical intervals are equal to zero, and the integral over $\varepsilon\gamma$ approaches 0 as ε does.

Now from the invariance of the 2-form $d\alpha$ and the commutativity of our diffeomorphism F with multiplication by numbers it follows that for any path γ

$$\int_{F\gamma} \alpha = \int_\gamma \alpha.$$

and thus the diffeomorphism F preserves the 1-form α. ☐

Definition. *The symplectification of a contact vector field is defined by the following construction. Consider the field as a velocity field of a one-parameter group of contact diffeomorphisms. Symplectify the diffeomorphisms. Consider the velocity field of this group. It is called the symplectification of the original field.*

Theorem. *The symplectification of a contact vector field is a hamiltonian vector field. The hamiltonian can be chosen to be homogeneous of first order with respect to the action of multiplication by the group of real numbers:*

$$H(\lambda x) = \lambda H(x).$$

Conversely, every hamiltonian field on a symplectified contact manifold, having a hamiltonian which is homogeneous of degree 1, projects onto the underlying contact manifold as a contact vector field.

PROOF. The fact that symplectifications of contact diffeomorphisms are symplectic implies that the symplectification of a contact field is hamiltonian. The homogeneity of the hamiltonian follows from the homogeneity of symplectic diffeomorphisms (from commutativity with multiplication by λ). Thus the first assertion of the theorem follows from the theorem on symplectifications of contact diffeomorphisms. The second part follows in the same way from the theorem on homogeneous symplectic diffeomorphisms. $\square$

Corollary. *Symplectification of vector fields is an isomorphic map of the Lie algebra of contact vector fields onto the Lie algebra of all locally hamiltonian vector fields with hamiltonians which are homogeneous of degree 1.*

The proof is clear.

H *Darboux's theorem for contact structures*

Darboux's theorem is a theorem on the local uniqueness of a contact structure. It can be formulated in any of the following three ways.

Theorem. *All contact manifolds of the same dimension are locally contact diffeomorphic (i.e., there is a diffeomorphism of a sufficiently small neighborhood of any point of one contact manifold onto a neighborhood of any point of the other which carries the noted point of the first neighborhood to the noted point of the second and the field of planes in the first neighborhood to the field of planes in the second).*

Theorem. *Every contact manifold of dimension $2m - 1$ is locally contact diffeomorphic to the manifold of contact elements of m-dimensional space.*

Theorem. *Every differential 1-form defining a nondegenerate field of hyperplanes on a manifold of dimension $2n + 1$, can be written in some local coordinate system in the "normal form"*

$$\omega = x\, dy + dz,$$

where $x = (x_1, \ldots, x_n)$, $y = (y_1, \ldots, y_n)$ and z are the local coordinates.

It is clear that the first two theorems follow from the third. We will deduce the third one from an analogous theorem of Darboux on the normal form of the 2-form giving a symplectic structure (cf. Section 43).

PROOF OF DARBOUX'S THEOREM. We symplectify our manifold. On this new $(2n + 2)$-dimensional symplectic manifold there are a canonical 1-form α, a nondegenerate 2-form $d\alpha$, a projection π onto the underlying contact manifold and a vertical direction at every point.

The given differential 1-form ω on the contact manifold defines a contact form at every point. These contact forms form a $(2n + 1)$-dimensional submanifold of the symplectic manifold. The projection π maps this submanifold diffeomorphically onto the underlying contact manifold, and the verticals intersect this submanifold at a nonzero angle.

Consider a point in the surface just constructed (in the symplectic manifold) lying over the point of the contact manifold we are interested in. In the symplectic manifold we can choose a local system of coordinates near this point such that

$$d\alpha = dp_0 \wedge dq_0 + \cdots + dp_n \wedge dq_n$$

and such that the coordinate surface $p_0 = 0$ coincides with our $(2n + 1)$-dimensional manifold (cf. Section 43, where in the proof of the symplectic Darboux's theorem the first coordinate may be chosen arbitrarily).

We note now that the 1-form $p_0\, dq_0 + \cdots p_n\, dq_n$ has derivative $d\alpha$. Thus, locally,

$$\alpha = p_0\, dq_0 + \cdots + p_n\, dq_n + dw,$$

where w is a function which can be taken to be zero at the origin. In particular, on the surface $p_0 = 0$ the form α takes the form

$$\alpha|_{p_0 = 0} = p_1\, dq_1 + \cdots + p_n\, dq_n + dw.$$

The projection π allows us to carry the coordinates $p_1, \ldots, p_n; q_0; q_1, \ldots, q_n$ and the function w onto the contact manifold. More precisely, we define functions x, y, and z by the formulas

$$x_i(\pi A) = p_i(A) \qquad y_i(\pi A) = q_i(A) \qquad z(\pi A) = w(A),$$

where A is a point on the surface $p_0 = 0$.

Then we obtain

$$\omega = x\, dy + dz$$

and it remains only to verify that the functions $(x_1, \ldots, x_n; y_1, \ldots, y_n; z)$ form a coordinate system. For this it is sufficient to verify that the partial derivative of w with respect to q_0 is not zero, or in other words that the 1-form α is not zero on a vector of the coordinate direction q_0. The latter is equivalent to the 2-form $d\alpha$ being nonzero on the pair of vectors: the basic vector in the direction of q_0 and the vertical vector.

But a vector in the coordinate direction q_0 is skew-orthogonal to all vectors of the coordinate plane $p_0 = 0$. If it was also skew-orthogonal to the vertical vector, then it would be skew-orthogonal to all vectors, which contradicts the nondegeneracy of $d\alpha$. Thus $\partial w/\partial q_0 \neq 0$, and the theorem is proved. $\qquad\square$

I Contact hamiltonians

Suppose that the contact structure of a contact manifold is given by a differential 1-form ω, and that this form is fixed.

Definition. The ω-*embedding* of the contact manifold into its symplectification is the map associating to a point of the contact manifold the restriction of the form ω on the tangent plane at this point.

Definition. The *contact hamiltonian function* of a contact vector field on a contact manifold with fixed 1-form ω is the function K on the contact manifold whose value at each point is the value of the homogeneous hamiltonian H of the symplectification of the field on the image of the given point under the ω-embedding:

$$K(A) = H(\omega|_A).$$

Theorem. *The contact hamiltonian function K of a contact vector field X on a contact manifold with a given 1-form ω is equal to the value of the form ω on this contact field:*

$$K = \omega(X).$$

PROOF. We use the expression for the increment of the ordinary hamiltonian function over a path in terms of the vector field and the symplectic structure (Section 48C). For this we draw a vertical interval $\{\lambda B\}$, $0 < \lambda \leq 1$, through the point B of the symplectification at which we want to calculate the hamiltonian function. The translations of this interval over small time τ under the action of the symplectified flow defined by our field X, fill out a two-dimensional region $\sigma(\tau)$. The value of the hamiltonian at the point B is equal to the limit

$$H(B) = \lim_{\tau \to 0} \tau^{-1} \iint_{\sigma(\tau)} d\alpha,$$

since $H(\lambda B) \to 0$ as $\lambda \to 0$. But the integral of the form $d\alpha$ over the region is the integral of the 1-form α along the edge formed by the trajectory of the point B (the other parts of the boundary give zero integrals). Therefore, the double integral is simply the integral of the 1-form α along the interval of trajectories, and the limit is the value of α on the velocity vector Y of the symplectified field. Thus $K(\pi B) = H(B) = \alpha(Y) = \omega(X)$, as was to be shown. $\square$

J Computational formulas

Suppose now that we make use of the coordinates in Darboux's theorem in which the form ω has the normal form

$$\omega = x\,dy + dz, \qquad x = (x_1, \ldots, x_n), \, y = (y_1, \ldots, y_n).$$

PROBLEM. Find the components of the contact field with a given contact hamiltonian function $K = K(x, y, z)$.

ANSWER. The equations of the contact flow have the form

$$\begin{cases} \dot{x} = -K_y + xK_z \\ \dot{y} = K_x \\ \dot{z} = K - xK_x. \end{cases}$$

Solution. A point of the symplectification can be given by the $2n + 2$ numbers x_i, y_i, z, and λ, where (x, y, z) are the coordinates of a point of the contact manifold and λ is the number by which we must multiply ω to obtain the given point of the symplectified space.

In these coordinates $\alpha = \lambda x\,dy + \lambda\,dz$. Therefore, in the coordinate system $\mathbf{p}, \mathbf{q}$, where

$$\mathbf{p} = (p, p_0), \qquad p = \lambda x, \, p_0 = \lambda$$
$$\mathbf{q} = (q, q_0), \qquad q = y, \, q_0 = z,$$

the form α takes the standard form:

$$\alpha = \mathbf{p}\,d\mathbf{q} \qquad d\alpha = d\mathbf{p} \wedge d\mathbf{q}.$$

The action T_μ of the multiplicative group is now reduced to multiplication of $\mathbf{p}$ by a number:

$$T_\mu(\mathbf{p}, \mathbf{q}) = (\mu\mathbf{p}, \mathbf{q}).$$

The contact hamiltonian K can be expressed in terms of the ordinary hamiltonian $H = H(p, q, p_0, q_0)$ by the formula

$$K(x, y, z) = H(x, y, 1, z).$$

The function H is homogeneous of degree 1 in $\mathbf{p}$. Therefore, the partial derivatives of K at the point (x, y, z) are related to the derivatives of H at the point $(p = x, p_0 = 1, q = y, q_0 = z)$ by the relations

$$H_q = K_y \qquad H_{q_0} = K_z,$$

$$H_p = K_x \qquad H_{p_0} = K - xK_x.$$

Hamilton's equations with hamiltonian function H therefore have the following form at the point under consideration:

$$\dot{x} + x\dot{\lambda} = -K_y \qquad \dot{\lambda} = -K_z,$$

$$\dot{y} = K_x, \qquad \dot{z} = K - xK_x.$$

from which we obtain the answer above.

PROBLEM. Find the contact hamiltonian of the Poisson bracket of two contact fields with contact hamiltonians K and K'.

ANSWER. $(K, K') + K_z EK' - K'_z EK$, where the brackets denote Poisson bracket in the variables x and y and E is the Euler operator $EF = F - xF_x$.

Solution. In the notation of the solution of the preceding problem we must express the ordinary Poisson bracket of the homogeneous hamiltonians H and H' at the point $(p = x, p_0 = 1, q = y, z_0 = z)$ in terms of the contact hamiltonians K and K'. We have

$$(H, H') = H_q H'_\mathbf{p} - H_\mathbf{p} H'_\mathbf{q} = H_q H'_p - H_p H'_q + H_{q_0} H'_{p_0} - H_{p_0} H'_{q_0}.$$

Substituting the values of the derivatives from the preceding problem, we find at the point under consideration

$$(H, H') = K_y K'_x - K_x K'_y + K_z(K' - xK'_x) - K'_z(K - xK_x).$$

K Legendre manifolds

The lagrangian submanifolds of a symplectic phase space correspond in the contact case to an interesting class of manifolds which may be called *Legendre manifolds* since they are closely related to Legendre transformations.

Definition. A *Legendre submanifold* of a $(2n + 1)$-dimensional contact manifold is an n-dimensional integral manifold of the field of contact planes.

In other words, it is an integral manifold of the highest possible dimension for a nondegenerate field of planes.

EXAMPLE 1. The set of all contact elements tangent to a submanifold of any dimension in an m-dimensional manifold is an $(m - 1)$-dimensional Legendre submanifold of the $(2m - 1)$-dimensional contact manifold of all contact elements.

EXAMPLE 2. The set of all planes tangent to the graph of a function $f = \varphi(x)$ in an $(n + 1)$-dimensional euclidean space with coordinates $(x_1, \ldots, x_n; f)$ is a Legendre submanifold of the $(2n + 1)$-dimensional space of all non-vertical hyperplane elements in the space of the graph (the contact structure is given by the 1-form

$$\omega = p_1 \, dx_1 + \cdots + p_n \, dx_n - df;$$

the element with coordinates (p, x, f) passes through the point with co-ordinates (x, f) parallel to the plane $f = p_1 x_1 + \cdots + p_n x_n$).

The Legendre transformation can be described in these terms in the following way.

Consider a second $(2n + 1)$-dimensional contact space with coordinates (P, X, F) and contact structure given by the form

$$\Omega = P \, dX - dF.$$

The *Legendre involution* is the map taking a point of the first space with coordinates (p, x, f) to the point of the second space with coordinates

$$P = x \qquad X = p \qquad F = px - f.$$

The Legendre involution, as can be easily calculated, carries the first contact structure to the second. Clearly, we have

Theorem. *A diffeomorphism of one contact manifold onto another which carries contact planes to contact planes, carries every Legendre manifold to a Legendre manifold.*

In particular, under the action of the Legendre involution the Legendre manifold of plane elements tangent to the graph of a function is carried into a new Legendre manifold. This new manifold is called the *Legendre transform of the original manifold.*

The projection of the new manifold onto the space with coordinates (X, F) (parallel to the P-direction) is in general not a smooth manifold, but has singularities. This projection is called the *Legendre transform of the graph of the function* φ.

If the function φ is convex, then the projection is itself the graph of a function $F = \Phi(X)$. In this case Φ is called the *Legendre transform of the function* φ.

As another example we consider the motion of oriented contact elements under the action of the geodesic flow on a riemannian manifold. As the "initial wave front" we take some smooth submanifold of our riemannian manifold (the dimension of the submanifold is arbitrary). The oriented contact elements tangent to this submanifold form a Legendre manifold in the space of all contact elements. From the preceding theorem we obtain

Corollary. *The family of all elements tangent to a wave front is transformed under the action of the geodesic flow after time t to a Legendre manifold of the space of all contact elements.*

It should be noted that this new Legendre manifold may not be the family of all elements tangent to some smooth manifold, since a wave front may develop singularities.

The *Legendre singularities* which arise in this way can be described in a manner similar to lagrangian singularities (cf. Appendix 12). A *Legendre fibration* of a $(2n + 1)$-dimensional contact manifold is a fibration all of whose fibers are n-dimensional Legendre manifolds. A *Legendre singularity* is a singularity of the projection of an n-dimensional Legendre submanifold of a $(2n + 1)$-dimensional contact manifold onto the $(n + 1)$-dimensional base of the Legendre fibration.

Consider the space $\mathbb{R}^{2n+1}$ with contact structure given by the form $\alpha = x\, dy + dz$, where $x = (x_1, \ldots, x_n)$ and $y = (y_1, \ldots, y_n)$. The projection $(x, y, z) \to (y, z)$ gives a Legendre fibration.

An *equivalence* of Legendre fibrations is a diffeomorphism of the total spaces of the fibrations carrying the contact structure and fibers of the first bundle to the contact structure and fibers of the second bundle. It can be shown that every Legendre bundle is equivalent to the special bundle just described in a neighborhood of every point of the space of the bundle.

The contact structure of the total space of fibration gives the fibers a local structure of a projective space. Legendre equivalence preserves this structure, i.e., defines locally projective fiber transformations.

The following theorem allows us to locally describe Legendre submanifolds and maps by using generating functions.

Theorem. *For any partition $I + J$ of the set of indices $(1, \ldots, n)$ into two disjoint subsets and for any function $S(x_I, y_J)$ of n variables x_i, $i \in I$, $j \in J$, the formulas*

$$y_I = \frac{\partial S}{\partial x_I} \qquad x_J = -\frac{\partial S}{\partial y_J} \qquad z = S - x_I \frac{\partial S}{\partial x_I}$$

define a Legendre submanifold of $\mathbb{R}^{2n+1}$. Conversely, every Legendre submanifold of $\mathbb{R}^{2n+1}$ is defined in a neighborhood of every point by these formulas for at least one of the 2^n possible choices of the subset I.

The proof is based on the fact that, on a Legendre manifold, $dz + x\, dy = 0$, so $d(z + x_I y_I) = y_I\, dx_I - x_J\, dy_J$. ☐

In the formulas of the preceding theorem, we replace S by a function from the list of the simple lagrangian singularities given in Appendix 12. We obtain Legendre singularities which are preserved under small deformations of the Legendre mapping $(x, y, z) \to (y, z)$ (i.e., are carried to equivalent singularities for small deformations of the function S). Every Legendre mapping for $n < 6$ can be approximated by a map, all of whose singularities are locally equivalent to singularities from the list A_k $(1 \leq k \leq 6)$, D_k $(4 \leq k \leq 6)$, E_6.

In particular, we obtain a list of the singularities of a wave front in general position in spaces of dimension less than 7.

In ordinary three-space this list is as follows:

$$A_1: S = \pm x_1^2 \qquad A_2: S = \pm x_1^3 \qquad A_3: S = \pm x_1^4 + x_1^2 y_2$$

where $I = \{1\}$, $J = \{2\}$, and $n = 2$.

The projections of the Legendre manifolds indicated here onto the base of the Legendre bundle (i.e., onto the space with coordinates y_1, y_2, and z) are: a simple point in the case of A_1, a cuspidal edge in the case of A_2, and a swallowtail (cf. Figure 246) in the case of A_3.

Thus a wave front in general position in three-space has only cusps and "swallowtail" points as singularities. At isolated moments of time during the motion of the front we can observe transitions of the three types A_4, D_4^- and D_4^+ (cf. Appendix 12, where the corresponding caustics filled out by the singularities of the front during its motion are drawn).

PROBLEM 1. Lay out an interval of length t on every interior normal to an ellipse in the plane. Draw the curve obtained and investigate its singularities and its transitions as t changes.

PROBLEM 2. Do the same thing for a triaxial ellipsoid in three-dimensional space.

L *Contactification*

Along with symplectification of contact manifolds, there is a contactification of symplectic manifolds with symplectic structure cohomologous to zero.

The contactification E^{2n+1} of the symplectic manifold (M^{2n}, ω^2) is constructed as the space of a bundle with fiber $\mathbb{R}$ over M^{2n}. Let U be a sufficiently small neighborhood of a point x in M, so that there is a canonical coordinate system p, q on U with $\omega = dp \wedge dq$. Consider the direct product $U \times \mathbb{R}$ with coordinates p, q, z. Let $V \times \mathbb{R}$ be the same kind of product constructed on another (or the same) neighborhood V, with coordinates $P, Q, Z; dP \wedge dQ = \omega$. If the neighborhoods U and V on M intersect, then we identify the fibers above the points of intersection in both representations so that the form $dz + p\, dq = dZ + P\, dQ = \alpha$ is defined on the whole (this is possible since $P\, dQ - p\, dq$ is a total differential on $U \cap V$).

It is easy to verify that after this pasting together we have a bundle E^{2n+1} on M^{2n} and that the form α defines a contact structure on E. The manifold E is called the *contactification* of the symplectic manifold M. If the cohomology class of the form ω^2 is integral, then we can define a contactification with fiber S^1.

M *Integration of first-order partial differential equations*

Let M^{2n+1} be a contact manifold, and E^{2n} a hypersurface in M^{2n+1}. The contact structure on M defines some geometric structure on E—in particular, the field of so-called characteristic directions. An analysis of this geometric structure can reduce the integration of general first-order nonlinear partial differential equations to the integration of a system of ordinary differential equations.

We assume that the manifold E^{2n} is transverse to the contact planes at all its points. In this case, the intersection of the tangent plane to E^{2n} at each of its points with the contact plane has dimension $2n - 1$, so that we have a field of hyperplanes on E^{2n}. Furthermore, the contact structure on M^{2n+1} defines on E^{2n} a field of lines lying in these $(2n - 1)$-dimensional planes.

In fact, let α be a 1-form on M^{2n+1} locally giving the contact structure; let $\omega = d\alpha$ and let $\mathbb{R}^{2n}$ be a contact plane at the point x in E^{2n}. Let $\Phi = 0$ be the local equation of E^{2n} (so $d\Phi$ is not zero at x). The restriction of $d\Phi$ to $\mathbb{R}^{2n}$ defines a nonzero linear form on $\mathbb{R}^{2n}$. The 2-form ω gives $\mathbb{R}^{2n}$ the structure of a symplectic vector space and thus an isomorphism of this space with its dual. The nonzero 1-form $d\Phi|_{\mathbb{R}^{2n}}$ corresponds to a nonzero vector ξ of $\mathbb{R}^{2n}$, so that $d\Phi(\cdot) = \omega(\xi, \cdot)$. The vector ξ is called the *characteristic vector* of the manifold E^{2n} at the point x. The characteristic vector ξ lies in the intersection of $\mathbb{R}^{2n}$ with the tangent plane to E^{2n}, so that $d\Phi(\xi) = 0$.

The vector ξ is not uniquely defined by the manifold E^{2n} and the contact structure on M, but only up to multiplication by a nonzero number. In fact, like the 2-form ω on $\mathbb{R}^{2n}$, the 1-form $d\Phi$ on $\mathbb{R}^{2n}$ is defined only up to multiplication by a nonzero number.

The direction of the characteristic vector (i.e., the line containing it) is determined uniquely by the contact structure at every point of the manifold E. Thus we have a field of characteristic directions on the hypersurface E of the contact manifold M. The integral curves of this field of directions are called the *characteristics*.

Now suppose we are given an $(n - 1)$-dimensional submanifold I of our hypersurface E^{2n}, which is integral for the contact field (so that the tangent plane to I at each point is contained in the contact plane).

Theorem. *If at a point x of I the characteristic on E^{2n} is not tangent to I, then in a neighborhood of the point x the characteristics on E^{2n} passing through points of I form a Legendre submanifold L^n in M^{2n+1}.*

PROOF. Let ξ be a vector field on E^{2n} made up of characteristic vectors. By the homotopy formula (cf. Section 36G) we have on E^{2n}

$$L_\xi \alpha = d i_\xi \alpha + i_\xi \, d\alpha.$$

But $i_\xi \alpha = 0$ since the characteristic vector belongs to the contact plane. Therefore, on E^{2n} we have $L_\xi \alpha = i_\xi \omega$. But the 1-form $i_\xi \omega$ is zero on the intersection of the tangent plane to E^{2n} with the contact plane (since on the contact plane $i_\xi \omega = d\Phi$, and on the tangent plane $d\Phi = 0$). Therefore, on the tangent plane to E^{2n} we have $i_\xi \omega = c\alpha$. Thus on the hypersurface E,

$$L_\xi \alpha = c\alpha$$

(where c is a function smooth in a neighborhood of x).

Now let $\{g^t\}$ be the (local) phase flow of the field ξ and η a vector tangent to E^{2n}. Set $\eta(t) = g^t_* \eta$ and $y(t) = \alpha(\eta(t))$. Then the function y satisfies the linear differential equation

$$\frac{dy}{dt} = c(t)y(t).$$

If $\eta(0)$ is tangent to I, then $y(0) = \alpha(\eta(0)) = 0$. This means $y(t) = \alpha(\eta(t)) = 0$, i.e., for all t, $\eta(t)$ lies in the contact plane. Therefore, $g^t I$ is an integral manifold of the contact field. Therefore the manifold formed by all $\{g^t I\}$ for small t is a Legendre manifold. $\qquad\square$

EXAMPLE. Consider $\mathbb{R}^{2n+1}$ with coordinates $x_1, \ldots, x_n; p_1, \ldots, p_n; u$ with contact structure defined by the 1-form $\alpha = du - p\, dx$. A function $\Phi(x, p, u)$ defines a differential equation $\Phi(x, \partial u/\partial x, u) = 0$ and a submanifold $E = \Phi^{-1}(0)$ in the space $\mathbb{R}^{2n+1}$ (called the *space of 1-jets* of functions on $\mathbb{R}^n$).

An initial condition for the equation $\Phi = 0$ is an assignment of a value f to the function u on an $(n-1)$-dimensional hypersurface Γ in the n-dimensional space with coordinates $x_1, \ldots, x_n$.

An initial condition determines the derivatives of u in the $n-1$ independent directions at each point of Γ. The derivative in a direction transverse to Γ can generally be found from the equation; if the conditions of the implicit function theorem are fulfilled, then the initial condition is called *noncharacteristic*.

A noncharacteristic initial condition defines an $(n-1)$-dimensional integral submanifold I of the form α (the graph of the mapping $u = f(x), p = p(x)$, $x \in \Gamma$). The characteristics on E intersecting I form a Legendre submanifold of $\mathbb{R}^{2n+1}$, the graph of the mapping $u = u(x)$, $p = \partial u/\partial x$. The function $u(x)$ is a solution of the equation $\Phi(x, \partial u/\partial x, u) = 0$ with initial condition $u|_\Gamma = f$.

Note that to find the function u we need only solve the system of $2n$ first-order ordinary differential equations for the characteristics on E, and perform a series of "algebraic" operations.

Appendix 5: Dynamical systems with symmetries

By the theorem of E. Noether, one-parameter groups of symmetries of a dynamical system determine first integrals. If a system admits a larger group of symmetries, then there are several integrals. Simultaneous level manifolds of these first integrals in the phase space are invariant manifolds of the phase flow. The subgroup of the group of symmetries mapping such an invariant manifold into itself acts on the manifold. In many cases, we can look at the quotient manifold of an invariant manifold by this subgroup. This quotient manifold, called the reduced phase space, has a natural symplectic structure. The original hamiltonian dynamical system induces a hamiltonian system on the reduced phase space.

The partition of the phase space into simultaneous level manifolds generally has singularities. An example is the partition of a phase plane into energy level curves.

In this appendix we will briefly discuss dynamical systems in reduced phase space and their relationship with invariant manifolds in the original space. All these questions were investigated by Jacobi and Poincaré ("elimination of the nodes" in the many-body problem, "reduction of order" in systems with symmetries, "stationary rotations" of rigid bodies, etc.). A detailed presentation in current terminology can be found in the following articles: S. Smale, "Topology and mechanics," Inventiones Mathematicae 10:4 (1970) 305–331, 11:1 (1970), 45–64; and J. Marsden and A. Weinstein, "Reduction of symplectic manifolds with symmetries," Reports on Mathematical Physics 5 (1974) 121–130.

A Poisson action of Lie groups

Consider a symplectic manifold (M^{2n}, ω^2) and suppose a Lie group G acts on it as a group of symplectic diffeomorphisms. Every one-parameter subgroup of G then acts as a locally hamiltonian phase flow on M. In many important cases, these flows have single-valued hamiltonian functions.

EXAMPLE. Let V be a smooth manifold and G some Lie group of diffeomorphisms of V. Since every diffeomorphism takes 1-forms on V to 1-forms, the group G acts on the cotangent bundle $M = T^*V$.

Recall that on the cotangent bundle there is always a canonical 1-form α ("pdq") and a natural symplectic structure $\omega = d\alpha$. The action of the group G on M is symplectic since it preserves the 1-form α and hence also the 2-form $d\alpha$.

A one-parameter subgroup $\{g^t\}$ of G defines a phase flow on M. It is easy to verify that this phase flow has a single-valued hamiltonian function. In fact, the hamiltonian function is given by the formula from Noether's theorem:

$$H(x) = \alpha\left(\frac{d}{dt}\bigg|_{t=0} g^t x\right), \quad \text{where } x \in M.$$

We now assume that we are given a symplectic action of a Lie group G on a connected symplectic manifold M such that, to every element a of the Lie algebra of G, there corresponds a one-parameter group of symplectic diffeomorphisms with a single-valued hamiltonian H_a. These hamiltonians

371

are determined up to the addition of constants which can be chosen so that the dependence of H_a upon a is linear. To do this, it is sufficient to choose arbitrarily the constants in the hamiltonians for a set of basis vectors of the Lie algebra of G, and to then define the hamiltonian function for each element of the algebra as a linear combination of the basis functions.

Thus, given a symplectic action of a Lie group G and a single-valued hamiltonian on M, we can construct a linear mapping of the Lie algebra of G into the Lie algebra of hamiltonian functions on M. The function $H_{[a,b]}$ associated to the commutator of two elements of the Lie algebra is equal to the Poisson bracket (H_a, H_b), or else it differs from this Poisson bracket by a constant:

$$H_{[a,b]} = (H_a, H_b) + C(a, b).$$

Remark. The appearance of the constant C in this formula is a consequence of an interesting phenomenon: the existence of a two-dimensional cohomology class of the Lie algebra of (globally) hamiltonian fields.

The quantity $C(a, b)$ is a bilinear skew-symmetric function on the Lie algebra. The Jacobi identity gives us

$$C([a, b], c) + C([b, c], a) + C([c, a], b) = 0.$$

A bilinear skew-symmetric function on a Lie algebra with this property is called a *two-dimensional cocycle* of the Lie algebra.

If we choose the constants in the hamiltonian functions differently, then the cocycle C is replaced by C'', where

$$C''(a, b) = C(a, b) + \rho([a, b])$$

where ρ is a linear function on the Lie algebra. Such a cocycle C' is said to be *cohomologous* to the cocycle C. A class of cocycles which are cohomologous to one another is called a *cohomology class* of the Lie algebra.

Thus, a symplectic action of a group G for which single-valued hamiltonians exist defines a two-dimensional cohomology class of the Lie algebra of G. This cohomology class measures the deviation of the action from one in which the hamiltonian function of a commutator can be chosen equal to the Poisson bracket of the hamiltonian functions.

Definition. An action of a connected Lie group on a symplectic manifold is called a *Poisson action* if the hamiltonian functions for one-parameter groups are single-valued, and chosen so that the hamiltonian function depends linearly on elements of the Lie algebra and so that the hamiltonian function of a commutator is equal to the Poisson bracket of the hamiltonian functions:

$$H_{[a,b]} = (H_a, H_b).$$

In other words, a Poisson action of a group defines a homomorphism from the Lie algebra of this group to the Lie algebra of hamiltonian functions.

EXAMPLE. Let V be a smooth manifold and G a Lie group acting on V as a group of diffeomorphisms. Let $M = T^*V$ be the cotangent bundle of the manifold V with the usual symplectic structure $\omega = d\alpha$. The hamiltonian functions of one-parameter groups are defined as above:

$$(1) \qquad H_z(x) = \alpha\left(\left.\frac{d}{dt}\right|_{t=0} g^t x\right), \qquad x \in T^*V.$$

Theorem. *This action is Poisson.*

PROOF. By definition of the 1-form α, the hamiltonian functions H_a are linear "in p" (i.e., on every cotangent space). Therefore, their Poisson brackets are also linear. Thus the function $H_{[a,b]} - (H_a, H_b)$ is linear in p. Since it is constant, it is equal to zero. $\qquad\square$

In the same way, we can show that the symplectification of any contact action is a Poisson action.

EXAMPLE. Let V be three-dimensional euclidean space and G the six-dimensional group of its motions. The following six one-parameter groups form a basis of the Lie algebra: the translations with velocity 1 along the coordinate axes q_1, q_2, and q_3 and the rotations with angular velocity 1 around these axes. By formula (1), the corresponding hamiltonian functions are (in the usual notation) p_1, p_2, p_3; M_1, M_2, M_3, where $M_1 = q_2 p_3 - q_3 p_2$, etc. The theorem implies that the pairwise Poisson brackets of these six functions are equal to the hamiltonian functions of the commutators of the corresponding one-parameter groups.

A Poisson action of a group G on a symplectic manifold M defines a mapping of M into the dual space of the Lie algebra of the group

$$P: M \to \mathfrak{g}^*.$$

That is, we fix a point x in M and consider the function on the Lie algebra which associates to an element a of the Lie algebra the value of the Hamiltonian H_a at the fixed point x:

$$p_x(a) = H_a(x).$$

This p_x is a linear function on the Lie algebra and is the element of the dual space to the algebra associated to x:

$$P(x) = p_x.$$

Following Souriau (*Structure des systèmes dynamiques*, Dunod, 1970), we will call the mapping P the *momentum*. Note that the value of the momentum is always a vector in the space $\mathfrak{g}^*$.

EXAMPLE. Let V be a smooth manifold, G a Lie group acting on V as a group of diffeomorphisms, $M = T^*V$ the cotangent bundle and H_a the hamiltonian functions constructed above of the action of G on M (cf. (1)).

Then the "momentum" mapping $P: M \to \mathfrak{g}^*$ can be described in the following way. Consider the map $\Phi: G \to M$ given by the action of all the elements of G on a fixed point x in M (so $\Phi(g) = gx$). The canonical 1-form α on M induces a 1-form $\Phi^*\alpha$ on G. Its restriction to the tangent space at the identity of G is a linear form on the Lie algebra.

Thus to every point x in M we have associated a linear form on the Lie algebra. It is easy to verify that this mapping is the momentum of our Poisson action.

In particular, if V is euclidean three-space and G is the group of rotations around the point 0, then the values of the momentum are the usual vectors of angular momentum; if G is the group of rotations around an axis, then the values of the momentum are the angular momenta relative to this axis; if G is the group of parallel translations, then the values of the momentum are the vectors of linear momentum.

Theorem. *Under the momentum mapping P, a Poisson action of a connected Lie group G is taken to the co-adjoint action of G on the dual space $\mathfrak{g}^*$ of its Lie algebra (cf. Appendix 2), i.e., the following diagram commutes:*

$$
\begin{array}{ccc}
M & \xrightarrow{\; g \;} & M \\
\downarrow{\scriptstyle P} & & \downarrow{\scriptstyle P} \\
\mathfrak{g}^* & \xrightarrow{\; Ad^*_g \;} & \mathfrak{g}^*
\end{array}
$$

Corollary. *Suppose that a hamiltonian function $H: M \to \mathbb{R}$ is invariant under the Poisson action of a group G on M. Then the momentum is a first integral of the system with hamiltonian function H.*

PROOF OF THE THEOREM The theorem asserts that the hamiltonian function H_a of the one-parameter group h^t is carried over by the diffeomorphism g to the hamiltonian function $H_{Ad_g a}$ of the one-parameter group $gh^t g^{-1}$.

Let g^s be a one-parameter group with hamiltonian function H_b. It is sufficient to show that the derivatives with respect to s (for $s = 0$) of the functions $H_a(g^s x)$ and $H_{Ad_{g^s} a}(x)$ are the same. The first of these derivatives is the value at x of the Poisson bracket (H_a, H_b). The second is $H_{[a, b]}(x)$. Since the action is Poisson, the theorem is proved. ☐

PROOF OF THE COROLLARY. The derivative, in the direction of the phase flow with hamiltonian function H, of each component of the momentum is zero, since it is equal to the derivative of function H in the direction of the phase flow corresponding to a one-parameter subgroup of G. ☐

B The reduced phase space

Suppose that we are given a Poisson action of a group G on a symplectic manifold M. Consider a level set of the momentum, i.e., the inverse image of some point $p \in \mathfrak{g}^*$ under the map P. We denote this set by M_p, so that (Figure 238)

$$M_p = P^{-1}(p).$$

In many important cases the set M_p is a manifold. For example, this will be so if p is a regular value of the momentum, i.e., if the differential of the map P at each point of the set M_p maps the tangent space to M onto the whole tangent space to $\mathfrak{g}^*$.

In general, a Lie group G acting on M takes the sets M_p into one another. However, the stationary subgroup of a point p in the co-adjoint representation (i.e., the subgroup consisting of those elements g of the group G for which $Ad^*_g p = p$) leaves M_p fixed. We denote this stationary subgroup by

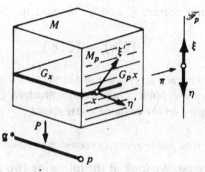

Figure 238   Reduced phase space

G_p. The group G_p is a Lie group, and it acts on the level set M_p of the momentum.

The reduced phase space is obtained from M_p by factoring by the action of the group G_p. In order for such a factorization to make sense, it is necessary to make several assumptions. For example, it is sufficient to assume that

1. p is a regular value, so that M_p is a manifold,
2. The stationary subgroup G_p is compact, and
3. The elements of the group G_p act on M_p without fixed points.

Remark. These conditions can be weakened. For example, instead of compactness of the group G_p we can require that the action be proper (i.e., that the inverse images of compact sets under the mapping $(g, x) \to (g(x), x)$ are compact). For example, the actions of a group on itself by left and right translation are always proper.

If conditions (1), (2), and (3) are satisfied, then it is easy to give the set of orbits of the action of G_p on M_p the structure of a smooth manifold. Namely, a chart on a neighborhood of a point $x \in M_p$ is furnished by any local transversal to the orbit $G_p x$, whose dimension is equal to the codimension of the orbit.

The resulting manifold of orbits is called the *reduced phase space of a system with symmetry.*

We will denote the reduced phase space corresponding to a value of the momentum by F_p. The manifold F_p is the base space of the bundle $\pi: M_p \to F_p$ with fiber diffeomorphic to the group G_p.

There is a natural symplectic structure on the reduced phase space F_p. Namely, consider any two vectors ξ and η tangent to F_p at the point f. The point f is one of the orbits of the group G_p on the manifold M_p. Let x be one of the points of this orbit. The vectors ξ and η tangent to F_p are obtained from some vectors ξ' and η' tangent to M_p at some point x by the projection $\pi: M_p \to F_p$.

Definition. The *skew-scalar product* of two vectors ξ and η which are tangent to a reduced phase space at the same point, is the skew-scalar product of

Appendix 5: Dynamical systems with symmetries

the corresponding vectors ξ' and η', tangent to the original symplectic manifold M:

$$[\xi, \eta]_p = [\xi', \eta'].$$

Theorem.[101] *The skew-scalar product of the vectors ξ and η does not depend on the choices of the point x and representatives ξ' and η', and gives a symplectic structure on the reduced phase space.*

Corollary. *The reduced phase space is even-dimensional.*

PROOF OF THE THEOREM. We look at the following two spaces in the tangent space to M at x:

$T(M_p)$, the tangent space to the level manifold M_p, and
$T(G_x)$, the tangent space to the orbit of the group G.

Lemma. *These two spaces are skew-orthogonal complements to one another in TM.*

PROOF. A vector ζ lies in the skew-orthogonal complement to the tangent plane of an orbit of the group G if and only if the skew-scalar product of the vector ζ with velocity vectors of the hamiltonian flow of the group G is equal to zero (by definition). But these skew-scalar products are equal to the derivatives of the corresponding hamiltonian functions in the direction ζ. Therefore, the vector ζ lies in the skew-orthogonal complement to the orbit of G if and only if the derivative of the momentum in the direction ζ is equal to zero, i.e., if ζ lies in $T(M_p)$. □

The representatives ξ' and η' are defined up to addition of a vector from the tangent plane to the orbit of the group G_p. But this tangent plane is the intersection of the tangent planes to the orbit Gx and to the manifold M_p (by the last theorem of part A). Consequently, the addition to ξ' of a vector from $T(G_px)$ does not change the skew-scalar product with any vector η' from $T(M_p)$ (since by the lemma $T(G_px)$ is skew-orthogonal to $T(M_p)$). Thus, we have shown the independence from the representatives ξ' and η'.

The independence of the quantity $[\xi, \eta]_p$ from the choice of the point x of the orbit f follows from the symplectic nature of the action of the group G on M and the invariance of M_p. Thus we have defined a differential 2-form on F_p:

$$\Omega_p(\xi, \eta) = [\xi, \eta]_p.$$

It is nondegenerate, since if $[\xi, \eta]_p = 0$ for every η, then the corresponding representative ξ' is skew-orthogonal to all vectors in $T(M_p)$. Therefore, ξ' must be the skew-orthogonal complement to $T(M_p)$ in TM. Then by the lemma $\xi' \in T(Gx)$, i.e., $\xi = 0$.

The form Ω_p is closed. In order to verify this we consider a chart, i.e., a piece of submanifold in M_p, transversally intersecting the orbit of the group G_p in one point.

The form Ω_f is represented in this chart by a 2-form induced from the 2-form ω which defines the symplectic structure in the whole space M, by means of the embedding of the submanifold piece. Since the form ω is closed, the induced form is also closed. The theorem is proved. □

[101] The theorem was first formulated in this form by Marsden and Weinstein. Many special cases have been considered since the time of Jacobi and used by Poincaré and his successors in mechanics, by Kirillov and Kostant in group theory, and by Faddeev in the general theory of relativity.

376

EXAMPLE 1. Let $M = \mathbb{R}^{2n}$ be euclidean space of dimension $2n$ with coordinates p_k, q_k and 2-form $\sum dp_k \wedge dq_k$. Let $G = S^1$ be the circle, and let the action of G on M be given by the hamiltonian of a harmonic oscillator

$$H = \tfrac{1}{2} \sum (p_k^2 + q_k^2).$$

Then the momentum mapping is simply $H \colon \mathbb{R}^{2n} \to \mathbb{R}$, a nonzero momentum level manifold is a sphere S^{2n-1}, and the quotient space is the complex projective space $\mathbb{C}P^{n-1}$.

The preceding theorem defines a symplectic structure on this complex projective space. It is easy to verify that this structure coincides (up to a multiple) with the one we constructed in Appendix 3.

EXAMPLE 2. Let V be the cotangent bundle of a Lie group, G the same group and the action defined by left translation. Then M_p is a submanifold of the cotangent bundle of G, formed by those vectors which, after right translation to the identity of the group, define the same element in the dual space to the Lie algebra.

The manifolds M_p are diffeomorphic to the group itself and are right-invariant cross-sections of the cotangent bundle. All the values p are regular.

The stationary subgroup G_p of the point p consists of those elements of the group for which left and right translation of p give the same result. The actions of elements different from the identity of G_p on M_p have no fixed points (since there are none by right translation of the group onto itself).

The group G_p acts properly (cf. remark above). Consequently, the space of orbits of the group G_p on M_p is a symplectic manifold.

But this space of orbits is easily identified with the orbit of the point p in the co-adjoint representation. Actually, we map the right-invariant section M_p of the cotangent bundle into the cotangent space to the group at the identity with *left* translations. We get a mapping

$$\pi \colon M_p \to \mathfrak{g}^*.$$

The image of this mapping is the orbit of the point p in the co-adjoint representation, and the fibers are the orbits of the action of the group G_p. The symplectic structure of the reduced phase space thus defines a symplectic structure in the orbits of the co-adjoint representation.

It is not hard to verify by direct calculation that this is the same structure which we discussed in Appendix 2.

EXAMPLE 3. Let the group $G = S^1$, the circle, and let it act without fixed points on a manifold V. Then there is an action of the circle on the cotangent bundle $M = T^*V$. We can define momentum level manifolds M_p (of co-dimension 1 in M) and quotient manifolds F_p (the dimension of which is 2 less than the dimension of M).

In addition, we can construct a quotient manifold of the configuration space V by identifying the points of each orbit of the group on V. We denote this quotient manifold by W.

Theorem. *The reduced phase space F_p is symplectic and diffeomorphic to the cotangent bundle of the quotient configuration manifold W.*

PROOF. Let $\pi: V \to W$ be the factorization map, and $\omega \in T^*W$ a 1-form on W at the point $w = \pi v$. The form $\pi^*\omega$ on V at the point v belongs to M_0 and projects to a point in the quotient F_0. Conversely, the elements of F_0 are the invariant 1-forms on V which are equal to zero on the orbits; they define 1-forms in W. We have constructed a mapping $T^*W \to F_0$; it is easy to see that this is a symplectic diffeomorphism.

The case $p \neq 0$ is reduced to the case $p = 0$ as follows. Consider a riemannian metric on V, invariant with respect to G. The intersection of M_p with the cotangent plane to V at the point v is a hyperplane. The quadratic form defined by the metric has a unique minimum point $S(v)$ in this hyperplane. Subtraction of the vector $S(v)$ carries the hyperplane $M_p \cap T^*V_v$ into $M_0 \cap T^*V_v$, and we obtain a possibly nonsymplectic diffeomorphism $F_p \to F_0$.

The difference between the symplectic structures on T^*W induced by that of F_p and F_0 is a 2-form, induced by a 2-form on W. $\square$

C *Applications to the study of stationary rotations and bifurcations of invariant manifolds*

Suppose that we are given a Poisson action of a group G on a symplectic manifold M; let H be a function on M invariant under G. Let F_p be a reduced phase space (we assume that the conditions under which this can be defined are satisfied).

The hamiltonian field with hamiltonian function H is tangent to every momentum level manifold M_p (since momentum is a first integral). The induced field on M_p is invariant with respect to G_p and defines a field on the reduced phase space F_p. This vector field on F_p will be called the *reduced field*.

Theorem. *The reduced field on the reduced phase space is hamiltonian. The value of the hamiltonian function of the reduced field at any point of the reduced phase space is equal to the value of the original hamiltonian function at the corresponding point of the original phase space.*

PROOF. The relation defining a hamiltonian field X_H with hamiltonian H on a manifold M with form ω

$$dH(\xi) = \omega(\xi, X_H) \quad \text{for every } \xi$$

implies an analogous relation for the reduced field in view of the definition of the symplectic structure on F_p. $\square$

EXAMPLE. Consider an asymmetric rigid body, fixed at a stationary point, under the action of the force of gravity (or any potential force symmetric with respect to the vertical axis).

The group S^1 of rotations with respect to a vertical line acts on the configuration space SO(3). The hamiltonian function is invariant under rotations, and therefore we obtain a reduced system on the reduced phase space.

The reduced phase space is, in this case, the cotangent bundle of the quotient configuration space (cf. Example 3 above). Factorization of the configuration space by the action of rotations around the vertical axis was done by Poisson in the following way.

We will specify the position of the body by giving the position of an orthonormal frame (e_1, e_2, e_3). The three vertical components of the basic vectors give a vector in three-dimensional euclidean space. The length of this vector is 1 (why?). This Poisson vector[102] γ determines the original frame up to rotations around a vertical line (why?).

Thus the quotient configuration space is represented by a two-dimensional sphere S^2, and the reduced phase space is the cotangent bundle T^*S^2 with a nonstandard symplectic structure. The reduced hamiltonian function on the cotangent bundle is represented as the sum of the "kinetic energy of the reduced motion," which is quadratic in the cotangent vectors, and the "effective potential" (the sum of the potential energy and the kinetic energy of rotation around a vertical line).

The transition to the reduced phase space in this case is almost by "elimination of the cyclic coordinate φ." The difference is that the usual procedure of elimination requires that the configuration or phase space be a direct product by the circle, whereas in our case we have only a bundle. This bundle can be made a direct product by decreasing the size of the configuration space (i.e., by introducing coordinates with singularities at the poles); the advantage of the approach above is that it makes it clear that there are no real singularities (except singularities of the coordinate system) near the poles.

Definition. The phase curves in M which project to equilibrium positions in the reduced system on the reduced phase space F_p are called the *relative equilibria* of the original system.

EXAMPLE. Stationary rotations of a rigid body which is fixed at its center of mass are relative equilibria. In the same way, rotations of a heavy rigid body with constant speed around the vertical axis are relative equilibria.

Theorem. *A phase curve of a system with a G-invariant hamiltonian function is a relative equilibrium if and only if it is the orbit of a one-parameter subgroup of G in the original phase space.*

PROOF. It is clear that a phase curve which is an orbit projects to a point. If a phase curve $x(t)$ projects to a point, then it can be expressed uniquely in the form $x(t) = g(t)x(0)$, and it is then easy to see that $\{g(t)\}$ is a subgroup. $\qquad\square$

[102] Poisson showed that the equations of motion of a heavy rigid body can be written in terms of γ in a remarkably simple form, the "Euler–Poisson equations":

$$\frac{d\mathbf{M}}{dt} - [\mathbf{M}, \omega] = \mu g[\gamma, \mathbf{l}] \qquad \frac{d\gamma}{dt} = [\gamma, \omega].$$

Corollary 1. *An asymmetrical rigid body in an axially symmetric potential field, fixed at a point on the axis of the field, has at least two stationary rotations (for every value of the angular momentum with respect to the axis of symmetry).*

Corollary 2. *An axially symmetric rigid body fixed at a point on the axis of symmetry, has at least two stationary rotations (for every value of the angular momentum with respect to the axis of symmetry).*

Both corollaries follow from the fact that a function on the sphere has at least two critical points.

Another application of relative equilibria is that they can be used to investigate modifications of the topology of invariant manifolds under changes of the energy and momentum values.

Theorem. *The critical points of the momentum and energy mapping*

$$P \times H: M \to \mathfrak{g}^* \times \mathbb{R}$$

on a regular momentum level set are exactly the relative equilibria.

PROOF. The critical points of the mapping $P \times H$ are the conditional extrema of H on the momentum level manifold M_p (since this level manifold is regular, i.e., for every x in M_p, we have $P_* T M_x = T \mathfrak{g}_p^*$).

After factorization by G_p, the conditional extrema of H on M_p define the critical points of the reduced hamiltonian function (since H is invariant under G_p). $\qquad\square$

The detailed study of relative equilibria and singularities of the energy-momentum mapping is not simple and has not been completely carried out, even in the classical problem of the motions of an asymmetrical rigid body in a gravitational field. The case when the center of gravity lies on one of the principal axes of inertia is treated in the supplement written by S. B. Katok to the Russian translation[103] of the article by S. Smale cited in the beginning of this appendix. In this problem the dimension of the phase space is six, and the group is the circle; the reduced phase space T^*S^2 is four-dimensional.

The nonsingular energy level manifolds in the reduced phase space are (depending on the values of momentum and energy) of the following four forms: S^3, $S^2 \times S^1$, $\mathbb{R}P^3$, and a "pretzel" obtained from the three-sphere S^3 by attaching two "handles" of the form

$$S^1 \times D^2 \qquad (D^2 = \text{the disc } \{(x, y) | x^2 + y^2 \leq 1\}).$$

---

[103] *Uspekhi Matematicheskikh Nauk* **27**, no. 2 (1972) 78–133.

Appendix 6: Normal forms of quadratic hamiltonians

In this appendix we give a list of normal forms to which we can reduce a quadratic hamiltonian function by means of a real symplectic transformation. This list was composed by D. M. Galin based on the work of J. Williamson in "On an algebraic problem concerning the normal forms of linear dynamical systems," Amer. J. of Math. 58, (1936), 141–163. Williamson's paper gives the normal forms to which a quadratic form in a symplectic space over any field can be reduced.

A Notation

We will write the hamiltonian as

$$H = \tfrac{1}{2}(Ax, x),$$

where $x = (p_1, \ldots, p_n; q_1, \ldots, q_n)$ is a vector written in a symplectic basis and A is a symmetric linear operator. The canonical equations then have the form

$$\dot{x} = IAx, \quad \text{where } I = \begin{pmatrix} 0 & -E \\ E & 0 \end{pmatrix}.$$

By the *eigenvalues* of the hamiltonian we will mean the eigenvalues of the linear infinitesimally-symplectic operator IA. In the same way, by a Jordan block we will mean a Jordan block of the operator IA.

The eigenvalues of the hamiltonian are of four types: real pairs $(a, -a)$, purely imaginary pairs $(ib, -ib)$, quadruples $(\pm a \pm ib)$, and zero eigenvalues.

The Jordan blocks corresponding to the two members of a pair or four members of a quadruple always have the same structure.

In the case when the real part of an eigenvalue is zero, we have to distinguish the Jordan blocks of even and odd order. There are an even number of blocks of odd order with zero eigenvalue and they can be naturally divided into pairs.

A complete list of normal forms follows.

B Hamiltonians

For a pair of Jordan blocks of order k with eigenvalues $\pm a$, the hamiltonian is

$$H = -a \sum_{j=1}^{k} p_j q_j + \sum_{j=1}^{k-1} p_j q_{j+1}.$$

For a quadruple of Jordan blocks of order k with eigenvalues $\pm a \pm bi$ the hamiltonian is

$$H = -a \sum_{j=1}^{2k} p_j q_j + b \sum_{j=1}^{k} (p_{2j-1} q_{2j} - p_{2j} q_{2j-1}) + \sum_{j=1}^{2k-2} p_j q_{j+2}.$$

381

For a pair of Jordan blocks of order k with eigenvalue zero the hamiltonian is

$$H = \sum_{j=1}^{k-1} p_j q_{j+1} \quad \text{(for } k = 1, H = 0).$$

For a Jordan block of order $2k$ with eigenvalue zero, the hamiltonian is of one of the following two inequivalent types:

$$H = \pm \frac{1}{2} \left(\sum_{j=1}^{k-1} p_j p_{k-j} - \sum_{j=1}^{k} q_j q_{k-j+1} \right) - \sum_{j=1}^{k-1} p_j q_{j+1}$$

(for $k = 1$ this is $H = \pm \frac{1}{2} q_1^2$).

For a pair of Jordan blocks of odd order $2k + 1$ with purely imaginary eigenvalues $\pm bi$, the hamiltonian is of one of the following two inequivalent types:

$$H = \pm \frac{1}{2} \left[\sum_{j=1}^{k} (b^2 p_{2j} p_{2k-2j+2} + q_{2j} q_{2k-2j+2}) \right.$$

$$\left. - \sum_{j=1}^{k+1} (b^2 p_{2j-1} p_{2k-2j+3} + q_{2j-1} q_{2k-2j+3}) \right] - \sum_{j=1}^{2k} p_j q_{j+1}.$$

For $k = 0$, $H = \pm \frac{1}{2} (b^2 p_1^2 + q_1^2)$.

For a pair of Jordan blocks of even-order $2k$ with purely imaginary eigenvalues $\pm bi$, the hamiltonian is of one of the following two inequivalent types:

$$H = \pm \frac{1}{2} \left[\sum_{j=1}^{k} \left(\frac{1}{b^2} q_{2j-1} q_{2k-2j+1} + q_{2j} q_{2k-2j+2} \right) \right.$$

$$\left. - \sum_{j=1}^{k-1} b^2 p_{2j+1} p_{2k-2j+1} + p_{2j+2} p_{2k-2j+2} \right]$$

$$- b^2 \sum_{j=1}^{k} p_{2j-1} q_{2j} + \sum_{j=1}^{k} p_{2j} q_{2j-1}$$

$$\left(\text{for } k = 1, H = \pm \frac{1}{2} \left(\frac{1}{b^2} q_1^2 + q_2^2 \right) - b^2 p_1 q_2 + p_2 q_1 \right).$$

Williamson's theorem. *A real symplectic vector space with a given quadratic form H can be decomposed into a direct sum of pairwise skew orthogonal real symplectic subspaces so that the form H is represented as a sum of forms of the types indicated above on these subspaces.*

C Nonremovable Jordan blocks

An individual hamiltonian in "general position" does not have multiple eigenvalues and reduces to a simple form (all the Jordan blocks are of first order). However, if we consider not an individual hamiltonian but a whole family of systems depending on parameters, then for some exceptional values of the parameters more complicated Jordan structures can arise. We can get rid of some of these by a small change of the family; others are non-removable and only slightly deformed after a small change of the family. If the number l of parameters of the family is finite, then the number of non-removable types in l-parameter families is finite. The theorem of Galin formulated below allows us to count all these types for any fixed l.

We denote by $n_1(z) \geq n_2(z) \geq \cdots \geq n_s(z)$ the dimensions of the Jordan blocks with eigenvalues $z \neq 0$, and by $m_1 \geq m_2 \geq \cdots \geq m_u$ and $\tilde{m}_1 \geq \tilde{m}_2 \geq \cdots \geq \tilde{m}_v$ the dimensions of the Jordan blocks with eigenvalues zero, where the m_i are even and the $\tilde{m}_i$ are odd (of every pair of blocks of odd dimension, only one is considered).

Theorem. *In the space of all hamiltonians, the manifold of hamiltonians with Jordan blocks of the indicated dimensions has codimension*

$$c = \frac{1}{2} \sum_{z \neq 0} \left[\sum_{j=1}^{s(z)} (2j - 1)n_j(z) - 1 \right] + \frac{1}{2} \sum_{j=1}^{u} (2j - 1)m_j$$

$$+ \sum_{j=1}^{v} [2(2j - 1)\tilde{m}_j + 1] + 2 \sum_{j=1}^{u} \sum_{k=1}^{v} \min\{m_j, \tilde{m}_k\}.$$

(Note that, if zero is not an eigenvalue, then only the first term in the sum is not zero.)

Corollary. *In l-parameter families in general position of linear hamiltonian systems, the only systems which occur are those with Jordan blocks such that the number c calculated by the formula above is not greater than l: all cases with larger c can be eliminated by a small change of the family.*

Corollary. *In one- and two-parameter families, nonremovable Jordan blocks of only the following 12 types occur:*

$$l = 1: (\pm a)^2, (\pm ia)^2, 0^2$$

(here the Jordan blocks are denoted by their determinants; for example, $(\pm a)^2$ denotes a pair of Jordan blocks of order 2 with eigenvalues a and $-a$, respectively;

$$l = 2: (\pm a)^3, (\pm ai)^3, (\pm a \pm bi)^2, 0^4, (\pm a)^2(\pm b)^2, (\pm ai)^2(\pm bi)^2,$$
$$(\pm a)^2(\pm bi)^2, (\pm a)^2 0^2, (\pm ai)^2 0^2$$

(the remaining eigenvalues are simple).

Galin has also computed the normal forms to which one can reduce any family of linear hamiltonian systems which depend smoothly on parameters, by using a symplectic linear change of coordinates which depends smoothly on the parameters. For example, for the simplest Jordan square $(\pm a)^2$, the normal form of the hamiltonian will be

$$H(\lambda) = -a(p_1 q_1 + p_2 q_2) + p_1 q_2 + \lambda_1 p_1 q_1 + \lambda_2 p_2 q_1$$

(λ_1 and λ_2 are the parameters).

Appendix 7: Normal forms of hamiltonian systems near stationary points and closed trajectories

In studying the behavior of solutions to Hamilton's equations near an equilibrium position, it is often insufficient to look only at the linearized equation. In fact, by Liouville's theorem on the conservation of volume, it is impossible to have asymptotically stable equilibrium positions for hamiltonian systems. Therefore, the stability of the linearized system is always neutral: the eigenvalues of the linear part of a hamiltonian vector field at a stable equilibrium position all lie on the imaginary axis.

For systems of differential equations in general form, such neutral stability can be destroyed by the addition of arbitrarily small nonlinear terms. For hamiltonian systems the situation is more complicated. Suppose, for example, that the quadratic part of the hamiltonian function at an equilibrium position (which determines the linear part of the vector field) is (positive or negative) definite. Then the hamiltonian function has a maximum or minimum at the equilibrium position. Therefore, this equilibrium position is stable (in the sense of Liapunov, but not asymptotically), not only for the linearized system but also for the entire nonlinear system.

On the other hand, the quadratic part of the hamiltonian function at a stable equilibrium position may not be definite. A simple example is supplied by the function $H = p_1^2 + q_1^2 - p_2^2 - q_2^2$. To investigate the stability of systems with this kind of quadratic part, we must take into account terms of degree ≥ 3 in the Taylor series of the hamiltonian function (i.e., the terms of degree ≥ 2 for the phase velocity vector field). It is useful to carry out this investigation by reducing the hamiltonian function (and, therefore, the hamiltonian vector field) to the simplest possible form by a suitable canonical change of variables. In other words, it is useful to choose a canonical coordinate system, near the equilibrium position, in which the hamiltonian function and equations of motion are as simple as possible.

The analogous question for general (non-hamiltonian) vector fields can be solved easily: there the general case is that a vector field in a neighborhood of an equilibrium position is linear in a suitable coordinate system (the relevant theorems of Poincaré and Siegel can be found, for instance, in the book, *Lectures on Celestial Mechanics*, by C. L. Siegel and J. Moser, Springer-Verlag, 1971.)

In the hamiltonian case the picture is more complicated. The first difficulty is that reduction of the hamiltonian field to a linear normal form by a canonical change of variables is generally not possible. We can usually kill the cubic part of the hamiltonian function, but we cannot kill all the terms of degree four (this is related to the fact that, in a linear system, the frequency of oscillation does not depend on the amplitude, while in a nonlinear system it generally does). This difficulty can be surmounted by the choice of a nonlinear normal form which takes the frequency variations into account. As a result, we can (in the "non-resonance" case) introduce action-angle variables near an equilibrium position so that the system becomes integrable up to terms of arbitrary high degree in the Taylor series.

This method allows us to study the behavior of systems over the course of large intervals of time for initial conditions close to equilibrium. However, it is not sufficient to determine whether an equilibrium position will be Liapunov stable (since on an infinite time interval the influence of the discarded remainder term of the Taylor series can destroy the stability). Such stability would follow from an exact reduction to an analogous normal form which did not disregard remainder terms. However, we can show that this exact reduction is generally not possible, and formal series for canonical transformations reducing a system to normal form generally *diverge*.

The divergence of these series is connected with the fact that reduction to normal form would imply simpler behavior of the phase curves (they would have to be conditionally periodic windings of tori) than that which in fact occurs. The behavior of phase curves near an equilibrium position is discussed in Appendix 8. In this appendix we give the formal results on normalization up to terms of high degree.

The idea of reducing hamiltonian systems to normal forms goes back to Lindstedt and Poincaré;[104] normal forms in a neighborhood of an equilibrium position were extensively studied by G. D. Birkhoff (G. D. Birkhoff, *Dynamical Systems*, American Math. Society, 1927).

Normal forms for degenerate cases can be found in the work of A. D. Bruno, "Analytic forms of differential equations," (Trudy Moskovskogo matematicheskogo obshchestva, v. 25 and v. 26).

A *Normal form of a conservative system near an equilibrium position*

Suppose that in the linear approximation an equilibrium position of a hamiltonian system with n degrees of freedom is stable, and that all n characteristic frequencies $\omega_1, \ldots, \omega_n$ are different. Then the quadratic part of the hamiltonian can be reduced by a canonical linear transformation to the form

$$H = \tfrac{1}{2}(\omega_1(p_1^2 + q_1^2) + \cdots + \tfrac{1}{2}\omega_n(p_n^2 + q_n^2)).$$

(Some of the numbers ω_k may be negative).

Definition. The characteristic frequencies $\omega_1, \ldots, \omega_k$ *satisfy a resonance relation of order K* if there exist integers k_l not all equal to zero such that

$$k_1\omega_1 + \cdots + k_n\omega_n = 0, \qquad |k_1| + \cdots + |k_n| = K.$$

Definition. A Birkhoff normal form of degree s for a hamiltonian is a polynomial of degree s in the canonical coordinates (P_l, Q_l) which is actually a polynomial (of degree $[s/2]$) in the variables $\tau_l = (P_l^2 + Q_l^2)/2$.

[104] Cf. H. Poincaré, *Les Méthodes Nouvelles de la Mécanique Céleste*, Vol. 1, Dover, 1957.

Appendix 7: Normal forms of hamiltonian systems near stationary points

For example, for a system with one degree of freedom the normal form of degree $2m$ (or $2m + 1$) looks like

$$H_{2m} = H_{2m+1} = a_1 \tau + a_2 \tau^2 + \cdots + a_m \tau^m, \qquad \tau = (P^2 + Q^2)/2,$$

and for a system with two degrees of freedom the Birkhoff normal form of degree 4 will be

$$H_4 = a_1 \tau_1 + a_2 \tau_2 + a_{11} \tau_1^2 + a_{12} \tau_1 \tau_2 + a_{22} \tau_2^2.$$

The coefficients a_1 and a_2 are characteristic frequencies, and the coefficients a_{ij} describe the dependence of the frequencies on the amplitude.

Theorem. *Assume that the characteristic frequencies ω_l do not satisfy any resonance relation of order s or smaller. Then there is a canonical coordinate system in a neighborhood of the equilibrium position such that the hamiltonian is reduced to a Birkhoff normal form of degree s up to terms of order $s + 1$:*

$$H(p, q) = H_s(P, Q) + R \qquad R = O(|P| + |Q|)^{s+1}.$$

PROOF. The proof of this theorem is easy to carry out in a complex coordinate system

$$z_l = p_l + iq_l \qquad w_l = p_l - iq_l$$

(upon passing to this coordinate system we must multiply the hamiltonian by $-2i$). If the terms of degree less than N entering into the normal form are not already killed, then the transformation with generating function $Pq + S_N(P, q)$ (where S_N is a homogeneous polynomial of degree N) changes only terms of degree N and higher in the Taylor expansion of the hamiltonian function.

Under this transformation the coefficient for a monomial of degree N in the hamiltonian function having the form

$$z_1^{\alpha_1} \cdots z_n^{\alpha_n} w_1^{\beta_1} \cdots w_n^{\beta_n} \qquad (\alpha_1 + \cdots + \alpha_n + \beta_1 + \cdots + \beta_n = N)$$

is changed into the quantity

$$s_{\alpha\beta}[\lambda_1(\beta_1 - \alpha_1) + \cdots + \lambda_n(\beta_n - \alpha_n)].$$

where $\lambda_l = i\omega_l$ and where $s_{\alpha\beta}$ is the coefficient for $z^\alpha w^\beta$ in the expansion of the function $S_N(P, q)$ in the variables z and w.

Under the assumptions about the absence of resonance, the coefficient of $s_{\alpha\beta}$ in the square brackets is not zero, except in the case when our monomial can be expressed in terms of the product $z_l w_l = 2\tau_l$ (i.e., when all the α_l are equal to the β_l). Thus we can kill all terms of degree N except those expressed in terms of the variables τ_l. Setting $N = 3, 4, \ldots, s$, we obtain the theorem.
$\square$

To use Birkhoff's theorem, it is helpful to note that a hamiltonian in normal form is integrable. Consider the "canonical polar coordinates" τ_l, φ_l, in which P_l and Q_l can be expressed by the formulas

$$P_l = \sqrt{2\tau_l} \cos \varphi_l \qquad Q_l = \sqrt{2\tau_l} \sin \varphi_l.$$

Since the hamiltonian is expressed in terms of only the action variables τ_l, the system is integrable and describes conditionally periodic notions on the tori $\tau = $ const with frequencies $\omega = \partial H/\partial \tau$. In particular, the equilibrium position $P = Q = 0$ is stable for the normal form.

B *Normal form of a canonical transformation near a stationary point*

Consider a canonical (i.e. area-preserving) mapping of the two-dimensional plane to itself. Assume that this transformation leaves the origin fixed, and that its linear part has eigenvalue $\lambda = e^{\pm i\alpha}$ (i.e., is a rotation by angle α in a suitable symplectic basis with coordinates p, q). We will call such a transformation *elliptic*.

Definition. A *Birkhoff normal form of degree s for a transformation* is a canonical transformation of the plane to itself which is a rotation by a variable angle which is a polynomial of degree not more than $m = [s/2] - 1$ in the action variable τ of the canonical polar coordinate system:

$$(\tau, \varphi) \to (\tau, \varphi + \alpha_0 + \alpha_1\tau + \cdots + \alpha_m\tau^m),$$

where

$$p = \sqrt{2\tau}\cos\varphi \qquad q = \sqrt{2\tau}\sin\varphi.$$

Theorem 2. *If the eigenvalue λ of an elliptic canonical transformation is not a root of unity of degree s or less, then this transformation can be reduced by a canonical change of variables to a Birkhoff normal form of degree s with error terms of degree $s + 1$ and higher.*

The multi-dimensional generalization of an elliptic transformation is the direct product of n elliptic rotations of the planes (p_l, q_l) with eigenvalues $\lambda_l = e^{\pm i\alpha_l}$. A Birkhoff normal form of degree s is given by the formula

$$(\tau, \varphi) \to \left(\tau, \varphi + \frac{\partial S}{\partial \tau}\right),$$

where S is a polynomial of degree not more than $[s/2]$ in the action variables $\tau_1, \ldots, \tau_n$.

Theorem 3. *If the eigenvalues λ_l of a multi-dimensional elliptic canonical transformation do not admit resonances*

$$\lambda_1^{k_1} \cdots \lambda_n^{k_n} = 1, \qquad |k_1| + \cdots + |k_n| \leq s,$$

then this transformation can be reduced to a Birkhoff normal form of degree s (with error in terms of degree s in the expansion of the mapping in a Taylor series at the point $p = q = 0$).

C *Normal form of an equation with periodic coefficients near an equilibrium position*

Let $p = q = 0$ be an equilibrium position of a system whose hamiltonian function depends 2π-periodically on time. Assume that the linearized equation can be reduced by a linear symplectic time-periodic transformation to an autonomous normal form with characteristic frequencies $\omega_1, \ldots, \omega_n$.

We say that a system is *resonant of order $K > 0$* if there is a relation

$$k_1 \omega_1 + \cdots + k_n \omega_n + k_0 = 0$$

with integers $k_0, k_1, \ldots, k_n$ for which $|k_1| + \cdots + |k_n| = K$.

Theorem. *If a system is not resonant of order s or less, then there is a 2π-periodic time-dependent canonical transformation reducing the system in a neighborhood of an equilibrium position to the same Birkhoff normal form of degree s as if the system were autonomous, with only the difference that the remainder terms R of degree $s + 1$ and higher will depend periodically on time.*

Finally, suppose that we are given a closed trajectory of an autonomous hamiltonian system. Then, in a neighborhood of this trajectory, we can reduce the system to normal form by using either of the following two methods:

1. Isoenergetic reduction: Fix an energy constant and consider a neighborhood of the closed trajectory on the $(2n - 1)$-dimensional energy level manifold as the extended phase space of a system with $n - 1$ degrees of freedom, periodically depending on time.
2. Surface of section: Fix an energy constant and value of one of the coordinates (so that the closed trajectory intersects the resulting $(2n - 2)$-dimensional manifold transversally). Then phase curves near the given one define a mapping of this $(2n - 2)$-dimensional manifold to itself, with a fixed point on the closed trajectory. This mapping preserves the natural structure on our $(2n - 2)$-dimensional manifold, and we can study it by using the normal form in Section B.

In investigating closed trajectories of autonomous hamiltonian systems, a phenomenon arises which contrasts with the general theory of equilibrium positions of systems with periodic coefficients. The fact is that the closed trajectories of an autonomous system are not isolated, but form (as a rule) one-parameter families. The parameter of the family is the value of the energy constant. In fact, assume that for some choice of the energy constant the closed trajectory intersects transversally the $(2n - 2)$-dimensional manifold described above in the $(2n - 1)$-dimensional energy level manifold. Then for nearby values of the energy, there will exist a similar closed trajectory. By the implicit function theorem we can even say that this closed trajectory depends smoothly on the energy constant.

If we now wish to use the Birkhoff normal form to investigate a one-parameter family of closed trajectories, we encounter the following difficulty. As the parameter describing the family varies, the eigenvalues of the linearized problem will generally change. Therefore, for some values of the parameter we will inevitably encounter resonances, obstructing reduction to the normal form.

Especially dangerous are resonances of low order, since they influence the first few terms of the Taylor series. If we are interested in a closed trajectory for which the eigenvalues nearly satisfy a resonance relation of low order, then the Birkhoff form must be somewhat modified. Namely, for resonance of order N some of the expressions

$$k_0 - [\omega_1(\beta_1 - \alpha_1) + \cdots + \omega_n(\beta_n - \alpha_n)], \qquad |\alpha| + |\beta| = N,$$

by which we must divide to kill the terms of order N in the hamiltonian function, may become zero. For non-resonant values of the parameter which are close to resonance, this combination of characteristic frequencies is generally not zero, but very small (this combination is therefore called a "small denominator").

Division by a small denominator leads to the following difficulties:

1. The transformation which reduces to normal form depends discontinuously on the parameter (it has poles for resonant values of the parameter);
2. The region in which the Birkhoff normal form accurately describes the system contracts to zero at resonance.

In order to get rid of these deficiencies, we must give up trying to annihilate some of the terms of the hamiltonian (namely, those which become resonant for resonance values of the parameter). Moreover, these terms must be preserved not only for resonance, but also for nearby values of the parameter.[105] The normal form thus obtained is somewhat more complicated than the usual normal form, but in many cases it gives us useful information on the behavior of solutions near resonance.

D Example: Resonance of order 3

As a simple example, we will study what happens to a closed trajectory of an autonomous hamiltonian system with two degrees of freedom, for which the period of oscillation (about the closed trajectory) of neighboring trajectories is three times the period of the closed trajectory itself. By what we said above, this problem may be reduced to an investigation of a one-parameter system of non-autonomous hamiltonian systems with one degree of freedom, 2π-periodically depending on time, in a neighborhood of an equilibrium position. This equilibrium position can be taken as the origin for all values of the parameter (to achieve this we must make a change of variables depending on the parameter).

Furthermore, the linearized system at the equilibrium position can be converted into a linear system with constant coefficients by a 2π-periodically time-dependent linear canonical change of variables. In the new coordinates the phase flow of the linearized system is represented as a uniform rotation

---

[105] The method indicated here is useful not only in investigating hamiltonian systems, but also in the general theory of differential equations. Cf., for example, V. I. Arnold, "Lectures on bifurcations and versal families," Russian Math. Surveys 27, No. 5, 1972, 54–123.

around the equilibrium position. The angular velocity ω of this rotation depends on the parameter.

At the resonance value of the parameter, $\omega = \frac{1}{3}$ (i.e., after time 2π, we have gone one-third of the way around the origin). The derivative of the angular velocity ω with respect to the parameter is generally not zero. Therefore, we can take as a parameter this angular velocity or, even better, its difference from $\frac{1}{3}$. We will denote this difference by ε. The quantity ε is called the *frequency deviation* or *detuning*. The resonance value of the parameter is $\varepsilon = 0$, and we are interested in the behavior of the system for small ε.

If we disregard the nonlinear terms in Hamilton's equations and disregard the frequency deviation ε, then all trajectories of our system become closed after making three revolutions (i.e., they have period 6π). We now want to study the influence of the nonlinear terms and frequency deviation on the behavior of the trajectories. It is clear that in the general case not all the trajectories will be closed. To study their behavior, it is useful to look at the normal form.

In the chosen coordinate system, $z = p + iq$, $\bar{z} = p - iq$, the hamiltonian function has the form

$$-2iH = -i\omega z\bar{z} + \sum_{\alpha+\beta=3} \sum_{k=-\infty}^{+\infty} h_{\alpha\beta k} z^\alpha \bar{z}^\beta e^{ikt} + \cdots,$$

where the dots indicate terms of order higher than three, and where $\omega = (\frac{1}{3}) + \varepsilon$.

In the reduction to normal form we can kill all terms of degree three except those for which the small denominator

$$\omega(\alpha - \beta) + k$$

becomes zero at resonance. These terms can be described also as those which are constant along trajectories of the periodic motion obtained by disregarding the frequency deviation and nonlinearity. They are called the *resonant terms*. Thus, for resonance $\omega = \frac{1}{3}$, the resonant terms are those for which

$$\alpha - \beta + 3k = 0.$$

Of the terms of third order, only $z^3 e^{-it}$ and $\bar{z}^3 e^{it}$ turn out to be resonant. Thus we can reduce the hamiltonian function to the form

$$-2iH = -i\omega z\bar{z} + hz^3 e^{-it} - \bar{h}\bar{z}^3 e^{it} + \cdots$$

(the conjugacy of h and $\bar{h}$ corresponds to the fact that H is real).

Note that, in order to reduce the hamiltonian function to this normal form, we made a 2π-periodic time-dependent smooth canonical transformation which depends smoothly on the parameter, even in the case of resonance. This transformation differs from the identity only by terms that are small of second order relative to the deviation from the closed trajectory (and its generating function differs from the generating function of the identity only by cubic terms).

Further investigation of the behavior of solutions of Hamilton's equations proceeds in the following way. First, we throw out of the hamiltonian function all terms of order higher than three and study the solutions of the resulting truncated system. Then we must see how the discarded terms can affect the behavior of the trajectories.

The study of the truncated system can be simplified by introducing a coordinate system in the complex z-plane which rotates uniformly with angular velocity $\frac{1}{3}$, i.e., by the substitution $z = \zeta e^{it/3}$. Then for the variable ζ we obtain an autonomous hamiltonian system with hamiltonian function

$$-2iH_0 = -i\varepsilon\zeta\bar{\zeta} + h\zeta^3 - \bar{h}\bar{\zeta}^3, \qquad \text{where } \varepsilon = \omega - (\tfrac{1}{3}).$$

The fact that, in a rotating coordinate system, the truncated system is autonomous is very good luck. The total system of Hamilton's equations (including terms of degree higher than three in the hamiltonian) is not only not autonomous in a rotating coordinate system, but is not even 2π-periodic (but only 6π-periodic) in time. The autonomous system with hamiltonian H_0 is essentially the result of averaging the original system over closed trajectories of the linear system with $\varepsilon = 0$ (where we disregard terms of degree higher than three).

The coefficient h can be made real (by a rotation of the coordinate system). Thus the hamiltonian function in the real coordinates (x, y) is reduced to the form

$$H_0 = \frac{\varepsilon}{2}(x^2 + y^2) + a(x^3 - 3xy^2).$$

The coefficient a depends on the frequency deviation ε as on a parameter. For $\varepsilon = 0$ this coefficient is generally not zero. Therefore, we can make this coefficient equal to 1 by a smooth change of coordinates depending on a parameter. Thus we must investigate the dependence on the small parameter ε of the phase portrait of the system with hamilton function

$$H_0 = \frac{\varepsilon}{2}(x^2 + y^2) + (x^3 - 3xy^2)$$

in the (x, y)-plane.

It is easy to see that this dependence consists of the following (Fig. 239).

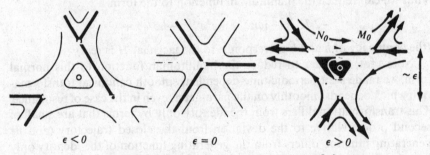

$\varepsilon < 0$ $\qquad\qquad\qquad$ $\varepsilon = 0$ $\qquad\qquad\qquad$ $\varepsilon > 0$

Figure 239   Passage through resonance 3:1

For $\varepsilon = 0$ the zero level set of the function H_0 consists of three straight lines through 0, intersecting at angles of 60°. Under a change of ε the level line always consists of three straight lines, where these three lines are moved forward as ε changes, always forming an equilateral triangle with center at the origin. The vertices of this triangle are saddle points of the hamiltonian function. As ε passes through zero (i.e., upon passage through resonance), the critical point at the origin changes from a minimum to a maximum.

Thus, for a system with hamiltonian function H_0, the origin is a stable equilibrium position for all values of the parameter except at resonance, and at resonance the origin is unstable. For values of the parameter close to resonance, the triangle close to the origin filled by closed phase curves is small (of order ε), so the "radius of stability" of the origin approaches zero as $\varepsilon \to 0$: a small (of order ε) perturbation of the initial condition is sufficient to make a phase point move outside the triangle and begin to go away from the equilibrium position.

Returning to the original problem of the periodic trajectory, we come to the following conclusions (which, of course, are not proven, since we threw out terms of degree higher than three, but which can be justified):

1. At the moment of passage through the resonance 3 : 1 a periodic trajectory generally loses its stability.
2. For values of the parameter close to resonance there is an unstable periodic trajectory near the periodic trajectory under consideration on the same energy level manifold. It is closed after making three circulations along the original trajectory and one revolution around it. For the resonance value of the parameter, this unstable trajectory merges with the original one.
3. The distance of this unstable periodic trajectory from the original decreases, as we approach resonance, to first order in the frequency deviation (i.e., as the first order of the difference of the parameter from the resonance value).
4. Through this unstable trajectory on the same three-dimensional energy level manifold there pass two two-dimensional invariant surfaces, filled with trajectories approximating this unstable periodic trajectory as $t \to \infty$ on one surface and as $t \to -\infty$ on the other.
5. The location of the separatrices is such that, by intersecting with a manifold transversal to the original trajectory, we obtain a figure close to the three sides of an equilateral triangle and their continuations. The vertices of the triangle are the points of intersection of the unstable periodic trajectory with the transversal manifold.
6. For initial conditions inside the triangle formed by the separatrices, a phase point stays near the original periodic trajectory (at a distance of order ε) for a long time (of order not less than $1/\varepsilon$), and for initial conditions outside the triangle it goes off quite rapidly to a distance which is large in comparison with ε.

E *Splitting of separatrices*

In reality, the separatrices we talked about in statements 4, 5, and 6 above have a very complicated structure (because of the influence of the terms of order higher than three which we disregarded in our approximation). In order to understand the situation, it is convenient to look at a two-dimensional surface transversally intersecting the original closed trajectory at some point on it (and lying entirely in one energy level manifold).[106] Trajectories beginning on this surface intersect it again after a time close to the time of circulation around the original closed trajectory. Thus we have a mapping of a neighborhood of the point of intersection of the closed trajectory with the surface onto a part of the surface. This mapping has a fixed point (at the point where the closed trajectory intersects the surface) and is approximately a rotation by 120° around this point, which we take for the origin in our surface.

We now consider the third power of the mapping indicated above. This is again a mapping of some neighborhood of the origin to a part of the surface, leaving the origin fixed. But now this mapping is approximately rotation by 360°, i.e., the identity: it is realized by the trajectories of our system after approximately three periods of our closed trajectory.

The calculations above give nontrivial information about the structure of this "mapping after three periods." In fact, by throwing out the terms of degree four and higher in the hamiltonian function, we change the terms of degree three and higher of the mapping. Therefore, the mapping after three periods which corresponds to the truncated hamiltonian function approximates (with cubic error) the actual mapping after three periods.

But we know the properties of the mapping after three periods corresponding to the truncated hamiltonian function, since it is the mapping of the phase flow of the system with hamiltonian function $H_0(x, y)$ after time 6π (the proof is based on the fact that after time 6π our rotating coordinate system returns to the original position). We now look at which of these properties are preserved for perturbations of third-order smallness relative to the distance from the fixed point, and which are not.

We let A_0 denote the mapping after three periods for the truncated system, and A the actual mapping after three periods.

1. The mapping A_0 is included in a flow: it is the transformation after time 6π in the phase flow with hamiltonian H_0.

   There is no reason to think that the mapping A is included in a flow.
2. The mapping A_0 is symmetric under a rotation by 120°: there is a nontrivial diffeomorphism g for which $g^3 = E$ and which commutes with A_0.

   There is no reason to think that the mapping A commutes with any nontrivial diffeomorphism g satisfying $g^3 = E$.

[106] Here we have the following general phenomenon: it is easier to think about mappings after a period, and easier to calculate with flows.

3. The mapping A_0 has three unstable fixed points at a distance ε from the origin, approximately the vertices of an equilateral triangle. For sufficiently small deviations from resonance (i.e., for sufficiently small ε) the mapping A also has three unstable fixed points near the vertices of an equilateral triangle. This follows from the implicit function theorem.

4. The separatrices of fixed points of the mapping A_0 form, for values of the parameter close to (but not at) resonance, a figure approximating the sides and extended sides of an equilateral triangle. If we begin with a point on one of the sides of the triangle, then after repeated applications of A_0 we obtain a sequence of points on the same side of the triangle approaching one of the vertices bounding the side, say M_0. Applying A_0^{-1}, we obtain a sequence approaching the other vertex, which we will denote by N_0.

Each of the three unstable fixed points of the mapping A also has separatrices approximating the sides of a triangle (Figure 240). Namely, those points of the plane which approach the fixed point M after applying the mappings A^n, $n \to +\infty$, form a smooth curve Γ^+ invariant under A, passing through M and, near M, close to the side $M_0 N_0$ of the separatrices of A_0. The points which approach N after applications of A^n, where $n \to -\infty$, form another smooth invariant curve Γ^-, passing through N and also near $M_0 N_0$ near N_0.

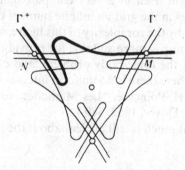

Figure 240   Splitting of separatrices

However the two curves Γ^+ and Γ^-, both near the line $M_0 N_0$, are not at all obliged to coincide. This is the phenomenon of splitting of separatrices, which accounts for the differing behavior of the trajectories of the truncated and total systems.

The magnitude of the splitting of separatrices is exponentially small for small ε; therefore it is easy to overlook the phenomenon of splitting in calculations in one or another scheme of "perturbation theory." However, this phenomenon is very important in fundamental questions. For example, its existence immediately implies the *divergence* of the series in numerous versions of perturbation theory (since if the series converged, there would be no splitting).

In general, the divergence of series in perturbation theory (while a good approximation is given by a few initial terms) is usually related to the fact that we are looking for an object which does not exist. If we try to fit a phenomenon to a scheme which actually contradicts the essential features of the phenomenon, then it is not surprising that our series diverge.

The Birkhoff series (which are obtained if one continues infinitely the normalizations of the initial terms of the Taylor series of the hamiltonian function) are one example of a formally convergent, but actually divergent, scheme of perturbation theory. If these series converged, then a general oscillating system with one degree of freedom with periodic coefficients would be reduced near an equilibrium position to an autonomous normal form and there would be no splitting of separatrices in it (whereas in fact there is).

Returning to the original closed trajectory, we see that the three unstable fixed points of the mapping A correspond to an unstable closed trajectory near the original triple. There is a family of trajectories approaching this unstable trajectory as $t \rightarrow +\infty$, and another family of trajectories approaching the unstable one as $t \rightarrow -\infty$. The points of the trajectories of each of these families form a smooth surface containing our unstable trajectory.

These two surfaces are also the separatrices we talked about in statements 4, 5, and 6 of Section D. By intersecting them with our transversal surface we obtain the invariant curves Γ^+ and Γ^- of the mapping A. The intersections of these two curves form a complicated network about which H. Poincaré, who first discovered the phenomenon of splitting of separatrices, wrote, "The intersections form a type of lattice, tissue, or grid with infinitely fine mesh. Neither of the two curves must ever cut across itself again, but it must bend back upon itself in a very complex manner in order to cut across all of the squares in the grid an infinite number of times.

"One will be struck by the complexity of this figure, which I shall not even attempt to draw. Nothing is more suitable for providing us with an idea of the complex nature of the three-body problem, and of all the problems of dynamics in general, where there is no uniform integral and where the Bohlin series are divergent." (H. Poincaré, "Les Méthodes Nouvelles de la Mécanique Céleste," Vol. III, Dover, 1957, 389.)

We should note that much is still unclear about the picture of intersecting separatrices.

F *Resonances of higher order*

Resonances of higher order can also be studied using a normal form. In this connection, we note that resonances of order higher than 4 do not usually induce instability, since in the normal form terms of degree 4 appear, guaranteeing a minimum or maximum of the function H_0 even at resonance.

In the case of resonance of order $n > 4$, the typical development of the phase portrait of the system with hamiltonian function H_0 is given by the formula

$$H_0 = \varepsilon\tau + \tau^2\alpha(\tau) + a\tau^{n/2}\sin n\varphi,$$

$$2\tau = p^2 + q^2, \qquad \alpha(0) = \pm 1,$$

and consists of the following (Figure 241).

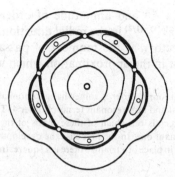

Figure 241   Averaged hamiltonian of phase oscillations near resonance 5:1

For small (of order ε) deviations of the frequency from resonance, and at a small (of order $\sqrt{|\varepsilon|}$) distance from the equilibrium position at the origin, the function H_0 has $2n$ critical points near the vertices of a regular n-gon with center at the origin. Half of these critical points are saddle points, and the other half are maxima if the origin is a minimum or minima if the origin is a maximum. The saddle points and stable points alternate. All n saddle points lie on one level of the function H_0; their separatrices, connecting successive saddle points, form n "islands," each of which is filled with closed phase curves encircling a stable point. The width of the islands is of order $\varepsilon^{(n/4)-(1/2)}$. The closed phase curves inside each island are called "phase oscillations" (since what varies essentially is the phase of the oscillations around the origin). The period of the phase oscillations grows with decreasing frequency deviation ε like $\varepsilon^{-n/4}$.

Inside the narrow ring formed by the islands, closer to the origin, there are closed phase curves encircling the origin; outside the ring the phase curves are closed, but motion along them proceeds in the direction opposite to that inside the ring. We note that the radius of the ring has order $\sqrt{|\varepsilon|}$ independently of the order of resonance, if this order is greater than 4. Also, the ring of islands exists for only one of the two signs of ε.

If we pass from the truncated system with hamiltonian H_0 to the total system, the separatrices split in a way similar to that described above for resonance of order 3. The size of the splitting of the separatrices is exponentially small (or order $e^{-1/\varepsilon^{n/4}}$), but the splitting is of fundamental importance for investigating stability, especially in the multi-dimensional case.

Returning to our original closed trajectory, we have the following picture. As we approach resonance along the ε axis from one side,[107] two periodic trajectories split off from our periodic trajectory: a stable one and an unstable one. These new trajectories close up after n circulations along the original trajectory and lie at a distance of order $\sqrt{|\varepsilon|}$ from the original trajectory. Near the stable trajectory there is a zone of slow phase oscillations

[107] Unlike resonance of order 3, for which there is an unstable periodic trajectory branching off from both sides of the resonance.

with period of order $\varepsilon^{-n/4}$ and amplitude of order π/n in the azimuthal direction and of order $\varepsilon^{(n/4)-(1/2)}$ in the radial direction. Loss of stability of the original periodic trajectory at the moment of passage through resonance does not occur, at least in the approximation which we have considered.

The case of resonance of fourth order is somewhat exceptional. In this case, in the normal form there are both resonant and non-resonant terms of order 4. The shape of the phase curves of the truncated system depends on which of these terms of the normal form dominates, a resonant one or a non-resonant one. In the first case the development is the same as for third-order resonance, except that in place of a triangle there is a square. In the second case the development is the same as for $n > 4$.

In conclusion, we remark that the given normal form becomes a better approximation as we get closer to resonance ($\varepsilon \ll 1$) and as the deviation of the initial point from the periodic trajectory gets smaller. That is, as the period of the closed trajectory and the period of oscillation of neighboring trajectories near it become more exactly commensurable, and as the initial condition approaches the closed trajectory, the interval of time grows on which our approximation accurately describes the behavior of the phase curves.

No conclusion about the behavior of non-closed phase curves on infinite intervals of time (for example, about the Liapunov stability of the original periodic trajectory) follows from our arguments, since the terms of higher order which were thrown out in reducing to normal form can, over an infinite period of time, completely change the character of the motion. Actually, under the conditions considered, the original periodic trajectory is Liapunov stable, but the proof requires substantially new techniques beyond the Birkhoff normal form (cf. Appendix 8).

Appendix 8: Theory of perturbations of conditionally periodic motion, and Kolmogorov's theorem

The collection of solvable "integrable" problems which we have at our disposal is not large (one-dimensional problems, motion of a point in a central field, eulerian and lagrangian motions of a rigid body, the problem of two fixed centers, and motion along geodesics on the ellipsoid). However, with the help of these "integrable cases," we can obtain meaningful information about motions of many important systems by considering an integrable problem as a first approximation.

An example of such a situation is the problem of motion of the planets around the sun under the law of universal gravitation. The mass of the planets is approximately 0.001 of the mass of the sun, so in a first approximation we can disregard the interaction of the planets on one another and consider only the attraction by the sun. As a result, we obtain the exactly integrable problem of the motion of non-interacting planets around the sun; each planet will describe its keplerian ellipse independently of the others, and the motion of the system as a whole will be conditionally periodic. If we now consider the interactions of the planets on one another, the keplerian motion of each planet will be slightly changed.

We call upon the theory of perturbations from celestial mechanics to study this interaction. It is clear that calculations for time of the order of 1,000 years do not present any fundamental difficulties. However, if we want to study longer intervals of time, and especially if we are interested in qualitative questions about the behavior of exact solutions of the equations of motion on an infinite time interval, then such difficulties arise. The accumulation of perturbations after an interval of time which is large in comparison to 1,000 years could cause a complete change in the character of the motion: for example, the planets could fall into the sun, escape from it, or collide with one another.

Note that the question of the behavior of solutions of the equations of motion on an infinite time interval has only an indirect relation to the problem of the motion of real planets. The reason is that, after intervals of billions of years, small non-conservative effects not considered in Newton's equations become important. Thus, the effects of the gravitational interaction of the planets are of real importance only when they seriously change the picture of motion within a finite time which is small in comparison with the time of development of non-conservative effects.

In calculating motion over such finite times, computers prove to be very useful, quickly determining the motion of the planets for many thousands of years in the future or past. However, we should note that even the application of modern calculating methods may be insufficient to predict the influence of perturbations if a phase point falls in the zone of exponential instability.

Asymptotic and qualitative methods have even greater value for the study of charged particles in magnetic fields, since in this situation a particle outstrips the computer and makes so many orbits that mechanical calculation of its trajectory is impossible even in the absence of exponential instability.

A whole series of methods has been devised for calculating perturbations in celestial mechanics. (A detailed analysis of them can be found in the book,

"Les Méthodes Nouvelles de la Mécanique Céleste," by H. Poincaré, Dover, 1957.)

A difficulty with all of these methods is that they lead to *divergent* series and therefore give no information about the behavior of motion as a whole over infinite intervals of time. The reason for the divergence of series in the theory of perturbations is "small denominators": integral linear combinations of frequencies of unperturbed motions by which it is necessary to divide in calculating the influence of perturbations. For exact resonance (i.e., for commensurable frequencies) these denominators vanish, and the corresponding term of the series in the theory of perturbations becomes infinitely large. Close to resonance, this term of the series is very large.

Thus, for example, in their motion around the sun, Jupiter and Saturn, in one day, go through approximately 299 and 120.5 seconds of arc respectively. Therefore, the denominator $2\omega_J - 5\omega_S$ is very small in comparison with each of their frequencies. This amounts to a large long-period perturbation of the planets on one another (its period is about 800 years); the study by Laplace of this effect was one of the first successes of the theory of perturbations.

We note that the difficulty caused by small denominators is essential. The rational numbers form a dense set; thus in the phase space of an unperturbed problem, initial conditions for which we have resonance and the small denominators vanish form a dense set. Hence, the functions given by the series of perturbation theory have a dense set of singular points.

The difficulty mentioned here is characteristic not only for problems of celestial mechanics, but for all problems which are close to integrable (for instance, for the problem of an asymmetrical rigid top under very fast rotation). Poincaré himself called the problem of studying perturbations of conditionally-periodic motions in a system given by the hamiltonian

$$H = H_0(I) + \varepsilon H_1(I, \varphi), \qquad \varepsilon \ll 1,$$

in action-angle variables I and φ, the *fundamental problem of dynamics*. Here H_0 is the hamiltonian of the unperturbed problem, and εH_1 a perturbation which is a 2π-periodic function of the angle variables $\varphi_1, \ldots, \varphi_n$. In the unperturbed problem ($\varepsilon = 0$) the angles φ change uniformly with constant frequencies

$$\omega_k = \frac{\partial H_0}{\partial I_k},$$

and all the action variables are first integrals.

We must investigate the phase curves of Hamilton's equations

$$\dot{I} = -\frac{\partial H}{\partial \varphi} \qquad \dot{\varphi} = \frac{\partial H}{\partial I}$$

in a phase space which is a direct product of a region in n-dimensional space with coordinates I and the n-dimensional torus with angular coordinates φ.

A substantial advance in the study of phase curves of this perturbed problem was begun in 1954 with the work of A. N. Kolmogorov in "On conservation of conditionally-periodic motions for a small change in Hamilton's function," Dokl. Akad. Nauk SSSR 98:4 (1954) 525–530 (Russian). In this appendix we present the basic results obtained since then in this area. The proofs can be found in the following works:

V. I. Arnold, "Small denominators I, Mapping the circle onto itself," Izv. Akad. Nauk SSSR Ser. Mat. 25 (1961), 21–86.

V. I. Arnold, "Small denominators II, Proof of a theorem of A. N. Kolmogorov on the preservation of conditionally periodic motions under a small perturbation of the Hamiltonian," Russian Math. Surveys 18:5 (1963).

V. I. Arnold, "Small denominators III. Small denominators and problems of stability of motion in classical and celestial mechanics." Russian Math. Surveys 18:6 (1963).

V. I. Arnold, A. Avez, Ergodic problems of classical mechanics, New York, Benjamin, 1968.

J. Moser, On invariant curves of area-preserving mappings of an annulus (Nachr. Akad. Wiss. Göttingen, Math. Phys. Kl IIa, (1962) 1–20).

J. Moser, A rapidly converging iteration method and nonlinear differential equations, (Annali della Scuola Norm. Sup. di Pisa, (3), 20 (1966), 265–315: (1966), 499–535.

J. Moser, Convergent series expansions for quasi-periodic motions, Math. Ann. 169 (1967), 136–176.

C. L. Siegel, J. K. Moser, Lectures on Celestial Mechanics, Springer-Verlag, 1971.

S. Sternberg, Celestial Mechanics, I, II, New York, Benjamin, 1969.

Before formulating our results, we will briefly discuss the behavior of phase curves in the unperturbed problem already studied in Chapter 10.

A Unperturbed motion

The system with hamiltonian $H_0(I)$ has n first integrals in involution (the n action variables). Every level set of all these integrals is an n-dimensional torus in $2n$-dimensional phase space. This torus is invariant with respect to the phase flow of the unperturbed system: every phase curve starting at a point of our torus remains on it.

The motion of a phase point on the invariant torus $I = $ const is conditionally-periodic. The frequencies of this motion are the derivatives of the unperturbed hamiltonian with respect to the action variables:

$$\dot{\varphi}_k = \omega_k(I), \quad \text{where } \omega_k = \frac{\partial H_0}{\partial I_k}.$$

Therefore, the phase curve densely fills a torus whose dimension is equal to the number of frequencies ω_k which are arithmetically independent.

We note that the frequencies depend on which torus we are looking at; i.e., which values of the first integrals we have fixed. A system of n functions ω of n variables I is generally functionally independent; in such a case we can simply number the tori by their frequencies, choosing the variables ω for coordinates in a neighborhood of the point under consideration in the space of action variables I.

The case when the frequencies are functionally independent will be called the *nondegenerate case*. The conditions for nondegeneracy have the form

$$\det \left| \frac{\partial \omega}{\partial I} \right| = \det \left| \frac{\partial^2 H_0}{\partial I^2} \right| \neq 0.$$

Thus, in the nondegenerate case, the unperturbed problem determines on the different invariant tori in phase space conditionally-periodic motions with different frequencies. In particular, the invariant tori on which the number of frequencies is maximal (i.e., n) form a dense set in phase space; such tori are called non-resonant tori.

It can be shown that the non-resonant tori form a set of full measure, i.e., the Lebesgue measure of the union of all invariant resonant tori of the unperturbed non-degenerate system is equal to zero. Nevertheless, invariant resonant tori exist and are mixed in with the non-resonant tori in such a way that they too form a dense set. Furthermore, the set of resonant tori with any number of independent frequencies from 1 to $n - 1$ is dense. In particular, the invariant tori on which all phase curves are closed (the number of independent frequencies is 1) form a dense set. Nevertheless, we note that the probability of landing on a resonant torus by a random choice of initial point in the phase space of the unperturbed system, is equal to zero (since the probability of landing on a rational number by a random choice of a real number is zero). Thus, by disregarding sets of measure zero, we can say that almost all invariant tori in a nondegenerate unperturbed system are non-resonant and have a total set of n arithmetically independent frequencies.

On a non-resonant torus, the trajectory of a conditionally-periodic motion is dense. Thus, for almost all initial conditions, a phase curve of a non-degenerate unperturbed system densely fills an invariant torus whose dimension is equal to the number of degrees of freedom (i.e., half the dimension of the phase space).

To better understand the whole picture, we consider the case of two degrees of freedom ($n = 2$). In this case, the phase space is four-dimensional so each energy level set is three-dimensional. We fix one such level set. This three-dimensional manifold, fibered by two-dimensional tori, can be represented in ordinary three-dimensional space as a family of concentric tori lying inside one another (Figure 242).

Figure 242   Invariant tori in a three-dimensional energy level manifold

The phase curves are windings of these tori; both frequencies of circulation change from torus to torus. In general, not only both frequencies but also their ratio will change from torus to torus. If the derivative of the ratio of frequencies with respect to the action variable numbering the tori on the given level set of the function H_0 is not zero, then we say that our system is *isoenergetically nondegenerate*. The condition for isoenergetic nondegeneracy has (as is easy to calculate) the form

$$\det \begin{vmatrix} \dfrac{\partial^2 H_0}{\partial I^2} & \dfrac{\partial H_0}{\partial I} \\[2ex] \dfrac{\partial H_0}{\partial I} & 0 \end{vmatrix} \neq 0.$$

The conditions for nondegeneracy and isoenergetic nondegeneracy are independent from one another; i.e., a nondegenerate system could be isoenergetically degenerate, and an iso-energetically nondegenerate system could be degenerate. In the many-dimensional case ($n > 2$) isoenergetic nondegeneracy means nondegeneracy of the following mapping of the $(n-1)$-dimensional level manifold of the function H_0 of n action variables to the projective space of dimension $n - 1$:

$$I \to (\omega_1(I) : \omega_2(I) : \cdots : \omega_n(I)).$$

Now consider an isoenergetically nondegenerate system with two degrees of freedom. It is easy to construct a two-dimensional plane in the three-dimentional energy level set transversally intersecting the two-dimensional tori of our family (in a family of concentric circles in the model in three-dimensional euclidean space).

A phase curve beginning in such a plane returns to it after making a circuit around the torus. As a result we obtain a new point on the same circle in which the torus intersects the plane. In this way there arises a mapping of the plane to itself.

This mapping of the plane to itself fixes the concentric meridian circles in which the plane intersects the invariant tori. Every circle is rotated through some angle, namely through that fraction of an entire revolution that the frequency along the meridian constitutes of the frequency along the equator.

If the system is isoenergetically nondegenerate, the angle of revolution of invariant circles in the plane of intersection changes from one circle to another. Therefore, on some circles this angle will be commensurable with a whole revolution, and on others it will be incommensurable. Each of these classes of circles will form a dense set, but on almost all circles (in the sense of Lebesgue measure) the angle of rotation will be incommensurable with a whole revolution.

The commensurability or incommensurability is manifested in the following way on the behavior of points of a circle under the mapping of the region to itself. If the angle of rotation is commensurable with a whole rotation, then after several iterations of the mapping the point will return to its initial position (the number of iterations will be larger as the denominator of the fraction expressing the angle of rotation is larger). If the angle of rotation is incommensurable with a whole rotation, the successive images of the point under repetitions of the mapping will densely fill up the meridian circle.

We note further that commensurability corresponds to resonant tori and incommensurability to non-resonant tori. Also, the existence of resonant tori implies the following property. Consider some power of the mapping of our region to itself induced by the phase curves. Let the exponent be the denominator of the fraction expressing the ratio of the frequencies on one of the resonant tori. Then the mapping raised to the indicated power has a whole circle consisting entirely of fixed points (namely, the meridian of the resonant torus under consideration).

Such behavior of fixed points is unnatural for mappings in any sort of general form, even canonical mappings (fixed points are usually isolated). In the given case, a whole circle of fixed points arises because we have considered an unperturbed integrable system. For arbitrarily small perturbations of general form, this property of the mapping (having a whole circle of fixed points) must fail. The circle of fixed points must be dispersed so that only a finite number remain.

In other words, under small perturbations of our integrable system, we expect a change in the qualitative picture of the phase curves, if only in the respect that entire invariant tori filled out by closed phase curves will disintegrate so that there remain only a finite number of closed curves, near those for the unperturbed system, and the remaining phase curves will be more complicated. We have already encountered such a case in Appendix 7 in investigating phase oscillations near resonance.

We now consider what happens to non-resonant invariant tori under a small perturbation of a hamiltonian function. Formal application of the principle of averaging (i.e., the first approximation of the classical theory of perturbations, cf. Section 52) leads us to the conclusion that a non-resonant torus does not undergo any evolution.

We note that the fact that the perturbations are hamiltonian is essential, since for non-conservative perturbations it is clear that the action variables may evolve. In celestial mechanics, their evolution means a secular change in the major semi-axes of the keplerian ellipses, i.e., the planets falling into the sun, colliding, or escaping to a large distance in a time which is inversely proportional to the size of the perturbation. If conservative perturbations led to evolutions in a first approximation, this would manifest itself in the fate of the planets after a time on the order of 1,000 years. Fortunately, the order of magnitude of the non-conservative perturbations is much less.

The theorem of Kolmogorov, formulated below, furnishes one justification for the conclusion, drawn from the non-rigorous theory of perturbations, about the absence of evolution of action variables.

B *Invariant tori in a perturbed system*

Theorem. *If an unperturbed system is nondegenerate, then for sufficiently small conservative hamiltonian perturbations, most non-resonant invariant tori do not vanish, but are only slightly deformed, so that in the phase space of the perturbed system, too, there are invariant tori densely filled with phase curves winding around them conditionally-periodically, with a number of independent frequencies equal to the number of degrees of freedom.*

These invariant tori form a majority in the sense that the measure of the complement of their union is small when the perturbation is small.

A. N. Kolmogorov's proof of this theorem is based on the following two observations.

1. We fix a non-resonance set of frequencies of the unperturbed system so that the frequencies are not only independent, but do not even approximately satisfy any resonance conditions of low order. More precisely, we fix a set of frequencies ω for which there exist C and ν such that $|(\omega, k)| > C|k|^{-\nu}$ for all integral vectors $k \neq 0$.

It can be shown that, if ν is sufficiently large (say $\nu = n + 1$), then the measure of the set of such vectors ω (lying in a fixed bounded region) for which the indicated condition of non-resonance is violated, is small when C is small.

Next, near a non-resonant torus of the unperturbed system corresponding to a fixed value of the frequencies, we will look for an invariant torus of the perturbed system on which there is conditionally-periodic motion with exactly the same frequencies as the ones we fixed, and which necessarily satisfy the condition of being non-resonant described above.

In this way, instead of the variations of frequency customary in perturbation schemes (consisting of the introduction of frequencies depending on the perturbation), we must hold constant the non-resonant frequencies, while selecting initial conditions depending on the perturbation in order to guarantee motion with the given frequencies. This can be done by a small (when the perturbation is small) change of initial conditions, because the frequencies change with the action variables according to the non-degeneracy condition.

2. The second observation is that, to find an invariant torus, instead of using the usual series expansion in powers of the perturbation parameter, we can use a rapidly convergent method similar to Newton's method of tangents.

Newton's method of tangents for finding roots of algebraic equations with initial error ε gives, after n approximations, an error of order ε^{2^n}. Such *super-convergence* allows us to paralyze the influence of the small denominators appearing in every approximation, and in the end succeeds not only in carrying out an infinite number of approximations, but also in showing the convergence of the entire procedure.

The assumption under which all this can be done is that the unperturbed hamiltonian function $H_0(I)$ is analytic and nondegenerate, and the perturbing hamiltonian function $\varepsilon H_1(I, \varphi)$ is analytic and 2π-periodic in the angle variables φ. The presence of the small parameter ε is immaterial: it is important only that the perturbation be sufficiently small in some complex neighborhood of radius ρ of the real plane of the variables φ (less than some positive function $M(\rho, H_0)$).

As J. Moser showed, the requirement of analyticity can be changed to differentiability of sufficiently high order if we combine Newton's method with an idea of J. Nash, the application of a smoothing operator at each approximation.

The resulting conditionally-periodic motions of the perturbed system with fixed frequencies ω turn out to be smooth functions of the parameter ε of perturbation. Therefore, they could have been sought, without Newton's method, in the form of a series in powers of ε. The coefficients of this series, called the Lindstedt series, can actually be found; however, we can prove its convergence only indirectly, with the help of newtonian approximations.

C Zones of instability

The presence of invariant tori in the phase space of the perturbed problem means that, for most initial conditions in a system which is nearly integrable, motion remains conditionally periodic with a maximal set of frequencies.

The question naturally arises of what happens to the remaining phase curves, with initial conditions falling into the gaps between the invariant tori which replace the resonant invariant tori of the non-perturbed problem.

The disintegration of a resonant torus on which the number of frequencies is one less than the maximum is easy to investigate in a first-order perturbation theory. To do this, we must average the perturbation over the $(n-1)$-dimensional invariant tori into which the resonant invariant torus is decomposed and which are densely filled out by phase curves of the unperturbed system. After averaging, we obtain a conservative system with one degree of freedom (cf. the investigation of phase oscillations near resonance in Appendix 7), which is easy to study.

In the approximation under consideration we have, near the n-dimensional reducible torus, stable and unstable $(n-1)$-dimensional tori, with phase oscillations around the stable ones. The corresponding conditionally-periodic motions have a full set of n frequencies, of which $n-1$ are the fast frequencies of the original oscillations and one is the slow (of order $\sqrt{\varepsilon}$) frequency of the phase oscillations.

However, one must not conclude that the only difference between motions in the unperturbed and perturbed systems is the appearance of "islands" of phase oscillations. In fact, the actual phenomena are much more complicated than the first approximation described above. One manifestation of this complicated behavior of the phase curves of the perturbed problem is the splitting of separatrices discussed in Appendix 7.

Appendix 8: Theory of perturbations of conditionally periodic motion

To study motions of a perturbed system outside of the invariant tori we must distinguish the cases of two and higher degrees of freedom. For two degrees of freedom, the dimension of the phase space is four, and an energy level manifold is three-dimensional. Therefore, the invariant two-dimensional tori divide each energy level set. Thus, a phase curve beginning in the gap between two invariant tori of the perturbed system remains forever confined between those tori. No matter how complicated this curve appears, it does not leave its gap, and the corresponding action variables remain forever near their initial conditions.

If the number n of degrees of freedom is greater than two, the n-dimensional invariant tori do not divide the $(2n - 1)$-dimensional energy level manifold but are arranged in it like points on a plane or lines in space. In this case the "gaps" corresponding to different resonances are connected to one another, so the invariant tori do not prevent phase curves starting near resonance from going far away. Hence, there is no reason to expect that the action variables along such a phase curve will remain close to their initial values for all time.

In other words, under sufficiently small perturbations of systems with two degrees of freedom (satisfying the generally fulfilled condition of iso-energetic nondegeneracy), not only do the action variables along a phase trajectory have no secular perturbations in any approximation of perturbation theory (i.e., they change little in a time interval on the order of $(1/\varepsilon)^N$ for any N, where ε is the magnitude of the perturbation), but these variables remain forever near their initial values. This is true, both for non-resonant phase curves conditionally-periodically filling out two-dimensional tori (and comprising most of the phase space), and for the remaining initial conditions.

At the same time, there exist systems with more than two degrees of freedom satisfying all the nondegeneracy conditions, in which, although for most initial conditions motion is conditionally periodic, for some initial conditions a slow drift of the action variables away from their initial values occurs. The average velocity of this drift in known examples[108] is on the order of $e^{-1/\sqrt{\varepsilon}}$, i.e., this velocity decreases faster than any power of the perturbation parameter. Thus it is not surprising that this drifting away does not appear in any approximation of perturbation theory. (By average velocity, we mean the ratio of the increase of action variables to time, so that we are actually dealing with an increase of order 1 after a time of order $e^{1/\sqrt{\varepsilon}}$).

An upper bound on the average velocity of the drift of the action variables in general nearly integrable systems of hamiltonian equations with n degrees of freedom is included in the recent work of N. N. Nekhoroshev.[109]

---

[108] Cf. V. I. Arnold, Instability of dynamical systems with many degrees of freedom. *Soviet Mathematics* **5**:3 (1964) 581–585.

[109] N. N. Nekhoroshev, The behavior of hamiltonian systems that are close to integrable ones, Functional Analysis and Its Applications, 5:4 (1971); Uspekhi Mat. Nauk 32:6 (1977).

This bound, like the lower bound mentioned above, has the form e^{-1/ε^d}; thus the increase of the action variables is small while the time is small in comparison with e^{1/ε^d}, if $\varepsilon < \varepsilon_0$. Here ε is the magnitude of the perturbation, and d is a number between 0 and 1 defined, like ε_0, by the properties of the unperturbed hamiltonian H_0. In addition, a nondegeneracy condition is imposed on the unperturbed hamiltonian (this condition has a long formulation, but is generally satisfied; in particular, strong convexity of the unperturbed hamiltonian is sufficient, i.e., positive or negative definiteness of the second differential of H_0).

From this upper bound it is clear that secular changes of the action variables are not detected by any approximation of perturbation theory, since the average velocity of these changes is exponentially small. We note also that secular changes of the action variables obviously have no directional character, but are represented by more or less random wandering in the resonant regions between the invariant tori. A more detailed discussion of the questions arising here can be found in the article, "Stochastic instability of nonlinear oscillations," by G. M. Zaslavski and B. V. Chirikov, Soviet Physics Uspekhi, v. 105, no. 1 (1971), 3–39.

D Variants of the theorem on invariant tori

Statements analogous to the theorem on conservation of invariant tori in an autonomous system have been proved for non-autonomous equations with periodic coefficients and for symplectic mappings. Analogous statements are valid in the theory of small oscillations in a neighborhood of an equilibrium position of an autonomous system or a system with periodic coefficients, as well as in a neighborhood of a closed phase curve of a phase flow or in a neighborhood of a fixed point of a symplectic mapping.

The nondegeneracy conditions necessary in the various cases are different. For reference, we will now give these nondegeneracy conditions. We will limit ourselves to the simplest requirements of nondegeneracy, which are all fulfilled by systems in "general position." In many cases, the requirements of nondegeneracy can be weakened, but the advantage gained by this is offset by the complication of the formulas.

1. *Autonomous systems.* The hamiltonian function is

$$H = H_0(I) + \varepsilon H_1(I, \varphi), \qquad I \in G \subset \mathbb{R}^n, \varphi \bmod 2\pi \in T^n.$$

The nondegeneracy condition

$$\det \left| \frac{\partial^2 H_0}{\partial I^2} \right| \neq 0$$

guarantees preservation[110] of most invariant tori under small perturbations ($\varepsilon \ll 1$).

---

[110] It is understood that the tori are slightly deformed under perturbations.

The condition for isoenergetic nondegeneracy

$$\det \begin{vmatrix} \dfrac{\partial^2 H_0}{\partial I^2} & \dfrac{\partial H_0}{\partial I} \\[2ex] \dfrac{\partial H_0}{\partial I} & 0 \end{vmatrix} \neq 0$$

guarantees the existence on every energy level manifold of a set of invariant tori whose complement has small measure. The frequencies on these tori generally depend on the size of the perturbation, but the ratios of frequencies are preserved under changes in ε.

If $n = 2$, then the condition for isoenergetic nondegeneracy also guarantees stability of the action variables, in the sense that they remain forever close to their initial values for sufficiently small perturbations.

2. *Periodic systems.* The hamiltonian function is

$$H = H_0(I) + \varepsilon H_1(I, \varphi, t), \qquad I \in G \subset \mathbb{R}^n, \varphi \bmod 2\pi \in T^n;$$

the perturbation is 2π-periodic not only in φ, but also in t. It is natural to look at the unperturbed system in the $(2n + 1)$-dimensional space $\{(I, \varphi, t)\} = \mathbb{R}^n \times T^{n+1}$. The invariant tori have dimension $n + 1$. The nondegeneracy condition

$$\det \left| \frac{\partial^2 H_0}{\partial I^2} \right| \neq 0$$

guarantees the preservation of most $(n + 1)$-dimensional invariant tori under a small perturbation ($\varepsilon \ll 1$).

If $n = 1$, this nondegeneracy condition also guarantees stability of the action variable, in the sense that it remains forever near its initial value for sufficiently small perturbations.

3. *Mappings $(I, \varphi) \to (I', \varphi')$ of the "2n-dimensional annulus."* The generating function is

$$S(I', \varphi) = S_0(I') + \varepsilon S_1(I', \varphi), \qquad I' \in G \subset \mathbb{R}^n, \varphi \in T^n.$$

The nondegeneracy condition

$$\det \left| \frac{\partial^2 S_0}{\partial I^2} \right| \neq 0$$

guarantees the preservation of most invariant tori of the unperturbed mapping $(I, \varphi) \to (I, \varphi + (\partial S_0 / \partial I))$ under small perturbations ($\varepsilon \ll 1$).

If $n = 1$, we obtain an area-preserving mapping of the ordinary annulus to itself. The unperturbed mapping is represented on each circle $I = \text{const}$ as a rotation. In this case the nondegeneracy condition means that the angle of rotation changes from one circle to another.

The invariant tori in the case $n = 1$ are ordinary circles. In this case, the theorem guarantees that under iterations of the mapping all the images of a point will remain near the circle on which the original point lay, if the perturbation is sufficiently small.

4. *Neighborhoods of equilibrium positions* (*autonomous case*). An equilibrium position is assumed to be stable in a linear approximation so that n characteristic frequencies $\omega_1, \ldots, \omega_n$ are defined. We assume that there are no resonance relations among the characteristic frequencies, i.e., no relations

$$k_1\omega_1 + \cdots + k_n\omega_n = 0 \quad \text{with integers } k_i \text{ such that } 0 < \sum |k_i| \leq 4.$$

Then the hamiltonian function can be reduced to the Birkhoff normal form (cf. Appendix 7)

$$H = H_0(\tau) + \cdots,$$

where $H_0(\tau) = \sum \omega_k \tau_k + \frac{1}{2} \sum \omega_{kl} \tau_k \tau_l$ and the dots denote terms of degree higher than four with respect to the distance from the equilibrium position.

The nondegeneracy condition

$$\det |\omega_{kl}| \neq 0$$

guarantees the existence of a set of invariant tori of almost full measure in a sufficiently small neighborhood of the equilibrium position.

The condition for isoenergetic nondegeneracy,

$$\det \begin{vmatrix} \omega_{kl} & \omega_k \\ \omega_l & 0 \end{vmatrix} \neq 0,$$

guarantees the existence of such a set of invariant tori on every energy level set (sufficiently close to the critical point).

In the case $n = 2$, the condition for isoenergetic nondegeneracy is satisfied if the quadratic part of the function H_0 is not divisible by the linear part. In this case, isoenergetic nondegeneracy guarantees Liapunov stability of the equilibrium position.

5. *Neighborhoods of equilibrium positions* (*periodic case*). Here again we assume stability in a linear approximation, so that n characteristic frequencies $\omega_1, \ldots, \omega_n$ are defined. We assume that there are no resonance relations

$$k_1\omega_1 + \cdots + k_n\omega_n + k_0 = 0 \quad \text{with } 0 < \sum_{i=1}^{n} |k_i| \leq 4$$

among the characteristic frequencies and the frequency of the time-dependence of the coefficients (which we will assume equal to 1).

Then the hamiltonian function can be reduced to a Birkhoff normal form in the same way as in the autonomous case, but with 2π-periodicity with respect to time in the remainder term.

The nondegeneracy condition

$$\det |\omega_{kl}| \neq 0$$

guarantees the existence of $(n + 1)$-dimensional invariant tori in the $(2n + 1)$-dimensional extended phase space, near the circle $\tau = 0$ representing the equilibrium position.

In the case $n = 1$ the nondegeneracy condition reduces to the non-vanishing of the derivative of the period of small oscillations with respect to the square of the amplitude of small oscillations. In this case, nondegeneracy guarantees that the equilibrium position is Liapunov stable.

6. *Fixed points of mappings.* Here we assume that all $2n$ eigenvalues of the linearization of a canonical mapping at a fixed point have modulus 1 and do not satisfy any low-order resonance relations of the form:

$$\lambda_1^{k_1} \cdots \lambda_n^{k_n} = 1, \qquad |k_1| + \cdots + |k_n| \leq 4$$

(where the $2n$ eigenvalues are $\lambda_1, \ldots, \lambda_n, \bar{\lambda}_1, \ldots, \bar{\lambda}_n$).

Then if we disregard terms of higher than third order in the Taylor series at the fixed point, the mapping can be written in Birkhoff normal form

$$(\tau, \varphi) \to (\tau, \varphi + \alpha(\tau)), \quad \text{where } \alpha(\tau) = \frac{\partial S}{\partial \tau},$$

$S = \sum \omega_k \tau_k + \frac{1}{2} \sum \omega_{kl} \tau_k \tau_l$ (the usual coordinates in a neighborhood of the equilibrium position are $p_k = \sqrt{2\tau_k} \cos \varphi_k, q_k = \sqrt{2\tau_k} \sin \varphi_k$).

The nondegeneracy condition

$$\det|\omega_{kl}| \neq 0$$

guarantees the existence of n-dimensional invariant tori (close to the tori $\tau = $ const), forming a set of almost full measure in a sufficiently small neighborhood of the equilibrium position.

If $n = 1$, we have a mapping of the ordinary plane to itself, and the invariant tori become circles. The nondegeneracy condition means that, for the normal form, the derivative of the angle of rotation of a circle with respect to the area bounded by the circle is not zero (at the fixed point and, therefore, in some neighborhood of it).

In the case $n = 1$ the nondegeneracy condition guarantees Liapunov stability of the fixed point of the mapping. We note that in this case the condition of absence of lower resonance has the form

$$\lambda^3 \neq 1 \qquad \lambda^4 \neq 1.$$

Thus a fixed point of an area-preserving mapping of the plane to itself is Liapunov stable if the linear part of the mapping is rotation through an angle which is not a multiple of $90°$ or $120°$ and if the coefficient ω_{11} in the normal Birkhoff form is not zero (guaranteeing nontrivial dependence of the angle of rotation on the radius).

We have not gone into the smoothness conditions assumed in these theorems. The minimal smoothness needed is not known in even one case.

For example, we point out that the last assertion about stability of fixed points of a mapping of the plane to itself was first proved by J. Moser under the assumption of 333-times differentiability, and only later (by Moser and Rüssman) was the number of derivatives reduced to 6.

E *Applications of the theorem on invariant tori and its generalizations*

There are many mechanical problems to which we can apply the theorem formulated above. One of the simplest of these problems is the motion of a pendulum under the action of a periodically changing exterior field or under the action of vertical oscillations of the point of suspension.

It is well known that, in the absence of parametric resonance, the lower equilibrium position of a pendulum is stable in the linear approximation. The stability of this position with regard to nonlinear effects (under the further assumption of the absence of resonances of order 3 and 4) can be proved only with the help of the theorem on invariant tori.

In an analogous way we can use the theorem on invariant tori to investigate conditionally-periodic motions of a system of interacting nonlinear oscillators.

Another example is the geodesic flow on a convex surface close to an ellipsoid. There are two degrees of freedom in this system, and we can show that most geodesics on a three-dimensional near-ellipsoidal surface oscillate between two "caustics" close to the lines of curvature of the surface, densely filling out the ring between them. At the same time, we can arrive at theorems on the stability of the two closed geodesics obtained, after deforming the surface, from the two ellipses containing the middle axis of the ellipsoid (in the absence of resonances of orders 3 and 4).

As one more example, we can look at closed trajectories on a billiard table of any convex shape. Among the closed billiard trajectories are those which are stable in the linear approximation, and we can conclude that in the general case they are actually stable. An example of such a stable billiard trajectory is the minor axis of an ellipse; therefore, a closed billiard trajectory, close to the minor axis of an ellipse on a billiard table which is almost the ellipse, is stable.

Application of the theorem on invariant tori to the problem of rotations of an asymmetric heavy rigid body allows us to consider the nonintegrable case of a rapidly rotating body. The problem of rapid rotation is mathematically equivalent to the problem of motion with moderate velocity in a weak gravitational field: the essential parameter is the ratio of potential to kinetic energy. If this parameter is small, then we can use eulerian motion of a rigid body as a first approximation.

By applying the theorem on invariant tori to the problem with two degrees of freedom obtained after eliminating cyclic coordinates (rotations around the vertical) we come to the following conclusion about the motion of a rapidly rotating body: if the kinetic energy of rotation of a body is sufficiently large in comparison with the potential energy, then the length of the vector of angular momentum and its angle with the horizontal remain forever close to their initial values.

It follows from this that the motion of the body will forever be close to a combination of Euler-Poinsot motion and azimuthal procession, except in the case when the initial values of kinetic energy and total momentum are close to those for which the body can rotate around the middle principal axis. In this last case, realized only for special initial conditions, the splitting of separatrices near the middle axis implies a more complicated undulation about the middle axis than in Euler-Poinsot motion.

One generalization of the theorem on invariant tori leads to the theorem on the adiabatic invariance for all time of the action variable in a one-dimensional oscillating system with periodically changing parameters. Here we must assume that the rule for changing parameters is given by a fixed smooth periodic function of "slow time," and the small parameter of the problem is the ratio of the period of characteristic oscillations and the period of change of parameters. Then, if the period of change of parameters is sufficiently large, the change in the adiabatic invariant of a phase point remains small in the course of an infinite interval of time.

In an analogous way we can prove the adiabatic invariance for all time of the action variable in the problem of a charged particle in an axially-symmetric magnetic field. Violation of axial symmetry in this problem increases the number of degrees of freedom from two to three, so that the invariant tori cease to divide the energy level manifolds, and the phase curve wanders about the resonance zones.

Finally, applying the theory to the three- (or many-) body problem, we succeed in finding conditionally periodic motions of "planetary type." To describe these motions, we must say a few words about the next approximation after the keplerian one in the problem of the motion of the planets. For simplicity we will limit ourselves to the planar problem.

For each keplerian ellipse, consider the vector connecting the focus of the ellipse (i.e., the sun) to the center of the ellipse. This vector, called the *Laplace vector*, characterizes both the magnitude of the eccentricity of the orbit and the direction to the perihelion.

The interaction of the planets on one another causes the keplerian ellipse (and therefore the Laplace vector) to change slowly. In addition, there is an important difference between changes in the major semi-axis and changes in the Laplace vector. Namely, the major semi-axis has no secular perturbations, i.e., in the first approximation it merely oscillates slightly around its average value ("Laplace's theorem"). The Laplace vector, on the other hand, performs both periodic oscillations and secular motion. The secular motion may be obtained if we spread each planet over its orbit proportionally to the time spent in travelling each piece of the orbit, and replace the attraction of the planets by the attraction of the rings obtained, that is, if we average the perturbation over the rapid motions. The true motion of the Laplace vector is obtained from the secular one by the addition of small oscillations; these oscillations are essential if we are interested in small intervals of time (years), but their effect remains small in comparison to the effect of the secular motion if we consider a large interval of time (thousands of years).

Calculations (carried out by Lagrange) show that the secular motion of the Laplace vector of each of n planets moving in one plane consists of the following (if we ignore the squares of the eccentricities of the orbits which are small in comparison with the eccentricities themselves). In the orbital plane of a planet we must arrange n vectors of fixed lengths, each rotating uniformly with its angular velocity. The Laplace vector is their sum.

This description of the motion of the Laplace vector is obtained because the hamiltonian system averaged with respect to rapid motions, which describes the secular motion of the Laplace vector, has an equilibrium position corresponding to zero eccentricities. The described motion of the Laplace vector is the decomposition of small oscillations near this equilibrium position into characteristic oscillations. The angular velocities of the uniformly rotating components of the Laplace vector are the characteristic frequencies, and the lengths of these components determine the amplitudes of the characteristic oscillations.

We note that the motion of the Laplace vector of the earth is, apparently, one of the factors involved in the occurrence of ice ages. The reason is that, when the eccentricity of the earth's orbit increases, the time it spends near the sun decreases, while the time it spends far from the sun increases (by the law of areas); thus the climate becomes more severe as the eccentricity increases. The magnitude of this effect is such that, for example, the amount of solar energy received in a year at the latitude of Leningrad (60°N) may attain the value which now corresponds to the latitudes of Kiev (50°N) (for decreased eccentricity) and Taimir (80°N) (for increased eccentricity). The characteristic time of variation of the eccentricity (tens of thousands of years) agrees well with the interval between ice ages.

The theorems on invariant tori lead to the conclusion that for planets of sufficiently small mass, there is, in the phase space of the problem, a set of positive measure filled with conditionally periodic phase curves such that the corresponding motion of the planets is nearly motion over slowly changing ellipses of small eccentricities, and the motion of the Laplace vectors is almost that given by the approximation described above. Furthermore, if the masses of the planets are sufficiently small, then motions of this type fill up most of the region of phase space corresponding in the keplerian approximation to motions of the planets in the same direction over non-intersecting ellipses of small eccentricities.

The number of degrees of freedom in the planar problem with n planets is equal to $2n$ if we take the sun to be fixed. The integral of angular momentum allows us to eliminate one cyclic coordinate; however, there are still too many variables for the invariant tori to divide an energy level manifold (even if there are only two planets this manifold is five-dimensional, and the tori are three-dimensional). Therefore, in this problem we cannot draw any conclusions about the preservation of the large semi-axes over an infinite interval of time for all initial conditions, but only for most initial conditions.

A problem with two degrees of freedom is obtained by further idealization. We replace one of the two planets by an "asteroid" which moves in the field of the second planet ("Jupiter"), not perturbing its motion.

The problem of the motion of such an asteroid is called the *restricted three-body problem*. The planar restricted three-body problem reduces to a system with two degrees of freedom, periodically depending on time, for the motion of the asteroid. If, in addition, the orbit of Jupiter is circular, then in a coordinate system rotating together with it we obtain, for the motion of the asteroid, an autonomous hamiltonian system with two degrees of freedom — called the planar restricted circular three-body problem.

In this problem, there is a small parameter — the ratio of the masses of Jupiter and the sun. The zero value of the parameter corresponds to unperturbed keplerian motion of the asteroid, represented in our four-dimensional phase space as a conditionally-periodic motion on a two-dimensional torus (since the coordinate system is rotating). One of the frequencies of this conditionally-periodic motion is equal to 1 for all initial conditions; this is the angular velocity of the rotating coordinate system, i.e., the frequency of the revolution of Jupiter around the sun. The second frequency depends on the initial conditions (this is the frequency of the revolution of the asteroid around the sun) and is fixed on any fixed three-dimensional level manifold of the hamiltonian function.

Therefore, the nondegeneracy condition is not fulfilled in our problem, but the condition for isoenergetic nondegeneracy is fulfilled. Kolmogorov's theorem applies, and we conclude that most invariant tori with irrational ratios of frequencies are preserved in the case when the mass of the perturbing planet (Jupiter) is not zero, but sufficiently small.

Furthermore, the two-dimensional invariant tori divide the three-dimensional level manifolds of the hamiltonian function. Therefore, the magnitude of the major semi-axis and the eccentricity of the keplerian ellipse of the asteroid will remain forever near their initial values if, at the initial moment, the keplerian ellipse does not intersect the orbit of the perturbing planet, and if the mass of this planet is sufficiently small.

In addition, in a stationary coordinate system, the keplerian ellipse of the asteroid could slowly rotate, since our system is only isoenergetically nondegenerate. Therefore under perturbations of an invariant torus frequencies are not preserved, but only their ratios. As a result of a perturbation, the frequency of azimuthal motion of the perihelion of the asteroid in a stationary coordinate system could be slightly different from Jupiter's frequency, and then in the stationary system the perihelion would slowly rotate.

Appendix 9: Poincaré's geometric theorem, its generalizations and applications

In his study of periodic solutions of problems in celestial mechanics, H. Poincaré constructed a very simple model which contains the basic difficulties of the problem. This model is an area-preserving mapping of the planar circular annulus to itself. Mappings of this form arise in the study of dynamical systems with two degrees of freedom. In fact, a mapping of a two-dimensional surface of section to itself is defined as follows: each point p of the surface of section is taken to the next point at which the phase curve originating at p intersects the surface (cf. Appendix 7). Thus, a closed phase curve corresponds to a fixed point of the mapping or of a power of the mapping. Conversely, every fixed point of the mapping or of a power of the mapping determines a closed phase curve.

In this way, a question about the existence of periodic solutions of problems in dynamics is reduced to a question about fixed points of area-preserving mappings of the annulus to itself. In studying such mappings, Poincaré arrived at the following theorem.

A *Fixed points of mappings of the annulus to itself*

Theorem. *Suppose that we are given an area-preserving homeomorphic mapping of the planar circular annulus to itself. Assume that the boundary circles of the annulus are turned in different directions under the mapping. Then this mapping has at least two fixed points.*

The condition that the boundary circles are turned in different directions means that, if we choose coordinates $(x, y \bmod 2\pi)$ on the annulus so that the boundary circles are $x = a$ and $x = b$, then the mapping is defined by the formula

$$(x, y) \to (f(x, y), y + g(x, y)),$$

where the functions f and g are continuous and 2π-periodic in y, with $f(a, y) \equiv a$, $f(b, y) \equiv b$, and $g(a, y) < 0$, $g(b, y) > 0$ for all y.

The proof of this theorem, announced by Poincaré not long before his death, was given only later by G. D. Birkhoff (cf. his book, *Dynamical Systems*, Amer. Math. Soc., 1927).

There remain many open questions related to this theorem; in particular, attempts to generalize it to higher dimensions are important for the study of periodic solutions of problems with many degrees of freedom. The argument Poincaré used to arrive at his theorem applies to a whole series of other problems. However, the intricate proof given by Birkhoff does not lend itself to generalization. Therefore, it is not known whether the conclusions suggested by Poincaré's argument are true beyond the limits of the theorem on the two-dimensional annulus. The argument in question is the following.

416

B *The connection between fixed points of a mapping and critical points of the generating function*

We will define a symplectic diffeomorphism of the annulus

$$(x, y) \rightarrow (X, Y)$$

with the help of the generating function $Xy + S(X, y)$, where the function S is 2π-periodic in y. For this to be a diffeomorphism we need that $\partial X/\partial x \neq 0$. Then

$$dS = (x - X)dy + (Y - y)dX,$$

and, therefore, the fixed points of the diffeomorphism are critical points of the function $F(x, y) = S(X(x, y), y)$. This function F can always be constructed by defining it as the integral of the form $(x - X)dy + (Y - y)dX$. The gradient of this function is directed either inside the annulus or outside on both boundary circles at once (by the condition on rotation in different directions).

But every smooth function on the annulus whose gradient on both boundary circles is directed inside the annulus (or out from it) has a critical point (maximum or minimum) inside the annulus. Furthermore, it can be shown that the number of critical points of such a function on the annulus is at least two. Therefore, we could assert that our diffeomorphism has at least two critical points if we were sure that every critical point of F is a fixed point of the mapping.

Unfortunately, this is true only under the condition that $\partial X/\partial x \neq 0$, so that we can express F in terms of X and y. Thus our argument is valid for mappings which are not too different from the identity. For example, it is sufficient that the derivatives of the generating function S be less than 1.

A refinement of this argument (with a different choice of generating function[111]) shows that it is even sufficient that the eigenvalues of the Jacobi matrix $D(X, Y)/D(x, y)$ never be equal to -1 at any point, i.e., that our mapping never flips the tangent space at any point. Unfortunately, all such conditions are violated at some points for mappings far from the identity. The proof of Poincaré's theorem in the general case uses entirely different arguments.

The connection between fixed points of mappings and critical points of generating functions seems to be a deeper fact than the theorem on mappings of a two-dimensional annulus into itself. Below, we give several examples in which this connection leads to meaningful conclusions which are true under some restrictions whose necessity is not obvious.

[111]
$$d\Phi = \tfrac{1}{2} \begin{vmatrix} X - x & Y - y \\ dX + dx & dY + dy \end{vmatrix}$$

417

C *Symplectic diffeomorphisms of the torus*

Consider a symplectic diffeomorphism of the torus which fixes the center of gravity

$$(x, y) \to (x + f(x, y), y + g(x, y)) = (X, Y),$$

where x and y mod 2π are angular coordinates on the torus, "symplectic" means the Jacobian $D(X, Y)/D(x, y)$ is equal to 1, and the condition on preserving the center of gravity means that the average values of the functions f and g are equal to zero.

Theorem. *Such a diffeomorphism has at least four fixed points, counting multiplicity, and at least three geometrically different ones, at least under the assumption that the eigenvalues of the Jacobi matrix are not equal to -1 at any point.*

The proof is based on consideration of the function on the torus given by the formula

$$\Phi(x, y) = \tfrac{1}{2} \int (X - x)(dY + dy) - (Y - y)(dX + dx),$$

and on the fact that a smooth function on the torus has at least four critical points (counting multiplicity) of which at least three are geometrically different.

Attempts at proving this theorem without restrictions on the eigenvalues meet with difficulties very similar to those encountered by Poincaré in the theorem about the annulus.

We note that the theorem about the annulus would follow from the theorem about the torus if in the latter we could throw out the condition on the eigenvalues. In fact, we can put together a torus from two copies of our annulus, inserting a narrow connecting annulus along each of the two boundary circles.

Then we can extend our mapping of the annulus to a symplectic diffeomorphism of the torus such that: (1) on each of the two large annuli the diffeomorphism coincides with the original, (2) on each of the connecting annuli the diffeomorphism has no fixed points, and (3) the center of gravity remains fixed.

The construction of such a diffeomorphism of the torus uses the property that the boundary circles rotate in different directions. On each connecting annulus all points are translated in the same direction as on both circles bounding the connecting annulus. Since the translations on the connecting annuli are in opposite directions, the size of the translations can be chosen to ensure preservation of the center of gravity.

Now out of four fixed points on the torus, two must lie in the original annulus, and we obtain the theorem on annuli from the theorem on tori.

The theorem on tori formulated above can be generalized to other symplectic manifolds, both two-dimensional and many-dimensional. To formulate these generalizations, we must first reformulate the condition of preservation of the center of gravity.

Let $g: M \to M$ be a symplectic diffeomorphism. We say that g is *homologous to the identity* if it can be connected to the identity diffeomorphism by a smooth curve g_t consisting of symplectic diffeomorphisms such that the field of velocities $\dot{g}_t$ at each moment of time t has a single-valued hamiltonian function. It can be shown that the symplectic diffeomorphisms homologous to the identity form the commutator subgroup of the connected component of the identity in the group of all symplectic diffeomorphisms of the manifold.

In the case when our manifold is the two-dimensional torus, the symplectic diffeomorphisms homologous to the identity are exactly those which preserve the center of gravity.

Thus we come to the following generalization of Poincaré's theorem.

Theorem. *Every symplectic diffeomorphism of a compact symplectic manifold, homologous to the identity, has at least as many fixed points as a smooth function on this manifold has critical points (at least if this diffeomorphism is not too far from the identity).*[112]

We note that the condition of the mapping being homologous to the identity is essential, as we see already from the example of a translation on the torus, which has no fixed points at all.

As to the last restriction (that the diffeomorphism be not too far from the identity), it is not clear whether it is essential.[112a] In the case that our manifold is the two-dimensional torus, it is sufficient that none of the eigenvalues of the Jacobi matrix of the diffeomorphism (in any global symplectic coordinate system on $\mathbb{R}^{2n}$) be equal to minus one.

A restriction of this sort may be necessary in higher-dimensional problems. It is not impossible that Poincaré's theorem is due to an essentially two-dimensional effect, as is the following theorem of A. I. Shnirel'man and N. A. Nikishin: every area-preserving diffeomorphism of the two-dimensional sphere to itself has at least two geometrically different fixed points.

The proof of this theorem is based on the fact that the index of the gradient vector field of a smooth function of two variables at an isolated critical point cannot be greater than 1 (although it can be equal to $1, 0, -1, -2, -3, \ldots$), and the sum of the indices of all the fixed points of an orientation-preserving diffeomorphism of the two-dimensional sphere to itself is equal to 2. On the other hand, the index of the gradient of a smooth function of a large number of variables at a critical point can take any integer value.

D Intersections of lagrangian manifolds

Poincaré's argument can be given a slightly different form if on every radius of the annulus we consider the points shifted only radially. There are such points on every radius, since the boundary circles of the annulus turn

[112] [For a proof, see V. Arnold, Sur les propriétés topologiques des applications globalement canoniques de la mécanique classique, C. R. Acad. Sci. Paris, 1965 and A. Weinstein, Symplectic manifolds and their lagrangian submanifolds, Advances in Math. 6 (1971) 329–346.]

[112a] [Recently, Conley and Zehnder, followed by others, have proved the theorem for tori, surfaces, and other manifolds, without the restriction of closeness to the identity.]

in different directions. Assume that we can make a smooth curve of radially shifting points, separating the interior and exterior circles of the annulus. Then the image of this curve under our mapping must intersect the curve (since the regions into which the curve divides the annulus are carried to regions of equal area).

If this curve and its image each intersect each radius once, then the points of intersection of the curve with its image are obviously fixed points of the mapping.

Part of this argument can be carried out in higher dimensions, and this gives useful results about periodic solutions of problems in dynamics. The role of the annulus in the many-dimensional case is played by the phase space: the direct product of a region in euclidean space with a torus of the same dimension (the annulus is the product of an interval with the circle). A symplectic structure on the phase space is defined in the usual way, i.e., it has the form $\Omega = \sum dx_k \wedge dy_k$, where the x_k are action variables and y_k are angle variables.

It is not difficult to explain which symplectic diffeomorphisms of our phase space are homologous to the identity. Namely, a symplectic diffeomorphism A is homologous to the identity if it can be obtained from the identity by a continuous deformation and if

$$\oint_\gamma x \, dy = \oint_{A\gamma} x \, dy$$

for any closed contour γ (not necessarily homologous to zero). The condition that the transformation be homologous to the identity prohibits systematic shifts along the x-direction ("evolution of the action variables"), but permits shifts along the tori.

We consider one of the n-dimensional tori $x = c = $ const and apply to it our symplectic diffeomorphism homologous to the identity. It turns out that the original torus intersects its image in at least 2^n points (counting multiplicities), of which at least $n + 1$ are geometrically different, at least under the assumption that the image torus has an equation of the form $x = f(y)$, where f is smooth.

For $n = 1$, this assertion means that each of the concentric circles constituting the annulus intersects its image in at least two points. This also follows from the preservation of area, so that the assumption that the image has equation $x = f(y)$ is not necessary.

Whether or not this assumption is necessary in higher dimensions is not known. If we make this assumption, the proof proceeds in the following way.

We note that the original torus is a lagrangian submanifold of phase space. Our diffeomorphism is symplectic, so the image torus is also lagrangian. Therefore, the 1-form $(x - c)dy$ on it is closed. Furthermore, this form on the torus is the total differential of some *single-valued* smooth function F, since our diffeomorphism is homologous to the identity, and therefore for any closed contour γ we have

$$\oint_{A\gamma} (x - c)dy = \oint_{A\gamma} x\, dy - \oint_{A\gamma} c\, dy = \oint_{\gamma} x\, dy - \oint_{A\gamma} c\, dy$$

$$= c \oint_{\gamma} dy - c \oint_{A\gamma} dy = 0.$$

We note that points of intersection of the torus with its image are critical points of the function F (since at them $dF = (x - c)dy = 0$).

From the condition of single-valued projection of the image torus (i.e., from the fact that the image torus has equation $x = f(y)$) it follows that, conversely, all critical points of the function F are points of intersection of our tori. In fact, under these conditions y can be taken for local coordinates on the torus, and therefore the fact that dF is zero for all vectors tangent to the image torus implies $x = c$.

A smooth function on an n-dimensional torus has at least 2^n critical points, counting multiplicities, of which at least $n + 1$ are geometrically different (cf., for example, Milnor, "Morse Theory," Princeton University Press, 1967).

Therefore, our tori intersect in at least 2^n points (counting multiplicities), and there are at least $n + 1$ geometrically different points of intersection.

Exactly the same argument shows that any lagrangian torus intersects its image in at least 2^n points (of which at least $n + 1$ are geometrically different), under the assumption that both the original torus and its image project single-valued onto the y-space, i.e., are given by equations $y = f(x)$ and $x = g(y)$, respectively. Besides, this statement reduces to the previous one by the canonical transformation $(x, y) \rightarrow (x - f(y), y)$.

E *Applications to determining fixed points and periodic solutions*

We now consider a symplectic transformation, homologous to the identity, of the special form which arises in integrable problems in dynamics, i.e., of the form

$$A_0(x, y) = (x, y + \omega(x)), \quad \text{where } \omega = \frac{\partial S}{\partial x}.$$

Here $x \in \mathbb{R}^n$ is the action variable and $y \bmod 2\pi \in T^n$ is the angular coordinate.

We assume that on the torus $x = x_0$ all the frequencies are commensurable:

$$\omega_i(x_0) = \frac{k_i}{N} 2\pi \quad \text{with integers } k_i, N; \omega(x_0) \neq 0,$$

and that the nondegeneracy condition

$$\det \left| \frac{\partial \omega}{\partial x} \right|_{x_0} \neq 0$$

is satisfied.

Appendix 9: Poincaré's geometric theorem, its generalizations and applications

Theorem. *Every symplectic diffeomorphism A homologous to the identity and sufficiently close to A_0 has, near the torus $x = x_0$, at least 2^n periodic points ξ of period N (such that $A^N \xi = \xi$), counting multiplicity.*

The proof could be reduced to investigating the intersection of two lagrangian submanifolds of a $4n$-dimensional space ($\mathbb{R}^n \times T^n \times \mathbb{R}^n \times T^n$) with $\Omega = dx \wedge dy - dX \wedge dY$, one of which is the diagonal ($X = x$, $Y = y$) and the other the graph of the mapping A^N.

However, it is easier to directly construct a suitable function on the torus. In fact, the mapping A_0^N has the form

$$(x, y) \to (x, y + \alpha(x)), \quad \text{where } \alpha(x_0) = 0, \det\left|\frac{\partial \alpha}{\partial x}\right|_{x_0} \neq 0.$$

By the implicit function theorem, the mapping A^N has, near the torus $x = x_0$, a torus which is displaced only radially ($(x, y) \to (X, Y)$) and is given by an equation of the form $x = f(y)$; its image is also given by an equation $x = g(y)$ of the same form. In this notation, $X(f(y), y) = g(y)$, $Y(f(y), y) = y$.

Since A is homologous to the identity, it follows that A^N has a single-valued global generating function of the form $Xy + S(X, y)$, where S has period 2π in the variable y.

The function $F(y) = S(X(f(y), y), y)$ has at least 2^n critical points y_k on the torus. All the points $\xi_k = (f(y_k), y_k)$ are fixed points for A^N. In fact,

$$dF = (x - X)\,dy + (Y - y)\,dX = (x - X)\,dy = (f(y) - g(y))\,dy.$$

Therefore, since $dF|_{y_k} = 0$, it follows that $f(y_k) = g(y_k)$, i.e., $A^N \xi_k = \xi_k$, as was to be shown.

We turn now to closed orbits of conservative systems. Using the terminology of Appendix 8, we can formulate the result as follows.

Corollary. *Upon disintegration of an n-dimensional torus, entirely filled up by closed trajectories of an isoenergetically nondegenerate system, at least 2^{n-1} closed trajectories of the perturbed problem are formed (counting multiplicities), among which at least n are geometrically distinct, at least if the perturbation is sufficiently small.*

The proof is reduced to the preceding theorem with the help of a $(2n - 2)$-dimensional surface of section. We must first choose angular coordinates y such that the closed trajectories of the unperturbed problem on the torus are given by the equations $\dot{y}_2 = \cdots = \dot{y}_n = 0$, and then define a surface of section by $y_1 = 0$.

In the case of two degrees of freedom we can apply Poincaré's theorem to the annuli formed by intersecting invariant tori with a two-dimensional intersecting surface. We obtain the following result:

In the gap between two two-dimensional invariant tori of a system with two degrees of freedom there are always at least two closed phase trajectories, if the ratio of the frequencies of conditionally-periodic motions on these tori are different.

In this way we obtain many periodic solutions in all problems with two degrees of freedom, where invariant tori are found (for example, in the bounded circular three-body problem, in the problem of closed geodesics, etc.). There is even a conjecture that in hamiltonian systems of "general form" with compact phase spaces, the closed phase curves form a dense set.[113] However, if this is true, the closedness of most of these curves has little importance since their periods are extremely large.

As an example of applying Poincaré's methods to systems with more than two degrees of freedom, we have a theorem of Birkhoff about the existence of infinitely many periodic solutions close to a given linearly stable periodic solution of general form (or about the existence of infinitely many periodic points in a neighborhood of a fixed point of a linearly stable nondegenerate symplectic mapping of a space to itself). In the proof, the mapping is first approximated by its normal form, and then the connection between fixed points of a mapping and critical points of the generating function is used.

Knowing periodic solutions allows us, among other things, to prove the *nonexistence of first integrals* (other than the classical ones) in many problems in dynamics. Assume, for example, that on some level manifold of known integrals we discover a periodic trajectory which is unstable. Its separatrices, in general, form a complicated network, which we considered in Appendix 7. If this phenomenon of splitting of separatrices is discovered, and if we can show that the separatrices are not contained in any manifold of lower dimension than the level manifold we are considering, then we can be sure that the system has no new first integrals.

The complicated behavior of phase curves, which obstructs the existence of first integrals, can often be detected without the help of periodic solutions by one simple glance at the picture, obtained by a computer, formed by the intersection of the phase curves with the surface of section.

F *Invariance of generating functions*

We have already noted the discouraging noninvariance of generating functions with respect to the choice of a canonical coordinate system on a symplectic manifold. On the other hand, we repeatedly used the connection between fixed points of a mapping and critical points of the generating function.

It turns out that, although generally the generating function is not invariantly associated to the mapping, near a fixed point there is an invariant connection. More precisely, suppose we are given a symplectic diffeomorphism fixing some point. In a neighborhood of this point, we define a "generating function"

$$\Phi = \tfrac{1}{2} \int \sum \begin{vmatrix} X_k - x_k & Y_k - y_k \\ dX_k + dx_k & dY_k + dy_k \end{vmatrix}$$

---

[113] A proof of this density in the C^1-topology has been announced by C. Pugh and C. Robinson. [Editor's note]

with the help of some symplectic coordinate system (x, y).[114] Using another symplectic coordinate system (x', y'), we construct a generating function Φ' in the same way.

Theorem. *If the linearization of the symplectic diffeomorphism at the fixed point has no eigenvalues equal to -1, then the functions Φ and Φ' are equivalent in a neighborhood of the fixed point, in the sense that there is a diffeomorphism g (in general not symplectic) such that*

$$\Phi(z) = \Phi'(g(z)) + \text{const.}$$

For the proof see the article: A. Weinstein, The invariance of Poincaré's generating function for canonical transformations, Inventiones Mathematicae, 16, No. 3 (1972), 202–214.

It should be noted that two diffeomorphisms with generating functions which are equivalent in a neighborhood of a fixed point are not necessarily equivalent in the class of symplectic diffeomorphisms (for example, rotation and rotation through an angle which depends on the radius, with non-degenerate quadratic parts of the generating function at zero).

Since the first edition of this book had appeared in 1974, the content of this Appendix has grown into a new branch of mathematics: symplectic topology. To describe this development (triggered by the conjectures in this Appendix, which still remain, for general manifolds, neither proved, nor disproved) one would need a book longer than the present one.

The interested reader might follow this development using the (incomplete) bibliography on pages 503–509.

---

[114] The increase of this function along any arc is equal to the integral of the form defining the symplectic structure over the band formed by the rectilinear intervals connecting each point with its image. Therefore, the function Φ is associated to the mapping invariantly with respect to linear canonical changes of coordinates.

Appendix 10: Multiplicities of characteristic frequencies, and ellipsoids depending on parameters

Several times in this course we have encountered families of ellipsoids in euclidean space. For example, in studying the dependence on parameters of characteristic frequencies of small oscillations, we encountered equipotential surfaces which were ellipsoids in euclidean space, depending upon the degree of rigidity of the system, (the metric of the space was defined by the kinetic energy). Another example was the ellipsoid of inertia of a rigid body (the parameter here was the shape of the rigid body and its distribution of mass).

Here we will consider the general problem of describing the values of the parameter for which the spectrum of eigenvalues degenerates, i.e., the corresponding ellipsoid becomes an ellipsoid of revolution. We note that the eigenvalues of a quadratic form on euclidean space (or the lengths of the axes of an ellipsoid) change continuously under continuous changes of the parameters of a system (the coefficients of the form). It seems natural to expect that in a system depending on one parameter, under changes of the parameter, at certain moments one of the eigenvalues would collide with another, so that for these values of the parameter the system would have a multiple spectrum.

Suppose, for example, that we want to make the ellipsoid of inertia of a rigid body into an ellipsoid of revolution by movement of an adjustable mass along an arc rigidly attached to the body so that there is one parameter at our disposal. The three major axes a, b, and c will be continuous functions of this parameter, and at first glance it seems that for a suitable value of the parameter (p) we can achieve equality of two of the axes, say $a(p) = b(p)$. It turns out, however, that this is not so, and that generally we need to attach *at least two* adjustable masses to make the ellipsoid of inertia an ellipsoid of revolution.

In general, a multiple spectrum in typical families of quadratic forms is observed only for two or more parameters, while in one-parameter families of general form the spectrum is simple for all values of the parameter. Under a change of parameter in the typical one-parameter family, the eigenvalues can approach closely, but when they are sufficiently close, it is as if they begin to repel one another. The eigenvalues again diverge, disappointing the person who hoped, by changing the parameter, to achieve a multiple spectrum.

In this appendix we consider the reasons for this seemingly strange behavior of the eigenvalues, and we discuss briefly analogous questions for systems with various groups of symmetries.

A *The manifold of ellipsoids of revolution*

Consider the set of all possible quadratic forms on the n-dimensional euclidean space $\mathbb{R}^n$. This set has itself a natural structure of a vector space of dimension $n(n + 1)/2$. For example, the quadratic forms on the plane form a three-dimensional space (a form $Ax^2 + 2Bxy + Cy^2$ has as coordinates the three numbers A, B, and C).

The positive-definite forms form an open region in this space of all quadratic forms (for example, in the case of the plane this is the inside of one nappe of the cone $B^2 = AC$ of degenerate forms).

Every ellipsoid centered at the origin defines a positive-definite quadratic form, for which it is the level set of 1; conversely, the set of level 1 of any positive-definite quadratic form is an ellipsoid. We can therefore identify the sets of positive-definite quadratic forms and ellipsoids centered at the origin. In this way we give the set of ellipsoids with center 0 in $\mathbb{R}^n$ the structure of a smooth manifold of dimension $n(n + 1)/2$ (this manifold is covered by one chart: a region in the space of quadratic forms).

Now consider the set of all *ellipsoids of revolution*. We claim that this set has codimension 2 in the space under consideration, i.e., it is given by *two* independent equations, rather than one as it would seem at first glance. More precisely, we have

Theorem 1. *The set of ellipsoids of revolution is a finite union of smooth submanifolds of codimension 2 and higher in the manifold of all ellipsoids.*

The *codimension* of a manifold is the difference between the dimension of the ambient space and the dimension of the submanifold.

PROOF. We first consider an ellipsoid in n-dimensional space which has two equal axes, and whose other axes are distinct. Such an ellipsoid is defined by the directions of the distinct axes, which gives

$$(n - 1) + (n - 2) + \cdots + 2 = \frac{(n + 1)(n - 2)}{2}$$

different parameters, and also by the magnitudes of the axes, which gives $n - 1$ parameters. Thus the total number of parameters is

$$\frac{n^2 - n - 2 + 2n - 2}{2},$$

which is two less than the dimension of the space of all ellipsoids (which is $n(n + 1)/2$). This count of parameters also shows that the set of ellipsoids with exactly two equal axes is a manifold.

As for ellipsoids with a larger number of equal axes, it is clear that they form a set of even smaller dimension. A rigorous proof follows from the following lemma.

Lemma. *The set of all ellipsoids with v_2 double, v_3 triple, v_4 four-fold axes, etc. is a smooth submanifold of the manifold of all ellipsoids, with codimension*

$$2v_2 + 5v_3 + 9v_4 + \cdots = \sum \tfrac{1}{2}(i - 1)(i + 2)v_i.$$

The proof of this theorem reduces to the same kind of parameter count as in the special case analyzed above (which corresponds to $v_2 = 1$, $v_3 = v_4 = \cdots = 0$). The reader can easily carry out this calculation, noting first that the dimension of the manifold of all k-dimensional subspaces in an n-dimensional vector space is equal to $k(n - k)$ (since a k-dimensional plane in general position in an n-dimensional space can be thought of as the graph of a mapping from a k-dimensional space to an $(n - k)$-dimensional space, and such a mapping is given by a rectangular $k \times (n - k)$ matrix).

EXAMPLE. Consider the case $n = 2$, i.e., ellipses in the plane. An ellipse is determined by three parameters (e.g., the lengths of the two axes and the angle giving the direction of one of them). Thus the manifold of ellipses in the plane is three-dimensional, as it must be by our formula.

A circle, however, is determined by one parameter (the radius). Thus the manifold of circles in the space of ellipses is a line in a three-dimensional space, and not a surface as it would seem at first glance.

This "paradox" becomes, perhaps, clearer from the following calculation. The quadratic forms $Ax^2 + 2Bxy + Cx^2$ with different eigenvalues form a submanifold of the three-dimensional space with coordinates A, B, and C, given by *one* equation $\lambda_1 - \lambda_2 = 0$, where $\lambda_{1, 2}(A, B, C)$ are the eigenvalues. However, the left-hand side of this equation is the sum of two squares, as is clear from the formula for the discriminant of the characteristic equation:

$$\Delta = (A + C)^2 - 4(AC - B^2) = (A - C)^2 + 4B^2.$$

Thus the single equation $\Delta = 0$ determines a line in the three-dimensional space of quadratic forms ($A = C$, $B = 0$), and not a surface.

A simple consequence of the fact that the manifold of ellipsoids of revolution has codimension 2 is that this manifold *does not divide* the space of all ellipsoids (and the manifold of quadratic forms with a multiple spectrum does not divide the space of quadratic forms), as a line does not divide a three-dimensional space. Therefore, we can assert not only that in an ellipsoid in "general position" all the axes share different lengths, but also that *any two such ellipsoids can be connected by a smooth curve in the space of ellipsoids consisting entirely of ellipsoids with axes of different lengths.* Furthermore, if two ellipsoids in general position are connected by a smooth curve in the space of ellipsoids which contains a point which is an ellipsoid of revolution, then by an arbitrarily small displacement of the curve we can remove it from the set of ellipsoids of revolution, so that on the new curve all the points will be ellipsoids without multiple axes.

One consequence of what we have said is a simple proof of the theorem that characteristic frequencies increase when the rigidity of a system is increased. The derivative of a non-multiple eigenvalue of a quadratic form with respect to a parameter is determined by the derivative of the quadratic form in the corresponding characteristic direction. If the rigidity is increased, the potential energy increases in every direction, including the characteristic directions. Thus the characteristic frequencies also increase. Hence we have proved the theorem on the growth of frequencies in the case when it is possible to go from the original system to a more rigid system, avoiding multiple spectra. The proof in the presence of multiple spectrum is now obtained by a passage to the limit, based on the fact that the interior of the path from the original system to the more rigid system can be removed by an arbitrarily small perturbation from the set of systems with multiple spectra.

In summary, we can say that a typical one-parameter family of ellipsoids (or quadratic forms in euclidean space) does not contain ellipsoids of revolution (quadratic forms with multiple spectra). Applying this to an ellipsoid of inertia we obtain the conclusion above about the necessity for two adjustable masses.

We turn now to two-parameter systems. It follows from our calculations that, in a typical two-parameter system, ellipsoids of revolution are encountered only at isolated points of the parameter plane.

Consider, for example, a convex surface in three-dimensional euclidean space. The second fundamental form of the surface determines an ellipse in the tangent space at every point. Therefore, we have a two-parameter family of ellipses (which can be translated to one plane by choosing a local coordinate system near a point on the surface). We come to the conclusion that, at every point of the surface except at certain isolated points, the ellipse has axes of different lengths. Therefore, on surfaces of general form, there are two orthogonal fields of directions (the major and minor axes of the ellipses) with isolated singular points. In differential geometry these directions are called the directions of principal curvature, and these singular points are called *umbilical points*. For example, on the surface of an ellipsoid there are four umbilical points: they lie on the ellipse containing the major and minor axes, and two of them are clearly visible in the picture of the geodesics on an ellipsoid (cf. Figure 207).

In exactly the same way, in a typical three-parameter family, ellipsoids of revolution are encountered only on certain lines in the three-dimensional parameter space. For example, if at every point of three-dimensional euclidean space, we are given an ellipsoid (i.e., a symmetric two-index tensor), then the singularities of the fields of principal axes will be, in general, on certain lines (where two of the three fields of directions have discontinuities). These lines, like the umbilical points in the preceding example, are of several different types. Their classification (for typical fields of ellipsoids) can be obtained from the classification of singularities of lagrangian projections given in Appendix 12.

In a typical *four-parameter* family, ellipsoids of revolution occur on two-dimensional surfaces in the space of parameters. These surfaces have no singularities other than transverse intersections at isolated points of the parameter space; these values of the parameters correspond to ellipsoids with two (different) pairs of equal axes.

Triple axes appear first for five parameters, at isolated points of the parameter space. The values of the parameters corresponding to ellipsoids with a double axis form a three-dimensional manifold in the five-dimensional parameter space with two types of singularities: transversal intersections of two branches along some curve and conic singularities at isolated points (not lying on this curve), i.e., at points of the parameter space corresponding to ellipsoids with three equal axes. These conic singularities have the following structure: by intersecting the three-dimensional manifold of ellipsoids of revolution with a four-dimensional sphere of small radius with center at the singular point, we obtain two copies of the projective plane. The resulting embeddings of the projective plane in the four-dimensional sphere are diffeomorphic to the embedding given by the five spherical harmonics of degree two on the two-dimensional sphere (five linear combinations of the functions $x_i x_j$, orthonormal in the space of functions on the sphere $x_1^2 + x_2^2 + x_3^2 = 1$, orthogonal to the identity, give an even mapping of S^2 into S^4 and, therefore, an embedding $\mathbb{R}P^2 \to S^4$).

It remains to describe the behavior of the eigenvalues of a quadratic form in a typical two-parameter family as the parameter approaches a singular point where the two eigenvalues coincide. A little calculation shows that the graph of the pair of eigenvalues we are considering has, over the plane of parameters near the singular point, the form of a two-sheeted cone, whose vertex corresponds to the singular point, and each of its nappes to one of the eigenvalues (Figure 243).

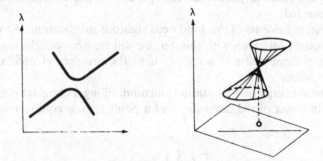

Figure 243   Characteristic frequencies of one- and two-parameter families of oscillating systems of general form

A typical one-dimensional subfamily of our two-dimensional family has the form of a curve in the plane of parameters which does not pass through any singular points. Every one-parameter family which contains a singular point can be removed from it by a small perturbation; the resulting one-parameter family will be a curve in the space of parameters passing near the singular point. The graph of the eigenvalues over a curve on the plane of parameters passing near a singular point consists of those points of the cone which project onto this curve. Therefore, this graph near the singular point is close to a hyperbola, resembling a pair of intersecting straight lines (a pair of straight lines would be obtained if our one-parameter family passed through the singular point).

This discussion of eigenvalues of two-parameter systems of quadratic forms explains the strange behavior of characteristic frequencies when a single parameter is varied: in general (except for completely singular cases), when a single parameter is varied the characteristic frequencies can approach one another but cannot collide; after approaching, they must again go off in different directions.

B *Application to the study of oscillations of continuous media*

The general argument above has numerous applications in the study of the dependence on parameters of the characteristic frequencies of various mechanical systems with finitely many degrees of freedom; however, the most interesting applications may be to systems with infinitely many degrees of freedom, describing oscillations of continuous media. These applications are based on the fact that the codimensions of manifolds of ellipsoids with given multiplicities of axes are determined by these multiplicities and do not depend on the dimension of the space.

For example, the codimension of the set of ellipsoids of revolution in the manifold of all ellipsoids is equal to two in a space of any dimension; therefore, it is natural to assume that in the infinite "manifold" of ellipsoids in infinite-dimensional hilbert space, the set of ellipsoids of revolution has codimension 2 (and, in particular, the space of ellipsoids without multiple axes is connected).

Of course, arguments of this kind need rigorous justification. We will not, however, occupy ourselves with this, but we will see what conclusions follow from the argument above if we apply it to the problem of oscillations in continuous media.

The kinetic energy of a continuous medium filling a compact region D is expressed in terms of the deviation u of a point x from equilibrium by the formula

$$T = \tfrac{1}{2} \int_D u_t^2 \, dx.$$

For definiteness, we can take the medium to be a membrane (in this case the region D is two-dimensional, and the deviation u one-dimensional). The kinetic energy defines a euclidean structure on the configuration space of the problem (i.e., in the space of functions u). The potential energy is given by the Dirichlet integral

$$U = \tfrac{1}{2} \int_D (\nabla u)^2 \, dx$$

(from the mathematical point of view these data constitute the definition of the membrane).

The squares of the characteristic frequencies of the membrane are the eigenvalues of the quadratic form U on the configuration space, whose metric is defined using the kinetic energy. We *assume* that a typical membrane corresponds to a typical quadratic form (this assumption means transversality of the manifold of quadratic forms corresponding to different membranes to the manifold of forms with multiple eigenvalues). If we believe in this property of general position, we come to the following conclusions.

1. For membranes in general position, all the characteristic frequencies are different. We can go from one membrane in general position to another by a continuous path consisting entirely of membranes with simple spectra. Furthermore, a typical path connecting any two membranes does not contain even one membrane with a multiple spectrum (except, possibly, the ends of the path).
2. By varying two parameters of the membrane we can make two characteristic frequencies coincide; to obtain a triple frequency, we must have at our disposal five independent parameters; for a four-fold frequency we need ten parameters, etc.
3. If, by starting from a membrane with a simple spectrum and continuously deforming it, we pass to another membrane with a simple spectrum along any path in general position, then as a result, the k-th largest characteristic frequency of the second membrane is always obtained independently of the path of deformation from the k-th largest characteristic frequency of the original membrane; continuations of characteristic functions, however, do generally depend on the path of deformation (i.e., by changing the path, the sign of the resulting characteristic function can be changed).

   In particular, if by starting from a membrane with a simple spectrum and deforming it we describe a closed path in the space of membranes and return to the original membrane, bypassing the set of membranes with multiple spectra (which has codimension 2), then the k-th characteristic frequency returns to its original value, while the k-th characteristic function may change sign. [Editor's note: Conclusions like this have been proven by K. Uhlenbeck (Amer. J. Math. 98 (1976), 1059–1078).]

C   The effect of symmetries on the multiplicity of the spectrum

A multiple spectrum is the exception in systems of general form, but it is not removable under small perturbations in cases when the given system is symmetric and the deformations preserve the symmetry.

Consider, for example, a system of three identical masses at the vertices of an equilateral triangle, connected to one another and to the center of the triangle by identical springs, and capable of moving in the plane of the triangle. The system has rotational symmetry of order 3. Therefore, there is a linear operator g acting on the configuration space (which has dimension 6), whose third power is equal to 1 and which leaves invariant both the euclidean structure of the configuration space and the ellipsoid in the configuration space giving the potential energy.

It follows that this ellipsoid must be an ellipsoid of revolution. If we let g be the indicated operator on the configuration space and ξ a vector on the major axis of the ellipsoid, then the axis in the direction $g\xi$ is also a major axis (since the rotation g takes the ellipsoid to itself).

There are two possibilities for the vector $g\xi$: either $g\xi = \xi$, or the vectors ξ and $g\xi$ are linearly independent. In the second case, the plane spanned by the vectors ξ and $g\xi$ consists entirely of major axes. Therefore, the eigenvalues corresponding to these axes are at least double. The space spanned by the three vectors ξ, $g\xi$, and $g^2\xi$ is invariant under g. It is either two dimensional (in which case g acts by a 120° rotation) or three dimensional (in which case g acts by the same rotation around $\xi + g\xi + g^2\xi$ as an axis). In the latter case, we may choose the direction of this sum for one of the principal axes of the ellipsoid, with the two other principal axes in the three-dimensional space perpendicular to it. It is therefore possible to choose the principal axes for an ellipsoid which is invariant under an orthogonal transformation of order three (in a space of any number of variables), so that each axis is either fixed under the transformation or is rotated by 120° in an invariant plane spanned by it and another axis (orthogonal to it, as well as to all other axes) of the same length. In what follows, we shall assume that the axes of ellipsoids and the directions of the corresponding characteristic oscillations have been chosen in the manner just described.

Our argument shows that characteristic oscillations of a system with third-order rotational symmetry can be of two types: those invariant under rotation by 120° ($g\xi = \xi$) and those passing under such a rotation to independent characteristic oscillations with the same frequency ($g\xi$ and ξ independent). In the second case, there actually arise three forms of characteristic oscillations with the same frequency (ξ, $g\xi$, and $g^2\xi$), but only two of them are independent:

$$\xi + g\xi + g^2\xi = 0$$

since the sum of three vectors of equal length on the plane forming angles of 120° is equal to zero.

The number of characteristic oscillations of our system is generally equal to 6. To find out how many of them are of the first (symmetric) and second (nonsymmetric) type, we can use the following argument. Consider the limiting case, when each of the masses oscillates independently from the others. In this case, we can choose an orthonormal basis of the configuration space consisting of six characteristic oscillations, two for each point, for which that point moves and the other two do not. We denote by ξ_i and η_i the characteristic vectors corresponding to the i-th point with characteristic frequencies a and b, respectively, and let x_i, y_i be coordinates in the orthonormal basis ξ_i, η_i. Then the potential energy can be written in the form

$$U = \tfrac{1}{2}(a^2 x_1^2 + b^2 y_1^2) + \tfrac{1}{2}(a^2 x_2^2 + b^2 y_2^2) + \tfrac{1}{2}(a^2 x_3^2 + b^2 y_3^2).$$

The symmetry operator g permutes the coordinate axes:

$$g\xi_1 = \xi_2 \qquad g\xi_2 = \xi_3 \qquad g\xi_3 = \xi_1,$$
$$g\eta_1 = \eta_2 \qquad g\eta_2 = \eta_3 \qquad g\eta_3 = \eta_1.$$

We can now represent our six-dimensional space as the orthogonal direct sum of two straight lines and two two-dimensional planes, invariant under the symmetry operator g. That is, the invariant lines are defined by the directions of the vectors

$$\xi_1 + \xi_2 + \xi_3 \quad \text{and} \quad \eta_1 + \eta_2 + \eta_3,$$

and the invariant planes are their orthogonal complements in the spaces spanned by the vectors ξ_i and η_i, respectively. The first straight line is the direction of a symmetric characteristic oscillation with frequency a, and the second the direction of one with frequency b. In exactly the same way, every vector in the first plane is a direction of characteristic oscillation with frequency a which, under rotation by $120°$, goes to an independent oscillation of the same frequency; for all vectors in the second plane, the oscillation is also not symmetric, with frequency b.

Thus, in this degenerate case of three independent points, there are two independent characteristic oscillations of symmetric type, and four unsymmetric, of which the latter are divided into two pairs. In each pair the oscillations have the same eigenvalue and are obtained from one another by rotation of the plane of our points by $120°$.

We now claim that the conclusion above holds true for any law of interaction between our points if the interaction is symmetric, i.e., if the potential energy of the system is preserved under rotation of the plane by $120°$.

In fact, decompose the 6-dimensional configuration space into an orthogonal sum of the plane of invariant vectors of g and of its orthogonal complement. The potential energy will decompose into a sum of two quadratic forms—one in two variables, the other in four. Now consider characteristic oscillations in the two-dimensional and four-dimensional configuration spaces, with potential energy described above. The four-dimensional space decomposes into two g-invariant planes, orthogonal in the potential energy metric. We have obtained a system of six characteristic oscillations having the required properties.

Thus, in a system in general form of three points in the plane with rotational symmetry of order 3, there are four different characteristic frequencies, two of which are simple and two double. Each of the simple characteristic frequencies corresponds to a symmetric characteristic oscillation, and each of the double ones to three characteristic oscillations obtained from one another by rotation by $120°$ and summing to zero (so that only two of them are independent).

Appendix 10: Multiplicities of characteristic frequencies, and ellipsoids

PROBLEM. Classify the characteristic oscillations of a system with the symmetries of an equilateral triangle (allowing not only rotation by 120°, but also reflection through the altitude of the triangle).

PROBLEM. Classify the characteristic oscillations of a system whose group of symmetries is the group of 24 rotations of the cube.

ANSWER. The oscillations will be of five types. By rotations, from each oscillation one can obtain systems of 8, or 6, or 4, or 2, or 1 independent oscillations (in the last case the oscillations are entirely symmetric).

Remark. To classify oscillations in systems with any group of symmetries, a special apparatus has been developed (the so-called theory of group representations). Cf., for example. Michael Tinkham, *Group Theory and Quantum Mechanics*, McGraw-Hill, 1964.

D *The behavior of frequencies of a symmetric system under a variation of parameters preserving the symmetry*

We assume now that our symmetric system depends in a general way on some number of parameters, and that the symmetry is not disturbed when the parameters are varied. Then the characteristic frequencies of various multiplicities will also depend on the parameters, and the question arises of when the characteristic frequencies will collide. We will confine ourselves to formulating a result for the simplest case of systems with third-order rotational symmetry (for rotational symmetry of any order $n \geq 3$, the answer is the same). The details can be found in the following articles: V. I. Arnold, Modes and quasi-modes, Functional Analysis and Its Applications, 6:2 (1972), 94–101; V. N. Karpushkin, The asymptotic behavior of the eigenvalues of symmetric manifolds and the "most probable" representations of finite groups, Moscow Univ. Math. Bull. 29 (1974), no. 2, 136–139.

Characteristic oscillations of any system with rotational symmetry of order 3 are divided into two types: symmetric oscillations, and oscillations carried by rotation by 120° into independent ones. For a general system with third-order rotational symmetry (without, in particular, any additional symmetry) all the characteristic frequencies of the first type are simple, and of the second, double. In addition, it turns out that if a system depends in a general way on one parameter and is symmetric for all values of the parameter, then under variation of the parameter, the characteristic frequencies of symmetric oscillations do not collide with one another, and the double characteristic frequencies of asymmetric oscillations do not split. In addition, the double characteristic frequencies of asymmetric oscillations do not collide with one another under a change of parameters. However, the characteristic frequencies of symmetric and asymmetric oscillations move under changes of parameter independently from one another, so that for discrete values of the parameter the characteristic frequency of a symmetric oscillation and the (double) characteristic frequency of an asymmetric oscillation can collide (and pass through one another).

In order to make two characteristic frequencies of symmetric oscillations collide, we must vary at least two parameters; and to make two characteristic frequencies of asymmetric oscillations collide we must vary at least three.

In general, in the typical family of systems with third-order rotational symmetry, for the collision of i simple characteristic frequencies (i symmetric oscillations) and j double frequencies (j unsymmetric oscillations) to occur, the number of parameters of the family must be at least

$$\frac{(i-1)(i+2)}{2} + j^2.$$

We apply this to oscillations of symmetric membranes. Here we will assume that the membrane is of general form, admits rotation by 120°, and corresponds to an ellipsoid of general form in the space of ellipsoids of the configuration space admitting the transformation of the configuration space induced by the rotation of the membrane.

The exact formulation of this assumption is that, for all membranes except a set of infinite codimension, the mapping from the space of symmetric membranes into the space of symmetric ellipsoids is transverse to each of the manifolds of ellipsoids with a given number of multiple axes.

If we agree to this assumption, we come to the following conclusions about oscillations of symmetric membranes.

1. For membranes of general form admitting rotation by 120°, asymptotically one-third of the characteristic frequencies (counting them with multiplicities) are simple, and the corresponding characteristic oscillations admit rotation by 120°. The remaining characteristic frequencies are double; each double characteristic frequency corresponds to three eigenfunctions whose sum is zero and which are taken to one another under rotation by 120°.

2. In general one-parameter families of such symmetric membranes, for isolated values of the parameters there are collisions of a single frequency with a double frequency, but there are no collisions of single frequencies with one another or collisions of double frequencies with one another.

3. The minimal number of parameters of a family of membranes for which more complicated collisions of characteristic frequencies are realized (stably with respect to small perturbations preserving the symmetry) is given by the formula

$$\sum_{i,j} \left[\frac{(i-1)(i+2)}{2} + j^2 \right] v_{ij},$$

where v_{ij} is the number of points of collision of i single and j double frequencies.

435

In particular, for a typical small deformation of a circular membrane preserving rotational symmetry of order 3, a third of the eigenvalues (corresponding to eigenfunctions with azimuthal part $\cos 3k\varphi$ and $\sin 3k\varphi$) immediately disperse. Under further one-parameter deformation the simple and double characteristic frequencies can pass through one another, but two simple or two double frequencies cannot collide with one another.

E *Discussion*

The value of the concepts of general position and symmetry lies, in particular, in the fact that they allow us to obtain some information in those cases where we cannot find an exact solution of a problem. In particular, for almost no membranes do we know the forms of the characteristic oscillations. Nevertheless, from general arguments we can say something, for example, about the multiplicities of eigenvalues.

The study of high-frequency oscillations of continuous media is very important in many fields (optics, acoustics, etc.), and special methods have been developed for approximate determination of the form of characteristic oscillations. One of these methods (called the method of *quasi-classical asymptotics*) consists of seeking an oscillation which is locally close to a simple harmonic wave of short length, but which changes its amplitude and the direction of its front from point to point.

Analysis (which we will not go into here) shows that in some cases we can construct approximate solutions, with the indicated properties, of the equation for eigenfunctions. They are approximate solutions in the sense that they almost satisfy the equation for eigenfunctions (not in the sense that they are close to real eigenfunctions).

In particular, if the membrane has the form of an equilateral triangle with smoothed and strongly blunted corners, then we can construct an approximate solution of the type described which differs appreciably from zero only in a neighborhood of one of the altitudes of the triangle. (Physicists call this approximate solution the wave analogue of a beam moving along the altitude of the triangle; this beam is a stable[115] trajectory on a billiard table having the shape of our membrane; c.f. the following appendix on short wave asymptotics).

It follows from symmetry and general position arguments that typical membranes with rotational symmetry of third order have no real characteristic oscillations of the type described. Assume that one of the characteristic

---

[115] The condition for linear stability of a billiard trajectory has the form

$$(r_1 + r_2 - l)(r_1 - l)(r_2 - l) > 0,$$

where l is the length of the interval of the trajectory and r_1 and r_2 are the radii of curvature of the walls at its ends.

oscillations of the membrane is concentrated near an altitude (but not near the center of the membrane). Then, rotating it by 120° and 240° we obtain three characteristic oscillations with the same characteristic frequency. These three oscillations are independent (this follows from the fact that their sum is not zero). Therefore, the characteristic frequency has multiplicity 3, which does not occur in typical systems with third-order rotational symmetry.

From this argument it is clear that attempting to construct rigorous high-frequency asymptotics for eigenfunctions is a rather hopeless task; what we can hope to do is to obtain approximate formulas for almost characteristic oscillations. Such an almost characteristic oscillation can differ very strongly from real characteristic oscillations, but if we give the membrane the initial condition corresponding to it, then for a long time the oscillation will resemble a standing wave (characteristic oscillation).

An example of an almost characteristic oscillation is the motion of one of two identical pendulums connected by a very weak spring. If, at the initial moment, we set the first pendulum in motion and leave the second fixed, then for a long time it will appear that only the first pendulum is oscillating, and the oscillation will be almost characteristic. For true characteristic oscillations, both pendulums oscillate with the same amplitude.

The problem of connecting the geometry of a membrane with the properties of its characteristic oscillations has been intensively studied in recent years by many authors (including H. Weyl, S. Minakshisundaram and A. Pleijel, A. Selberg, J. Milnor, M. Kac, I. Singer, H. McKean, M. Berger, Y. Colin de Verdière, J. Chazarain, J. J. Duistermaat, V. F. Lazutkin, A. I. Shnirel'man, and S. A. Molchanov).

To the simplest question, "Can you hear the shape of a drum?" the answer turns out to be negative: there exist non-isometric riemannian manifolds with the same spectrum. On the other hand, several properties of a manifold can be recovered from the eigenvalues of the laplacian and from the properties of eigenfunctions (for example, the complete set of lengths of closed geodesics can be recovered).

Appendix 11: Short wave asymptotics

From the point of view of physical optics, the description of the propagation of light in geometric optics, using rays (i.e., Hamilton's canonical equations) or wave fronts (i.e., the Hamilton-Jacobi equation), is only an approximation. According to the ideas of physical optics, light is electromagnetic waves, and geometric optics is a first approximation, a good description of phenomena only when the length of the waves is small compared to the size of the objects being considered.

A mathematical version of these physical ideas consists of asymptotic formulas for solving the corresponding differential equations—formulas which give better approximations for higher-frequency oscillations (i.e., for shorter waves). These asymptotic formulas can be written in terms of rays (i.e., motions in some hamiltonian dynamical system) or fronts (i.e., solutions of the Hamilton-Jacobi equation).

Similar short wave asymptotics exist for solutions of many equations in mathematical physics, describing all wave processes. In different areas of physics and mathematics they are connected with different names. For example, in quantum mechanics, short wave asymptotics are called quasi-classical approximations; they are determined by the so-called WKBJ method (Wentzel, Kramers, Brillouin, Jeffreys), although these approximations were used much earlier by Liouville, Green, Stokes, Rayleigh and others.

The construction of short wave asymptotics is based on the idea that, locally, a series of almost strictly sinusoidal waves is observed at each place, although the amplitudes of these waves and the directions of their fronts change slowly from point to point. Formal substitution of a function of this form into the partial differential equations describing the wave process reduces us (in a first approximation for waves of small length) to the Hamilton-Jacobi equation for wave fronts. The higher-order approximations allow us to determine as well the dependence of the amplitude of oscillation on the point.

Of course, the entire procedure requires a mathematical foundation. The exact formulation and proof of the corresponding theorems are not at all easy. Particular difficulty is introduced by "caustics" (i.e., focal or conjugate points, or turning points).

Caustics are envelopes of families of rays; they can be seen on a wall illuminated by rays reflected from some smooth curved surface. If the rays orthogonal to the wave fronts intersect and form caustics, then near the caustics the formulas for short wave asymptotics must be slightly changed. Namely, the phase of oscillations along each ray undergoes a standard discontinuity (one-fourth of a wave) upon each passage of the ray through a caustic.

A precise description of all these phenomena may be conveniently developed in terms of the geometry of lagrangian submanifolds of the corresponding phase space and their projections onto the configuration space. Here, caustics are interpreted as singularities of the projection, from phase space to configuration space, of that lagrangian manifold which represents a

438

family of rays. Thus, the normal forms of singularities of lagrangian projections introduced in Appendix 12 supply a classification of singularities of caustics formed by systems of rays in "general position."

In this appendix we introduce (without proof) the simplest formulas of short wave asymptotics for the Schrödinger equation of quantum mechanics. A more detailed exposition can be found in the following places:

J. Heading, Introduction to phase integral methods, Methuen Co. Ltd., 1962. (Cf. especially Appendix II (by V. P. Maslov) in the Russian translation of Heading's book, Moscow 1965).

V. P. Maslov, Théorie des perturbations et méthodes asymptotiques, Pairs, Dunod, 1972 (Russian edition: Moscow University, 1965).

V. I. Arnold, On a characteristic class entering into conditions of quantization, Functional Analysis and its Applications, v. I (1967).

L. Hörmander, Fourier integral operators, Acta Math. 127 (1971), 79–183.

A *Quasi-classical approximation for solutions of Schrödinger's equation*

Schrödinger's equation for a particle in a field with potential energy U in euclidean space is an equation for a complex-valued function $\psi(q, t)$:

$$ih \frac{\partial \psi}{\partial t} = -\frac{h^2}{2} \Delta \psi + U(q)\psi, \qquad q \in \mathbb{R}^n, t \in \mathbb{R}.$$

Here, h is some real constant which is also a small parameter of the problem being considered, and Δ is the Laplace operator.

We assume that the initial condition has the short wave form

$$\psi|_{t=0} = \varphi(q)e^{(i/h)s(q)},$$

where the smooth function φ is nonzero only inside some bounded region. We will find below an asymptotic (as $h \to 0$) formula for the solution of Schrödinger's equation with such an initial condition.

First of all, we consider the motion of a classical particle in the field with potential energy U, i.e., we consider Hamilton's equations

$$\dot{q} = \frac{\partial H}{\partial p} \qquad \dot{p} = -\frac{\partial H}{\partial q}, \quad \text{where } H = \tfrac{1}{2}p^2 + U(q)$$

in $2n$-dimensional phase space. The solutions of these equations determine a phase flow (under some conditions on the potential, which we assume fulfilled; these conditions prevent the particle from going off to infinity in a finite time).

We associate to our short wave initial condition a lagrangian submanifold of the phase space (i.e., a manifold whose dimension is equal to the dimension of the configuration space and on which the 2-form $dp \wedge dq$ defining the symplectic structure on the phase space is identically zero). Namely, we define

439

the "momentum" corresponding to our initial condition as the gradient of the phase, i.e., we set

$$p(q) = \frac{\partial s}{\partial q}.$$

Lemma. *For any smooth function s, the graph of the function p(q) constructed by it in the phase space $\mathbb{R}^{2n} = \{(p, q)\}$ is a lagrangian manifold. Conversely, if a lagrangian manifold projects diffeomorphically onto the q-space (i.e., it is a graph), then it is given by some generating function s, according to the formula above.*

We denote the lagrangian manifold constructed from the initial condition (with the function s) by M. After time t the phase flow g^t carries the manifold M to another manifold $g^t M$. This new manifold is also lagrangian, since the phase flow preserves the symplectic structure.

For small t, the new lagrangian manifold, like the old, projects diffeomorphically onto the configuration space. However, for large t this is not necessarily true (Figure 244).

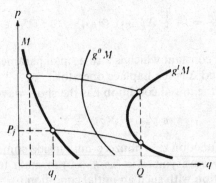

Figure 244 Transformation of lagrangian manifolds by the phase flow

In other words, several points of the new lagrangian manifold may project to one point Q of the configuration space. We assume that there are only finitely many of these points and that they are all nondegenerate (i.e., that at each of the points of the new lagrangian manifold which project onto Q, the derivative of the projection mapping onto the configuration space is non-degenerate).

The nondegeneracy condition is satisfied for almost all points Q. Those exceptional points Q for which it is not satisfied form a set of measure zero in the configuration space. In the general case, this set is a surface whose dimension is one less than the dimension of the configuration space. This surface, playing the role of a caustic in our problem, can itself have complicated singularities.

The points of the new lagrangian manifold projecting to the point Q arose under the phase flow transformation from several points of the original lagrangian manifold (constructed from the initial condition). In other words, after time t, several trajectories of classical particles, with initial conditions belonging to the original lagrangian manifold, arrive at Q.

We let (p_j, q_j) denote these initial points in the phase space, and S_j the action along the trajectories of the phase flow coming from the point (p_j, q_j). More precisely, we set

$$S_j(Q, t) = s(q_j) + \int_0^t L \, d\theta,$$

$$\text{where } L = \frac{\dot{q}^2}{2} - U(q) \text{ and } g^\theta(p_j, q_j) = (p(\theta), q(\theta)).$$

Then, as $h \to 0$, the solution of Schrödinger's equation with the oscillating initial condition given by the functions s and φ has asymptotic form

$$\psi(Q, t) = \sum_j \varphi(q_j) \left| \frac{DQ}{Dq_j} \right|^{-1/2} e^{(i/h)S_j(Q, t) - (i\pi/2)\mu_j} + O(h),$$

where μ_j is an integer (the Morse index) which will be defined below.

In order to explain this formula, we first consider the case when the time interval t is small. In this case, the sum is reduced to a single term, since the lagrangian manifold obtained from the original lagrangian manifold by the phase flow transformation after small time projects diffeomorphically onto the configuration space. In other words, of the family of particles corresponding to the initial condition for Schrödinger's equation, only one arrives at Q after the small time t.

For small t, the Morse index is equal to zero (as we will see below from its definition). In this way the function $\psi(Q, t)$ has, like the initial condition, a rapidly oscillating form. Thus, the function S defining the wave fronts at time t is none other than the value at time t of the solution of the Hamilton-Jacobi equation, the initial condition for which is given by the function s defining the wave front at the initial moment. The amplitude of the wave at time t at the point Q is obtained from the amplitudes, at the initial moment at the original point, of the trajectories coming to Q multiplied by a certain factor. This factor is chosen so that, under motions of the particles corresponding to our initial conditions, the integral of the square of the modulus of the function ψ, over a region of configuration space filled with particles, does not change with time. (Here we assume that at the initial moment, some region in the configuration space has been selected; then the phase points on the original lagrangian manifold are selected whose projections onto the configuration space lie in this region; their images under the action of the phase flow after time t are found; finally, the projections of these images onto the configuration space form the region "filled with particles at time t.")

B *The Morse and Maslov indices*

The number μ_j is defined as the number of focal points to the manifold M on the interval $[0, t]$ of the phase curve starting out at the point (p_j, q_j).

Focal points to the manifold M are defined as follows. We chose the point Q so that, under projection of the lagrangian manifold obtained from M at time t, a nondegeneracy condition is satisfied at this point. However, if we consider the entire phase curve coming from the point (p_j, q_j), then at some moments of time θ between 0 and t, the nondegeneracy condition may not be satisfied at the point $(p(\theta), q(\theta))$ of the lagrangian manifold $g^\theta M$. Such points are called *focal points* to the manifold M along this phase curve.

We note that the definitions of focal points to M and the Morse index do not depend on Schrödinger's equation, but relate simply to the geometry of the phase flow in the cotangent bundle to the configuration space (or to the calculus of variations, which is the same thing).

In particular, as our lagrangian manifold M we may take the fiber of the cotangent bundle passing through the point (p_0, q_0) (given by the condition $q = q_0$). In this case a focal point to M on the phase curve going out from (p_0, q_0) is called *conjugate* to the original point (more precisely, the projection of this focal point onto the configuration space is said to be conjugate to the point q_0 along the extremal in the configuration space starting at q_0 with momentum p_0). In the even more special case of motion along a geodesic on a riemannian manifold, a focal point to a fiber of the cotangent bundle is called conjugate to the initial point of the geodesic along this geodesic. For example, the south pole of a sphere is conjugate to the north pole along any meridian.

The Morse index of an interval of a geodesic, equal to the number of points conjugate to the initial point, plays an important role in the calculus of variations. Namely, we consider the second differential of the action as a quadratic form on the space of variations (with fixed endpoints) of the geodesic we are studying. Then the index of inertia of this quadratic form is equal to the Morse index (cf., for instance, J. Milnor, *Morse Theory*, Princeton University Press, 1967).

Thus the geodesic, up to the first conjugate point, is a minimum of the action, which justifies the name "principle of least action" for various variational principles of mechanics.

We note that in calculating the Morse index, the focal points must be counted with multiplicity (the multiplicity of a focal point in general position is equal to 1).

The Morse index is a particular case of the so-called *Maslov index*, which is defined independently of the phase flow for any curve on a lagrangian manifold of the cotangent bundle over the configuration space.

Consider the projection of our n-dimensional lagrangian manifold onto the n-dimensional configuration space. This is a smooth mapping of manifolds of the same dimension. It can have singular points, i.e., points at which the rank of the derivative mapping drops, and in a neighborhood of which the projection is not a diffeomorphism.

It turns out that in general the set of singular points has dimension $n - 1$ and consists of the union of a smooth manifold of dimension $n - 1$ made up of simple singular points at which the rank drops to 1, and a finite set of manifolds whose dimensions are $n - 3$ and smaller. Here, "in general" means that these properties can be attained by an arbitrarily small perturbation of the lagrangian manifold, under which it remains lagrangian.

We should point out that, among the pieces of various ranks into which the set of singular points is divided, there is no piece of dimension $n - 2$. After the simplest singular points, forming a manifold of dimension $n - 1$, there are the points where the rank drops by two; they form a manifold of dimension $n - 3$. The projection of the set of singular points onto the configuration space (the caustic) consists, in general, of pieces of all dimensions from 0 to $n - 1$ without omissions.

Furthermore, it turns out that the $(n - 1)$-dimensional manifold of the simplest singular points is two-sided in the lagrangian manifold; that is, we can coordinate the orientations of the normals at all points in the following way.

Consider some simple singular point on the lagrangian manifold. We take a system of coordinates $q_1, \ldots, q_n$ in a neighborhood of the projection of this point onto the configuration space. Let $p_1, \ldots, p_n$ be corresponding coordinates in the fiber of the cotangent bundle. In a neighborhood of our singular point, we can consider the lagrangian manifold as the graph of the vector function $(q_1, p_2, \ldots, p_n)$ of the variables $(p_1, q_2, \ldots, q_n)$ (or a vector function of an analogous form in which the role of the distinguished coordinate is played not by the first coordinate but by any of the remaining coordinates).

Singular points near the given one are then defined by the condition $\partial q_1/\partial p_1 = 0$. For lagrangian manifolds in general position, this derivative changes sign upon passing from one side of the manifold of singular points to the other in our neighborhood of the simple singular point. We will call the side where this derivative is positive the *positive side*.

We note that it is necessary to prove that the definitions of positive direction near different points agree with one another. Furthermore, it must be shown that the positive direction near one point is well defined, i.e., does not depend on the coordinate system. All this can be done by direct calculations (cf. the article cited above in "Functional Analysis"). For further development of these ideas, see V. I. Arnold, Sturm theorems and symplectic geometry, Funct. Anal. Appl. 19 (1985).

Now the Maslov index of an oriented curve on a lagrangian manifold is defined as the number of passages from the negative side of the manifold of singularities to the positive side, minus the number of passages in the other direction. In this we assume that the ends of the curve are nonsingular and that the curve intersects only the manifold of simple singular points and only with nonzero angles. Having defined the index for such curves, we can define it for an arbitrary curve connecting two nonsingular points: to do this it is sufficient to approximate the curve by one which intersects only the manifold of simple singular points and only with nonzero angles. It can be shown that the index does not depend on the choice of the approximating curve.

PROBLEM. Find the index of the circle $p = \cos t$, $q = \sin t$ oriented by the parameter t, $0 \leq t \leq 2\pi$, in the lagrangian manifold $p^2 + q^2 = 1$ of the phase plane.

ANSWER. $+2$.

Finally, the Morse index of a phase curve in $\mathbb{R}^{2n}$ can now be defined as the Maslov index of a curve in an $(n + 1)$-dimensional lagrangian manifold in a suitable $(2n + 2)$-dimensional phase space. As coordinates in this space we will take $(p_0, p; q_0, q)$ (where $(p, q) \in \mathbb{R}^{2n}$). If we set $q_0 = t$ and $p_0 = -H(p, q)$, and let the point (p, q) range over the n-dimensional lagrangian manifold in $\mathbb{R}^{2n}$ obtained from the original after time t by the action of the phase flow, then under change of t the points in $\mathbb{R}^{2n+2}$ form an $(n + 1)$-dimensional lagrangian manifold. The graph of the motion of a phase point under the action of the phase flow can be considered as a curve on this $(n + 1)$-dimensional lagrangian manifold. We can verify that the Maslov index of this graph agrees with the Morse index of the original phase curve.

C Indices of closed curves

The indices of closed curves on lagrangian submanifolds of a linear phase space can also be calculated with the help of a complex structure. In addition to the symplectic structure $dp \wedge dq$ on the linear phase space $\mathbb{R}^{2n} = \{(p, q)\}$, we introduce a euclidean structure (with scalar square $p^2 + q^2$) and a complex structure, in which multiplication by i is

$$I: \mathbb{R}^{2n} \to \mathbb{R}^{2n} \qquad I(p, q) = (-q, p) \qquad z = p + iq \qquad \mathbb{C}^n = \{z\}.$$

All three structures are connected by the relation

$$[x, y] = (Ix, y),$$

where the square brackets denote the skew-scalar product.

Linear transformations of the phase space preserving any two (and, therefore, all three) structures are called *unitary transformations*. Such transformations take lagrangian planes to lagrangian planes.

Every lagrangian plane can be obtained from any other (e.g., from the real plane $\mathbb{R}^n$ given by the equation $q = 0$) by a unitary transformation. In addition, any two unitary transformations A and B carrying the real plane to the same lagrangian plane differ by a unitary transformation which is a real orthogonal transformation:

$$B = AC, \quad \text{where } C\mathbb{R}^n = \mathbb{R}^n.$$

Conversely, any preliminary orthogonal transformation does not change the image of the plane under the action of a unitary transformation.

We now note that the determinant of an orthogonal transformation is equal to ± 1. Therefore the *square of the determinant* of a unitary transformation carrying the real plane to a given lagrangian plane depends only on the lagrangian plane itself and does not depend at all on the choice of unitary transformation.

After these preliminary remarks we return to our lagrangian manifold and closed oriented curve lying in it. At every point of the curve, there is a plane tangent to the lagrangian manifold in the symplectic vector space. The square of the determinant of the unitary transformation carrying the real plane to this tangent plane is a complex number with modulus one. As a point moves along our closed curve, this complex number changes. After an entire circuit of the curve, the square of the determinant makes some integral number of rotations around the origin on the plane of complex variables, oriented from 1 to i. This integer is the index of the closed curve.

The indices of closed curves enter into asymptotic formulas for stationary problems (characteristic oscillations). Assume that the phase flow corresponding to the potential U has an invariant lagrangian manifold lying on the energy level $H = E$. Then the equation

$$\tfrac{1}{2}\Delta\psi = \lambda^2(U(q) - E)\psi$$

has a series of eigenvalues $\lambda_N \to \infty$ with asymptotic form $\lambda_N = \mu_N + O(\mu_N^{-1})$ if, for every closed contour γ on the lagrangian manifold, we have the congruence

$$\frac{2\mu_N}{\pi} \oint_\gamma p \, dq \equiv \text{ind } \gamma \pmod 4.$$

In the one-dimensional case, the lagrangian manifold is a circle, its index is equal to 2, and the formula above reduces to the so-called "quantization condition"

$$\mu_N \oint_\gamma p \, dq = 2\pi(N + \tfrac{1}{2}).$$

The eigenfunctions corresponding to these eigenvalues are also associated with lagrangian manifolds, but this association is not so simple. In fact, we cannot write down asymptotic formulas for eigenfunctions, but only for functions approximately satisfying the equations of characteristic functions. These functions turn out to be small outside the projection of the lagrangian manifold onto the configuration space. The asymptotic formulas have singularities near the caustics formed by the projection.

The actual eigenfunctions, however, can behave entirely differently, at least if the eigenvalue is multiple or if there are eigenvalues close to it (cf. Appendix 10).

Appendix 12: Lagrangian singularities

Lagrangian singularities are singularities of projections of lagrangian manifolds onto configuration space. Such singularities are encountered in investigating global solutions to the Hamilton-Jacobi equation, in studying caustics, focal or conjugate points, in analyzing the propagation of discontinuities and shock waves in the mechanics of a solid medium, and also in problems of short wave asymptotics (cf. Appendix 11).

In order to describe lagrangian singularities we must first say a few words about singularities of smooth mappings in general. We begin with the simplest examples.

A *Singularities of smooth mappings of a surface onto a plane*

The mapping projecting a sphere onto a plane is singular on the equatorial circle (at points of the equator the rank of the derivative drops to one). As a result, a curve is formed on the plane of projection (the so-called apparent contour) bounding regions in which points have different numbers of pre-images: every point of the plane inside the apparent contour has two pre-images, and every point outside has none.

In more complicated cases of "apparent contours" there can be more complicated singularities. Consider, for example, the surface given in three-dimensional space with coordinates (x, y, z) by the equation (Figure 245)

$$x = yz - z^3$$

and the mapping of projection parallel to the z-axis onto the plane with coordinates (x, y).

The singular points of the projection form a smooth curve on the surface (with equation $3z^2 = y$). However, the image of this curve on the (x, y) plane is not a smooth curve. This image is a semi-cubical parabola with a cusp at the point $(0, 0)$ with equation

$$27x^2 = 4y^3.$$

Such a curve divides the plane into two parts: a smaller part (inside the cusp) and a larger part (outside). Over each point of the smaller part there are three points of our surface, and over each point of the larger part there is only one.

We now consider any small deformation of our surface. It turns out that, under projection of any surface close to ours, the apparent contour will always have a similar singularity (semi-cubical cusp) at some point close to the singularity of the apparent contour of the original surface. In other words, this singularity is not removable by a small perturbation of the surface.

Furthermore, in place of a deformation of the surface, we can arbitrarily deform the mapping itself of the surface to the plane (no longer caring whether it is a projection), as long as it remains smooth and the deformation is small. It turns out that, for these deformations too, the cusp does not disappear but is only slightly deformed.

The examples presented here exhaust all typical singularities of mappings of a surface to the plane. It can be shown that all more complicated singu-

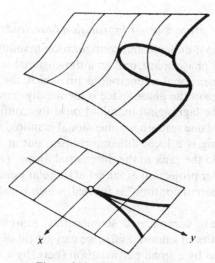

Figure 245   Whitney's tuck

larities are removable by a small perturbation. Therefore, by slightly deforming any smooth mapping, we can always arrange that in a neighborhood of any point of the surface, the mapping will be either nonsingular, or structurally similar to the projection mapping of a sphere onto a plane near the equator, or structurally similar to the projection mapping of the surface considered above with a cubic cusp on the apparent contour.

The words "structurally similar to" mean that, on the pre-image surface and the image plane, we can choose local coordinates (in a neighborhood of our point and its image) such that in these coordinates the mapping will be written in a special way. Namely, the normal forms to which the mapping of the surface to the plane will be reduced in a neighborhood of points of the three types indicated above will be

$y_1 = x_1 \qquad y_2 = x_2$ \qquad (nonsingular point)

$y_1 = x_1^2 \qquad y_2 = x_2$ \qquad (a fold, as on the equator of the sphere)

$y_1 = x_1 x_2 - x_1^3 \qquad y_2 = x_2$ \qquad (a "tuck" with a cusp on the apparent contour)

Here (x_1, x_2) are the local coordinates in the pre-image, and (y_1, y_2) are the local coordinates in the image.

The proof of this theorem (it is due to H. Whitney) and its multidimensional generalizations can be found in works on the theory of singularities of smooth maps, such as

V. I. Arnold, Singularities of smooth mappings, Russian Math. Surveys 23:1 (1968) 1–44.

Symposium on Singularities of Smooth Manifolds and Maps, Univ. of Liverpool, 1969–70. Proceedings. Springer, 1971. See especially the article of R. Thom and H. Levine.

Golubitsky and Guillemin, Stable Mappings and Their Singularities, Springer-Verlag, 1973.

B *Singularities of projection of lagrangian manifolds*

We now consider an n-dimensional configuration manifold, the corresponding $2n$-dimensional phase space, and an n-dimensional lagrangian submanifold (i.e., an n-dimensional submanifold on which the 2-form giving the symplectic structure of the phase space is identically zero).

By projecting the lagrangian manifold onto the configuration space, we obtain a mapping of one smooth n-dimensional manifold to another. At most points, this mapping is a local diffeomorphism, but at some points of the lagrangian manifold the rank of the differential drops. These points are said to be singular. Under projection of the set of singular points to the configuration space an "apparent contour" is formed, which is called a *caustic* in the lagrangian case.

Caustics can have complicated singularities; however, as in the usual theory of singularities of smooth maps, we can get rid of singularities which are too complicated by a small perturbation (here, by a small perturbation, we mean a small deformation of a lagrangian manifold in phase space under which this manifold remains lagrangian).

After this there remain only the simplest unremovable singularities, for which we can write out normal forms and which we can study once and for all. When considering problems in general position which do not satisfy any special properties of symmetry, it is natural to expect that only these simple unremovable singularities will appear.

Consider, for example, the caustics formed on a wall by light from a point source reflected from some smooth curved surface (here the four-dimensional phase space is formed by straight lines intersecting the surface of the wall in all possible directions, and the lagrangian submanifold by the rays of light coming from the source as they intersect the wall). By moving the source, we can see that generally the caustics have only simple singularities (semi-cubical cusps), while more complicated singularities appear only for special, exceptional positions of the source.

We will give below, for $n \leq 5$, normal forms for singularities of the projection of an n-dimensional lagrangian submanifold of $2n$-dimensional phase space onto an n-dimensional configuration space. There are a finite number of these normal forms, and their classification is related (in a rather mysterious way) with the classifications of simple Lie groups, simple degenerate critical points of functions, regular polyhedra, and many other objects. For $n \geq 6$, the normal forms of some singularities must inevitably contain parameters. For further details the reader is referred to the articles:

V. I. Arnold, Normal forms for functions near degenerate critical points, the Weyl groups of A_k, D_k, E_k, and lagrangian singularities, Functional Analysis and Its Applications 6:4 (1972) 254–272.

V. I. Arnold, Critical points of smooth functions and their normal forms, Uspekhi Mat. Nauk 30:5 (1975).

C Tables of normal forms of typical singularities of projections of lagrangian manifolds of dimension $n \leq 5$

We will use the following notation:

$(q_1, \ldots, q_n)$ are coordinates on the configuration space,

$(p_1, \ldots, p_n)$ are the corresponding impulses,

so that p and q together form a symplectic coordinate system in the phase space.

We will give a lagrangian manifold with the help of a generating function F by the formulas

$$q_i = \frac{\partial F}{\partial p_i} \qquad p_j = -\frac{\partial F}{\partial q_j},$$

where the index i runs over some subset of $\{1, \ldots, n\}$ and j runs over the remainder of $\{1, \ldots, n\}$. That is, $i = 1, j > 1$ for singularities denoted in the list by A_k, and $i = 1, 2, j > 2$ for singularities denoted by D_k and E_k.

With this notation, one and the same expression $F(p_i, q_j)$ can be considered as giving a lagrangian manifold in spaces of a different number of dimensions: we can add arbitrarily many arguments q_j, on which F does not actually depend.

The list of normal forms of typical singularities is now as follows: for $n = 1$

$$A_1: F = p_1^2 \qquad A_2: F = \pm p_1^3;$$

for $n = 2$, in addition to the two above, there is

$$A_3: F = \pm p_1^4 + q_2 p_1^2;$$

for $n = 3$, in addition to the three preceding, there are

$$A_4: F = \pm p_1^5 + q_3 p_1^3 + q_2 p_1^2,$$
$$D_4: F = \pm p_1^2 p_2 \pm p_2^3 + q_3 p_2^2;$$

for $n = 4$, in addition to the five preceding, there are

$$A_5: F = \pm p_1^6 + q_4 p_1^4 + q_3 p_1^3 + q_2 p_1^2,$$
$$D_5: F = \pm p_1^2 p_2 \pm p_2^4 + q_4 p_2^3 + q_3 p_2^2;$$

for $n = 5$, in addition to the seven preceding, there are

$$A_6: F = \pm p_1^7 \pm q_5 p_1^5 + \cdots + q_2 p_1^2,$$
$$D_6: F = \pm p_1^2 p_2 \pm p_2^5 + q_5 p_2^4 + q_4 p_2^3 + q_3 p_2^2,$$
$$E_6: F = \pm p_1^3 \pm p_2^4 + q_5 p_1 p_2^2 + q_4 p_1 p_2 + q_3 p_2^2.$$

D Discussion of the normal forms

A point of type A_1 is nonsingular. A singularity of type A_2 is a fold singularity. If we take $(p_1, q_2, \ldots, q_n)$ as coordinates on the lagrangian manifold, then the projection mapping may be written as

$$(p_1, q_2, \ldots, q_n) \to (\pm 3p_1^2, q_2, \ldots, q_n).$$

A singularity of type A_3 is a tuck with a semi-cubical cusp on the visible contour. To convince ourselves of this, it is enough to write out the corresponding mapping of the two-dimensional lagrangian manifold to the plane:

$$(p_1, q_2) \to (\pm 4p_1^3 + 2q_2 p_1, q_2).$$

A singularity of type A_4 first appears in the three-dimensional case, and the corresponding caustic is represented by a surface in three-dimensional space (Figure 246) with a singularity called a swallowtail (we already encountered this in Section 46).

The caustic of a singularity of type D_4 in three-dimensional space is represented as a surface with three cuspidal edges (of type A_3), tangent at one point; two of these cuspidal edges can be imaginary, so that there are two versions of the caustic of D_4.

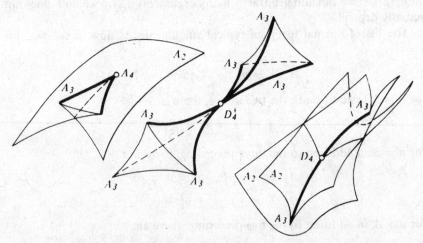

Figure 246   Typical singularities of caustics in three-dimensional space

E Lagrangian equivalence

We must now say in what sense the examples mentioned are normal forms of typical singularities of projections of lagrangian manifolds. First of all, we will define which singularities we will consider to have the "same structure."

A projection mapping of a lagrangian manifold onto configuration space will be called a *lagrangian mapping* for short. Suppose that we are given two

450

lagrangian mappings of manifolds of the same dimension n (the correspond-ing n-dimensional lagrangian manifolds lie, in general, in different phase spaces which are cotangent bundles of two different configuration spaces). We say that two such lagrangian mappings are *lagrangian equivalent* if there is a symplectic diffeomorphism of the first phase space to the second, taking fibers of the first cotangent bundle to fibers of the second, and taking the first lagrangian manifold to the second. The symplectic diffeomorphism itself is then called a *lagrangian equivalence mapping*.

We note that two lagrangian equivalent lagrangian mappings are taken one to the other with the help of diffeomorphisms in the pre-image space and the image space (or, as they say in analysis, are carried to one another by a change of coordinates in the pre-image and in the image). In fact, our sym-plectic diffeomorphism restricted to the lagrangian manifold gives a diffeo-morphism of the pre-images; a diffeomorphism of the configuration-space images arises because fibers are carried to fibers.

In particular, the caustics of the two lagrangian equivalent mappings are diffeomorphic, hence a classification up to lagrangian equivalence implies a classification of caustics. However, the classification up to lagrangian equiv-alence is finer than the classification of caustics, since a diffeomorphism of caustics does not in general give rise to a lagrangian equivalence of the map-pings. Furthermore, the classification up to lagrangian equivalence is finer then the classification up to diffeomorphisms of the pre-image and image, since not every such pair of diffeomorphisms is realized by a symplectic diffeomorphism of the phase space.

A lagrangian mapping considered in a neighborhood of some chosen point is called lagrangian equivalent at that point to another lagrangian mapping (also with a chosen point), if there is a lagrangian equivalence of the first mapping in some neighborhood of the first point onto the second in some neighborhood of the second point, carrying the first point to the second.

We can now formulate a classification theorem for singularities of lagrangian mappings in dimensions $n \leq 5$.

Theorem. *Every n-dimensional lagrangian manifold ($n \leq 5$) can, by an arbi-trarily small perturbation in the class of lagrangian manifolds, be made into one such that the projection mapping onto the configuration space will be lagrangian equivalent at every point to one of the lagrangian mappings in the list above.*

In particular, a two-dimensional lagrangian manifold can be put in "general position" by an arbitrarily small perturbation in the class of lagrangian manifolds, so that the projection mapping onto the configuration space (two-dimensional) will not have singularities other than folds (which can be reduced by a lagrangian equivalence to the normal form A_2) or tucks (which can be reduced by a lagrangian equivalence to the normal form A_3).

We note that this assertion about two-dimensional lagrangian mappings does not follow from the classification theorem for general (non-lagrangian) mappings. In the first place, lagrangian mappings make up a very restricted class among all smooth mappings, and therefore they can (and actually do for $n > 2$) have as typical, singularities which are not typical for mappings of general form. Secondly, the possibility of reducing a mapping to normal form by diffeomorphisms of the pre-image and image does not imply that this can be done using a lagrangian equivalence.

In this way, the caustics of a two-dimensional lagrangian manifold in general position have as singularities only semi-cubical cusps (and points of transversal intersection). All more complicated singularities break up under a small perturbation of the lagrangian manifold, the resulting cusps and self-intersection points of caustics are unremovable by small perturbations, and are only slightly deformed.

Normal forms of the singularities $A_4, D_4, \ldots$ can be used in a similar way for studying the caustics of lagrangian manifolds of higher dimensions, and also for studying the development of caustics of low-dimensional lagrangian manifolds, when parameters on which the manifold depends are varied.[116]

Other applications of the formulas of this section can be found in the theory of *Legendre singularities*, i.e., singularities of wave fronts. Legendre transforms, envelopes, and convex hulls (cf. Appendix 4). The theories of lagrangian and Legendre singularities have direct application, not only in geometric optics and the theory of asymptotics of oscillating integrals, but also in the calculus of variations, in the theory of discontinuous solutions of nonlinear partial differential equations, in optimization problems, pursuit problems, etc. R. Thom has suggested the general name *catastrophe theory* for the theory of singularities, the theory of bifurcations, and their applications.

[116] See, e.g., V. Arnold, Evolution of wavefronts and equivariant Morse lemma, Comm. Pure Appl. Math., 1976, No. 6.

Appendix 13: The Korteweg–de Vries equation

Not all first integrals of equations in classical mechanics are explained by obvious symmetries of a problem (examples are specific integrals of Kepler's problem, the problem of geodesics on an ellipsoid, etc.). In such cases, we speak of "hidden symmetry."[117]

Interesting examples of such hidden symmetry are furnished by the Korteweg–de Vries equation

$$(1) \qquad\qquad u_t = 6uu_x - u_{xxx}.$$

This nonlinear partial differential equation first arose in the theory of waves in shallow water; later it turned out that this equation is encountered in a whole series of problems in mathematical physics.

As a result of a series of numerical experiments, remarkable properties of solutions of this equation with zero boundary conditions at infinity were discovered: as $t \to \infty$ and $t \to -\infty$ these solutions decompose into "solitons"—waves of definite form moving with different velocities.

To obtain a soliton moving with velocity c, it is sufficient to substitute the function $u = \varphi(x - ct)$ into equation (1). Then we obtain the equation $\varphi'' = 3\varphi^2 + c\varphi + d$ for φ (d is a parameter). This is Newton's equation with a cubic potential. There is a saddle on the phase space (φ, φ'). The separatrix going from this saddle to the saddle for which $\varphi = 0$ determines a solution φ tending to 0 as $x \to \pm \infty$; it is a soliton.

When solitons collide, there is a complicated nonlinear interaction. However, numerical experiments showed that the sizes and velocities of the solitons do not change as a result of collision. And, in fact, Kruskal, Zabusky, Lax, Gardner, Green, and Miura succeeded in finding a whole series of first integrals for the Korteweg–de Vries equation. These integrals have the form $I_s = \int P_s(u, \ldots, u^{(s)})dx$, where P_s is a polynomial. For example, it is easy to verify that the following are first integrals of equation (1):

$$I_{-1} = \int u \, dx \qquad I_0 = \int u^2 \, dx \qquad I_1 = \int \left(\frac{u'^2}{2} + u^3\right) dx,$$

$$I_2 = \int \left(\frac{u''^2}{2} - \frac{5}{2}u^2 u'' + \frac{5}{2}u^4\right) dx.$$

The appearance of an infinite series of first integrals is easily explained by the following theorem of Lax.[118] We will denote the operator of multiplication by a function of x by the symbol for the function itself, and the operator of differentiation with respect to x by the symbol ∂. Consider the Sturm-Liouville operator $L = -\partial^2 + u$ depending on a function $u(x)$. We verify directly:

Theorem. The Korteweg–de Vries equation (1) is equivalent to the equation $\dot{u} = [L, A]$, where $A = 4\partial^3 - 3(u \partial + \partial u)$.

---

[117] The term "accidental symmetry" is frequently used in English. [Trans. note.]

[118] Lax, P. D., Integrals of nonlinear equations of evolution and solitary waves, *Comm. Pure Appl. Math.* **21** (1968) 467–490.

Appendix 13: The Korteweg–de Vries equation

Directly from this theorem of Lax, we have

Corollary. *The operators L constructed from a solution of equation* (1) *are unitarily equivalent for all t; in particular, each of the eigenvalues λ of the Sturm–Lionville problem $Lf = \lambda f$ with zero boundary conditions at infinity is a first integral of the Korteweg–de Vries equation.*

Gardner, V. E. Zakharov and L. D. Faddeev noted that equation (1) is a completely integrable infinite-dimensional hamiltonian system, and found the corresponding action-angle variables.[119] A symplectic structure on the space of functions vanishing at infinity is given by the skew-scalar product $\omega^2(\partial w, \partial v) = \frac{1}{2} \int (w \, \partial v - v \, \partial w)dx$, and the hamiltonian of equation (1) is the integral I_1. In other words, equation (1) can be written in the form of Hamilton's equation in the functional space of functions of x, $\dot{u} = (d/dx)(\delta I_1/\delta u)$.

Every integral I_s gives in this way a "higher Korteweg–de Vries equation" $\dot{u} = Q_s[u]$, where $Q_s = (d/dx)(\delta I_s/\delta u)$ is a polynomial in the derivatives $u, u', \ldots, u^{2s+1}$. The integrals I_s are in involution, and the flows corresponding to them on the functional space commute.

The explicit form of the polynomials P_s and Q_s, and also the explicit form of the action-angle variables (and therefore of solutions of equation (1)), is described in terms of solutions of the direct and inverse problems of scattering theory with potential u.

The explicit form of the polynomials Q_s can also be obtained from the following theorem of Gardner, generalizing Lax's theorem. In the space of functions of x, we consider a differential operator of the form $A = \sum p_i \partial^{m-i}$, where $p_0 = 1$, and the remaining coefficients p_i are polynomials in u and the derivatives of u with respect to x. It turns out that, for any s there is an operator A_s of order $2s + 1$ such that its commutator with the Sturm–Liouville operator L is the operator of multiplication by a function $[L, A_s] = Q_s$.

The operator A_s is defined by these conditions uniquely up to the addition of linear combinations of the A_r with $r < s$; in the same way, the polynomials Q_s are determined up to the addition of linear combinations of the preceding Q_r's.

V. E. Zakharov, A. B. Shabat, L. D. Faddeev, and others, using Lax's method and techniques of inverse scattering theory, have studied a whole series of physically important equations, including the equations $u_{tt} - u_{xx} = \sin u$ and $i\psi_t + \psi_{xx} \pm \psi|\psi|^2 = 0$.

Investigation of the problem with periodic boundary conditions for the Korteweg–de Vries equation led S. P. Novikov[120] to the discovery of an interesting class of completely integrable systems with a finite number of degrees of freedom. These systems are constructed in the following way.

Consider any finite linear combination of first integrals, $I = \sum c_i I_{n-i}$, and let $c_0 = 1$. The set of stationary points of the flow with hamiltonian I

[119] Zakharov, V. E. and Faddeev, L. D., The Korteweg–de Vries equation is a completely integrable hamiltonian system, Functional Analysis and Its Applications, 5:4 (1971) 280–287.

[120] Novikov, S. P., The periodic problem for the Korteweg–de Vries equation, Functional Analysis and Its Applications, 8:3 (1974) 236–246.

on the functional space is invariant under the phase flows with hamiltonians I_s, including the phase flow of equation (1).

On the other hand, these stationary points are determined from the equations $(d/dx)(\delta I/\delta u) = 0$, or $\delta I/\delta u = d$. The second equation is the Euler-Lagrange equation for the functional $I - dI_{-1}$, involving derivatives of order n. Therefore, it has order $2n$ and can be written as a hamiltonian system of equations in $2n$-dimensional euclidean space.

It turns out that this hamiltonian system with n degrees of freedom has n integrals in involution and can be integrated completely with the help of suitable action-angle coordinates. In this way, we obtain a finite-dimensional family of particular solutions of the Korteweg–de Vries equation depending on $3n + 1$ parameters ($2n$ phase coordinates and $n + 1$ further parameters $c_1, \ldots, c_n; d$).

These solutions have, as Novikov showed, remarkable properties; for example, in the periodic problem they give functions $u(x)$ for which the linear differential equation with periodic coefficients

$$-X'' + u(x)X = \lambda X$$

has a finite number of zones of parametric resonance (cf. Section 25) on the λ-axis.

After this book was written, much work was done on the subjects discussed in this appendix, in particular by Novikov, Dubrovin, Krichever, Manakov, Matveev, Its, Dikii, Manin, Drinfeld, Gelfand, Lax, Moser, McKean, Van Moerbeke, Adler, Perelomov, Olshanetskii, and many others. Among other things, Manakov solved the Euler equations of a rigid body in $\mathbb{R}^n$ for arbitrary n: these are completely integrable. For more details see the forthcoming book by Novikov and his collaborators. (Note added by author in translation.)

Appendix 14: Poisson structures

Along with the classical Poisson bracket of functions, one also encounters more general (degenerate) brackets. A typical example is the Poisson bracket of functions of the components M_i of the angular momentum vector: $\{F, G\} = \sum (\partial F/\partial M_i)(\partial G/\partial M_j)\{M_i, M_j\}$. Such degenerate brackets may be considered as families of ordinary Poisson brackets or families of sympletic manifolds. These families generally have singularities (they are not foliations): they consist of symplectic manifolds (leaves) of different dimensions, related to one another by the condition of smoothness for the given degenerate Poisson bracket structure on the ambient space. (In the angular momentum example above, the leaves are concentric spheres and their center at the origin.)

In this appendix, we shall present the simplest elementary properties of Poisson structures on finite-dimensional manifolds. One should keep in mind, though, that in applications (especially to the mathematical physics of continuous media) one frequently encounters Poisson structures on infinite-dimensional manifolds. In these cases, the symplectic leaves often (but not always) have finite dimension or codimension.

A Poisson manifolds

A *Poisson structure* on a manifold is a Lie algebra structure on its space of smooth functions (i.e., a bilinear skew-symmetric operation of "Poisson bracket" on functions, satisfying the Jacobi identity) such that the operator $\mathrm{ad}_a = \{a, \ \}$ (contraction of the Poisson bracket with any fixed function a) is an operator of differentiation by some vector field θ_a. The vector field θ_a is then called the *hamiltonian vector field* with hamiltonian function a. The mapping $a \mapsto \theta_a$ gives a homomorphism from the Lie algebra of functions to the Lie algebra of vector fields. A manifold with a given Poisson structure is called a *Poisson manifold*.

Two points on a Poisson manifold are called *equivalent* if they can be joined by a path consisting of segments of integral curves of hamiltonian vector fields. The equivalence classes under this relation are called the *leaves* of the Poisson manifold. The values of all possible hamiltonian vector fields at a given point of a Poisson manifold form a linear space which is just the tangent space of the leaf through that point. Thus the leaves are smooth manifolds, but they are in general not closed, and they have different dimensions.

The classical (explicitly described by S. Lie in 1890, but essentially considered already by Jacobi) example of a Poisson manifold is the dual space of a (finite-dimensional) Lie algebra. The elements of the algebra itself may be considered as linear functions on this space. The Poisson structure is defined as an extension of the Lie algebra structure from this finite-dimensional subspace to the entire space of smooth functions on the dual of the original Lie algebra. Such an extension exists and is unique: if $\omega_1, \ldots, \omega_n$ is a basis of the original Lie algebra, then

$$\{a, b\}_{\text{Poisson}} = \sum (\partial a / \partial \omega_i)(\partial b / \partial \omega_j)[\omega_i, \omega_j]_{\text{Lie}}.$$

In this example, the leaves are the orbits of the co-adjoint representation of the underlying Lie group in the dual of its Lie algebra.

Every leaf of a Poisson manifold carries a *natural symplectic structure* (closed nondegenerate 2-form), defined in the following way. Consider the values of two hamiltonian vector fields at a point of the leaf. The value of the 2-form on this pair of vectors is defined to be the value of the Poisson bracket of the hamiltonian functions at the given point (this value depends only on the two vectors and not on the choice of hamiltonian functions). The fact that the form is closed on the leaf follows from the Jacobi identity; nondegeneracy comes from the fact that, if the derivative of every function by a given tangent vector is zero, then the vector itself must be zero. The phase flow of every hamiltonian vector field preserves the symplectic structures on the leaves.

Thus, the leaves of a Poisson manifold are even dimensional, and the manifold may be considered as a union of sympletic manifolds (generally of different dimensions), whose symplectic structures are coordinated by the condition that the Poisson bracket on the ambient space be smooth.

For example, the co-adjoint orbits of SO(3) (spheres centered at the origin) may be organized according to local Darboux coordinates: in the neighborhood of any nonzero point, the Poisson structure in suitable local coordinates takes the form $\{x, y\} = 1, \{x, z\} = \{y, z\} = 0$. This normal form for the Poisson structure on the space of angular momenta is convenient in carrying out the process of elimination of the nodes in the many-body problem (see Section III.5.5 of the paper: V. I. Arnol'd, Small denominators and problems of stability of motion in classical and celestial mechanics, Russian Math. Surveys 18, No. 6 (1963), 85–191).

Jacobi realized that the (classical) Poisson brackets of the first integrals of any hamiltonian system could be considered as a Poisson structure (this structure is discussed in Section VI.1.3 of the author's paper cited above).

The construction of a Poisson structure on the dual space of a Lie algebra leads to a new Lie algebra. This construction may then be repeated, leading to a whole series of new (infinite-dimensional) Poisson structures. More generally, suppose that one is given any Poisson structure on a manifold. Then the space of functions on that manifold carries the structure of a Lie algebra. This implies that the dual space of this function space carries its own Poisson structure. Elements of this dual space may be interpreted as distribution densities on the original manifold. Thus, the space of distributions on a Poisson manifold (for example, on a symplectic phase space) has a natural Poisson structure. This structure makes it possible to apply the hamiltonian formalism to equations of Vlasov type, which describe the evolution of distributions of particles in phase space under the action of a field which is consistent with the particles themselves.

B *Poisson mappings*

A mapping from one Poisson manifold to another is called a Poisson mapping if it is consistent with the Poisson structures, i.e., if for any two functions on the second manifold, the Poisson bracket of their pullbacks to the first manifold coincides with the pullback of their Poisson brackets. For example, the embedding of each symplectic leaf in a Poisson manifold is a Poisson mapping.

The cartesian product of two Poisson manifolds has a natural Poisson structure, for which the projection on each factor is a Poisson mapping (the Poisson bracket of functions pulled back from different factors is zero).

S. Lie showed that every Poisson manifold is locally (in the neighborhood of a point where the dimension of the symplectic leaves is *locally constant*, for example, in the neighborhood of a generic point, where the rank is locally maximal) decomposable into the product of a symplectic leaf and a complementary space on which all Poisson brackets are zero.

On such a neighborhood, one may introduce coordinates p_i, q_i, c_i such that p and q have the usual symplectic Poisson brackets, while the Poisson bracket of each c_i with any function is equal to zero. In physics, the coordinates p_i and q_i are called *Clebsch variables*,[121] while the c_i's are called *Casimir functions*. Clebsch introduced his variables for the hamiltonian description of the hydrodynamics of ideal fluids, while Casimir considered the center of the Lie algebra of functions on the dual space of a given Lie algebra.

The dimension of the symplectic leaf through a nongeneric point of a Poisson manifold is less than that for nearby generic points. In the neighborhood of such a point, the Poisson manifold may still be represented as the product of a neighborhood of the point in its symplectic leaf and a neighborhood of a distinguished point in some Poisson manifold of complementary dimension. In other words, on a minimal transverse manifold to a symplectic leaf there arises a (unique up to diffeomorphism) local Poisson structure—the so-called *transverse Poisson structure* (cf. A. Weinstein, The local structure of Poisson manifolds, J. Diff. Geom. 18 (1983), 523–557).[122] In the transverse structure, the Poisson brackets of all functions are zero at the distinguished point (which may be taken as the origin of a coordinate system). The Taylor series for these brackets begin with

$$\{x_i, x_j\} = \sum c_{i,j}^k x_k + \cdots,$$

[121] Translator's note: The term *Clebsch variables* is also used to refer to canonical coordinates on a symplectic manifold which projects onto (rather than embeds into) a Poisson manifold.

[122] Warning: As A. B. Giventaľ has noted, Theorem 3.1 in this paper is incorrect. (Translator's note: For further discussion, see A. Weinstein, Lie algebras and Poisson structures, Astérisque, hors série (1985), 257–271.)

where $c_{i,j}^k$ are the structure constants of a finite-dimensional Lie algebra (the *linearized transverse structure*).

A natural question arises: Is it possible to annihilate the higher order terms in the Taylor series by a suitable change of coordinates?

The question of the form of transverse structures was already raised by the author in Section VI.1.3 of the previously cited article.

If the linearized algebra is semisimple and the Poisson structure is analytic, then one can eliminate the higher order terms of the Taylor series by an analytic change of coordinates: J. Conn, Linearization of analytic Poisson structures, Annals of Math. 119 (1984), 577–601. An analogous result is true for the C^∞ case, when the linearized algebra is of compact type: J. Conn, Linearization of C^∞ Poisson structures, Annals of Math. (1985).

A. Weinstein, along with his earlier proof of an analogous result for formal series, expressed the conjecture that semisimplicity was a necessary condition for the annihilation of nonlinear terms. The study of singularities of Poisson structures in the plane (or, more generally, structures with symplectic leaves of codimension 2) leads, however, to a different conclusion.

C Poisson structures in the plane

From the point of view of differential geometry, a Poisson structure is given by a smooth bivector field on a manifold. In fact, the Poisson brackets at each point associate a number to each pair of cotangent vectors. Therefore they define a section of the second exterior power of the tangent bundle, i.e., a bivector field.

The Jacobi identity expresses a sort of "closedness" of this bivector field. On a two-dimensional manifold, this closedness condition is automatically satisfied everywhere, so that every smooth bivector field on the plane gives a Poisson structure. This circumstance allows one to apply to the classification of Poisson structures in the plane the usual considerations of general position (transversality, etc.). In terms of coordinates x, y, a bivector field may be expressed in the form $f(\partial_x \wedge \partial_y)$, where f is a smooth function. The corresponding Poisson structure is defined by the condition

$$(1) \qquad \{x, y\} = f(x, y).$$

A Poisson structure on the plane may also be given by a differential 2-form $dx \wedge dy/f$. This form, like the bivector field, is invariantly connected with the Poisson structure; however, unlike the bivector field, it has pole singularities along the curve $f = 0$. The leaves in this case are the points of the curve $f = 0$ and the connected components of the complement of this curve in the plane. Points of the curve $f = 0$ are called *singular points* of the Poisson structure. In the neighborhood of a nonsingular point, any Poisson structure in the plane may be put into the normal form $\{x, y\} = 1$.

The following diagram shows the beginning of the hierarchy of singularities of Poisson structures on the plane in the neighborhood of a singular point.

459

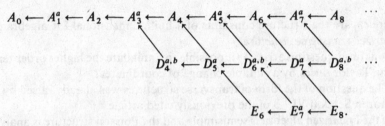

Each letter in the diagram represents a Poisson structure which, in suitable local coordinates with origin at the singular point under consideration, can be written in the form $\{x, y\} = f$, where the function f is given by Table 1.

Table 1

A_0	A_{2k}	A^a_{2k-1}	$D^{a,b}_{2k}$
y	$x^2 + y^{2k+1}$	$\dfrac{x^2 \pm y^{2k}}{1 + ay^{k-1}}$	$\dfrac{x^2 y \pm y^{2k-1}}{1 + ax + by^k}$

D^a_{2k+1}	E_6	E^a_7	E_8
$\dfrac{x^2 y + y^{2k}}{1 + ax}$	$x^3 + y^4$	$\dfrac{x^3 + xy^3}{1 + ay^2}$	$x^3 + y^5$

Theorem. *Given a Poisson structure on a two-dimensional manifold, it is either reducible in a neighborhood of each point to one of the normal forms in Table 1, or it belongs to a set of codimension 8 in the space of Poisson structures.*

Thus, a generic Poisson structure may be reduced in a neighborhood of each point to the normal form $\{x, y\} = 1$ (nonsingular point) or $\{x, y\} = y$ (point of type A_0). In a generic one-parameter family, one encounters for special values of the parameter structures of the type A_1: $\{x, y\} = b(x^2 \pm y^2)$, $b \neq 0$; in two-parameter families one finds A_2, etc.

Remark 1. In the two-dimensional case, the set of all Poisson structures forms a linear space, so that one may speak of a generic structure or family of structures (having in mind a structure [family] belonging to some open dense subset of the space of structures [families]). The problem of classifying generic Poisson structures in three or more dimensions is not uniquely posed, since the set of all such structures does not form a single manifold (one may find components of "different dimensions," as in the classification of Lie algebras).

Remark 2. The structure $\{x, y\} = y$ of type A_0 is the standard Poisson structure on the dual space of the Lie algebra of the group of affine transformations of the line. This structure was considered in 1965, in connection with the study of the Euler equations for left-invariant metrics on groups (in this case—the Lobachevskii metric on a half-plane), at which time it was already realized that the structure is stable and is locally equivalent to any structure of the form $\{x, y\} = y + \cdots$, where the dots designate higher order terms. This (evident) observation contradicts the previously mentioned conjecture of A. Weinstein, according to which the possibility of removing any higher order terms by a formal change of coordinates was characteristic of the linear Poisson structures on the dual spaces of *semisimple* Lie algebras.

Remark 3. The parameters a, b in the table above are *moduli* (invariants depending continuously on the structure). More precisely, structures equivalent to a given one are found only a finite number of times as the parameters are varied.

The rational functions in Table 1 may be replaced by polynomials, but it is not very convenient to do so. The number of moduli in the numerator is one less than the number of irreducible components of the curve $f = 0$. This is not merely a coincidence. One invariant of a Poisson structure on the plane is the residue constructed from the form $dx \wedge dy/f$ (initially, one constructs a residue-form on each component, then its residue at the origin). The sum of the residues corresponding to all the components is zero. Therefore the number of moduli is 1 less than the number of components.

D *Powers of volume forms*

The classification of Poisson structures on the plane may be considered as the classification of differential forms of the type $f(dx \wedge dy)^{-1}$, where f is a smooth (or holomorphic) function. More generally, it is natural to consider forms of the type

$$(2) \qquad f(dx)^{\alpha} = f(x_1, \ldots, x_n)(dx_1 \wedge \cdots \wedge dx_n)^{\alpha},$$

where α is a fixed number, generally complex. The classification of such forms and their deformations in the one-dimensional case, recently carried out by V. P. Kostov, revealed the role of resonance values of α (certain negative rational numbers).

For example, the resonance case $n = 1$, $\alpha = -1$ corresponds to the classification of the singularities and their bifurcations for vector fields on the line, i.e., singular points of differential equations $\dot{x} = v(x)$ and their bifurcations in finite-parameter families. A generic one-parameter family may be reduced by a smooth (holomorphic) change of the parameter and a smooth (holomorphic) change of the variable x, depending smoothly (holomorphically) on the parameter, to the form $\dot{x} = x^2 + \varepsilon + c(\varepsilon)x^3$. (For k parameters, the corresponding form is $\dot{x} = x^{k+1} + \varepsilon_1 x^{k-1} + \cdots + \varepsilon_k + c(\varepsilon)x^{2k+1}$.)

The nonresonance case was studied by S. Lando for all n and α: he showed that almost every versal deformation of the function f defines, after multiplication by $(dx)^{\alpha}$, a versal deformation of the form, as long as α is not a resonance value.

The case $\alpha = -1$, which is interesting in connection with Poisson structures, is generally a resonance case. Instead of powers of volume forms, as in (2), we may consider the differential forms

$$(3) \qquad\qquad f^\beta \, dx, \qquad \beta = 1/\alpha,$$

whose classification is obviously equivalent.

The hypersurface $f = 0$ is invariantly connected with the form (3). The classification therefore begins with the reduction to normal form of the singularity manifold $f = 0$. The beginning of the hierarchy of singular points of hypersurfaces is known. In suitable local coordinates, a hypersurface is given by one of the equations in the following list:

$$
\begin{aligned}
A_\mu: &\quad \pm x_1^{\mu+1} \pm x_2^2 \pm \cdots \pm x_n^2 = 0, \qquad \mu \geq 0; \\
D_\mu: &\quad x_1^2 x_2 \pm x_2^{\mu-1} \pm x_3^2 \pm \cdots \pm x_n^2 = 0, \qquad \mu \geq 4; \\
E_6: &\quad x_1^3 + x_2^4 \pm x_3^2 \pm \cdots \pm x_n^2 = 0; \\
E_7: &\quad x_1^3 + x_1 x_2^3 \pm x_3^2 \pm \cdots \pm x_n^2 = 0; \\
E_8: &\quad x_1^3 + x_2^5 \pm x_3^2 \pm \cdots \pm x_n^2 = 0.
\end{aligned}
$$

After we have brought the hypersurface into normal form, the classification of the forms (2) or (3) comes down to classifying forms of the type

$$(4) \qquad\qquad f^\beta h(x_1, \ldots, x_n) \, dx, \qquad h(0) \neq 0,$$

where $f = 0$ is the given equation of the singularity hypersurface and h is a smooth (holomorphic) function which remains to be put in normal form.

E *The quasi-homogeneous case*

We shall consider here the case in which the singularity hypersurface $f = 0$ is quasi-homogeneous (this condition holds for the cases A, D, E).

Definition. A function f is called *quasi-homogeneous of weight p*, with weights w_i attached to the variables x_i, if it is an eigenfunction with eigenvalue p for the quasi-homogeneous Euler vector field ε (or is zero):

$$\varepsilon f = pf, \quad \text{where } \varepsilon = \sum w_i x_i (\partial/\partial x_i).$$

A quasi-homogeneous polynomial is called *nondegenerate* if the critical point 0 has finite multiplicity (i.e., it is $\mathbb{C}$ isolated). From here on, we will take the weights w_i to be positive numbers.

Theorem. *Let f be a nondegenerate quasi-homogeneous polynomial of weight 1. Then the differential form $f^\beta h \, dx$ (where $dx = dx_1 \wedge \cdots \wedge dx_n$ and h is a holomorphic function on a neighborhood of 0) may be reduced by a biholomorphic coordinate change in a neighborhood of zero to the form $f^\beta (1 + \phi) \, dx$, where ϕ is a quasi-homogeneous polynomial of weight $-\beta - \sigma$, $\sigma = w_1 + \cdots + w_n$.*

The weight of ϕ is chosen so that the weight of the form $f^\beta \phi \, dx$ is zero.

An analogous theorem is true for smooth h (and smooth coordinate changes), except that in the real case one must replace $1 + \phi$ by $\pm 1 + \phi$.

EXAMPLE 1. If β is positive, then $\phi \equiv 0$, so that the complex form reduces to $f^\beta \, dx$.

More generally, $\phi \equiv 0$ if the (possibly complex) number β is not a negative rational number: in this case, a nonzero quasi-homogeneous polynomial of weight $-\beta - \sigma$ does not appear. If the polynomial f (or just its quasi-homogeneity type w) is fixed, then the resonance values of β form a finite set of arithmetic progressions in the negative rationals (for the remaining β, $f^\beta h \, dx$ reduces to the form $f^\beta \, dx$).

EXAMPLE 2. If $\beta = -1$, then the monomials occurring in ϕ may be enumerated by the interior integral points of the Newton diagram of f. The monomial $x^m = x_1^{m_1} \ldots x_n^{m_n}$ corresponds to the point $(m_1 + 1, \ldots, m_n + 1)$ of the diagram (i.e., the exponent of the form $x^m \, dx$).

EXAMPLE 3. Suppose that $\beta = -1, n = 3$, and f is one of the A, D, E polynomials introduced above, defining a simple singularity. Calculating weights, we find that $-\beta - \sigma < 0$; therefore $\phi \equiv 0$, from which we obtain:

Corollary 1. *The form with pole singularity*

$$\frac{h(x, y, z) \, dx \wedge dy \wedge dz}{f(x, y, z)}, \qquad h(0) \neq 0,$$

where f is one of the polynomials A, D, E, may be reduced to the form $dx \wedge dy \wedge dz/f$ by a holomorphic (smooth) change of coordinates.

In exactly the same way for any $n \geq 3$, a factor $h(x_1, \ldots, x_n)$ which does not vanish at the origin can be converted to unity.

Corollary 2. *A simple form (i.e., one not having moduli) of the type $dx_1 \wedge \cdots \wedge dx_n/f(x_1, \ldots, x_n)$, where f is a holomorphic (smooth) function near the origin and $n > 2$, may be reduced by a coordinate change in a neighborhood of the origin to a normal form in which f is either 1 or one of the A, D, E polynomials.*

Corollary 3. *A simple (not having moduli) n-vector field in n-dimensional space $(n > 2)$ is locally equivalent to a normal form $f \cdot (\partial_1 \wedge \cdots \wedge \partial_n)$, where f is either 1 or one of the A, D, E polynomials; $\partial_k = \partial/\partial x_k$.*

Corollary 4. *For $l \leq 6$, in generic l-parameter families of n-vector fields on n-dimensional space $(n > 2)$, the field in a neighborhood of each point and for each value of the parameters is equivalent to one of the simple fields in the preceding corollary.*

Appendix 14: Poisson structures

Corollary 5. *For $l \leq 6$, in generic l-parameter families of forms $dx \wedge dy \wedge dz/ f(x, y, z)$, one finds only forms which in the neighborhood of each point are locally equivalent to one of the following 24 types:*

$$\frac{dx \wedge dy \wedge dz}{1}, \quad \frac{dx \wedge dy \wedge dz}{x}, \quad \frac{dx \wedge dy \wedge dz}{x^2 + y^2 \pm z^2}, \quad \frac{dx \wedge dy \wedge dz}{x^3 + y^2 \pm z^2},$$

$$\frac{dx \wedge dy \wedge dz}{x^4 \pm y^2 \pm z^2}, \quad \frac{dx \wedge dy \wedge dz}{x^5 + y^2 \pm z^2}, \quad \frac{dx \wedge dy \wedge dz}{x^2 y \pm y^3 + z^2}, \quad \frac{dx \wedge dy \wedge dz}{x^6 \pm y^2 \pm z^2},$$

$$\frac{dx \wedge dy \wedge dz}{x^2 y + y^4 \pm z^2}, \quad \frac{dx \wedge dy \wedge dz}{x^7 + y^2 \pm z^2}, \quad \frac{dx \wedge dy \wedge dz}{x^2 y \pm y^5 + z^2}, \quad \frac{dx \wedge dy \wedge dz}{x^3 + y^4 \pm z^2}.$$

For $n = 2$ and $\beta = -1$, the theorem may be applied in the following way.

Corollary 6. *Let f be a nondegenerate quasi-homogeneous polynomial of weight 1 with argument weights w_1, w_2. Then the form*

$$\frac{h(x, y) dx \wedge dy}{f(x, y)}, \quad h(0, 0) \neq 0,$$

where h is a smooth (holomorphic) function in a neighborhood of 0, can be reduced by a suitable smooth (holomorphic) coordinate change to a form in which $h = \pm 1 + \phi$, where ϕ is a quasi-homogeneous polynomial of weight $1 - w_1 - w_2$.

Correspondingly, bivector fields and Poisson structures may be locally reduced to the form

$$\frac{f(x, y)(\partial_x \wedge \partial_y)}{\pm 1 + \phi(x, y)}, \quad \{x, y\} = \frac{f(x, y)}{\pm 1 + \phi(x, y)}.$$

Calculating the weights of the simple singularity types A, D, E for functions of two variables, we obtain Table 1 from the last corollary. For example, for A_1 we have $w_1 = w_2 = \frac{1}{2}$, the weight of ϕ equals 0, and so ϕ is constant.

The dimension of the space of equivalence classes of forms $h \, dx \wedge dy/f$, where $h(0) \neq 0$ and f is a fixed nondegenerate quasi-homogeneous polynomial, equals the dimension of the space of quasi-homogeneous polynomials of weight σ.

F Varchenko's theorem

A. N. Varchenko has proven a series of generalizations of the preceding theorem. Here we shall describe the simplest of these.

1. Let f be a quasi-homogeneous polynomial of weight 1 in the variables $x_1, \ldots, x_n$ with weights $w_1, \ldots, w_n$. Suppose that, for some set I of multi-indices, the residue classes of the monomials x^I generate (as a vector space) the factor algebra of the algebra of formal power series

$$\mathbb{C}[[x_1, \ldots, x_n]]/(\partial f/\partial x_1, \ldots, \partial f/\partial x_n).$$

Theorem. *Every germ $f^\beta h\, dx$ is equivalent to a germ of the form $f^\beta(1 + \sum \lambda_{m,l} x^m f^l)\, dx$, where the l's are nonnegative integers and the m's are elements of I such that the weight of each form $f^\beta x^m f^l\, dx$ is equal to zero.*

2. We define the *degree of non-quasi-homogeneity* of the germ f to be the dimension of the factor space $(f, \partial f/\partial x_1, \ldots, \partial f/\partial x_n)/(\partial f/\partial x_1, \ldots, \partial f/\partial x_n)$.

Theorem. *For almost all β, the number of moduli of the form $f^\beta h\, dx_1 \wedge \cdots \wedge dx_n$ (for fixed β and f and arbitrary h, $h(0) \neq 0$) is equal to the degree of non-quasi-homogeneity of the germ f. The exceptional (resonance) values of β consist of a finite number of arithmetic progressions of negative rational numbers, with difference -1. In particular, for any $\beta \geq 0$, the number of moduli equals the degree of non-quasi-homogeneity.*

3. EXAMPLE. For $\beta = 0$, we obtain:

Corollary. *The number of moduli of the form $h\, dx$ ($h(0) \neq 0$), relative to the group of diffeomorphisms preserving the germ of f, equals the degree of non-quasi-homogeneity of f (equal to zero, if the germ of f is equivalent to a quasi-homogeneous one).*

4. In the resonance cases, the result is more complicated.

EXAMPLE. Let $n = 2$, $\beta = -1$ (Poisson structures in the plane).

Theorem. *The number of moduli for a germ of a Poisson structure with given singular curve $f = 0$ equals the degree of non-quasi-homogeneity of the germ of f augmented by one less than the number of irreducible components of the germ of the curve $f = 0$.*

In resonance cases, the number of moduli behaves in a rather regular way along each arithmetic progression with difference -1. Namely, when β decreases by 1 the number of moduli increases (not necessarily strictly), but its maximal value does not exceed (for any $\beta > -n$) the "nonresonant" value (i.e., the degree of non-quasi-homogeneity of f) by more than the number of Jordan blocks associated with the eigenvalue $e^{2\pi i \beta}$ of the monodromy operator of the function f.

G Poisson structures and period mappings

An interesting source of Poisson structures is provided by the period mappings of critical points of holomorphic functions (A. N. Varchenko and A. B. Givental', Mapping of periods and intersection form, Funct. Anal. Appl. 16, (1982), 83–93).

Period mappings allow one to transfer to the base of a fibre bundle certain structures which live on the (co)homology spaces of the fibres. A Poisson

465

structure on the base arises in this way from the intersection form in the middle-dimensional homology of the fibres, when this form is skew-symmetric.

Period mappings are defined by the following construction. Suppose that one is given a locally trivial fibration. Associated to such a fibration are the bundles (over the same base) of homology and cohomology of the fibres with complex coefficients. These bundles are not only locally trivial, but they are locally trivialized in a canonical way (the integer cycles in a fibre are uniquely identifiable with integer cycles in the nearby homology fibres). A period mapping is defined as a section of the cohomology bundle.

Suppose now that one is given, on the total space of a differentiable fibre bundle, a differential form which is closed on each fibre. The *period mapping of this form* associates to each point of the base the cohomology class of the form on the fibre over this point.

If one is given a vector field on the base of the fibration, then any (smooth) period mapping may be differentiated along this vector field, and *the derivative is again a period mapping*. In fact, neighboring fibres of the cohomology bundle are identified with one another by the above-mentioned "integer" local trivialization, so a section may be considered (locally) as a map into one fibre and may be differentiated as an ordinary (vector-valued) function.

Suppose now that the base is a complex manifold having the same complex dimension as the fibres of the cohomology bundle. A period mapping is called *nondegenerate* if its derivatives along any $\mathbb{C}$-independent vectors at each point are linearly independent. In other words, a period mapping is nondegenerate if the corresponding local maps from the base to typical fibres are diffeomorphisms.

The derivative of a nondegenerate period mapping thus allows us to map the tangent bundle of the base isomorphically onto the cohomology bundle. The dual isomorphism goes from the homology bundle to the cotangent bundle of the base. This isomorphism transfers to the base any additional structures carried by the homology groups.

Suppose that the fibres of our original bundle are (real) oriented even dimensional manifolds, and consider their homology in the middle dimension. In this case, the homology of each fibre carries a bilinear form: the index of intersection. This form is symmetric if the dimension of the fibre is a multiple of 4; otherwise, it is skew-symmetric. The form is nondegenerate if the fibre is closed (i.e., compact and without boundary); otherwise, it may be degenerate. We shall suppose below that we are in the situation where the form is skew-symmetric.

In this situation *a nondegenerate period mapping induces a Poisson structure on the base*. In fact, the isomorphism described above, between the cotangent spaces of the base and the homology groups of the fibres (carrying their skew-symmetric intersection forms), defines a skew-symmetric bilinear form on pairs of cotangent vectors. The Poisson bracket of two functions on the base is defined as the value of this form on the differentials of the functions.

This bracket defines a Poisson structure (of constant rank) on the base.

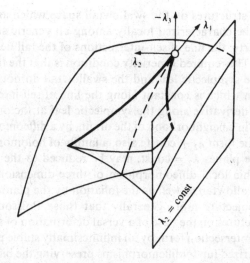

Figure 247   Poisson structure and the swallowtail

This is obvious from the fact that the local identification of the base with the cohomology of the typical fibre, given by the period mapping, provides the base with local coordinates whose Poisson brackets are constant.[123]

Varchenko and Givental' observed that if one constructs, in the way just described, using a generic 1-form, a Poisson structure on the complement of the discriminant locus in the base of a versal deformation of a critical point of a function of two variables, then this structure may be holomorphically extended across the discriminant locus. (One may replace the discriminant locus above by the wave front of a typical singularity.) We shall limit ourselves here to the simplest examples of Poisson structures arising in this way.

Consider the three-dimensional space of polynomials $\mathbb{C}^3 = \{x^4 + \lambda_1 x^2 + \lambda_2 x + \lambda_3\}$ with coordinates λ_k. The polynomials with multiple roots form therein the discriminant surface (a swallowtail; see Figure 247).

The Poisson structures arising from period mappings may be reduced (by diffeomorphisms preserving the swallowtail) to the following form: the symplectic leaves are the planes $\lambda_2 =$ const., and their symplectic structures are of the form $d\lambda_1 \wedge d\lambda_3$.

The fibration of interest here is formed by the complex curves $\{(x, y): y^2 = x^4 + \lambda_1 x^2 + \lambda_2 x + \lambda_3\}$, and the period mapping is given by, for example, the form $y\, dx$. (See V. I. Arnold, A. N. Varchenko, S. M. Gusein-Zade, "Singularities of Differentiable Mappings," Vol. 2: *Monodromy and the Asymptotics of Integrals*, Birkhäuser, 1988, §15, or Uspekhi Mat. Nauk 40, no. 5 (1985).)

---

[123] In the case where the intersection form is symmetric, the analogous construction defines on the base a flat pseudo-riemannian (possibly degenerate) metric.

The Poisson structures on the swallowtail space which arise from period mappings may be characterized locally among all generic structures by the following property: the line of self-intersections of the tail lies entirely in one symplectic leaf. The required genericity condition is that the tangent planes at the origin to the symplectic leaf and the swallowtail do not coincide. Every smooth function which is constant along the line of self-intersections of the tail, and whose derivative along the symplectic leaf at the origin is nonzero, may be reduced in a neighborhood of the origin, by a diffeomorphism preserving the tail, to the form $\lambda_2 + \text{const.}$; also, a family of holomorphic symplectic structures in the planes $\lambda_2 = \text{const.}$ may be reduced to the form $d\lambda_1 \wedge d\lambda_3$ by a holomorphic local diffeomorphism of three-dimensional space which preserves the swallowtail as well as the foliation by the planes.

One may conjecture more generally that those Poisson (in particular, symplectic) structures on the base of a versal deformation of a singularity, induced from the intersection form by an infinitesimally stable period mapping, may be characterized (up to diffemorphisms preserving the bifurcation set) by a natural condition on the rank of the restricted Poisson structure to the strata of the discriminant locus. The "natural condition" in the three-dimensional example above is that the line of self-intersections of the swallowtail be contained in a symplectic leaf. In four-dimensional space, an analogous role would apparently be played by the condition that a certain submanifold be lagrangian, namely, the manifold of polynomials having two critical points with critical value zero in the symplectic space of polynomials $x^5 + \lambda_1 x^3 + \lambda_2 x^2 + \lambda_3 x + \lambda_4$ (the ranks of the symplectic structure on the tangent spaces to the other strata may also be important).

Appendix 15: On elliptic coordinates

A system of Jacobi's elliptic coordinates is associated to each ellipsoid in euclidean space. These coordinates make it possible to integrate the equations of geodesics on the given ellipsoid, as well as certain other equations, such as the equations of motion for a point on a sphere under the influence of a force with quadratic potential, or for a point on a paraboloid under the influence of a uniform gravitational field.

These facts suggest that, even on an infinite-dimensional Hilbert space, there should be a class of integrable systems associated to each symmetric operator. To study these systems, it is necessary to extend the theory of elliptic coordinates to the infinite-dimensional case. To do this, it is first necessary to express the finite-dimensional theory of confocal quadric surfaces in coordinate free form.

In the transition to the infinite-dimensional case, symmetric operators on finite-dimensional euclidean spaces must be replaced by self-adjoint operators on Hilbert spaces. Since the elliptic coordinates are not really connected with the operator itself, but rather with its resolvent, the unboundedness of the original operator (which might be, for example, a differential operator) does not present a serious obstacle.

In some cases, the elliptic coordinates on Hilbert space obtained from a self-adjoint operator form a countable sequence; however, when the operator has a continuous spectrum, the coordinates form a continuous family. In this case, the transformation from the original point of the Hilbert space (thought of as a function space) to the continuous family of elliptic coordinates of the point may be considered as a nonlinear mapping between function spaces. This mapping, by analogy with the Fourier transform, might be called the Jacobi transform: the original function is transformed into a function which expresses the elliptic coordinates in terms of some continuous "index." (More precisely, the result of the transform is a measure on the spectral parameter axis.) The study of the functional analytic properties and the inversion of the Jacobi transform will probably be accomplished before too long.

Following an exposition of the general theory of elliptic coordinates, we shall describe below some of the applications of these coordinates to potential theory.

This appendix is based on the following papers by the author.

Some remarks on elliptic coordinates, Notes of the LOMI Seminar (volume dedicated to L. D. Faddeev on his 50th birthday), 133 (1984), 38–50.

Integrability of hamiltonian systems associated with quadrics (after J. Moser), Uspekhi 34, no. 5, 214.

Some algebro-geometrical aspects of the Newton attraction theory, Progress in Math. (I. R. Shafarevich volume), 36 (1983), 1–4.

Magnetic analogues of the theorem of Newton and Ivory, Uspekhi 38, no. 5 (1983), 145–146.

Further details on background material for the results in this appendix may be found in the following papers.

R. B. Melrose, Equivalence of glancing hypersurfaces, Invent. Math. 37 (1976), 165–191.

J. Moser, Various aspects of integrable Hamiltonian systems, in: J. Guckenheimer and S. E. Newhouse, eds. "Dynamical systems", CIME Lectures, Bressanone, Italy, June 1978, Cambridge, Mass., Birkhäuser, Boston, 1980, pp. 233–289.

V. I. Arnold, Lagrangian manifolds with singularities, asymptotical of rays, and unfoldings of the swallowtail, Funct. Anal. Appl. 15 (1981).

V. I. Arnold, Singularities in variational calculus, J. Soviet Mathematics 27 (1984), 2679–2713.

A. B. Givental', Polynomial electrostatic potentials (Seminar report, in Russian), Uspekhi Mat. Nauk 39, no. 5 (1984), 253–254.

V. I. Arnold, On the Newtonian potential of hyperbolic layers, Selecta Math. Sovietica 4 (1985), 103–106.

A. D. Vainshtein and B. Z. Shapiro, Higher-dimensional analogs of the theorem of Newton and Ivory, Funct. Anal. Appl. 19 (1985), 17–20.

A Elliptic coordinates and confocal quadrics

Elliptic coordinates in euclidean space are defined with the aid of confocal quadrics (surfaces of degree two). The geometry of these quadrics is obtained from the geometry of pencils of quadratic forms in euclidean space (i.e., from the theory of principal axes of ellipsoids or from the theory of small oscillations) by a passage to the dual space.

Definition 1. A *eucildean pencil of quadrics* (resp. *quadratic forms*) in a euclidean vector space V is a one-parameter family of surfaces of degree two

$$\tfrac{1}{2}(A_\lambda x, x) = 1$$

(resp. forms A_λ), where

$$A_\lambda = A - \lambda E \qquad (E = \text{"identity"}),$$

and where A is a symmetric operator

$$A: V \to V^*, \qquad A^* = A.$$

Definition 2. A *confocal family* of quadrics in a euclidean space W is a family of quadrics dual to the quadrics of a euclidean pencil in W^*:

$$\tfrac{1}{2}(A_\lambda^{-1} \xi, \xi) = 1.$$

Thus, quadrics which are confocal to one another form a one-parameter family, but the quadratic forms defining the family do not depend linearly on the parameter.

EXAMPLE. The family of plane curves which are confocal to a given ellipse consists of all those ellipses and hyperbolas with the same foci. In Figure 248,

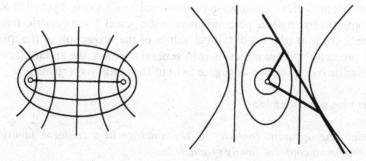

Figure 248 A confocal family and the corresponding euclidean pencil

the curves of a confocal family are shown on the left, and the curves of the corresponding euclidean pencil are shown on the right.

The *elliptic coordinates* of a point are the value of the parameter λ for which the corresponding quadrics of a fixed confocal family pass through the point. We fix an ellipsoid in eucildean space with all its axes of different lengths.

Theorem 1 (Jacobi). *Through each point of an n-dimensional euclidean space there pass n quadrics confocal to a given ellipsoid. Smooth confocal quadrics intersect at right angles.*

PROOF. Each point other than 0 in our space corresponds to an affine hyperplane in the dual space, consisting of those linear functionals whose value is 1 at the given point. In terms of the dual space, Theorem 1 means that every hyperplane not passing through 0 in an n-dimensional euclidean space is tangent to precisely n of the quadrics in a euclidean pencil, and the vectors from 0 to the points of tangency are pairwise orthogonal (Figure 248, right).

The proof of the property of euclidean pencils just stated is based on the fact that the aforementioned vectors define the principal axes of the quadratic forms $B = \frac{1}{2}(Ax, x) - \frac{1}{2}(l, x)^2$, where $(l, x) = 1$ is the equation of the hyperplane.

As a matter of fact, on a principal axis of any quadratic form B, corresponding to the proper value λ, the form $B - \lambda E$ reduces to 0 along with its gradient. The vanishing of this form at the point of intersection of the principal axis and the hyperplane means that the point of intersection lies on the quadric $\frac{1}{2}(Ax, x) = 1$, while the vanishing of the gradient means that the quadric and the hyperplane are tangent at the point. □

Theorem 2 (Chasles). *Given a family of confocal quadrics in n-dimensional euclidean space, a line in general position is tangent to $n - 1$ different quadrics in the family, and the planes tangent to the quadrics at the points of tangency are pairwise orthogonal.*

PROOF. We project the quadrics in the confocal family along a pencil of parallel lines onto the hyperplane perpendicular to the pencil. Each quadric defines an apparent contour (the set of critical values of the projection of the quadric). For a projection whose direction is in general position, the apparent contour is a quadric (i.e., a surface of degree two) in the image hyperplane.

Here we need a lemma.

Lemma. *The apparent contours of the quadrics in a confocal family form themselves a confocal family of quadrics.*

PROOF. On passage to the dual, sections become projections and vice versa. The apparent contours of the projections of confocal quadrics along a pencil of parallel lines are therefore dual to the sections of the dual quadrics by a hyperplane passing through the origin.

The sections of the quadrics in a euclidean pencil by a hyperplane through 0 form a euclidean pencil of quadrics in the hyperplane. The lemma now follows by duality. □

Returning to the proof of Theorem 2, we apply the lemma above to the projections along the line in the statement of the theorem. According to the lemma, the apparent contours of the projections of the confocal quadrics in Theorem 2 form a confocal family of quadrics in a hyperplane. By Theorem 1, $n - 1$ of these apparent contours pass through each point, where they intersect at right angles. This completes the proof of Theorem 2. □

Theorem 3 (Jacobi and Chasles). *Given a geodesic on a quadric Q in n-dimensional space, there is a set of $n - 2$ quadrics confocal to Q such that all the tangent lines to the geodesic are also tangent to the quadrics in the set.*

PROOF (Beginning). We consider the manifold of oriented lines in euclidean space. This manifold has a natural symplectic structure as the manifold of characteristics in the hypersurface $p^2 = 1$ in the phase space of a free particle moving under its own inertia in our euclidean space.

(The characteristics on a hypersurface in a symplectic manifold are the integral curves of the field of characteristic directions, i.e., the field of directions which are skew-orthogonal to the tangent spaces of the hypersurface. In other words, the characteristics of the hypersurface are the phase curves for any hamiltonian flow whose hamiltonian function vanishes to first order on the hypersurface.

The symplectic structure on the manifold of characteristics on a hypersurface in a symplectic manifold is defined in such a way that the skew-scalar product of any two vectors tangent to the hypersurface is equal to the skew-scalar product of their projections in the manifold of characteristics.

Note, finally, that the notion of characteristics is equally well defined for any submanifold of a symplectic manifold on which the induced 2-form has constant nullity. The characteristics then have dimension equal to that nullity, and the manifold of characteristics still inherits a symplectic structure.) ☐

Lemma A. *Each characteristic of the manifold of lines tangent to a given hypersurface in euclidean space consists of all the lines tangent to a single geodesic on the hypersurface.*

PROOF OF LEMMA A. For efficiency of expression, we will identify the cotangent vectors to euclidean space with tangent vectors by using the euclidean structure, so that our original phase space is represented as the space of vectors based at points of euclidean space (i.e., momenta are identified with velocities). The unit vectors to the given hypersurface form a submanifold of odd codimension (equal to 3) in phase space. The characteristics of this submanifold define the geodesic flow on the hypersurface.

The map which assigns to each vector the line in which it lies takes the codimension 3 submanifold just described to the manifold of lines tangent to the hypersurface. Under this mapping, characteristics are transformed to characteristics (with respect to the symplectic structure on the space of lines). This proves the lemma. ☐

[*Remark.* The preceding argument may be easily extended to the following general situation, first considered by Melrose. Let Y and Z be a pair of hypersurfaces in a symplectic manifold X which intersect transversally along a submanifold W. We consider the manifolds of characteristics B and C of the hypersurfaces Y and Z together with the canonical quotient fibrations $Y \twoheadrightarrow B$ and $Z \twoheadrightarrow C$; the manifolds B and C inherit symplectic structures from X.

In the intersection W, there is a distinguished hypersurface (of codimension 3 in X) consisting of points at which the restriction to W of the symplectic structure on X is degenerate. This hypersurface Σ in W may also be defined as the set of critical points of the composed mapping $W \hookrightarrow Y \twoheadrightarrow B$ (or $W \hookrightarrow Z \twoheadrightarrow C$ if one wishes). These objects form the following commutative diagram:

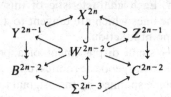

The analogue to Lemma A in this situation is the assertion that the characteristics on the images of the mappings $\Sigma \to B$ and $\Sigma \to C$ are the images of one and the same curve on Σ (namely, the characteristics of Σ considered as a submanifold of the symplectic manifold X).

Lemma A itself is the special case of the assertion above in which $X = \mathbb{R}^{2n}$

(the phase space of a free particle in $\mathbb{R}^n$), the hypersurface Y consists of the unit vectors (given by the condition $p^2 = 1$, i.e., a level surface of the hamiltonian for a free particle), and the hypersurface Z consists of those vectors which are based at the points of the given hypersurface in $\mathbb{R}^n$. In this case, B is the manifold of all oriented lines in euclidean space, and Σ is the manifold of unit vectors tangent to the hypersurface. The mapping $\Sigma \rightarrow B$ assigns to each unit vector the line which contains it. The manifold C is the (co)tangent bundle of the given hypersurface. $\Sigma \rightarrow C$ is the embedding into this bundle of its unit sphere bundle (in other words, the embedding of a level surface of the kinetic energy, i.e., the hamiltonian for motion constrained to the hypersurface).

It is always useful to keep the diagram above in mind when one is dealing with constraints in symplectic geometry.]

PROOF OF THEOREM 3 (Middle). We suppose given a smooth function on euclidean (configuration) space whose restriction to a certain line has a non-degenerate critical point. In this situation, the function will also have a critical point when restricted to each nearby line; i.e., on each nearby line, there will be a nearby point where the line is tangent to a level surface of the function. The value of the function at the critical point is thus a function (defined locally) on the space of lines. We call this function of lines the *induced line function* (from the original *point function*). □

Lemma B. *If two point functions in euclidean space are such that the tangent planes to their level surfaces are orthogonal at the points where a given line is tangent to these surfaces (these points being in general different for the two functions), then the Poisson bracket of the induced line functions is zero at the given line (considered as a point in the space of lines).*

PROOF OF LEMMA B. We calculate the derivative of the second induced line function along the phase flow whose hamiltonian is the first induced function. The phase curves for the first induced function, which lie on its level surfaces, are the characteristics of those surfaces. A level surface for the first induced function consists of those lines which are tangent to a single level surface of the first point function. Each characteristic of this surface, according to Lemma A, consists of the lines which are tangent to a single geodesic on the level surface of the first point function.

For an infinitesimally small displacement of a point on a geodesic in a surface, the tangent line to the geodesic rotates (up to infinitesimal quantities of higher order) in the plane spanned by the original tangent and the normal to the surface. By hypothesis, the tangent plane to the level surface of the second function at the point where this surface is tangent to our line is perpendicular to the tangent plane of the level surface of the first function. Therefore, under the above-mentioned infinitesimally small rotation, the line remains tangent to the same level surface of the second function (up to infinitesimals of higher order). It follows that the rate of change of the second induced function under the action of the phase flow given by the first is zero at the element in question of the space of lines, which proves Lemma B. □

PROOF OF THEOREM 3 (End). We fix a line in general position in $\mathbb{R}^n$. According to Theorem 2, this line is tangent to $n - 1$ quadrics in the confocal family, at $n - 1$ points. We construct in the neighborhood of each of these points a smooth function, without critical points, whose level surfaces are the quadrics of our confocal family.

We fix one of these quadrics (the "first") and consider the hamiltonian system on the space of lines whose hamiltonian function is the first induced line function. Each of its phase curves on a fixed level surface of the hamiltonian function consists of the tangent lines to one geodesic of that quadric (Lemma A). The remaining induced functions have zero Poisson bracket with the hamiltonian, by Lemma B (since the planes tangent to the confocal surfaces at the points where they touch one line are orthogonal, by Theorem 2).

Thus all the induced functions are first integrals for the hamiltonian system generated by any one of them. Since the lines tangent to a geodesic on the first quadric form a phase curve of the first system, all the induced functions take constant values on this curve. That proves Theorem 3, as well as the following result. □

Theorem 4. *The geodesic flow on a central surface of degree 2 in euclidean space is a completely integrable system in the sense of Liouville (i.e., it has as many independent integrals in involution as it has degrees of freedom).*

Remark. Strictly speaking, we proved Theorem 3 only for lines in general position, but the result extends by continuity to the exceptional cases (in particular, to asymptotic lines of our quadrics). In the same way, Theorem 4 was initially proved just for quadrics with unequal principal axes, but passage to a limit extends the result to more symmetric quadrics of revolution (as well as to noncentral "paraboloids").

B Magnetic analogues of the theorems of Newton and Ivory

Elliptic coordinates make it possible to extend Newton's well-known theorem on the gravitational attraction of a sphere to the case of attraction by an ellipsoid.

Definition. A *homeoidal density* on the surface of an ellipsoid E is the density of a layer between E and an infinitely nearby ellipsoid which is homothetic to E (with the same center).

The following is a well-known result.

Ivory's Theorem. *A finite mass, distributed on the surface of an ellipsoid with homeoidal density, does not attract any internal point; it attracts every* *external point the same way as if the mass were distributed with homeoidal density on the surface of a smaller confocal ellipsoid.*

The attraction in Ivory's theorem is defined by the law of Newton or Coulomb: in n-dimensional space, the force is proportional to r^{1-n} (as prescribed by the fundamental solution of Laplace's equation).

Newton's theorem on the (non)attraction of an internal point carries over to the case of a hyperbolic homeoidal layer and to the case of an attracting mass distributed on a level hypersurface of a hyperbolic polynomial of any degree. (A polynomial of degree m, $f(x_1,\ldots,x_n)$ is called *hyperbolic* if its restriction to any line through the origin has all its roots real.)

A homeoidal charge density on the zero hypersurface $f = 0$ of a hyperbolic polynomial is defined as the density of a homogeneous infinitesimally thin layer between the hypersurfaces $f = 0$ and $f = \varepsilon \to 0$ (the signs of the charges being chosen so that successive ovaloids have opposite charges).

[*A homeoidal charge does not attract the origin* (*nor any other point within the innermost ovaloid*), *and this property is preserved if the charge density is multiplied by any polynomial of degree at most $m - 2$.*

Generalization: *If a homeoidal charge density is multiplied by any polynomial of degree $m - 2 + r$, then the potential inside the innermost ovaloid is a harmonic polynomial of degree r* (A. B. Givental', 1983).]

When one attempts to find a version for hyperboloids of Ivory's theorem on the attraction of confocal ellipsoids, it turns out that an essential role is played by the topology of the hyperboloids. When passing to hyperboloids of different signatures, one must consider, instead of homeoidal densities, harmonic forms of different degrees, and instead of the Newton or Coulomb potential, the corresponding generalized forms-potentials given by the Biot–Savart law.

In the simplest nontrivial case of a hyperboloid of one sheet in three-dimensional euclidean space, the result is as follows.

The hyperboloid divides space into two parts: "internal" and "external," the latter being nonsimply connected. We consider elliptic coordinate curves from the system whose level surfaces are the quadrics confocal to the given hyperboloid.

The elliptic coordinate curves on our hyperboloid, which are obtained by intersecting with the confocal ellipsoids (closed lines of curvature on the hyperboloid), are called the *parallels* of the hyperboloid. The orthogonal curves, obtained by intersection with the two-sheeted hyperboloids, are called the *meridians*.

Although the elliptic coordinate system has singularities (on each symmetry plane of the quadrics in the family), the hyperboloid is smoothly fibred by the parallels (diffeomorphic to the circle) and meridians (diffeomorphic to the line).

The region inside the hyperboloidal tube is also smoothly fibred by meridians (orthogonal to the ellipsoids in the confocal family), while the annular

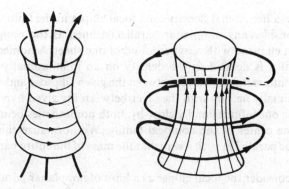

Figure 249   Magnetic fields generalizing the theorems of Newton and Ivory

region outside the hyperboloid is smoothly fibred by parallels (orthogonal to the hyperboloids of two sheets).

Theorem. *A current with a suitable density, flowing along the meridians of a hyperboloid, produces a magnetic field which is zero inside the hyperboloidal tube, while the field in the annular exterior region is directed along the parallels. A current with a suitable density, flowing along the parallels of a hyperboloid, produces a magnetic field which is zero in the exterior annular region, while the field inside the hyperboloidal tube is directed along the meridians. (See Figure 249.)*

The current densities giving rise to such magnetic fields, which generalize the homeoidal charge densities on ellipsoids, may be described in the following way. There are associated to each family of confocal quadrics in three-dimensional euclidean space two "focal curves": an ellipse and a hyperbola. (See Figure 250.) The focal ellipse is the boundary of the limiting ellipsoid of the family in which the shortest axis shrinks to zero; the focal hyperbola arises in a similar way from the hyperboloids of one or two sheets.

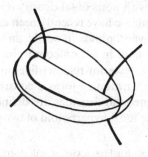

Figure 250   Focal ellipse and focal hyperbola

We define a homeoidal density on a focal ellipse in the following way. To begin we consider any nonplanar parallel, defined as the nonplanar intersection of an ellipsoid with a hyperboloid of one sheet. A homeoidal density on this parallel is defined as the density on an infinitesimally thin "wire," obtained by intersecting the layer between the given ellipsoid and a homothetic one infinitesimally nearby with the layer between the given hyperboloid and a homothetic one infinitesimally close by, both homotheties being taken with respect to the center of the confocal family. We normalize this homeoidal density on the parallel in such a way that the mass of the entire parallel is equal to 1.

Now we consider the focal ellipse as a limit of nonplanar parallels. It turns out that the normalized homeoidal densities on the parallels have a well-defined limit as the parallels approach the focal ellipse. This limiting density is called the homeoidal density on the focal ellipse.

The homeoidal density on a focal hyperbola is defined in an analogous way.

We may now describe the current densities referred to as "suitable" in the theorem above on magnetic fields. The surface of a hyperboloid of one sheet is fibred over the focal ellipse (the fibre over a point is the meridian which lies on the same hyperboloid of two sheets as that point).

The flux of the meridianal current suitable for the theorem, through any curve on the hyperboloid, equals the integral of the homeoidal density form on the focal ellipse over the projection of that curve onto the focal ellipse (along the hyperboloids of two sheets).

The density of the flow along the parallels is induced in an analogous way from the homeoidal density on the focal hyperbola.

Remark. The magnetic field of the parallel flow with the indicated density, inside the hyperboloidal tube, coincides outside each confolal ellipsoid (up to sign) with the newtonian or coulombian field produced by a charge which is distributed with homeoidal density on that ellipsoid.[124]

In exactly the same way, the magnetic field in the annular domain outside the hyperboloid of one sheet coincides (up to sign), in the region between the sheets of each confocal hyperboloid of two sheets, with the coulombian field produced by two equal charges with opposite signs distributed on the two sheets of the hyperboloid with homeoidal density (O. P. Shcherbak).

The results formulated above have recently been extended by B. Z. Shapiro and A. D. Vainshtein to hyperboloids in euclidean spaces of any number of dimensions. For a hyperboloid in $\mathbb{R}^n$, diffeomorphic to $S^k \times \mathbb{R}^l$, a harmonic k-form is constructed on the exterior region (diffeomorphic to the product of S^k with a half-space) and a harmonic l-form is constructed on the interior.

The corresponding homeoidal densities are defined on the focal ellipsoid with codimension k and the focal hyperboloid of two sheets with codimension

---

[124] This is actually the density with which a charge will distribute itself on the surface of a conducting ellipsoid.

l by the same limiting procedure that we described above for $k = l = 1$, using the intersections of layers between infinitesimally close and homothetic quadrics.

Noncomputational proofs of these geometric theorems are unknown, even for the special case of magnetic fields in three-dimensional space.

Remark. The presence of distinguished harmonic forms on hyperboloids and in their complementary domains suggests that one might try to find filtrations, analogous to those arising in the theory of mixed Hodge structures, in spaces of differential forms on noncompact (and possibly even singular) algebraic and semialgebraic real manifolds.

Appendix 16: Singularities of ray systems

The simplest example of a ray system is the system of normals to a surface in euclidean space.

In a neighborhood of a smooth surface, its normals form a smooth fibration, but at some distance from the surface various normals begin to intersect one another (Figure 251). The complicated figures which are thereby formed were already investigated by Archimedes, but their full details were not revealed until the discovery in 1972 of the relation between singularities of ray systems and the theory of groups generated by reflections.

This relation, for which there is no evident *a priori* reason (and which is as surprising as, say, the relation between the problems of tangents and areas), has turned out to be a powerful instrument for the study of critical points of functions. By 1978, it had become clear that the theory of reflection groups also governs the singularities of the Huygens evolvents.

Huygens (1654) discovered that the evolvent of a plane curve has a cusp singularity at each point where it meets the curve (Figure 252). Evolents of plane curves and their higher-dimensional generalizations are wave fronts on manifolds with boundary. Singularities of wave fronts, like those of ray systems, are classified in terms of reflection groups.

While rays and fronts on manifolds without boundary are related to the Weyl groups in the A, D, and E series, singularities of evolvents are described by the groups of types B, C, and F (the ones with double connections in their Dynkin diagrams).

The remaining reflection groups ($I_2(p)$, H_3, H_4) continued for some time to have no visible relation to the theory of singularities. This situation changed in the fall of 1982 when it was discovered that the symmetry group H_3 of the icosahedron governs the singularities of evolvent systems in the neighborhood of inflection points of plane curves.

The appearance of the icosahedron at an inflection point of a curve looks as mystical as the icosahedron in Kepler's law of planetary distances. But the presence of the icosahedron here is not an accident: upon the investigation in 1984 of more complicated systems of rays and fronts, the remaining group H_4 appeared.

We shall give in this appendix a brief description of the theory of singularities of ray systems. Further details may be found in the following references:

V. I. Arnold, Singularities of ray systems, Russian Math. Surveys 38 (1983).

V. I. Arnold, Singularities in variational calculus, J. Soviet Math. 27 (1984), 2679–2713.

O. V. Lyashko, Classification of critical points of functions on a manifold with singular boundary, Funct. Anal. Appl. 17 (1983), 187–193.

O. P. Shcherbak, Singularities of families of evolvents in the neighborhood of an inflection point of the curve, and the group H_3, generated by relections, Funct. Anal. Appl. 17 (1983), 301–303.

A. N. Varchenko and S. V. Chmutov, Finite irreducible groups, generated by relections, are monodromy groups of suitable singularities, Funct. Anal. Appl. 18 (1984), 171–183.

Figure 251   A caustic as the envelope of rays

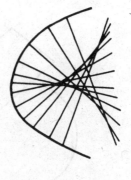

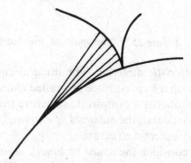

Figure 252   An evolvent of a curve

V. I. Arnold, Singularities of solutions of variational problems (Seminar report, in Russian), Uspekhi Mat. Nauk 39, no. 5 (1984), 256.

O. P. Shcherbak, Wave fronts and reflection groups. Russian Math. Surveys, 43, no. 3 (1988).

Itogi Nauki i Tekhniki, Sovremennye Problemy matematiki, Noveishie dostijenia, Moscow, VINITI, vol. 33 (1988). English translation: J. Sov. Math. 27 (1984).

Many of the results which we will describe concern such simple geometric objects that it is surprising that they were not already known in classical times. For instance, the local classification of projections of generic surfaces in three-dimensional space was not discovered until 1981. The number of equivalence classes of germs of projections turned out to be finite—namely 14: neighborhoods of points on generic surfaces can have that many different appearances when viewed from different points in space.

A *Symplectic manifolds and ray systems*

1. The *space of oriented lines* in euclidean space may be identified with the (co)tangent bundle of the sphere (Figure 253), and it thereby obtains a symplectic structure.

2. More generally, we consider any hypersurface in a symplectic manifold. The skew-orthogonal complement to its tangent space at each point is called

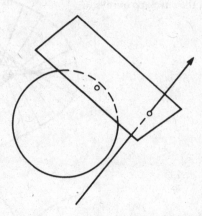

Figure 253   The space of oriented lines in euclidean space

the *characteristic direction*. The integral curves of the field of characteristic directions on a hypersurface are called *characteristics*. The manifold of characteristics inherits a symplectic structure from the original manifold.

3. In particular, the *manifold of extremals* of a general variational problem carries a symplectic structure.

4. We consider the *space of binary forms* (homogeneous polynomials in two variables) of a particular odd degree. The group of linear transformations of the plane acts on this even dimensional linear space. Up to multiplication by a constant, there is a unique nondegenerate skew-symmetric form on this space which is invariant under the action of the group SL(2) of linear transformations with determinant equal to 1. This form gives a natural symplectic structure on the manifold of binary forms of each odd degree.

5. The binary forms in x and y for which the coefficient of x^{2k+1} is unity form a hypersurface in the space of all forms. The manifold of characteristics of this hypersurface is naturally identified with the *manifold of monic polynomials of even degree* $x^{2k} + \cdots$ in x. We have thereby defined a natural symplectic structure on this space of polynomials.

6. The one-parameter group of translations along the x-axis preserves the symplectic structure just introduced. The hamiltonian function for this group is a quadratic polynomial (found already by Hilbert (1893)). The manifold of characteristics for any level surface of this hamiltonian function may be identified with the *manifold of monic polynomials of degree* $2k - 1$ *in x for which the sum of the roots is zero*. Thus we have a natural symplectic structure on this space of polynomials.

B *Submanifolds of symplectic manifolds*

The restriction of a symplectic structure to a submanifold is a closed 2-form, but it is not necessarily nondegenerate. For submanifolds in euclidean space there is, in addition to the intrinsic geometry, an extensive theory of extrinsic curvatures. In symplectic geometry, the situation is simpler:

Theorem (A. B. Givental', 1981). *The restriction of the symplectic form to a germ of a submanifold in a symplectic manifold determines the germ up to a symplectic diffeomorphism of the ambient manifold.*

An intermediate theorem, in which one uses the values of the symplectic form at all vectors based on the submanifold, not just those tangent to it, was proved earlier by A. Weinstein (1971). Unlike Weinstein's theorem, Givental's theorem makes it possible to classify generic submanifold germs in symplectic manifolds: it is sufficient to use the classification of degenerate symplectic structures obtained by J. Martinet (1970) and his successors.

EXAMPLES. 1. A generic two-dimensional surface in symplectic space is symplectically diffeomorphic in a neighborhood of each point with the surface $p_2 = p_1^2, p_3 = q_3 = \cdots = 0$ (in Darboux coordinates). 2. On four-dimensional submanifolds, one finds stable curves of elliptic and hyperbolic Martinet singular points with normal forms

$$p_2 = p_1 p_3 \pm q_1 q_2 + q_3^3/6, \qquad p_3 = 0, \qquad p_4 = q_4 = \cdots = 0.$$

[The ellipticity or hyperbolicity of a singular point is determined by the nature of the dynamical system invariantly attached to the submanifold. The divergence-free vector fields in three-dimensional space which arise have entire curves of singular points. The classification of singular lines turns out to be less pathological than the classification of singular points (which is almost as difficult as all of celestial mechanics).]

This concludes a description of the first steps in the theory of symplectic singularities on smooth manifolds.

C *Lagrangian submanifolds in the theory of ray systems*

We recall that a *lagrangian submanifold* is a submanifold of symplectic space on which the symplectic structure pulls back to zero and which has the highest possible dimension consistent with this property (equal to half the dimension of the ambient manifold).

EXAMPLES. 1. Each *fibre of a cotangent bundle* is lagrangian. 2. The *manifold of all oriented normals* to a smooth submanifold (of any dimension) in euclidean space is a lagrangian submanifold of the space of lines. 3. The manifold of all *polynomials $x^{2m} + \cdots$ divisible by x^m* is lagrangian.

A *lagrangian fibration* is a fibration all of whose fibres are lagrangian.

EXAMPLES. 1. The *cotangent fibration* is lagrangian. 2. The *Gauss fibration* from the space of lines in euclidean space to the unit sphere of directions is lagrangian.

All lagrangian fibrations of a fixed dimension are locally (on a neighborhood of a point in the total space) symplectically diffeomorphic.

A *lagrangian mapping* is the projection of a lagrangian submanifold to the base of a lagrangian fibration, i.e., a triple $V \to E \to B$, where the first arrow is an immersion onto a lagrangian manifold and the second arrow is a lagrangian fibration.

EXAMPLES. 1. *A gradient mapping* $q \mapsto \partial S/\partial q$ is lagrangian. 2. The *normal mapping* which maps each normal vector of a submanifold in euclidean space to its tip is lagrangian. 3. The *Gauss mapping* which takes each point of a transversely oriented hypersurface in euclidean space to the unit vector at the origin in the direction of the normal is lagrangian. (The corresponding lagrangian manifold consists of the normals themselves.)

An *equivalence* of lagrangian mappings is a fibre-preserving symplectic diffeomorphism of the total spaces of the fibrations which takes the first lagrangian manifold to the second.

The set of critical values of a lagrangian mapping is called a *caustic*. The caustics of equivalent mappings are diffeomorphic.

EXAMPLE. The caustic of the normal mapping of a surface is the envelope of the family of normals, i.e., the *focal surface* (surface of centers of curvature).

Every lagrangian mapping is locally equivalent to a gradient (or normal, or Gauss) mapping. The singularities of generic gradient (or normal, or Gauss) mappings are the same as those for arbitrary generic lagrangian mappings. The simplest of these are classified by the reflection groups A_k, D_k, E_6, E_7, E_8 (see Appendix 12).

EXAMPLE. We consider a medium of dust particles moving inertially, with their initial velocities forming a potential field. After time t, the particle at x moves to $x + t(\partial S/\partial x)$. We thereby obtain a one-parameter family of smooth mappings $\mathbb{R}^3 \to \mathbb{R}^3$.

These mappings are lagrangian. In fact, a potential field of velocities gives a lagrangian section of the cotangent bundle. The phase flow of Newton's equations preserves the lagrangian property. For large t, though, our lagrangian manifold is no longer a section: its projection on the base develops singularities. The caustics of the corresponding lagrangian mappings are places where the density of particles has become infinite.[125] According to Ya. B. Zel'dovich (1970) an analogous model (taking into account gravity and the expansion of

---

[125] The relation between caustics and dust-like media was first discovered by Lifshitz, Sudakov, and Khalatnikov: see the survey by E. M. Lifshitz and I. M. Khalatnikov, Investigations in relativistic cosmology, Adv. Phys. 12 (1963), 185.

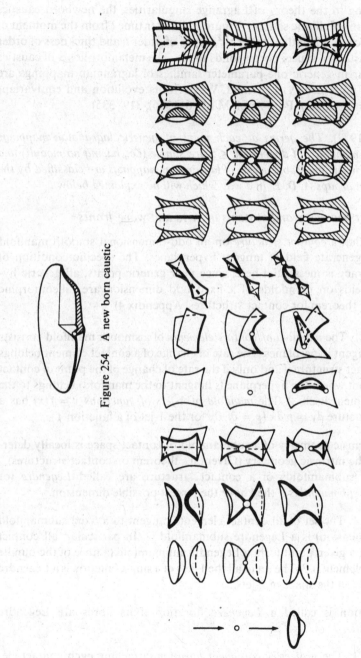

Figure 254 A new born caustic

Figure 255 Perestroikas of caustics in 3-space

the universe) describes the formation of large scale nonhomogeneities in the distribution of matter in the universe.

According to the theory of Lagrange singularities, the newborn caustics have the form of elliptic saucers (Figure 254) (after time t from the moment of birth, a saucer has length of order $t^{1/2}$, depth of order t, and thickness of order $t^{3/2}$). The birth of a saucer corresponds to A_3. The metamorphoses of caustics which occur in generic one-parameter families of lagrangian mappings are shown in Figure 255 (V. I. Arnold, Wave fronts evolution and equivariant Morse lemma, Comm. Pure Appl. Math. 6 (1976), 319–335).

Theorem (1972). *The germs at each point of generic lagrangian mappings between manifolds of dimension ≤ 5 are simple (i.e., having no moduli) and stable. The simple stable germs of lagrangian mappings are classified by the reflection groups A, D, E, in a way which will be explained below.*

D *Contact geometry and systems of rays and wave fronts*

We recall that a *contact structure* on an odd-dimensional smooth manifold is a nondegenerate field of tangent hyperplanes. The specific condition of nondegeneracy is inessential here, since near generic points, all generic hyperplane fields on manifolds of a fixed odd dimension are diffeomorphic (Darboux's theorem for contact structures, Appendix 4).

EXAMPLES. 1. The *manifold of contact elements* of a smooth manifold consists of all its tangent hyperplanes. The rate of change of a contact element belongs to the contact structure if and only if the rate of change of the point of contact (i.e., the point where the hyperplane is tangent to the manifold) belongs to the contact element itself. 2. The *manifold of 1-jets of functions* $y = f(x)$ has a contact structure $dy = p\,dx$ ($p = \partial f/\partial x$ for the 1-jet of a function f).

The extrinsic geometry of a submanifold of contact space is locally determined by the intrinsic geometry (Givental's theorem on contact structures).

Integral submanifolds of a contact structure are called *Legendre* (or *legendrian*) *submanifolds* if they have the largest possible dimension.

EXAMPLES. 1. The set of all contact elements tangent to a fixed submanifold (of any dimension) is a Legendre submanifold. 2. In particular, all contact elements at a given point form a Legendre submanifold (a fibre of the bundle of contact elements). 3. The set of all the 1-jets of a single function is a Legendre submanifold in the space of 1-jets.

A fibration is called a *Legendre* fibration if its fibres are Legendre submanifolds.

EXAMPLES. 1. The *projective cotangent fibration* (attaching each contact element to its point of contact) is Legendre. 2. The *fibration of 1-jets of functions over the 0-jets* (forgetting the derivative) is Legendre.

All Legendre fibrations of a fixed dimension are locally contact diffeomorphic (in a neighborhood of a point in the total space of the fibration).

The projection of a Legendre submanifold on the base of a Legendre fibration is called a *Legendre mapping*. The image of a Legendre mapping is called its *front*.

EXAMPLES. 1. The *Legendre transformation*: A hypersurface in projective space may be lifted to the space of contact elements of projective space as a Legendre submanifold. The manifold of contact elements of projective space is also fibred over the dual projective space. (The fibration assigns to each contact element the plane containing it.) This is a Legendre fibration. The projection of the lifted Legendre submanifold maps it onto the hypersurface which is projectively dual to the original one. Thus, *the projective dual of a smooth hypersurface is the front of a Legendre mapping*. 2. *Frontal mappings*: Laying out a segment of length t on each normal to a hypersurface in euclidean space, we obtain a Legendre mapping whose front is *equidistant* from the given hypersurface.

Every Legendre mapping is locally equivalent to a Legendre transformation, as well as to a frontal mapping. The theory of Legendre singularities thus coincides exactly with the theory of singularities of Legendre transformations and of frontal mappings. Equivalence, stability, and simplicity of Legendre mappings are defined just as the lagrangian case.

Theorem (1973). *The germs, at all points, of generic Legendre mappings between manifolds of dimension ≤ 5 are simple and stable. The simple and stable germs of Legendre mappings are classified by the groups A, D, E: their fronts are locally diffeomorphic (in the complex domain) to the manifolds of nonregular orbits of the corresponding reflection groups.*

EXAMPLE. The only singularities of a typical wave front in three-dimensional space are (semicubic) cuspidal curves (A_2) and "swallowtails" (A_3, Figure 256; near such a point, the front is diffeomorphic to the surface formed by the polynomials with multiple roots in the space of polynomials $x^4 + ax^2 + bx + c$).

Figure 256   Singularities of wave fronts

Of course, there may also be transverse intersections of branches of fronts of the types just described.

Remark. The real forms of simple singularities of fronts may also be described in terms of reflection groups. E. Looijenga has shown that the real components in the complement of a simple germ of a front may be identified with the conjugacy classes of involutions (elements of order 2) in the normalizer of the reflection group, conjugacy being taken with respect to the reflection group itself. (See E. Looijenga, The discriminant of a real simple singularity, Compositio Math. 37 (1978), 51–62.)

E *Applications of contact geometry to symplectic geometry*

All lagrangian singularities may be obtained from Legendre singularities, if one realizes the latter by projections of Legendre submanifolds of the space of 1-jets of functions onto the space of 0-jets. If one forgets the value of each function, the space of 1-jets is projected onto phase space (i.e., the cotangent bundle); a Legendre submanifold in the first space projects to a lagrangian submanifold in the second. In particular, *the caustic of a lagrangian mapping is the image of the cuspidal edge of the front of a Legendre mapping* under a projection with one-dimensional fibres.

Theorem (O. V. Lyashko, 1979). *All holomorphic vector fields transverse to the front of a simple singularity are locally equivalent under holomorphic diffeomorphisms preserving the front.*

EXAMPLE. A generic vector field in the neighborhood of the most singular point of a swallowtail $\{x^4 + ax^2 + bx + c = (x + d)^2 \dots\}$ is equivalent, by a holomorphic diffeomorphism preserving the swallowtail, to the normal form $\partial/\partial c$ (Figure 257).

The reduction of various objects to normal form, by a diffeomorphism preserving a wave front or caustic, is a basic technique for studying the geometry of systems of rays and fronts. For instance, the study of the meta-

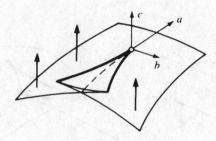

Figure 257   The normal form of a vector field at the swallowtail

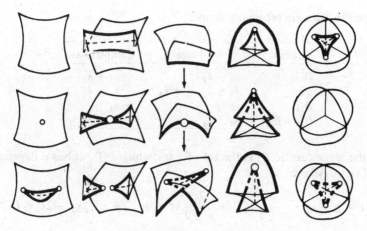

Figure 258   Perestroikas of wave fronts

morphoses of moving wave fronts is based on the following result, which is "dual" to the previous one.

Theorem (1976). *All generic holomorphic functions equal to zero at the most singular point of a simple singularity of a front are locally equivalent under holomorphic diffeomorphisms which preserve the front.*

EXAMPLE. In a neighborhood of the most singular point of a swallowtail, a generic function may be reduced, by a diffeomorphism preserving the swallowtail, to the normal form a.

This theorem is a special case of the equivariant Morse lemma. It is applied in the following way. The instantaneous wave fronts together form a "large front" in space–time. "Time" is a function on space–time. We reduce this function to normal form by a diffeomorphism which preserves the front, and we thereby obtain a normal form for the metamorphoses of the instantaneous fronts. The metamorphoses of fronts in $\mathbb{R}^3$ are shown in Figure 258. The problem of describing the metamorphoses of caustics in generic one-parameter families (Figure 255) is solved in exactly the same way. In this case, the time function is reduced to normal form by a transformation of space–time which preserves the "large caustic." If the dimension of space–time is no larger than 4, then all the singularities of the large caustic are of types A and D.

The caustics of lagrangian singularities in the A series differ from the wave fronts in the A series only by a shift of 1 unit in the index. The same is therefore true for their metamorphoses.

The caustics in the D series are not the same as the fronts. The normal forms for a generic time function in the neighborhood of a caustic singularity of type D were found by V. M. Zakalyukin (1975). The topological normal forms for the time function are especially simple:

Caustic	Real case	Complex case
D_4^-	$\lambda_1 + \lambda_2$	$\lambda_1 + \lambda_2$
D_4^+	$\lambda_1 \pm \lambda_2, \lambda_1 + \lambda_4$	$\lambda_1 + \lambda_2$
D_{2k+1}	$\pm \lambda_1$	λ_1
$D_{2k}, k \geq 3$	$\lambda_1 \pm \lambda_2$	$\lambda_1 + \lambda_2$

Here, the large caustic D_μ is the set of λ for which $\mathcal{F}(\cdot, \lambda)$ has a degenerate critical point, where

$$\mathcal{F}(x, \lambda) = \pm x_2^2 x_1 + \frac{1}{\mu - 1} x_1^{\mu-1} + \frac{\lambda_1}{\mu - 2} x_1^{\mu-2} + \cdots + \lambda_{\mu-2} x_1 + 2\lambda_\mu x_2 \ (\mu \geq 4).$$

The reduction to normal form of the germ of the time function is accomplished by a local homeomorphism of the space $\mathbb{R}^{\mu-1}$ ($\mathbb{C}^{\mu-1}$), which preserves the large caustic and which is smooth everywhere except at 0 (V. I. Bakhtin, 1984).

J. Nye (1984) has noticed that not all metamorphoses of caustics and fronts may be realized by the motion of a front under an equation of eikonal (or Hamilton–Jacobi) type. For example, the caustic of a ray system cannot have the form of "lips" with two cusps (although this is possible for lagrangian caustics). The point is that the inclusion of a lagrangian or Legendre manifold in the hypersurface given by a Hamilton–Jacobi or eikonal equation imposes topological restrictions on the coexistence, and thus on the metamorphoses, of singularities, even though the individual singularities may be realized on hypersurfaces. This is namely the case when the level surface of the hamiltonian is locally nondegenerately convex in the momentum variables.

The vector fields generating the diffeomorphisms preserving a front are those which are tangent to it. The study of these vector fields leads to an unusual "convolution" operation on the invariants of a reflection group. To a pair of invariants (functions on the orbit space) we associate a new invariant—the scalar product of the gradients of the functions (pulled back from the orbit space to the original euclidean space).

The linearization of this operation defines a symmetric bilinear mapping from each cotangent space of the orbit space into itself.

Theorem (1979). *The linearized convolution of invariants of a reflection group is isomorphic as a bilinear operation to the operation on the local algebra of the corresponding singularity given by the formula $(p, q) \mapsto S(p \cdot q)$, where $S = D + (2/h)E$, D is Euler's quasi-homogeneous derivation, and h is the Coxeter number.*

In 1981, A. N. Varchenko and A. B. Givental' (who also proved the theorem above for the exceptional groups) found a far-reaching generalization of this result. They replaced the euclidean structure by the intersection form of the underlying period mapping, which arises from a family of holomorphic differential forms on the fibres of the Milnor fibration of a versal family of functions. A nondegenerate intersection form defines (depending on the parity of the number of variables) either a locally flat pseudo-euclidean metric with a standard singularity on the Legendre front or a symplectic structure which extends holomorphically to the front.

EXAMPLE. The space of monic polynomials with odd degree and sum of the roots equal to zero acquires yet another symplectic structure. Relative to this structure, the submanifold of polynomials with the maximal number of double roots turns out to be lagrangian.

When the intersection form is indefinite, the symplectic structure is replaced by a Poisson structure (see Appendix 14).

F Tangential singularities

The first applications of the theory of lagrangian and Legendre singularities, around which the theory itself developed (~ 1966), concerned short wave asymptotics in the form of the asymptotics of oscillatory integrals. A survey of these applications (including the determination of uniform estimates for oscillatory integrals when saddle points meet, the calculation of asymptotics using Newton polyhedra, the construction of mixed Hodge structures, applications to number theory and the theory of convex polyhedra, and estimates of the index of singular points of vector fields and the number of singular points of algebraic surfaces) may be found in the book:

V. I. Arnold, A. N. Varchenko, and S. M. Gusein-Zade, "Singularities of Differentiable Mappings," Vol. 2, *Monodromy and Asymptotics of Integrals*, Moscow, Nauka, 1984. English translation: Birkhäuser, 1988.

and in the paper

V. I. Arnold, Singularities of ray systems, Proceedings of the International Congress of Mathematicians, August 16–24, 1983, Warsaw.

Here we shall present other applications of the theory of lagrangian and Legendre singularities to the study of the configurations of projective manifolds and tangential planes of various dimensions. One is led to such problems from variational problems with one-sided constraints (such as the obstacle problem), as well as from the study of Nekhoroshev's exponent of roughness for unperturbed hamiltonian functions (see Appendix 8).

We consider a generic surface in three-dimensional projective space (Figure 259). The curve of parabolic points (p) divides the surface into a domain of elliptic points (e) and a domain of hyperbolic points (h); the latter domain contains the curve of inflection points of the asymptotic lines (f), with its

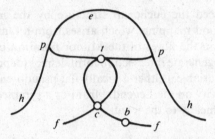

Figure 259   Projective classification of points of a surface

points of biinflection (b), self-intersection (c), and tangency to the parabolic curve (t),

From this classification of points, one may derive both estimates of curvature exponents and the following classification of projections.

Theorem (O. A. Platonova and O. P. Shcherbak, 1981). *Every projection from a point outside a generic surface in $\mathbb{R}P^3$ is locally equivalent at each point of the surface to the projection along lines parallel to the x-axis of a surface $z = f(x, y)$, where f is one of the following 14 functions:*

$$x, x^2, x^3 + xy, x^3 \pm xy^2, x^3 + xy^3, x^4 + xy,$$

$$x^4 + x^2y + xy^2, x^5 \pm x^3y + xy, x^3 \pm xy^4, x^4 + x^2y + xy^3, x^5 + xy.$$

By a projection we mean here a diagram $V \to E \to B$ consisting of an embedding and a fibration; an equivalence of projections is then a 3×2 commutative diagram whose vertical arrows are diffeomorphisms.

The only singularities of the projection from a generic center are folds and Whitney tucks. The tucks appear when the projection is along an asymptotic direction. The remaining singularities are visible only from special points. The finiteness of the number of singularities of projections (and therefore the number of singularities of apparent contours) was not obvious before the result above was obtained, since there is a continuum of inequivalent singularities for generic three-parameter families of mappings from a surface to the plane.

The regions of space from which the generic surface has a different appearance, as well as the corresponding views of the surface, are shown in Figure 260 (for the most complicated cases).

The hierarchy of tangential singularities becomes more comprehensible when it is reformulated in terms of symplectic and contact geometry. R. Melrose (1976) observed that the rays tangent to a surface are described by a pair of hypersurfaces in symplectic phase space: one of them, $p^2 = 1$, is defined by the metric; the other is defined by the surface.

A significant part of the geometry of asymptotic lines may be reformulated in terms of this pair of hypersurfaces. In this way, we may transfer concepts from the geometry of surfaces to the more general case of arbitrary pairs of

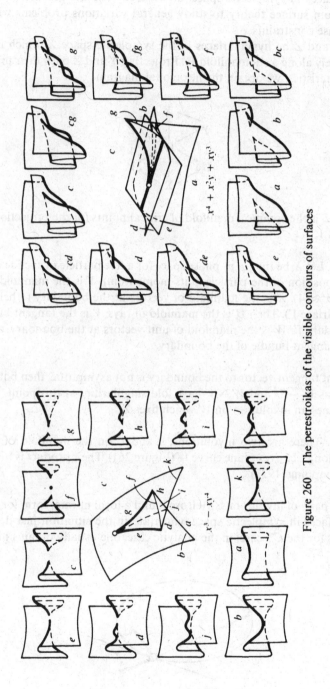

Figure 260 The perestroikas of the visible contours of surfaces

493

hypersurfaces in symplectic space, and thereby use the geometric intuition gained from surface theory to study general variations problems with one-sided phase constraints.

Let Y and Z be hypersurfaces in the symplectic space X which intersect transversely along a submanifold W. Projecting Y and Z onto their manifolds of characteristics, we obtain the hexagonal diagram

in which Σ is the common manifold of critical points for the projections of W on U and V.

EXAMPLE. Let X be the $\{q, p\}$ phase space for a free particle in euclidean space (q is the position of the particle, p its momentum). Y is the manifold of unit vectors ($p^2 = 1$). Z is the manifold of vectors at the boundary (q belongs to a hypersurface Γ). Then U is the manifold of rays, V is the tangent bundle of the boundary Γ, W is the manifold of unit vectors at the boundary, and Σ is the unit tangent bundle of the boundary.

If a unit tangent vector to the boundary is not asymptotic, then both of the projections $W \to U$ and $W \to V$ have fold singularities at this point. Each of them defines an involution on W which fixes Σ.

EXAMPLE. There are two involutions, σ and τ, on the manifold of tangent vectors along a convex plane curve W (Figure 261). Their product is Birkhoff's billiard mapping (1927).

Using pairs of involutions, Melrose found a local normal form for pairs of hypersurfaces in symplectic space which are in the situation just described. (This was for the C^∞ case; in the analytic case, one usually obtains divergent

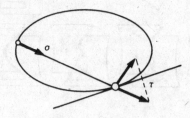

Figure 261   The two involutions generating the billiard mapping

series, just as in the theory of Ecalle (1975) and Voronin (1981) on resonant dynamical systems.)

For more complicated singularities (for example, near asymptotic directions), pairs of hypersurfaces have moduli. For the two simplest singularity types after the fold, it is possible to put in normal form (at least formally) the pair consisting of the first hypersurface and its intersection with the second. This allows us to study, in a neighborhood of an asymptotic or biasymptotic unit tangent vector to the boundary, the mapping which assigns the ray containing it to each unit vector at the boundary. The critical values of this mapping in the symplectic space of lines are described by the following result, since the manifold of tangent rays is locally diffeomorphic near a biasymptotic ray to the product of a swallowtail and a line.

Theorem (1981). *All the generic symplectic structures in the neighborhood of a point in the direct product of a swallowtail and a linear space are formally diffeomorphic by local diffeomorphisms preserving the product structure.*

G The obstacle problem

We consider an obstacle bounded by a smooth surface in euclidean space. The obstacle problem consists of the study of the singularities of the function defined outside the obstacle whose value at each point is the length of the shortest path remaining outside the obstacle and joining the point to a fixed initial set. This variational problem on a manifold with boundary is unsolved even in three-dimensional space.

Each minimizing path consists of segments of straight lines and segments of geodesics on the surface of the obstacle (Figure 262). We consider therefore a system of geodesics on the surface of the obstacle, orthogonal to a fixed front. The system of all rays tangent to these geodesics forms a lagrangian variety in the symplectic space of lines, just as any system of extremals for a variational problem. But while in an ordinary variational problem this lagrangian variety is a smooth manifold (even at caustics), the lagrangian variety arising in the obstacle problem has singularities. From the last theorem (in the previous section), one obtains:

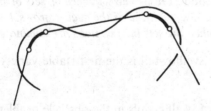

Figure 262   An extremal of the obstacle problem

Figure 263   The open ("unfurled") swallowtail

Corollary (1981). *The lagrangian variety of rays in a generic obstacle problem has a semicubic cuspidal edge along each asymptotic ray and a singularity diffeomorphic to an open swallowtail at each biasymptotic ray.*

The open swallowtail is the surface in the four-dimensional space of monic polynomials $x^5 + Ax^3 + Bx^2 + Cx + D$ formed by the polynomials with triple roots. Differentiation of the polynomials turns the open swallowtail into an ordinary one; when the swallowtail is opened, the cuspidal edge is retained, but the self-intersection disappears (Figure 263).

Theorem (1981). *In the generic motion of a wave front, the cuspidal edges of the instantaneous fronts sweep out an open swallowtail in four-dimensional space–time (over the usual swallowtail caustic).*

Theorem (O. P. Shcherbak, 1982). *Consider a generic one-parameter family of space curves and suppose that, for some value of the parameter (time), one of the curves has a point of double flatness (of type 1, 2, 5). Then the projective duals of these curves form a surface in space–time which is locally diffeomorphic to the open swallowtail.*

The open swallowtail is the first member of a whole series of singularities. Consider, in the space of monic polynomials $x^n + \lambda_1 x^{n-1} + \cdots + \lambda_{n-1}$, the set of polynomials with a root of fixed comultiplicity k, $(x - \alpha)^{n-k}(x^k + \cdots)$. Differentiation of polynomials preserves the comultiplicity of roots.

Theorem (A. B. Givental', 1981). *The sequence of sets of polynomials of fixed comultiplicity becomes stabilized as the degree grows, beginning with degree $n = 2k + 1$ (i.e., when the self-intersections are eliminated).*

EXAMPLE. The open swallowtail is the first stable variety over the ordinary swallowtail.

The appearance of swallowtails in the obstacle problem was axiomatized by Givental' (1982) in his theory of triads.

Definition. A *symplectic triad* (H, L, l) consists of a smooth hypersurface H in a symplectic manifold and a lagrangian submanifold L which is tangent to H to first order along a hypersurface l of L.

The lagrangian variety generated by the triad is the image of L in the manifold of characteristics of the hypersurface H.

EXAMPLE 1. Consider, in the problem of bypassing an obstacle with boundary $\Gamma \subset \mathbb{R}^n$, the distance along geodesics from an initial front as a function $s: \Gamma \to \mathbb{R}$. The manifold L consisting of all extensions of the 1-form ds from Γ to $\mathbb{R}^n$, together with the hypersurface $H: p^2 = 1$, forms a triad. The lagrangian variety generated by this triad is precisely the variety of rays tangent to the geodesics in our system of extremals on Γ.

EXAMPLE 2. In the symplectic manifold of monic polynomials $\mathscr{F} = x^d + \lambda_1 x^{d-1} + \cdots + \lambda_d$ with even degree $d = 2m$, the polynomials divisible by x^m form a lagrangian submanifold L.

Consider the hamiltonian for translation along the x-axis. [This polynomial in λ is equal to

$$h = \sum (-1)^\lambda \mathscr{F}^{(i)} \mathscr{F}^{(j)}, \qquad i + j = d, \quad \mathscr{F}^{(i)} = d^i \mathscr{F}/dx^i.$$

The hypersurface $h = 0$ is tangent to the lagrangian submanifold L along the subspace l of polynomials divisible by x^{m+1}, thus forming a triad. The lagrangian variety generated by this triad is an open swallowtail of dimension $m - 1$ (the set of polynomials $x^{d-1} + a_1 x^{d-3} + \cdots + a_{d-2}$ having a root of multiplicity greater than half the degree).]

Theorem (A. B. Givental', 1982). *The triads in Example 2 are stable. Every germ of a generic triad is diffeomorphic to a germ of a triad in Example 2.*

Corollary. *The variety of rays tangent to the geodesics in the system of extremals of a generic obstacle problem is locally symplectically diffeomorphic to a lagrangian open swallowtail.*

In contact geometry, there are two kinds of Legendre varieties associated to obstacle problems: varieties of contact elements of fronts and varieties of 1-jets of time functions. The first of these are diffeomorphic to lagrangian open swallowtails; the second are diffeomorphic to cylinders over the first.

EXAMPLE. Consider the problem of bypassing an obstacle in the plane which is bounded by a curve with an inflection point. The fronts, which are the evolvents of the curve, have two kinds of singularities: ordinary cusps (of order 3/2) on the curve itself and singularities of order 5/2 on the tangent line through the inflection point (Figure 264). Over points of the boundary curve,

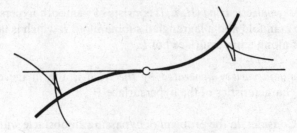

Figure 264   The evolvents of a cubical parabola

the Legendre variety is nonsingular, while over points on the tangent line through the inflection point it has a cuspidal edge of order 3/2.

Theorem (1978). *In the space of contact elements to the plane, fibered over the plane itself, the surface consisting of the contact elements of the evolvents of a generic curve near a point of inflection is locally equivalent by a fiber-preserving diffeomorphism to the surface consisting of all polynomials with multiple roots in the space of polynomials $x^3 + ax^2 + bx + c$, fibered into lines parallel to the b-axis.*

This surface (Figure 265), together with the surface $c = 0$ representing the contact elements along the boundary curve, forms a variety which is diffeomorphic to the set of irregular orbits for the reflection group B_3. This observation led to the theory of boundary singularities (1978).

EXAMPLE (I. G. Shcherbak, 1982). Consider a generic curve on a surface in three-dimensional euclidean space. At certain points, the direction of the curve coincides with principal curvature directions of the surface. It follows from the theory of lagrangian boundary singularities that the Weyl group F_4 is

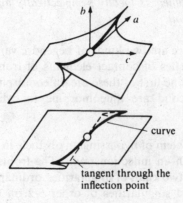

curve

tangent through the
inflection point

Figure 265   The surface of contact elements of the evolvents

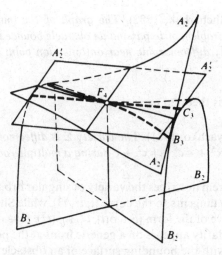

Figure 266   The caustic singularity F_4

connected with each such point: the focal points of the surface (A_2), focal points of the curve (A_2'), and normals to the surface at points of the curve (B_2) together form an F_4 caustic near the center of curvature (Figure 266).

We will not dwell here on the theory of boundary singularities, but it is worth mentioning the "Lagrange duality" relating a function and its restriction to the boundary (up to stable equivalence): this may be thought of as a modern version of the Lagrange multiplier rule (I. G. Shcherbak, 1982).

Returning to inflection points of plane curves, we consider the graph of the multiple-valued time function in an obstacle problem. The level curves of this function are the evolvents of the obstacle boundary. Therefore, the graph of this function has the form (shown in Figure 267) of a surface with two cuspidal edges (of orders 3/2 and 5/2). When I showed this surface to A. B. Giventalʹ, he recognized O. V. Lyashko's drawing of the singular orbit Σ of the group H_3 (symmetries of the icosahedron). Givental's conjecture was soon verified:

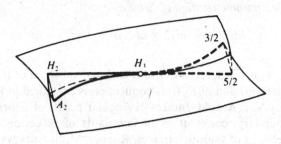

Figure 267   The discriminant of H_3

Theorem (O. P. Shcherbak, 1982). *The graph of the (multiple-valued) time function in the problem of bypassing an obstacle bounded by a generic plane curve is formally diffeomorphic near an inflection point of the curve to the variety Σ.*

The proof of this theorem uses:

Theorem (O. V. Lyashko, 1981). *The variety Σ is diffeomorphic to the variety of polynomials $x^5 + ax^4 + bx^2 + c$ having a multiple root.*

Lyashko's theorem describes the variety of singular orbits for the group H_3 as the union of the tangents to the curve (t, t^3, t^5), while Shcherbak's theorem applies to any curve of the form $(t + o(t), t^3 + o(t^3), t^5 + o(t^5))$.

The same singularity appears on a generic front at the point of tangency of a asymptotic ray with the bounding surface of an obstacle in $\mathbb{R}^3$.

Finally, we describe a variational problem leading to the singularity H_4 (after O. P. Shcherbak).

The *group H_4* consists of the symmetries of a regular polyhedron in $\mathbb{R}^4$. Its 120 vertices lie on $S^3 \approx SU(2)$ and form the binary icosahedral group (the binary group being the inverse image of the symmetry group of the icosahedron under the double covering $S^3 \to SO(3)$).

Consider the problem of bypassing an obstacle bounded by a smooth surface in three-dimensional euclidean space. The extremals beginning at a fixed point outside the obstacle generate a pencil (one-parameter family) of geodesics on the surface. A *time function* is the distance from a fixed initial manifold (e.g., a point) along stationary (not necessarily minimizing) paths consisting of arcs of geodesics and their tangents, considered as a (multiple-valued) function of the terminal point in space (solution of the Hamilton–Jacobi equation).

Theorem (O. P. Shcherbak, 1984). *For a generic obstacle, the graph of the time function at a point which is focal for the pencil along an asymptotic tangent at a parabolic point of the surface is locally diffeomorphic to the variety Σ of singular orbits of the group H_4.*

An explicit parametrization of Σ is:

$$(a, b^2/2 + ac, c^2/2 + ab^3, b^5/5 + c^3/3 + ab^3c).$$

The group H_4 is related to a four-dimensional subspace of the base space of the versal deformation of E_8 (this connection is explained in Remark 7, §9 of the paper by V. I. Arnold, Indices of singular points of 1-forms on manifolds with boundary, convolution of invariants of reflection groups, and singular projections of smooth surfaces, Russian Math. Surveys 34:2 (1979), 1–42).

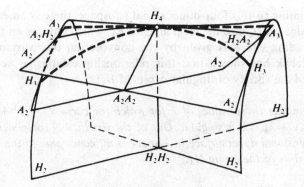

Figure 268   The caustic singularity H_4

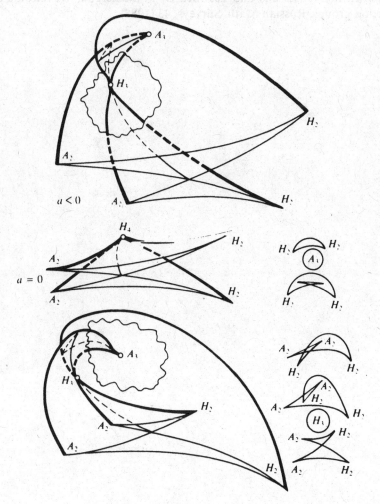

Figure 269   The front perestroika H_4

Corresponding to this four-dimensional subspace, there is an embedding of the local algebra D_4 into the local algebra E_8, which induces on the former the same grading which is given by the convolution of invariants of H_4. O. P. Shcherbak has shown that this relationship establishes yet another description of the variety of singular orbits of H_4:

Theorem. *Consider those values of λ for which the curve $x^5 + y^3 + \lambda_1 x^3 y + \lambda_2 x^3 + \lambda_3 y + \lambda_4 = 0$ is singular. One of the irreducible components of this three-dimensional hypersurface in λ-space is diffeomorphic to the variety of singular orbits of the group H_4.*

The caustic and three typical sections of the variety of singular orbits of H_4 are shown in Figures 268 and 269. See O. P. Shcherbak, Wavefronts and reflection groups, Russian Math. Surveys, 43 (1988).

Bibliography of Symplectic Topology

Arnold, V.I. Sur une propriété topologique des applications globalement canoniques de la mécanique classique. *C.R. Acad. Sci. Paris* **261** (1965), 3719–3722.

Arnold, V.I. On a characteristic class entering the quantization conditions. *Funct. Anal. Appl.* **1:1** (1967), 1–14.

Arnold, V.I. A comment on "Sur un théorème de géométrie". In: *Izbrannye trudy A. Puankaré*. Moscow, Nauka, 1972, vol. 2, pp. 987–989.

Arnold, V.I. Lagrange and Legendre cobordisms. *Funct. Anal. Appl.* **14:3** (1980), 1–13; **14:4** (1980), 8–17.

Arnold, V.I. The Sturm theorems and symplectic geometry. *Funct. Anal. Appl.* **19** (1985), 251–259.

Arnold, V.I. First steps in symplectic topology. *Russian Math. Survey* **41:6** (1986), 1–21.

Arnold, V.I. On functions with mild singularities. *Funct. Anal. Appl.* **23:3** (1989), 1–10.

Arnold, V.I., and Givental, A.B. Symplectic geometry. In: *Dynamical Systems IV* (Enc. of Math. Sc. vol. 4). Berlin-Heidelberg-New York, Springer, 1990, pp. 1–136.

Arnold, V.I. Sur les propriétés topologiques des projections lagrangiennes en géométrie symplectique des caustiques. Preprint 9320, CEREMADE, Université Paris-Dauphine, 1993, pp. 1–9 (Cahiers de Mathématiques de la Décision, 14/6/93).

Arnold, V.I. Some remarks on symplectic monodromy of Milnor fibration. In: *Progress in Math., A Floer Memorial Volume*. Basel-Boston, Birkhäuser, 1993.

Arnold, V.I. Invariants and perestroikas of plane fronts. *Trudy (Proceedings) Steklov Math. Inst., Russ. Acad. of Sc.*, Vol. 209, 1985.

Arnold, V.I. On topological properties of Legendre projections in contact geometry of wave fronts. Algebra and Analysis. *S. Petersbourg Math. J.* **6:3** (1994).

Arnold, V.I. Symplectic geometry and topology. In: *Trends and Perspectives in Modern Mathematics*. Cambridge Univ. Press, to appear (Preprint MIT, 1993, 68 pp).

Arnold, V.I. Topological Invariants of Plane Curves and Caustics. *J.B. Lewis Memorial Lectures*, Rutgers, 1993, 106 pp; AMS University Lecture Series, Vol. 5, Providence, AMS, 1994, 60 pp.

Arnold, V.I., ed. *Singularities and Curves* (Advances in Sov. Math.), Providence, AMS, 1994.

Atiyah, M. New invariants of 3- and 4-manifolds. In: *The Mathematical Heritage of H. Weyl.* Durham, NC, 1987 (Sympos. Pure Math., vol. 48). Providence, AMS, 1988, pp. 285–289.

Audin, M. Quelques calculs en cobordisme lagrangien. *Ann. Inst. Fourier* **35:3** (1985), 159–194.

Audin, M. *Cobordismes d'immersions lagrangiennes et legendriennes* (Travaux en Cours, vol. 20). Hermann, 1987, 203 pp.

Audin, M. Fibrés normaux d'immersion en dimension double, points doubles d'immersions lagrangiennes et plongements totalement réels. *Comm. Math. Helvet.* **63** (1988), 593–623.

Audin, M. Hamiltoniens périodiques sur les variétés symplectiques compactes de dimension 4. In: *Lect. Notes in Math.* **1416**. Berlin-Heidelberg-New York, Springer, 1990, pp. 1–25.

Audin, M. *The Topology of Torus Actions on Symplectic Manifolds.* Basel, Birkhäuser, 1991.

Banyaga, A. Sur la structure du groupe des difféomorphismes qui préservent une forme symplectique. *Comm. Math. Helv.* **53** (1978), 174–227.

Bennequin, D. Entrelacements et équations de Pfaff. *Astérisque* **107–108** (1983), 83–161.

Bennequin, D. Quelques remarques simples sur la rigidité symplectique. In: *Géométrie Symplectique et de Contact: Autour du Théorème de Poincaré-Birkhoff*, P. Dazord and N. Desolneux-Moulis, eds. Paris, Herman, 1984, pp. 1 50.

Bialy, M.L., and Polterovich, L.V. Lagrangian singularities of invariant tori of hamiltonian systems with two degrees of freedom. *Invent. Math.* **97:2** (1989), 291 303.

Bialy, M., and Polterovich, L. Hamiltonian diffeomorphisms and Lagrangian distributions. *Geom. Funct. Anal.* **2** (1992), 173–21.

Bialy, M., and Polterovich, L. Optical Hamiltonian functions. Preprint 1992, 20 p.

Boothby, W.M., and Wang, H.C. On contact manifolds. *Ann. Math.* **68** (1958), 721–734.

Calabi, E. On the group of automorphisms of a symplectic manifold. In: *Problems in Analysis (Symposium in honour of S. Bochner).* Princeton Univ. Press, 1970, 1–26.

Chaperon, M. Quelques questions de géométrie symplectique [d'après, entre autres, Poincaré, Arnold, Conley et Zehnder], Séminaire Bourbaki 1982 83. *Astérisque* **105–106** (1983), 231 249.

Chaperon, M. Une idée du type "géodésiques brisées" pour les systèmes hamiltoniens. *C.R. Acad. Sci. Paris* **298** (1984), 293 296.

Chaperon, M. An elementary proof of the Conley-Zehnder theorem in symplectic geometry. In: *Dynamical Systems and Bifurcations*, B.L.J. Braaksma, H.W. Broer, F. Takens, eds. (Lecture Notes in Math. **1125**) Berlin-Heidelberg-New York, Springer, 1985, 1–8.

Chaperon, M. Familles génératrices. Cours à l'école d'été Erasmus de Samos (1990), Publication Erasmus, 1993.

Chekanov, Yu.V. Lejandrova teoriya Morsa. *Uspekhi Mat. Nauk* **42:4** (1987), 139–141.

Chekanov, Yu.V. Caustics in geometrical optics. *Funct. Anal. Appl.* **20** (1986), 223–226.

Chekanov, Yu.V. Lagrangian tori in a symplectic vector space and global symplectomorphisms. Bochum Preprint 169, 1993, 13 p (to appear in *Math. Z*).

Conley, C., and Zehnder, E. The Birkhoff-Lewis fixed point theorem and a conjecture of V.I. Arnold. *Invent. Math.* **73** (1983), 33–49.

Duistermaat, J.J. On the Morse index in variational calculus. *Adv. Math.* **21** (1976), 173–195.

Duistermaat, J.J. On global action-angle variables. *Comm. Pure Appl. Math.* **33** (1980), 687–706.

Ekeland, I., and Hofer, H. Symplectic topology and Hamiltonian dynamics. *Math. Z.* **200** (1988), 355–378.

Ekeland, I., and Hofer, H. Symplectic topology and Hamiltonian dynamics II. *Math. Z.* **203** (1990), 553–567.

Eliashberg, Y. Rigidity of symplectic and contact structures, Preprint, 1981.

Eliashberg, Y. Cobordisme des solutions de relations différentielles. In: *Sem. Sud-Rhodanien de Géom.*, tome 1, P. Dazord and N. Desolneux-Moulis, eds. Hermann, 1984, pp. 17–32.

Eliashberg, Y. The complexification of contact structures on a 3-manifold. *Uspekhi Mat. Nauk* **6:40** (1985), 161–162.

Eliashberg, Y. Classification of overtwisted contact structures on 3-manifolds. *Invent. Math.* **98** (1989), 623–637.

Eliashberg, Y. Filling by holomorphic discs and its applications. In: *Geometry of Low-Dimensional Manifolds*, Vol. 2, S.K. Donaldson and C.B. Thomas, eds. (London Math. Soc. Lect. Notes Ser. **151**) Cambridge Univ. Press, 1990, pp. 45–67.

Eliashberg, Y., and Gromov, M. Convex symplectic manifolds. *Proceedings of Symposia in Pure Mathematics*, E. Bedford *et al.* (eds), **52:2** (1991), 135–162.

Eliashberg, Y., and Polterovich, L. Bi-invariant metrics on the group of Hamiltonian diffeomorphisms. Preprint, 1991.

Eliashberg, Y. New invariants of open symplectic and contact manifolds. *J. Amer. Math. Soc.* **4** (1991), 513–520.

Eliashberg, Y., and Ratiu, T. The diameter of the symplectomorphism group is infinite. *Invent. Math.* **103** (1991), 327–340.

Eliashberg, Y. On symplectic manifolds with some contact properties. *J. Diff. Geometry* **33** (1991), 233–238.

Eliashberg, Y., and Hofer, H. Unseen symplectic boundaries. Preprint, 1992, 16 pp.

Eliashberg, Y., and Polterovich, L. Unknottedness of Lagrangian surfaces in symplectic 4-manifolds. Preprint, 1992, 9 pp.

Eliashberg, Y., and Polterovich, L. New applications of Luttinger's surgery. Preprint, 1992, 12 pp.

Eliashberg, Y. Contact 3-manifolds twenty years since J. Martinet's work. *Ann. Inst. Fourier* **42** (1992), 165–191.

Eliashberg, Y., and Hofer, H. An energy-capacity inequality for the symplectic holonomy of hypersurfaces flat at infinity. Preprint, 1992, 8 pp.

Eliashberg, Y. Topology of 2-knots in $\mathbb{R}^4$ and symplectic geometry. In: *Progress in Math.*, A. Floer Memorial Volume. Boston-Basel, Birkhäuser, 1993.

Eliashberg, Y. Legendrian and transversal knots in tight contact 3-manifolds. In: *Topological Methods in Modern Mathematics*. Houston, Publish or Perish, 1993, pp. 171–193.

Eliashberg, Y. Classification of contact structures on $\mathbb{R}^3$. *Duke Math. J. Intern. Math. Res. Notes* N°3 (1993), 87–91.

Eliashberg, Y., and Hofer, H. Towards the definition of symplectic boundary. Preprint, 1993.

Floer, A. Proof of the Arnold conjecture and generalizations to certain Kaehler manifolds. *Duke Math. J.* **53** (1986), 1–32.

505

Floer, A. Morse theory for Lagrangian intersections. *J. Diff. Geom.* **28** (1988), 513–547.

Floer, A. The unregularized gradient flow for the symplectic action. *Comm. Pure Appl. Math.* **41** (1988), 775–813.

Floer, A. A relative Morse index for the symplectic action. *Comm. Pure Appl. Math.* **41** (1988), 393–407.

Floer, A. An instanton invariant for 3-manifolds. *Comm. Math. Phys.* **118:2** (1988), 215–240.

Floer, A. Witten's complex in infinite dimensional Morse theory. *J. Diff. Geom.* **30** (1989), 207–221.

Floer, A. Cuplength estimates for Lagrangian intersections. *Comm. Pure Appl. Math.* **42** (1989), 335–356.

Floer, A. Symplectic fixed points and holomorphic spheres. *Comm. Math. Phys.* **120** (1989), 575–611.

Floer, A., and Hofer, H. Symplectic homology I: open sets in $\mathbb{C}^n$. Preprint, 1992.

Floer, A., Hofer, H., and Wysocki, K. Applications of symplectic homology I. Preprint, 1992.

Fortune, B., and Weinstein, A. A symplectic fixed point theorem for complex projective spaces. *Bull. Am. Math. Soc.* **12:1** (1985), 128–130.

Fuchs, D.B. Maslov-Arnold characteristic classes. *Sov. Math. Dokl.* **9** (1968), 96–99.

Ginzburg, V.L. Calculation of contact and symplectic cobordism groups. *Topology* **31:4** (1992), 757–762.

Ginzburg, V.L., and Khesin, B.A. Steady fluid flows and symplectic geometry. Preprint IHES, October 1992, 20 pp. (to appear in: *J. Geom. Phys.*).

Giroux, E. Convexité en topologie de contact. *Comm. Math. Helvet.* **66** (1991), 637–677.

Givental, A.B. Lagrangian embeddings of surfaces and the open Whitney umbrella. *Funct. Anal. Appl.* **20:3** (1986), 35–41.

Givental, A.B. Periodic mappings in symplectic topology. *Funct. Anal. Appl.* **23:4** (1989), 287–300.

Givental, A.B. Nonlinear generalization of the Maslov index. In: *Singularity Theory and Its Applications*, V. Arnold, ed. (Advances in Soviet Math., vol. 1), Providence, AMS, 1990, pp. 71–103.

Givental, A.B. A symplectic fixed point theorem for toric manifolds. In: *Progress in Math., A. Floer Memorial Volume*. Boston-Basel, Birkhäuser, 1993.

Gray, J.W. Some global properties of contact structures. *Ann. Math.* **69** (1959), 421–450.

Gromov. M. *Partial Differential Relations*. Berlin-Heidelberg-New York, Springer, 1996.

Gromov, M. Pseudo holomorphic curves in symplectic manifolds. *Invent. Math.* **82** (1985), 307–347.

Guillemin, V., and Sternberg, S. Birational equivalence in symplectic category. *Invent. Math.* **97** (1989), 485–522.

Harlamov, V., and Eliashberg, Y. On the number of complex points of a real surface in a complex surface. *Proc. LITC–82* (1982), 143–148.

Hofer, H., and Zehnder, E. A new capacity for symplectic manifolds. In: *Analysis Et Cetera*, Boston, Academic Press, 1990, 405–428.

Hofer, H. On the topological properties of symplectic maps. *Proc. Roy. Soc. Edinburgh, Ser. A.* **115** (1990), 25–38.

Hofer, H. Symplectic Invariants. In: *Proceedings ICM Kyoto 1990*. Berlin–Heidelberg–New York, Springer, 1991.

Hofer, H. Symplectic capacities. In: *Durham Conferences*, S.K. Donaldson and C.B. Thomas, eds. London Math. Soc., 1992.

Hofer, H., and Salamon, D. Floer homology and Novikov rings. Preprint, 1992. 39 pp.

Hofer, H. Estimates for the energy of a symplectic map. *Comm. Math. Helvet.* **68** (1993), 48–72.

Kazarian, M.È. Umbilical characteristic number of Lagrangian mappings of 3-dimensional pseudo-optical manifolds. Preprint, Ruhr-Univ. Bochum, 1993, 12 pp.

Kuksin, S. Infinite-dimensional symplectic capacities and a squeezing theorem for Hamiltonian PDE's. Preprint Forschungsinstitut für Mathematik ETH Zürich, August 25, 1993.

Lalonde, F., and Sikorav, J.-C. Sous-variétés lagrangiennes exactes des fibrés cotangents. *Comm. Math. Helvet.* (1991), 18–33.

Lalonde, F. Isotopy of symplectic balls, Gromov's radius and the structure of ruled symplectic 4-manifolds. Preprint, 1992.

Lalonde, F., and McDuff, D. The geometry of symplectic enérgy. Preprint #1993/6 IMS SUNY Stony Brook, June 1993, 26 pp.

Laudenbach, F., and Sikorav, J.-C. Persistence d'intersection avec la section nulle au cours d'une isotopie hamiltonienne dans un fibré cotangent. *Invent. Math.* **82:2** (1985), 349–358.

Laudenbach, F., and Sikorav, J.C. Disjonction hamiltonienne et limites de sous-variétés lagrangiennes. Preprint, Centre de Math., Ecole Polytechnique, septembre 1993.

Lee, Yng-Ing. Nonlagrangian limits of Lagrangian discs. *Duke Math. J. Intern. Math. Res. Notes.* N°2, 1993.

Luttinger, K. Lagrangian tori in $\mathbb{R}^4$. Preprint, 1992.

Lutz, R. Structures de contact sur les fibrés principaux en cercles de dimension 3. *Ann. Inst. Fourier* **3** (1977), 1–15.

Martinet, J. Formes de contact sur les variétés de dimension 3. In: *Lect. Notes in Math.* **209**. Berlin-Heidelberg-New York, Springer, 1971, pp. 142–163.

Meckert, C. Formes de contact sur la source connexe de deux variétés de contact. IRMA, Strasbourg, 1980.

McDuff, D. The structure of rational and ruled symplectic 4-manifolds. *JAMS* **3:1** (1990), 679–712.

McDuff, D. Elliptic methods in symplectic geometry. *Bull. Amer. Math. Soc.* **23** (1990), 311–358.

McDuff, D. Symplectic manifolds with contact-type boundaries. *Invent. Math.* **103** (1991), 651–671.

McDuff, D. Blow-ups and symplectic embeddings in dimension 4. *Topology* **30** (1991), 409–421.

McDuff, D. Singularities of J-holomorphic curves. *J. Geom. Anal.* **3** (1992), 249–266.

McDuff, D. Notes on ruled symplectic 4-manifolds. Preprint, 1992 (to appear in *Trans. Amer. Math. Soc.*).

McDuff, D., and Polterovich, L. Symplectic packing and algebraic geometry. Preprint, 1992.

McDuff, D. Remarks on the uniqueness of symplectic blowing-up. *Proceedings of 1990 Warwick Symposium*, Cambridge Univ. Press, 1993.

McDuff, D., and Salamon, D. Notes on J-holomorphic curves. Stony Brook preprint, 1993.

McDuff, D., and Traynor, L. The 4-dimensional symplectic camel and related results. (London Math. Soc. Lect. Notes Series). Cambridge Univ. Press (to appear).

McDuff, D., and Salamon, D. *Symplectic Topology* (in preparation).

Moser, J. On the volume elements on a manifold. *Trans. Amer. Math. Soc.* **120** (1965), 286–294.

Oh, Y.-G. A symplectic fixed point theorem on $T^{2n} \times \mathbb{C}P^k$. *Math. Z.* **203:4** (1990), 535–552.

Polterovich, L. New invariants of embedded totally real tori and one problem of Hamiltonian mechanics. In: *Methods of Qualitative Theory and the Theory of Bifurcations*, Gorki, 1988, pp. 84–90.

Polterovich, L. Strongly optical Lagrange manifolds. *Math. Notes Ac. Sc. USSR* **45** (1989), 152–158.

Polterovich, L. Symplectic displacement energy for Lagrangian submanifolds. Preprint, 1991.

Polterovich, L. The surgery of Lagrange submanifolds. *Geom. Funct. Anal.* **2** (1991), 213–246.

Polterovich, L. The Maslov class of Lagrange surfaces and Gromov's pseudoholomorphic curves. *Trans. Amer. Math. Soc.* **325** (1991), 241–248.

Rabinowitz, P. Critical points of indefinite functionals and periodic solutions of differential equations. In: *Proceedings ICM Helsinki 1978*. Acad. Sci. Fennica, Helsinki, 1980, pp. 791–796.

Sato, H. Remarks concerning contact manifolds. *Tôhoku Math. J.* **29** (1977), 577–584.

Siegel, C.L. Symplectic geometry. *Amer. J. Math.* **65:1** (1943).

Sikorav, J.C. Problèmes d'intersections et de points fixes en géométrie hamiltonienne. *Comm. Math. Helvet.* **62:1** (1987), 62–73.

Sikorav, J.-C. Rigidité symplectique dans le cotangent de T^n. *Duke Math. J.* **59** (1989), 227–231.

Sikorav, J.-C. *Systèmes hamiltoniens et topologie symplectique*. Pisa, ETS Editrice, 1990.

Sikorav, J.-C. Quelques propriétés des plongements lagrangiens. Preprint, 1990.

Tabachnikov, S.L. Calculation of the generalized Bennequin invariant of a Legendrian curve from the geometry of its front. *Funct. Anal. Appl.* **22:3** (1988), 246–248.

Tabachnikov, S. Around four vertices. *Russian Math. Surveys* **45:1** (1990), 229–230.

Tabachnikov, S. Geometry of Lagrangian and Legendrian 2-web. Preprint, Arkansas Univ., 1992, 22 pp.

Traynor, L. Symplectic embedding trees for generalized camel spaces. Preprint 034-93 MSRI Berkeley, January 1993, 19 pp.

Traynor, L. Symplectic packing constructions. Preprint, October 1993, 20 pp.

Vasil'ev, V.A. Characteristic classes of Lagrangian and Legendre manifolds dual to singularities of caustics and wave fronts. *Funct. Anal. Appl.* **15** (1981), 164–173.

Vasil'ev, V.A. Self-intersections of wave fronts and Legendre (Lagrangian) charactristic numbers. *Funct. Anal. Appl.* **16** (1982), 131–133.

Vassilyev, V.A. *Lagrange and Legendre Characteristic Classes*. New York, Gordon and Breach, 1988.

Vasil'ev, V.A. Topology of spaces of functions having no complicated singularities. *Funct. Anal. Appl.* **23:4** (1989), 24–36.

Viterbo, C. Capacités symplectiques et applications. Séminaire Bourbaki, n°714, *Astérisque* **177–178** (1989), 345–362.

Viterbo, C. A new obstruction to embedding Lagrangian tori. *Invent. Math.* **100** (1990), 301–320.

Viterbo, C. Plongement lagrangiens et capacités symplectiques des tores dans $\mathbb{R}^{2n}$. *C.R. Acad. Sci. Paris, Sér. I, Math.* **311** (1990), 487–490.

Viterbo, C. Symplectic topology as the geometry of generating functions. *Math. Ann.* **292** (1992), 685–710.

Weinstein, A. Lectures on symplectic manifolds. *C.B.M.S. Regional Conf. Ser. in Math.* vol. 29, Providence, AMS, 1977.

Weinstein, A. Periodic orbits for convex hamiltonian systems. *Ann. Math.* **108** (1978), 507–518.

Weinstein, A. On the hypotheses of Rabinowitz's periodic orbit theorems. *J. Diff. Eq.* **33** (1979), 353–358.

Weinstein, A. Contact surgeries and symplectic handlebodies. *Hokkaido Math. J.* **20** (1991), 241–251.

Weinstein, A. Symplectic manifolds and their lagrangian submanifolds. *Adv. Math.* **6** (1971), 329–346.

Index

Graduate Texts in Mathematics

(continued from page ii)